THE
COACH-MAKERS' ILLUSTRATED
HAND-BOOK

SECOND EDITION, 1875

CONTAINING COMPLETE INSTRUCTIONS
IN ALL THE DIFFERENT BRANCHES OF

CARRIAGE BUILDING

ADAPTED TO THE WANTS OF EVERY PERSON DIRECTLY OR
INDIRECTLY CONNECTED WITH THE MANUFACTURE
OF CARRIAGES.

WRITTEN AND REVISED BY PRACTICAL MEN OF ACKNOWLEDGED
ABILITY AND LONG EXPERIENCE IN THEIR SEVERAL
DEPARTMENTS

ASTRAGAL PRESS
Mendham, New Jersey

Copyright © 1995 by The Astragal
Press. All rights reserved.

Reprinted from an original copy
of the 1875 edition
in the collection of the
Carriage Museum of America

Library of Congress Catalog Card Number
95-75706 International Standard Book Number:
978-1-879335-61-5

Published by
The Astragal Press
5 Cold Hill Road, Suite #12
P.O. Box 239
Mendham NJ 07945-0239

PREFACE.

The design of this work is to place in the hands of carriage-makers a clear, concise, and perfectly reliable book of reference, containing instructions in the most important matters connected with the four several branches of the trade.

That the book is needed by a large number of those who are working at the trade of coach-making, we have abundant evidence, and the interest that its announcement awakened furnished positive proof that its advent, or something similar to it, had been long and anxiously awaited.

In presenting it to the trade for their patronage, we therefore have the satisfaction of feeling that we are carrying out the repeatedly expressed wishes of a number of carriage artisans, and as this is the first edition of a work of the kind published in the United States, we may lay claim to a share of the gratitude which is generally accorded to those who aim to furnish valuable information at a cheap rate.

A perusal of the book will disclose to the reader that each department is complete in itself, and arranged in the order that a carriage is carried forward in the factory thus resolving the matter in each department into the simplest and most convenient form that could be devised; and as an aid to the speedy finding of any article in either of the departments, a copious index is added.

Great care also has been exercised to avoid errors that would tend to mislead the student in his search after truth and safe methods of proceeding in his every-day work.

Trusting that our labor has been directed toward a good purpose, and that the "Hand Book" may dispel darkness by the introduction of light, we send it out on its mission to stand or fall on its own merits.

THE PUBLISHER.

CONTENTS.

PART I.—WOOD DEPARTMENT.

Scale Drafting and Coloring; French or Square Rule; Application of the Square Rule to all Styles of Carriages in general use; Construction of Carriage Parts and Wheels; Drawing Instruments; Scroll Patterns; Corner Bevels; Iron Planes; Round and Ribbed Boots; Wood Carving; Elevating Landau Seats; Canvasing Outside, etc.

PART II.—SMITH DEPARTMENT.

Styles and Manner of Ironing Platform and Perch Carriages; Hanging Off; Hooping Wheels; Welding Iron and Steel; Axle Centers; C Springs; Folding Steps; Drafting Joints; Patterns for Stays; Setting Axles; Steel Axles; Tempering Springs; Weight of Springs; Axles, Iron and Steel; Making Tools, etc.

PART III.—PAINT DEPARTMENT.

The Art of Coach Painting; Principles of Coloring; Complete Instructions from the Priming to the last coat of Varnish; Designs for Monograms and Ornaments; Compounding of Colors; Causes of Varnish Pitting; Complementary Colors; Mixing Paints; Varnish Brushes; Enlarging Letters and Ornaments; Gilding; Imitation Cane Work; Light and Ventilation; Oil Colors; Transparent Colors, etc.

PART IV.—TRIMMING DEPARTMENT.

Illustrations of different Styles of Trimming; Instructions for Standing and Fall Top Work; Setting Bows; Quantity of Stock required; Covering Dashes; Cutting Stock; Drafting Tops; Method in the Trimming Shop; Spring Cushions; Softening Buggy Tops, etc.

PART V.—MISCELLANEOUS DEPARTMENT.

Apprentices; Success in Business; System; Credit; Sharpers; Convenient Work Shops; Order and Disorder; Out of Work; Competition; Carriage Materials; Box Wood; Wood Engraving; Electrotyping; Grinding of Colors, together with items of general interest to Manufacturers and all engaged in the Carriage Business.

INDEX.

PART I.

WOOD DEPARTMENT.

A Word on Plugs and Brads	93
Back and Front Pillars, (*Illustration*)	59
Blackboard Paint	95
Brette, (*ill.*)	65
Bristol Board	21
Brushes	22
Buck Board Wagon, (*ill.*)	82
Cant of Brette, (*ill.*)	65
" Circular Hearse, (*ill.*)	63
" Coal Box Body, (*ill.*)	81
" Coupelette, (*ill*)	61
" English Phaeton, (*ill.*)	70
" Five-Glass Landau, (*ill.*)	62
" Four-in-Hand Drag, (*ill.*)	66
" Four-seat Phaeton, No. 1, (*ill.*)	77
" Four-seat Phaeton, No. 2, (*ill.*)	78
" French Cabriolet, (*ill.*)	70
" Glass Front Landau, (*ill.*)	62
" Jump Seat Rockaway, (*ill.*)	79
" Landau, (*ill.*)	56
" Landaulette, (*ill.*)	60
" Physician's Carriage, (*ill.*)	81
" Rockaway Landaulette, (*ill*)	64
" Six-seat Extension Top, (*ill.*)	67
" Six-seat Rockaway, *ill.*)	64
" Stanhope Buggy, (*ill.*)	80
" Victoria Phaeton, No. 1, (*ill.*)	73
" Victoria Phaeton, No. 2, (*ill.*)	73
" Victoria Phaeton, No. 3, (*ill.*)	75
Canvasing Outside	94
Carriage Parts, (2 *ills.*)	95
Centering Square, (*ill.*)	103
Circular Hearse	63
Coal Box Body, (*ill.*)	81
Corner Bevels	88
Corner Bevels, No. 1, (*ill.*)	88
Corner Bevels, No. 2, (*ill.*)	89
Corner Bevels No. 3, (*ill.*)	90
" " No. 4, (*ill.*)	91
" " No. 5, (*ill.*)	92
Designs for Carved Body Blocks, (*ill.*)	101
Diagram No. 1, Scale Drafting, (*ill.*)	23
" 2, " " (*ill.*)	24
" 3, " " (*ill.*)	26
" 4, " " (*ill.*)	29
" 5, " " (*ill.*)	31
" 6, " " (*ill.*)	33
" 7, " " (*ill.*)	34
" 8, " " (*ill.*)	35
" 9, " " (*ill.*)	36
" 10, " " (*ill.*)	39
" 1, Square Rule, (*ill.*)	42
" 2, " " (*ill.*)	43
" 3, " " (*ill.*)	45
" 4, " " (*ill.*)	48
" 5, " " (*ill.*)	48
" 6, " " (*ill.*)	50
" 7, " " (*ill.*)	52
" 8, " " (*ill.*)	52
Door Rocker, (*ill.*)	59
Dotting Pen, (*ill.*)	21
Drawing Board	18
" Pen, (*ill.*)	18
Drop Front, Close Top Buggy, (*ill.*)	26
English Phaeton, (*ill.*)	70
Express Carriage Part (2 *ills.*)	98
Fastening Tacks, (*ill.*)	21
Five-Glass Landau, (*ill.*)	57
Folding Front Seat, (*ill.*)	83
Four-in-hand Drag, (*ill.*)	66
Four-seat Phaeton, (*ill.*)	30
" " " No. 1, (*ill.*)	78
" " " No. 2, (*ill.*)	78
French Cabriolet Americanized, (*ill.*)	71

(v)

INDEX.

French or Square Rule	42
Glass Front Landau, (*ill.*)	62
Glue Brush	94
How to make a cheap Glue Brush	94
How to make It	87
India Ink	21
Insertion of Screws in Wood	94
Iron Planes	93
Irregular Curves, (*ill.*)	19
Its advantages over the wooden plane	94
Jump Seat Rockaway, (*ill.*)	79
Landau, (*ill.*)	40, 56
Landau Bows, (*ill.*)	86
Landaulette, (*ill.*)	60
Monitor Buggy, (*ill.*)	22
Origin and Composition	116
Panel Quarter, (*ill.*)	84
Parallel Ruler, (*ill.*)	20
Pear Wood for Sweep Patterns	27
Philadelphia Ribbed Boot, (*ill.*)	84
Physician's Carriage, (*ill*)	81
Plain Dividers, (*ill.*)	20
Platform Carriage for pole and shafts, (*ill.*)	97
Plugs and Brads	93
Portland Cutter	87
Practice with the Pen, (*ill.*)	25
Price of Drawing Instruments	25
Road Drawing Pen, (*ill.*)	19
Rockaway Landaulette, (*ill.*)	64
Scale Drafting	17
Scroll Patterns for Bar Ends, (4 *ills.*)	100
Six-seat Extension Top Phaeton, (*ill.*)	67
Six-seat Rockaway, (*ill.*)	36, 64
Six-seat Sociable, (*ill.*)	34
Sizes for Wheels	102
Slate Finish	94
Spring Bow Pen with Pencil Point, (*ill.*)	19
Square Rule	42
Standing Boot Piece, (*ill.*)	59
Stanhope Buggy, (*ill.*)	80
Steel Spacing Dividers, (*ill.*)	19
Sweep Patterns, (*ill.*)	27, 28
Sweep Patterns, (3 *ills.*)	32
The Door	63
The Draft	87
The Length of the Curb	87
The Manner of Obtaining Corner Bevels, (*ill.*)	88
The Plain Dividers, (*ill.*)	20
The Use of Glue	93
To Imitate Mahogany	94
To Lower the Top	63
Trammel for Ovals, (*ill.*)	85
Triangular Scale of Boxwood, (*ill.*)	20
T-square, (*ill.*)	18
Victoria Phaeton, No. 1, (*ill.*)	73
Victoria Phaeton, No. 2, (*ill.*)	73
Victoria Phaeton No. 3, (*ill.*)	75
Water Colors	22
Wood Carving	99
Working Draft of Buck Board Wagon, (*ill.*)	82
Working Draft of Brett, (*ill.*)	65
" " Circular Hearse, (*ill.*)	63
" " Coal Box Buggy, (*ill.*)	81
" " Coupelette, (*ill.*)	61
" " English Phaeton, (*ill.*)	70
" " Five-Glass Landau, (*ill.*)	57
Working Draft of Four-in-hand Drag, (*ill.*)	66
Working Draft of Four-seat Phaeton, No. 1, (*ill.*)	77
Working Draft of Four-seat Phaeton, 2, (*ill.*)	78
Working Draft of French Cabriolet, (*ill.*)	70
Working Draft of Glass Front Landau, (*ill.*)	62
Working Draft of Jump Seat Rockaway, (*ill.*)	79
Working Draft of Landau, (*ill.*)	56
Working Draft of Landaulette, (*ill.*)	60
Working Draft of Physician's Carriage, (*ill.*)	81
Working Draft of Rockaway Landaulette, (*ill.*)	64
Working Draft of Six-seat Extension, (*ill.*)	67
Working Draft of Six-seat Rockaway, (*ill.*)	64
Working Draft of Stanhope Buggy, (*ill.*)	80
Working Draft of Victoria Phaeton, No. 1, (*ill.*)	73
Working Draft of Victoria Phaeton, No. 2, (*ill.*)	73
Working Draft of Victoria Phaeton, No. 3, (*ill.*)	75
Working Drafts	56

PART II.

SMITH DEPARTMENT.

A cheap way to make, etc.	163
A Mandrel, (*ill.*)	111
Application of Body Loops	133
A Rule applicable to fifth wheels, (*ill.*)	146
Axle Centers	162
Back Bar, (*ill.*)	127
Back Quarter Wings, Landau, (*ill.*)	144
Bob Punch, (*ill.*)	109
Body Loops	133
Bottom Bed, (*ill*)	122
Brett, Hanging off, (*ill.*)	115
Carriage Part on "C" Springs, (*ill.*)	132
" Springs	157
Coach Steps, (*ill.*)	146
Cold Chisel	163
Cross Springs	158
"C" Springs	118
Curved Dash, (*ill.*)	142
Device for Elevating Landau Seat, (*ill.*)	149
Die Plate and Screw, (3 *ill.*)	113
Direction of a Given Force	116
" " Bearing Straps	118
Double folding Coach Step, (*ill.*)	146
Drafting Joints, (*ill.*)	148
Drop Pole, (*ill.*)	122
Fifth Wheels, (*ill.*)	146
Flatter, (*ill.*)	109
" for Clips, (*ill.*)	108
Folding Coach Step, (*ill.*)	246
Force of Power	116
Front Running Part, (*ill.*)	129
Fuller, (*ill.*)	107
Fulling Block, (*ill.*)	107
Gauge Chisel, (*ill.*)	108
Gauges, (*ill.*)	112
Gearing, No. 1, (*ill.*)	119
" " 2, (*ill.*)	120
" " 3, (3 *ills.*)	122
" " 4, (*ill.*)	123
" " 5, (2 *ills.*)	125
" " 6, (*ill.*)	126
" " 7, (*ill.*)	127
" " 8, (*ill.*)	129
" " 9, (*ill.*)	129
" " 10, (4 *ills.*)	129
" " 11, (*ill.*)	129
" " 12, (2 *ills.*)	132
" " 13, (*ill.*)	134
Hand Hammer, (2 *ills.*)	106
Hooping Wheels	150
How to Find the Right Sweep, etc., (*ill.*)	150
How Tools for Lamp Sockets are Made, (*ill.*)	169
Iron Back Bar, (*ill.*)	143
Iron Back Bar for Landaulette, (*ill.*)	140
Iron for Axle Centers,	162
Ironing Platform Express, (4 *ills.*)	139
Ironing Sulkeys	149
Landau Seat, (*ill.*)	149
Landau, 8 Springs, (*ill.*)	115
Light Platform, (*ill.*)	134
Line of Draft, (*ill.*)	143
Lower part with King or Body Bolt, (*ill.*)	126
Lower part with Springs attached, (*ill.*)	123
Lower Tools, (*ill.*)	113
Mandrel	111
Material	161
Monitor Dashes, 5 (*ills.*)	141
Origin and Composition	116
Parallel Forces	116
Patterns of Straight Stays and Steps, (*ill.*)	115
Platform Spring Carriage	119
" " " with Iron Perch, (*ill.*)	132
Power or Strength of Springs	118
Riveting Hammer, (*ill.*)	107
Rocker Plates	162
Round Corner Dash, (*ill.*)	141
Self-shutting Door Step, (2 *ills.*)	145
Setting Axles, (*ill.*)	154
" " when Cold, (*ill.*)	156
Shifting Seat Rail, (2 *ills.*)	144
Short and Easy Turning, (*ill.*)	135
Simultaneous Forces	116
Skeleton Boots, (*ill.*)	137
Small Flat Iron	163
Square Dash—plain, (*ill.*)	142
" Puncheon, (*ill.*)	107
Stay for Perch Carriage, (*ill.*)	111, 113
" Four-seat Rockaway, (*ill.*)	131
Steel	152
" Axles	152
Swage Hammer, (*ill.*)	107
" Iron, (*ill.*)	107

Tempering Drills,	162
" Springs	158
" Thin Tools	162
The Blacksmith	104
" Height of Wheels, etc., (*ill.*)	143
" Ironing of a carriage part, (*ill.*)	136
" Manner of producing Scrolls	134
" Mode of Making Lamp Sockets	109
" Tidy Blacksmith	105
Tillbury Shaft, (*ill.*)	122
To hang Brett or 8-Spring Landau, (*ill.*)	115
Tools, (12 *ills.*)	106 to 109
" for making Lamp Sockets, (6 *ills.*)	109 to 112
Tools for welding pins in Shifting Rail, (4 *ills.*)	113
Top Bed, (*ill.*)	123
" Part of Carriage, (*ill.*)	124
Value of Iron	163
Wear Irons for Concave Bodies, (*ill.*)	147
Weight of Bar Iron	163
" Common Axles	164
" Elliptic Springs	164
" Round Iron	164
" Square "	164
" Wrought-iron and Steel	165
Wings of Landau, (*ill.*)	141

PART III.

PAINT DEPARTMENT.

A B C Monogram, (*ill.*)	210
A C G Monogram, (*ill.*)	215
Action of Water on Varnish	285
All Sharp Edges	280
Alphabets	295
A Mantle, (*ill.*)	210
American and English Painting	257
A Natural Palette	259
An Old Body	256
Another Method	274
Apaumee, (*ill.*)	227
A P Monogram, (*ill.*)	212
Art of Coach Painting	171
Asphaltum	248
Axle Bed, Ornament and Stripe, (*ill.*)	235
Barred Helmet, (*ill.*)	229
Basket Work	263
B German Text, (*ill.*)	207
Bismarck	187, 252
Blacks	170
Black Turning Green	285
Blenders	174
Blistering	285
Blues	169, 203
Breast Plate, (*ill.*)	229
Brilliant Yellow	253
Broad Stripe, (*ill.*)	233
Browns	169, 203
Browns Follow Yellow	253
Bronze Powder	254
Brushes	173
Buck's Head, (*ill.*)	224
Buff	187, 202
Burnt Sienna	248
Burnt Umber	248
Cadmium Yellow	247
Canary Color	187, 203
Cane "	203
" Work, (*ill.*)	244
Cans for Oil, Varnish	181
Care of Varnish Brushes	282
Carmine	248, 253, 254
" Carriage	253
Causes of Varnish Pitting	281
C. A. V. Monogram, (*ill.*)	214
C. B. " (*ill.*)	213
Chairos—Curo. and Flat Tints	168
Chamois	178
Changeable colors	187
Chatamuck Lake	248
Chocolate	187
Chrome Green	184, 248
" Yellow	247
" " and Black	184
C. I. N. Monogram, (*ill.*)	214
Cinnamon	203
Citron	187
Claret Color	249
Coagulation of Mixed Paints	266
Colored Grays	187

INDEX.

Color Items	253	Glazing	255
Colors	182, 247	Goat's Head, (*ill*.)	238
" and the Eye	248	Gold Beating	272
" from Wolfram	255	" Color	187, 254
Complementary Colors	167	" Leaf	271, 272
Compound Colors	186	" Paint for Striping	249
" Striping Colors	202	" Striping	252
Concerning Slush	261	Grass Green	203
Corn Color	202	Green Leaf	203
Corner Patterns, (9 *ills*.)	231 to 233	Greens	169, 187
" Piece, (*ill*.)	233	G Roman Ornate, (*ill*.)	208
" Scroll, (*ill*.)	232	Ground and Striping Colors, etc.	199
Cream Color	257	" Colors	18
Cremnitz White	247	" for Ultramarine Blue	253
Crimson Lake	248, 254	Hairing Off	284
C. R. Monogram, (*ill*.)	214	Heavy Felt	256
C. T. B. " (*ill*.)	213	" Varnishing	255
Dark Rich Brown	187	H R G Monogram, (*ill*.)	215
Deep Buff	187, 202	Imitation Cane Work	244
" Sea Green	252	Indian Red	183, 247
D. E. F. Monogram, (*ill*.)	210	Initial Letters	206
D. German Text, (*ill*.)	207	" Letter B, (*ill*.)	207
Drabs	203	" " D, (*ill*.)	207
Dragon's Blood	247	" " G, (*ill*.)	208
Drop Black	184, 248, 254	" " P, (*ill*.)	208
Dusters	174	" " R, (*ill*.)	209
Dutch Pink	247, 254	" " S, (*ill*.)	209
Emerald Green	248	" " V, (*ill*.)	209
English Black Japan	261	Irregularities on Surface of Varnishes	190
" Purple Lake	248	Is the Milky Appearance, etc.	285
Enlarging Letters	207	Ivory Black	248
" Ornaments	246	Japans	188, 255
Enriched Panel Stripe, (*ill*.)	237	J E C Monogram, (*ill*.)	215
Enrichment of a Hub, (*ill*.)	243	J K " (*ill*.)	210
E. P. Monogram, (*ill*.)	215	King's Yellow	247
Escutcheon, (*ill*.)	222	Lake	185, 248
Extra Colors	170	Lamp Black	184, 248, 253
Facing a Body	267	Laying Gold Leaf	272
F. E. S. Monogram, (*ill*.)	215	Lead Color	187
Finished Scroll, (*ill*.)	229	Leafing, (*ill*.)	229
Finishing Coat	196	Length of Time, etc.	285
Fitch Hair	283	Less Japan and more Oil	260
Flake White	247	Lettering Pencils	175
Flesh Color	187, 203	Leveling Varnish	196, 279
Florentine Lake	253	Light and Ventilation	172
French Gray	187, 202, 252	" Buff	187, 202
From Priming to the Finish	274	" Red	247
Gall Stone	247	Lilac	187, 203
Garter	221	Linen Color	187
Gearings	265, 276	Little Things	290
German Text	206	L M Monogram, (*ill*.)	210
G H I Monogram, (*ill*.)	210		

INDEX.

Magenta Lake	248
Maroon	187
" Lake	248
Massicote	247
Mediums	170
Menhaden Oil	291
Method of Varnishing	277
Milky appearance on Varnish	285
Milori Green	250
Miscellaneous Items	255
Mixture of Striping Colors	201
Monograms	209
Monogram A B C (*ill.*)	210
" D E F "	210
" G H I "	210
" J K "	210
" L M "	210
" N O P "	211
" Q R S "	211
" T U V "	211
" W X Y "	212
" Z & "	212
" A P "	212
" P S "	212
" P S A "	212
" U S A "	213
" F R A "	213
" S A P "	213
" C B "	213
" C T B "	213
" N I B "	214
" C R "	214
" C A V "	214
" C I N "	214
" E P "	215
" F E S "	215
" H R G "	215
" S N G "	215
" A C G "	215
" J E C "	215
" S K J "	216
" M E V "	216
" M S "	216
" N P "	216
" O M L "	217
" O V I "	217
" O T S "	217
" O M T "	217
" V A T "	217
" T R A "	217
" S M T "	217
" S W B "	218
More Oil	260
Naples Yellow	247
Neglect of the Person	289
New Apprentices	256
New Colors	254
New York Red	187
N I B Monogram, (*ill.*)	214
Noble & Hoare's Private Mark	279
" " Varnish	285
N O P Monogram, (*ill.*)	211
Normal or Pure Gray	187
N P Monogram, (*ill.*)	216
Oil Colors in Patent Tubes	169
Oils, Japan and Varnish	188
Olive	187
Olive Green	249, 250
O M L Monogram, (*ill.*)	217
O M T Monogram, (*ill.*)	217
Opera Board Scroll, (*ill.*)	240
Orange	187, 203
Orange Mineral	254
Ornamental Panel Striping, (*ill.*)	236
Ornamental Work for Sleighs	236
Ornamented Alphabet	205
Ornamented Broad Lines, (7 *ills.*)	234, 236
Ornament for Sleigh Panel, (*ill.*)	240
Ornamenting Pencils	175
Ornaments, (41 *ills.*)	218 to 230
Ornaments composed of Scrolls, (5 *ills.*)	238 to 241
O T S Monogram, (*ill.*)	217
O V I Monogram, (*ill.*)	217
Paint Brushes	174
Paint Cracking	257
Paint Cracks	267
Painters' Colic	257
Painting Bodies	263
" Buggies	269
" Coach	192
" Irons	266
Paint Mills	176
" Pots	175
" Shop	171
" Stone	177
Palette Knife	178
Panel Colors	182, 185
" Stripe	237
Patched On	195
Patent Yellow	247
Pea Green	187, 203
Pearl Color	187
Pencil Grease	256
Pencils	174

INDEX.

Pencils for Striping	253
Permanent Wood Filling	265
Perspective	168
Pigment	253
Pink	203
Pitting	191
" Checked	282
Practical application of Oil Colors	170
Practice with the Pen	271
Principles of Coloring	166
Prussian Blue	247
P S A Monogram, (*ill.*)	212
P, Two Styles, (*ill.*)	208
Pulverized Pumice Stone	259
Pumice Stone	255, 256, 260
Pure Gray	187
Pure Toned Straw Color	250
Purple	187, 203, 254, 255
" Lake	248
Putty	268
Puttying	275
Putty Knives	178
Q R S Monogram, (*ill.*)	211
Quick Drying Color	254
" " Putty	269
" " Rough Stuff	265
Raw Umber	183, 248, 254
Red Lead	247
" " and Umber	267
Reds	169, 203
Removing Paint	262
Resolvents for Removing Paint	262
Rigid or Stiff Ruler	245
R Ornamented, (*ill.*)	209
Rose Lake	248
" Pink	247
Rough Stuff	257, 265, 275
Royal Purple Lake	248
" Red	254
Rubbing Coats	197
Running Gear	198
Salmon Color	187
S A P Monogram, (*ill.*)	213
Saxon Green	248
Screens	180, 263
Scroll and Fine Lines, (4 *ills.*)	236
Scrolled Ornament, (*ill.*)	241
Scroll, (4 *ills.*)	223
Sea Green	187, 252
Secondaries	166
Secondary Hues	166
Shammy	178
Sharp Edges	280
Shield and Monogram, (*ill.*)	229
Sienna	247
Silver Color	202, 254
Size of Rulers	245
Sizing for Gold Leaf	273
Skeleton, (*ill.*)	229
Sleigh Painting	256
Slush	261
S M T Monogram, (*ill.*)	217
S N G Monogram, (*ill.*)	215
S Ornamented, (*ill.*)	209
Spoke Face Ornaments, (11 *ills.*)	241 to 243
Sponges	177
Spontaneous Combustion	269
Stone Color	187
Straw Color	187, 202, 254
Striping Colors	199, 201
" in Gold	273
" Pencils	268
" Pens	270
Styles of Striping	201
S U B Monogram, (*ill.*)	218
Surface of Varnish	190
Tan Color	187
Terre Verte	248
Test for Japan	255
The Art of Coach Painting	171
The Captive Fly	286
The Carriage Part	198, 276
The Coloring	238
The Excessive Use of Tobacco	290
The First Coat	279
The Paint Shop	171
The Paint Stone	177
The Palette Knife	178
The Principles of Coloring	166
The Rubbing or Leveling Varnish	279
The Shammy	256
Tiger Scroll, (*ill.*)	239
Tobacco	290
To Color a Body	251
To Fasten Hair in Pencils	267
To Lay Out Imitation Cane Work	244
To Mix Gold Bronze	251
" Paint a Light Carriage, three weeks	259
" " Carmine	264
" " Light Buggies	274
" Prevent Gold Leaf Adhering	274
" Remove Varnish Cracks	281
T R A Monogram, (*ill.*)	213, 217
Transparent Colors	251, 253
Treatment of Spoke Face	241

XII   INDEX.

Trestles... 179	Warm Feet...................................... 256
Tuscan Red........................248, 254	Washing Windows........................ 256
T U V Monogram, (*ill*.)...................... 211	" with Turpentine............. 256
Ultramarine Blue..............186, 248, 254	Water Buckets................................. 177
Umber Toned Drabs........................ 187	" on Varnish............................. 285
Unique Alphabet............................. 205	Wheel Board or Prop....................... 181
U S A Monogram, (*ill*.).................... 213	Where should Varnish be kept........ 284
	White and Straw Color, etc............ 250
Vandyke Brown................................ 248	" Broad Line............................... 202
Varnish.. 189	" Chalk..................................... 247
" Brushes...........................173, 283	" Lead............................. 247, 255
" Cracks................................... 281	Whites... 169
Varnishing....................................... 201	Wine Color....................................... 187
" Body.. 195	Winged Eclipse, (*ill*.)...................... 220
" Carriage Part......................... 278	Wire Stand....................................... 187
Varnish Items.................................. 286	Worthy a Passing Thought............ 288
" Pitting.................................... 281	W X Y Monogram, (*ill*.)................ 212
V A T Monogram,(*ill*.)................... 217	Wyvern, (*ill*.)................................. 224
Venetian Red................................... 247	Yellow Arsenic................................ 247
Ventilation.............................172, 287	" Ochre..................................... 247
Verdigris................................186, 248	" Orpiment............................... 247
Vermilion.. 248	Zanzibar Gum................................. 257
Violet.. 187	Zinc White...................................... 247
V Ornamented, (*ill*.)...................... 209	Z & Monogram, (*ill*.)..................... 212

PART IV.

TRIMMING DEPARTMENT.

About Stock..................................... 340	Cloth Broad Lace, etc., (2 *ills*.)........ 331
Apron.. 321	Corded Straps, (2 *ills*.)................... 330
Arm Piece and Quarter, (*ill*.).......... 299	Covering Dashes............................. 328
A Simple Cutting Machine, (*ill*.)..... 339	Cushion for No-Top Buggy............ 325
Back, (*ill*.)..................................... 307	" Hook...................................... 322
" Boot, (*ill*.)............................... 338	" Top with Folds....................... 325
" Cushion and Fall, (*ill*.)............ 307	" and Fall, (*ill*.).............. 307, 308
" for Landau Sleigh, (*ill*.)........... 302	" " Sewed together,(*ill*.)......... 316
" of Light Top, (*ill*.)................... 333	Cutting Machine, (*ill*.)................... 339
" Straps..................................... 330	" Stock...................................... 340
" Supporter................................ 337	Dash.. 321
" Valence, (*ill*.).......................... 330	Door Fall, (*ill*.)............................... 299
" with Iron Armpiece, (*ill*.)......... 309	Drab Cloth....................................... 337
Black Leather Varnish..................... 337	Draft of an Extension Top, (*ill*.)..... 336
Bows, To Prevent from Springing.... 323	Drag Cushion and Fall, (*ill*.).......... 313
Broad Lace...................................... 331	Driving Cushion for Dog Cart, (*ill*.)... 317
Buggy Seat and Fall, (*ill*.).............. 311	Drop Backs, (*ill*.)................ 305, 307, 309
Card Pocket..................................... 300	End Finish of Drop Backs............... 313
Cheap Back Straps for Hack............ 330	Extension Top, (*ill*.)....................... 336
Clarence Doors, (2 *ills*.).................. 300	Fall for Dickey Seat, (*ill*.).............. 323

Fall for Landau Sleigh, (*ill.*).............. 303
" Coach or Phaeton Door........ 302
Fall Pattern, (*ill.*)............................ 321
Fitting on Top after Stuffing............ 334
" Side Quarters...................... 331
Form with which to Stitch Knob Holes, (*ill.*)................................... 339
Head Lining Close Tops.................... 330
Horse Shoe Back, (*ill.*)..................... 309
How to Clean Drab Cloth................ 337
" Make a Back Boot, (*ill.*)........ 338
" " the Cushion................ 316
" " the Fall...................... 316
" " the Front Facing........ 326
" " the Valence, (*ill.*)........ 332
" " Trimming for Lazy Back 331
Improvement, Trimming Turn-over Seat, (*ill.*)................................... 314
Inside Round-front Clarence, (*ill.*) 301
Knife for Cutting Cord,(*ill.*)............. 322
Knob Holes... 339
Landau Back Quarter, etc., (*ill.*).......... 298
" Sleigh Trimming, (*ill.*).......... 302
Lazy Backs, (*ill.*)........................312, 331
Leather Blacking................................ 337
Leather Varnish................................. 337
Light Buggy Seat, (*ill.*)..................... 310
" Top, (*ill.*)............................... 333
Method Applied to Close Top Gig, etc 297
" " " No Top Wagon..... 295
" " " Top Buggy.......... 295
" in Trimming Shop...................... 294
Mixing and Cooking Paste................ 337
New Box Loop.................................... 313
Park Phaeton Seat, (*ill.*).................. 306
Paste That Will Keep...................... 337
Patent Leather................................... 341
Patterns for Stick Seat Rolls, (*ill.*)...... 326

Plaited Welting, (*ill.*)....................... 313
Roll Stick... 321
Round Cross Straps, (*ill.*)................. 329
Rustic Trimming, (*ill.*)..................... 320
Scolloped Back Valence, (*ill.*)............ 343
Setting Bows...................................... 335
Setting Tubular Bows by Draft, (*ill.*).. 335
Softening Buggy Tops....................... 336
Spring Cushion.................................. 324
Squab Top Cushion........................... 324
Stock.. 340
Strips for Cording............................. 337
Summer Tops..................................... 334
The Back Part of Close Tops............ 334
" Cushion Front........................ 304
" " Top and Fall............... 305
" Door Fall, (*ill.*)..................... 299
" Grecian Bend Drop Back, (*ill.*)... 310
" Phœnix Shaft Straps................ 320
" Pocket Fall, &c., (*ill.*).............. 324
" Saratoga, (*ill.*)........................ 303
" " Back............................. 304
" Star Top Cushion, &c., (*ill.*)......... 318
" Swinging Holders.................... 300
Three Edged Raiser Machine, (*ill.*)... 339
To Make Shell Work, &c................. 305
To Make the Back............................ 303
To Mend a Broken Bow................... 340
Top of Cushion, (*ill.*)....................... 315
To Prevent Front and Back Bows from Springing................................... 323
To Prevent Head Lining, &c............. 330
To Prevent Tops of Dashes, &c......... 328
Trestle, (*ill.*)..................................... 296
Trimming for Landau Sleigh, (*ill.*)..... 302
" " Light Road Wagon, (*ill.*) 319
Trimmer's Don't Use Sticks............. 336
Twisted Leather Seaming Cord, (*ill.*)... 329
Valence... 332

PART V.

MISCELLANEOUS DEPARTMENT.

A Safe and Profitable Investment...... 367
American Timber Interests............... 368
Apprentices....................................... 355
Are They Competent Judges............ 346
A Word to Apprentices..................... 355

Be Careful of Your Credit................ 346
Be Ready in Time............................. 351
Beware of Sharpers........................... 345
Blacksmithing in Germany............... 366
Boxwood 356

INDEX.

Carriage Materials............................ 362	System 344
Coldness.. 351	System in Carriage Shops................. 360
Competent Judges.............................. 346	The Cover... 350
Competition 357	The Electrotype................................. 356
Convenient and Comfortable Shops.... 344	The Grinding of Colors..................... 359
Credit 346	The Leather....................................... 350
Electrotyping 356	The Reins as Held by the Fair Sex... 364
From Farm Wagon to Landau........... 352	The Washing...................................... 350
Goddard Method of Painting............. 363	The First Operation........................... 358
Greasing... 350	The Sarven Patent Wheel.................. 365
History of Wood Engraving............... 355	To Employers..................................... 360
Order and Disorder............................ 342	Water-Proof Glue 367
Out of Work.. 353	Where Carriages Should be Kept...... 366
Sharpers... 345	Wood Engraving................................ 356
Sponges and Chamois........................ 350	Workshops 344
Success in Business............................ 343	Wrenches.. 350

COACH-MAKERS'
Illustrated Hand-Book.

PART FIRST.—WOOD-WORK DEPARTMENT.

SCALE DRAFTING.

THIS, to the progressive body-maker, or learner, is one of the most important subjects we could treat upon, and while we feel satisfied that we fully understand the subject on which we are writing, we are aware of the difficulty before us of making it plain to our readers. Our object is to give a clear and full illustration of the whole system from the beginning to the *end*, showing the kind of instruments necessary by cuts, with description, together with the kind of material required to make a perfect draft. New beginners will find much to discourage them, but by constant practice and perseverance the object can be obtained.

Scale drafting is reducing a carriage, or any other object, to any given scale and retaining its proportions—half inch to the foot being generally used by coach draftsmen. As a general rule, most of the present styles of carriages originate in the scale draft, and are then transferred to the blackboard to full size.

We have divided the sets of drawing instruments into three classes, viz.: brass, fine German silver, and extra fine Swiss. The brass instruments are intended for schools; the fine German silver and the extra fine Swiss instruments for practical carriage draftsmen.

Without the aid of some drawing instrument, a student cannot obtain a thorough knowledge of geometry or trigonometry; but as very few who go over these branches in youth ever make any practical use of them in after life, it is not necessary that the drawing instruments which are furnished to schools should be any finer in finish and quality than is sufficient for a clear demonstration of the problems.

But with the practical carriage draftsman his drawing instruments are next to his head and hands, and they must be of the best material, well and accurately finished. Being in constant use, and if they are not perfectly correct, the loss and delay occasioned by them, in one instance, will be much greater than the cost of a good set of instruments, which can be used his life-time.

The fine German silver drawing instruments meet the wants of the practical man. The extra fine Swiss drawing instruments are more nicely finished than the fine German silver; the metal of which they are made resembles more closely pure silver;

they are more substantial in construction, and consequently more durable. As a general rule, draftsmen give the preference to extra fine Swiss drawing instruments.

Having made these general remarks we will now proceed to describe each of the instruments required, their use and how to use them:

DRAWING BOARD

is a rectagular frame of walnut, with an open center, in which a soft pine board, carefully planed and perfectly smooth, is fitted and fastened with buttons. The frame is made of hard wood, so as not to wear easily and become incorrect, and the center of soft wood, so the fastening pins can be easily put in. The angles and edges of the frame should be as correct as possible, for resting the head of the T-square against.

T-SQUARE.

The T-square is usually made of hard (pear) wood, having the head permanently and securely fastened at right angles to the blade, and a secondary head of the same size attached to it with a clamp-screw, and thus, when other angles than right angles are to be made, the movable head can be fixed at the proper inclination to the blade, while a right angle is still maintained by the fixed head. The blade is attached between the two parts of the head, so that in using either the fixed or movable side, there is an edge to come against the drawing-board, while the blade rests on the board.

The T-square is always used in connection with a drawing-board, and with it all the straight and parallel lines of a drawing are very easily added—the head of the T-square being held against the edge of the board, and, by sliding the head along the edge of the drawing board, parallel lines can be drawn. The edges of the blade of the T-square are apt to get rough from constant use; to prevent this, and also to make the blade stiffer and less liable to warp, a thin strip of brass is set into the edges, and finished off smooth and true.

DRAWING PEN.

This is a most important instrument to every carriage draftsman, and should be well made and always kept in good order. It consists of two steel blades, attached to an ivory handle, and so bent that when the points are almost touching, there is a space between the blades for holding ink. One of the blades is hinged where it joins the handle, so that it can be opened away from the other blade when it is to be cleansed. A steel screw, having a German silver head, is passed through the hinged blade, and screws into the other blade; by turning this screw the points can be brought to the distance apart for making the required thickness of line. Size, $4\frac{1}{2}$ inches long, from the point of pen to the end of handle. To use the drawing pen, put the ink between the blades with a common writing pen, or a camel-hair pencil, drawing it down and out between the points of the blades; screw the blades to the proper distance apart for making a line the required thickness. In drawing the line the pen should be held firmly against the ruler, or pattern, slightly inclined in the direction the line is being drawn; the points of both blades must touch the paper. The handles of most drawing pens are made to unscrew, and a needle is fitted in the screw end, which can be used for pricking drawings from one paper to another.

ROAD DRAWING PEN.

For drawing close parallel lines, as moldings, a double drawing pen is used. It  consists of two drawing pens, attached parallel to each other on one handle; the distance of the two pens apart is regulated by the adjusting screw, between the end of the handle and the top of the pens.

IRREGULAR CURVES

Are made of wood or horn. A variety of curves are cut upon the outer edges, and pieces are cut from the body in such a manner that there is a curve for every side of the opening. These curves are used in designing carriages.

SPRING BOW PEN, WITH PENCIL POINT.

The leg, body and handle are made of one piece of German silver or brass, three inches long, for describing small circles from one-sixteenth of an inch to two inches in diameter, such as heads of screws, the hubs and tires of wheels, etc. The lower end 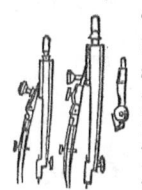 of the leg is finished with a small tube and clamp screw, for receiving and retaining a needle point; the body is almost twice the width of the leg, and a groove is cut the whole length of one of its sides; the pen or pencil point is attached to a tempered steel spring, the end of which is screwed fast into the upper end of the cut in the body; a steel wire, half an inch long, with a fine thread cut on it, is fastened into the body, and passes through the spring just above the pen or pencil point; a nut is screwed on the end of this wire, and bears against the spring, and forces it in or lets it out of the cut in the body, which brings the pen or pencil and needle point nearer together, or puts them farther apart.

STEEL SPACING DIVIDERS.

In carriage drawings it frequently occurs that a large number of small equal distances are to be set off, not only at one time, but repeatedly, upon the same drawing. For this purpose the ordinary dividers are too large and inconvenient to handle rapidly, and, having nothing but the joint to hold them in their position, are liable to get their extension altered. For such work there is used a pair of very delicate dividers, made altogether of steel, the two legs of which are united at the top by an arched spring, and drawn together or opened by the screw in the middle. On the top of the arched spring an ivory or German silver handle is attached by which the instrument can be quickly turned over and over when used in spacing off a number of equal distances. The size of the spacing dividers mostly used are three inches long, with the legs delicately rounded from the regulating screw to the points. The advantages by these spacing dividers are, greater nicety and accuracy of adjustment, and no liability of accidental change when once adjusted.

PARALLEL RULER.

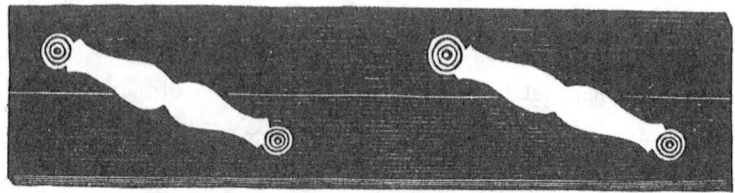

This consists of two straight edges, of ebony or metal, six inches long, by three-quarters of an inch to one and a half inches wide, joined together by two parallel strips of brass, which move upon pivots at the points where they are attached to the ruler; thus, when the bars are put apart they are always held parallel to each other by the brass strips, consequently, if the edge of one of the bars is brought to a line and firmly held there, and the other bar pushed away from it, a line or lines drawn by the second bar will be parallel to the original line.

TRIANGULAR SCALE OF BOXWOOD,

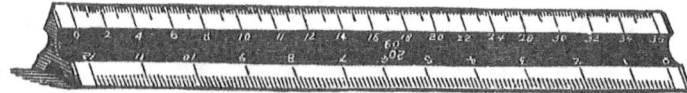

six inches long, has five edges, graduated with two scales on each edge, as follows: one edge has 3-32 of an inch, and 3-16 of an inch marked off; the 3-32 are numbered from one end, and the 3-16 from the other; one edge has $\frac{1}{8}$ of an inch, and $\frac{1}{4}$ of an inch; one edge has $\frac{3}{8}$ of an inch, and $\frac{3}{4}$ of an inch; one edge has $\frac{1}{2}$ of an inch, and 1 inch; one edge has $1\frac{1}{2}$ inches, and 3 inches; and one edge is divided into inches and 16ths of an inch. The first division of the 3-32 scale is divided into four equal parts; consequently, if the 3-32 represent one foot, each of the sub-divisions will represent 3 inches. The 3-16, $\frac{1}{8}$, $\frac{1}{4}$ and $\frac{3}{8}$ have the first division divided into twelve equal parts, therefore, if the primary division represents one foot, each of the sub-divisions will represent one inch. The $\frac{1}{2}$ and $\frac{3}{4}$ of an inch have the first division divided into twenty-four equal parts; therefore, if the primary divisions represent one foot, each of the sub-divisions will represent the half of an inch. The 1 inch and $1\frac{1}{2}$ inches have the first division divided into forty-eight equal parts; and if the primary division represents one foot, each of the sub-divisions will stand for $\frac{1}{4}$ of an inch. The 3 inches has the first division divided into ninety-six equal parts; and if the primary division represents one foot, each of the sub-divisions will represent the $\frac{1}{8}$ of an inch.

THE PLAIN DIVIDERS.

This instrument consists of two legs, the upper halves of which are made of brass or German silver, and the lower halves, or points, of tempered steel.

In the fine instruments, the joints about which the legs move should be framed of the two different metals—German silver and steel. By this arrangement the wear is much diminished, and greater uniformity and smoothness of motion is obtained. If this uniformity and smoothness be wanting, it is extremely difficult to set the legs quickly apart, at a desired distance; for, being opened and closed by the fingers of one hand, if the joint is not good they will move by fits and starts, and either go beyond or stop short of the point; but when they move evenly, the pressure can be so applied as to open the legs at once to the exact distance, and the joint must be sufficiently tight to hold them in this position, and not permit them to deviate from it in consequence of a small amount of pressure which is inseparable from their use. The joints of the dividers are tightened or loosened by inserting the two steel points of the key into the two small holes on one side of the head of the dividers, and turning from one to tighten it, and in the opposite direction to loosen it.

DOTTING PEN.

The dotting pen is made like the drawing pen, but has a finely-toothed wheel, which revolves between the points, and instead of a continuous ink line, it makes a dot for each tooth, and consequently a line of dots when drawn between two points. It is used when imaginary lines are to be shown on the drawing.

FASTENING TACKS

Are small nails used for fastening the paper to the drawing board. They have large flat heads, and very small, sharp points. The heads are round, and made of brass, German silver or steel, and the points of the best tempered steel, carefully sharpened.

In putting them into the drawing board, the point should be well started with the fingers, and the pin pushed home with a small bottle cork. If the thumb is used for pressing them in, there is danger of the upper part of the pin coming through the head and injuring the thumb.

A new form of fastening tack has just been introduced. It is a right-angled piece of metal, each side of which is one half an inch long, with three points. It is intended for fastening the paper at the corners.

Having completed our list of instruments, we will now proceed to give the material necessary for a perfect draft.

BRISTOL BOARD.

We can recommend Reynold's superfine drawing boards—the largest size being the thickest in quality.

Foolscap, 15 inches by 12.
Demy, 18 inches by 14.
Medium, 20½ inches by 15½.
Royal, 22½ inches by 17½.
Imperial, 28 inches by 20.

This you will cut to the size of scale required.

There is also a tinted paper that comes in various colors, and there is the tinted Bristol board, which comes the same size, and is preferred by some, being not so easily soiled by handling. You will also need the finest French vegetable tracing paper; this is used to lay over your draft, and tracing from the original for transmitting in letter.

Faber's pencils, No. 4.

INDIA INK.

Which comes: Lion Head, Round Gilt Ink, Large Square Gilt Ink, Large Octagon Ink, Liquid Gold Ink.

BRUSHES.

For brushes, you will require fine brown sables in quills, brown dyed sables in tin ferrules, with handles.

WATER-COLORS.

In water-colors we give the entire list, which comes in whole and half cakes; or moist water-colors, which retain, from processes and treatment known only to themselves, their solubility and dampness for an unlimited period, and a box of them, though laid aside for two or three years, will be found, when required again, equally moist and serviceable as when purchased.

Antwerp Blue.	Lamp Black.	Verditer.	
Bistre.	Light Red.	Yellow Ochre.	Green Oxide of Chromium.
Burnt Sienna.	Neutral Tint.	Yellow Lake.	Lemon Yellow.
Brown Pink.	Naples Yellow.		French Blue.
Blue Black.	New Blue.	Sepia.	Pink Madder.
British Ink.	Olive Green.	Warm Sepia.	Rose Madder.
Brown Ochre.	Orpiment.	Roman Sepia.	Intense Blue.
Burnt Roman Ochre.	Prussian Blue.	Brown Madder.	
Burnt Umber.	Prussian Green.	Constant White.	Mars Orange.
Chrome Yellow, 1, 2 and 3.	Payne's Gray.	Chinese White.	Pure Scarlet.
Cologne Earth.	Raw Sienna.	Indian Yellow.	Burnt Carmine.
Dragon's Blood.	Raw Umber.	Crimson Lake.	Smalt.
Emerald Green.	Roman Ochre.	Scarlet Lake.	Purple Madder.
Gamboge.	Red Lead.	Purple Lake.	Ultramarine Ash.
Hooker's Green, No. 1.	Red Ochre.	Mars Brown.	Carmine.
Hooker's Green, No. 2.	Red Chalk.	Mars Yellow.	Gallstone.
Indigo.	Sap Green.	Scarlet Vermilion.	Cadmium Yellow.
Indian Red.	Terre Verte.	Chalon's Brown.	Orange Vermillion.
Italian Pink.	Vandyke Brown.	Black Lead.	
Ivory Black.	Venitian Red.		Ultramarine.
King's Yellow.	Vermilion.	Cobalt Blue.	Ditto. Quarter Cake.

DIAGRAMS.

Diagram No. 1 illustrates the manner of laying out for a correct scale draft of the "Monitor" pattern, being the simplest one we can treat upon.

First, we shall establish the base line, holding that the base is the foundation of all mechanical structure. Second, the height of the wheels; carriage-part to the top of spring-bar; distance of the same apart. Third, the laying out of body, also the top in two positions, demonstrating the points for the knuckles in the joints. Fourth, we illustrate the point where the front wheel will strike the body in turning, establishing the place for the wear-iron, also the space between the front and hind wheels on a turning point—having your drawing-board, with paper secured to it with fastening tacks. It will be necessary for the beginner to be well versed in the *scale rule* or triangular scale. The scale that we are using to this drawing is half inch to the foot (U. S. standard), the half inch being divided into twenty-four equal parts, each part representing a half inch, each two parts one inch.

We will now proceed and lay out the pencil sketch. (Use Faber's No. 4.) Place the T-square on the right hand edge of the board, and draw the base line A. Next establish the height of the wheels, which is 4 feet and 4 feet 4; also, the distance from center of wheels or axles, which is 54 inches. This distance allows a 50-inch body on bottom, and gives the right space for forming the body loops. Set your pencil compasses 26 inches, being half the height of the hind wheels from the center point established, and 24 inches being half the height of the front wheels; also, draw the hubs $3\frac{1}{4}$ inches diameter; bands, $2\frac{3}{8}$. Now proceed and lay out for the height of the spring bars—back bed, $2\frac{1}{4}$ sweep, $1\frac{3}{8}$ deep in center; springs, 11 inches over all, $1\frac{1}{4}$ wide, $1\frac{1}{2}$ inches deep for the spring bar, making in all $15\frac{5}{8}$ from center of the hind wheel to the top of spring bar. The front bed 2 inches sweep, $1\frac{3}{8}$ deep; fifth wheels, 1 inch; spring-block, $1\frac{1}{2}$; springs, $10\frac{3}{4}$ over all; spring-bar, $1\frac{1}{2}$ deep, making in all $17\frac{5}{8}$ from the center of the front wheel to the top of spring bar, which levels for the loops. We give these dimensions of the carriage, being a matter of taste, but the principle illustrated remains the same.

SCALE DRAFTING.

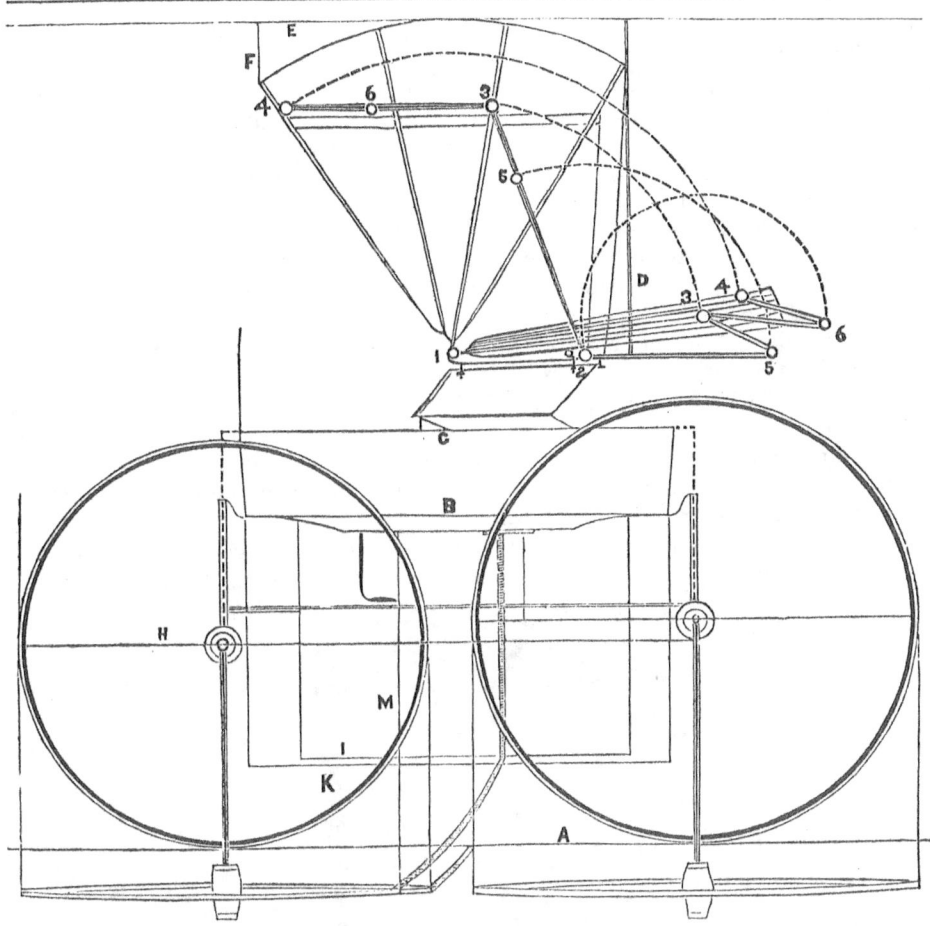

DIAGRAM NO. 1. HALF-INCH SCALE.

Proceed and lay out for the width of springs and draw them. Measure from base line A 39 inches; draw line B, which is the bottom line of body, which drops $2\frac{1}{2}$ inches from top of spring bar. Measure from B 10 inches, and draw line C, top line of body. Measure on line B, from front and back springs, $2\frac{1}{2}$ inches, and prick it off, which leaves the body 4 feet 2 inches long on the bottom line. Now proceed and draw the two ends, giving $\frac{3}{4}$ inch bevel to each.

When they are drawn square, the top line of body C will appear shorter than bottom line B. Measure from line B 2 inches, and draw the concave rocker; also, sweep the ends as represented by the diagram. You can now dot off the loops, steps, perch, and lay out the dash, which is 11 inches from top of body. Measure from front of body, on line C, 21 inches; this being the front of the seat, $15\frac{3}{4}$ for the depth, seat elevated sufficient to bring the bottom edge of skirting on top line of body. Measure $4\frac{3}{4}$ inches from seat frame for seat panel; give $\frac{1}{4}$ inch slope, and draw this line. Lay off the back and front bevels to suit taste. You will now lay out and draw the line for shifting rail; turn the gooseneck for the slat irons with the pencil string-bow pen. Measure from top line, back of seat, 4 inches; draw perpendicular line D; also measure 45 inches from top of seat frame; draw the top line E; measure from line D 42 inches, and draw line F, this being the height, length and back flare of top. The top sweep has 4½ inches round, 1 inch lower front than back; it is on a circle of 75 inches. The needle point of compasses is placed at the center point of gooseneck for receiving the slat irons; this will strike the sweep. Proceed and space off and draw the bows, and establish the props, Nos. 2, 3 and 4, and draw the joints for establishing the different points for the stub joints, illustrated in top laying down by Nos. 5 and 6.

We will now lay out for establishing the point where the front wheel will strike the concave rocker, and where the wear iron should be placed. You draw center line H, being the center of body, and the center point of the front wheel, representing the center of king bolt and the center of axles. Line I represents the outside of concave rocker, and K the body.

Draw the wheels below line A, as represented in the diagram; these represent half the track from center line H. Draw perpendicular line M; this takes the size of the wheel on a line with the concave rocker. Now, with the pencil compasses place the needle point in center of front wheel, being the center of body-bolt, which strikes the circle to line I, which represents the concave in the rocker. From this point, squared to the body, will give the exact place for the wear iron. The short circle shows the distance between the front and back wheels in turning.

This principle illustrates how close the front and back wheels can be brought together with safety.

DIAGRAM NO. 2. HALF-INCH SCALE.

SCALE DRAFTING.

In diagram No. 2 we illustrate the completion of the pencil sketch with India ink.

An India ink draft, with lines drawn and shaded correctly, will resemble a steel engraving. To acquire this art, considerable practice and the use of the best quality of drawing instruments will be necessary. We have before illustrated these instruments, but we here give the same list with the price attached to each instrument, subject to variation in cost of material.

Pair plain dividers, Swiss, $2.00; German silver, 85 cents.
Road drawing pen, Swiss, $3.75; German silver, $2.50.
Spring-bow pen with pencil point, Swiss, $3.80; German silver, $3.25.
Spacing dividers, Swiss, $1.75; German silver, $1.75.
Dotting pen, Swiss, $2.60; German silver, $1.00.
Drawing pen, Swiss, $1.50; German silver, $1.00.
Fastening tacks per dozen, 75 cents.
T-square, $1.75.
Parallel ruler, 35 cents.
Drafting scale, triangular boxwood, 6 inches, $2.00.
Irregular curves, 50 cents.
Drawing boards, $2.50.

In preparing for inking, neatness is the first thing in order, being different from penciling, as false lines cannot be erased without spoiling the drawing. You will require a small china saucer. Dip the ink in clear water, rub it well on your finger, which makes a pure liquid free from grit, and from your finger allow it to drop into the saucer. When you have a sufficient quantity prepared, take the camel-hair pencil and fill the spring-bow pen with the fluid for inking the wheels. You can now practice as illustrated, by holding the top of the spring-bow between the thumb and forefinger; this will give you the full control of the pen in making the circular lines. Having practiced sufficiently with this on a separate piece of paper to illustrate to you in regard to thickness and evenness of the line, you will be prepared to apply it to the drawing. The width of the line may be regulated by the set-screw on the pen.

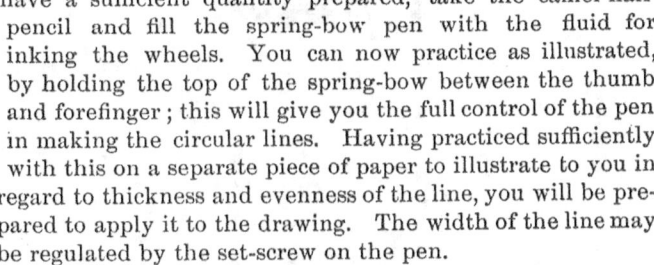

If there should be any imperfection in the line, you will open your pen and clean out with a piece of chamois skin, and if the pen should require sharpening, use a piece of fine clay stone with water; fill the pen again with ink, as before described, and practice until a perfect line or circle is obtained. Having this, you will now commence and strike the hubs of the wheels as laid out in the diagram; next strike the tire and rims; then, with your spring-bow pen, ink the props, prop nuts, knuckles to joints. Always ink *first* that which is shown on the outer surface. With the drawing pen filled with fluid, practice in the same manner as with the bow pen, using a straight edge for your guide. When the line is satisfactory, commence and ink the spokes and joints to top, as in the diagram, keeping the center of the hub for your guide. All other lines are inked in the same manner, using the T-square for the parallel and perpendicular lines. The circle of the top is described by dividers with changeable points, the props and knuckles with the spring-bow pen, as was illustrated in Diagram No. 1.

When your inking is complete, you will clean off by erasing with India rubber all pencil lines which were used for laying out for a correct draft. For shading you will require another saucer like the one for inking, for reducing or getting the required tints. With a small sable hair pencil reduce the ink with water to the shade you want. This can be obtained by practicing on a separate piece of paper in spaces of the size of panels of the body, also seat and top. All parts that stand in the foreground should be light; those that stand in the back, dark. The shading of this draft comes next in art to the coloring of a draft, which, to a beginner, might seem discouraging, but he will soon accomplish it by practice.

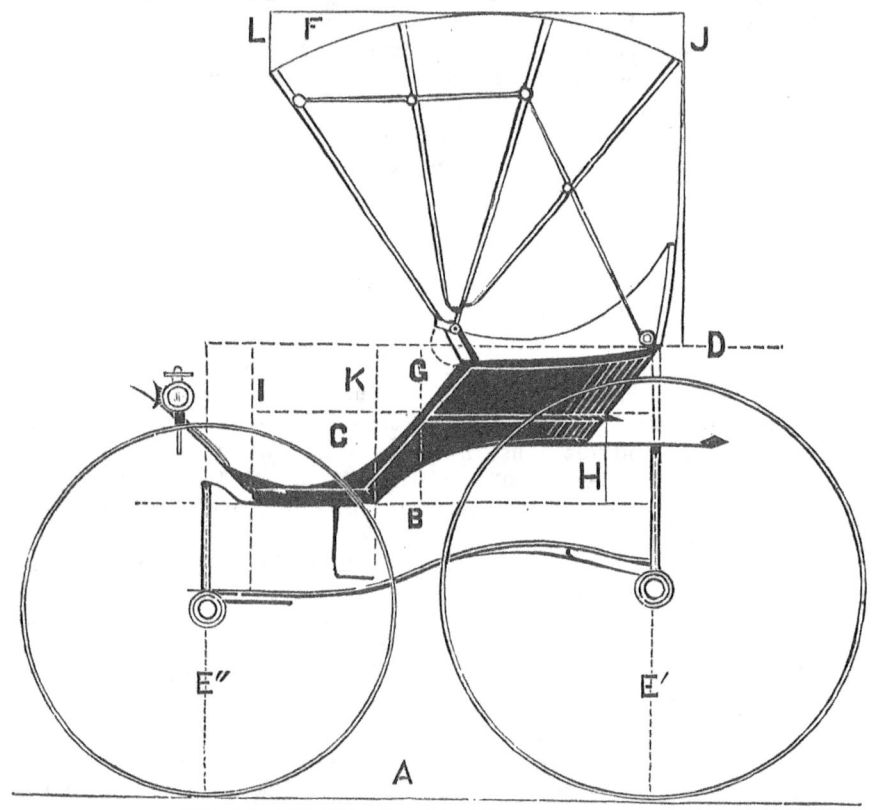

DIAGRAM NO. 3. HALF-INCH SCALE.

The accompanying diagram, No. 3, is a drop-front, close-top buggy, which, by its varying lines, requires the use of "sweep" patterns, which are represented and explained in this connection.

Commence by drawing base line A. Measure 35 inches from base line A, and produce dotted line B, being the full depth of the drop-front. From this line measure 11½ inches, and draw dotted line C, which represents the height of seat frame. From line C measure 8 inches, and draw dotted line D, being full height of seat. Measure 45½ inches from dotted line C, and draw line F, showing the square of the full height of top.

Having established the full height of the buggy from the base line A, noting the different points as we ascended to line F, we will proceed and lay out for the length. Measure 51½ inches, and draw the vertical dotted lines E' E'', which represent the centers of front and hind wheels, which are 3 feet 8 inches, and 4 feet 1 inch. Measure from base line A 24½ inches, being half the height of hind wheel; and also 22 inches from base line A, on line E″, being half the height of front wheel. Now, with your dividers carrying the pencil point, strike the wheels and afterward the hubs. (It should be borne in mind that we are making a pencil sketch.)

We will now proceed, laying out for the height of pump-handle, the back bed arched 3 inches, 1½ inches deep; springs, 11 inches over all; spring bar, 1½ inches deep; making in all 17 inches from the center of hind wheel to the bearing of pump-handle At this point draw a horizontal line, which is the bottom of pump-handle.

Next, lay out for height of front spring bar. The front axle bed drops 1½ inches, and is 1½ inches deep. Fifth wheels, ½ inch each; spring block, 1½ inches; springs, 11 inches over all; depth of front spring bar, 1½ inches; making in all 15 inches from center of wheel to the bottom of body loop. Measure from vertical dotted line E″ 4½ inches, and draw vertical line I. This establishes the front bottom corner of body. From line I measure 15 inches, and produce vertical dotted line K, which gives the front of the arch where it intersects base line B. From dotted line I measure 20 inches, and draw vertical dotted line G, which, at its intersection of dotted line C, establishes a point which is the front of seat.

Measure from line G 21 inches, and draw line H, which gives the depth of seat. This is intended for a spring back. Lay off for the bevel of the seat and body, inclining them as fancy may dictate.

On dotted horizontal line D, from its point of intersection with dotted vertical line E′, measure off 3½ inches; and from this point erect the line J, and from line J produce line L, 47 inches distant, which gives the width of the top.

Having now completed the establishment of the main points, we will proceed to apply the sweep patterns. But before we explain their use, let us remark that in using irregular sweeps, it is the better plan to always use the round side of the pattern as the pencil or pen follows it with more precision.

The carriage draftsman should make his own sweeps; his first care being to select suitable wood, as the evenness of the edge depends, in a great measure, on the closeness of the grain of the wood.

The patterns for body designing should be taken from pear-wood veneer, or fine grain apple tree, the pear wood being preferable for its fine grain, and is not liable to split. It is absolutely necessary to have a clean smooth edge on the sweep pattern, which cannot be obtained by the use of rosewood or any other coarse-grained veneers. If the edge varies by being crossed with the grain, the trace of the pen will show an imperfection which should be avoided.

This degree of nicety might not be apparent to the generality of persons, but should you be required to delineate for the engraver, whatever imperfections of lines were on the blocks would be left, and the "proof" would plainly show the defects.

Irregular sweeps are also made of horn, which presents a true, firm edge, but the body-designer should provide himself with veneers, and form his own sweeps, for there are lines to be drawn which cannot be obtained in any other manner.

In finishing the sweep to suit the eye, use fine sand-paper, rounding the edge a trifle. Test the accuracy of the edge of the pattern on a separate piece of paper, and in no case apply it to the draft in hand until it is uniform throughout.

The first to be drawn is the "arch" sweep. To produce this, apply so much of the pattern (Fig. 1) as is contained between the Nos. 1 and 2, the back part coming in above

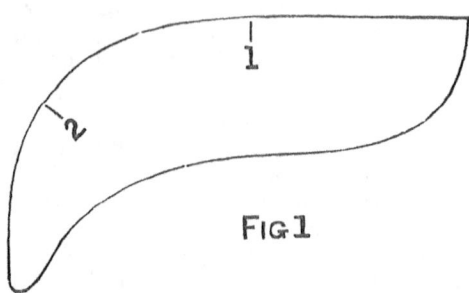

FIG. 1

the pump-handle, which leaves the pump-handle straight, giving the body a depth of 3 inches on a vertical line.

Fig. 2, between the Nos. 1 and 2, was made use of in producing the sweep on the

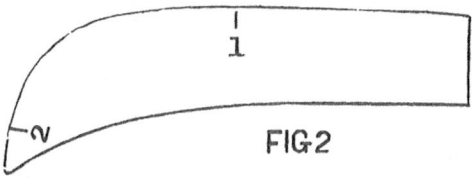

FIG. 2

front of body, and at other points on the pattern (not numbered); the dash and top line of seat were traced.

Fig. 3 is very useful for turning the sweep of slat-irons.

FIG. 3

Fig. 4, the perch sweep; line 1, center of hind spring; line 2, center of front spring.

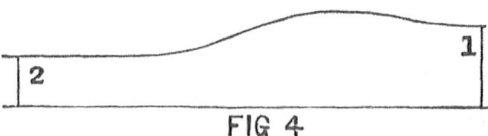

FIG. 4

Fig. 5, the loop sweep for bodies, varying to suit the different classes of work. Line

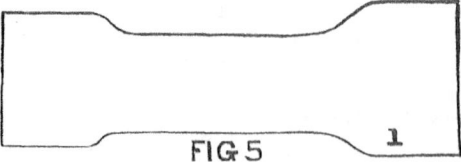

FIG. 5

1 is the sweep for front loop on the draft now under consideration.

Top is 5½ inches round; front bow 1 inch lower than the back one, and struck to a circle of 8 feet 2 inches.

The seat has a high back. The side arm-piece to which the trimming is attached is represented by the swept line shown above the top of seat, which is not seen from the outside with a close top.

SCALE DRAFTING. 29

To complete the draft, it is necessary only to add the joints, lamp and step.

As previously stated an India ink draft, when properly shaded, will resemble a steel-plate engraving. While the method of applying the ink is necessarily similar in every style of carriage sought to be represented, the student will discover that more skill is required to produce a clean and well-executed drawing on the outlines of a buggy, like the one here shown, in diagram No. 4, than on the straight, plain design in diagram No. 2.

DIAGRAM NO. 4. HALF-INCH SCALE.

It will be discovered by the beginner that he must use more dexterity in handling the sable-hair pencils, and there will be many failures before the hand will become trained to its work in giving those delicate touches which the experienced draftsman appears to perform without an effort. Practice alone will give that confidence which is requisite in making the drawing pen, or the inking pencils, obey the will. Nervousness, which is always experienced by the novice, will wear away gradually, and, after a time, what seemed to be a vexatious task will be the source of much pleasure.

The starting points in this draft, as in all others, will be to strike the rims of the wheels; next the tire line, which is the outside line that bounds the wheel; then follow with the hubs, circle of the top, prop nuts, knuckles of joints; last, the lamp.

We now refer you to the sweep patterns which were used in defining the pencil sketch.

The "arch" sweep, Fig. 1, may now be applied to the draft, using so much of it as is contained between the figures 1 and 2, using the pen instead of the pencil, as in the first instance. The remaining patterns are to be applied in the same manner, the only difference being to make sure work with the pen, for when ink becomes set in the grain of the paper it cannot be erased. In drawings of this kind it is necessary, in some instances, to change your pattern and piece the line. Here is a nice operation to perform, for the contact of the lines, where joined together, must not be perceptible. Practice alone, which gives a perfect control of the pen, will accomplish it. Another point to which we would direct the attention is, that where lines intersect, care should be taken to have them of the same size, also to avoid the defect of one line overreaching the other. It is preferable to run on a line starting from the end of the one already drawn, rather than to begin at a distant point on the pattern, and conclude the line at the intersecting point.

Having completed the inking of the lines laid down, erase all pencil marks, and then proceed with the shading, as previously explained. The draft now under consideration being a close-top buggy, it will require more skill to produce than the previous one, which was a roll-up-top. The larger the surface over which a tint is to be laid, the more difficult it will be found to lay on an even coating of ink, and graduate the shades so as to represent the roundness and receding portions. The joints must be white. To do this, the paper will be left clear throughout their outline, the ink to be worked up to the joints, forming them clean and sharp.

Lighting and shading is governed by the following principles: When light falls upon any object, it (the light) is reflected to a greater or less degree, according to the form of the object. A level surface would reflect light throughout its plane, with but a slight modification. But a globe, cylinder, or a rounded surface, as the top or back corner of this buggy, would appear lighter at the highest points; and as the planes recede they would gradually fall into shade. Any portion of a draft, then, which you wish to bring forward, must be represented lighter than those parts which are retiring.

We now take a four-seat and four-spring phaeton. (Diagram No. 5.)

Having begun by taking up a plain buggy, we next gave one with curved lines, and now we present to the student an extension-top phaeton.

By degrees we have led the way from plain work up to that requiring more skill, and without further comment will proceed to the explanation of the one in hand. Having secured the Bristol paper to the drawing board, draw base line A; from this line measure 37 inches, and draw dotted base line of body B. From line B measure $1\frac{1}{2}$ inches, and draw dotted line C, this being the base of the front. From line B measure $8\frac{1}{2}$ inches, and draw line D. From this measure 3 inches, and draw line F, this being full height of the back-seat frame. From F measure $1\frac{1}{2}$ inches, and produce line E, this giving the full height of frame on the front seat, elevating it $1\frac{1}{2}$ inches higher than the back seat. From E measure 44 inches, and draw line K, giving the full height of top.

Having established the different points in ascending, we will now lay out for width of door, by drawing dotted lines O and P, which are $17\frac{1}{2}$ inches apart. Next in order is the height of wheels. Hind ones 4 feet 2 inches. Measure from vertical line O 23 inches, and draw vertical line M; this allows the wheel to advance 2 inches front of door, and gives ample room for its opening.

Strike the hind wheel as heretofore given. Measure from center of wheel 12 inches for height of spring. From this point measure $1\frac{1}{8}$ inches for depth of back bar; this establishes the bottom of pump-handle, also the square cut in body, which we intend

SCALE DRAWING. 31

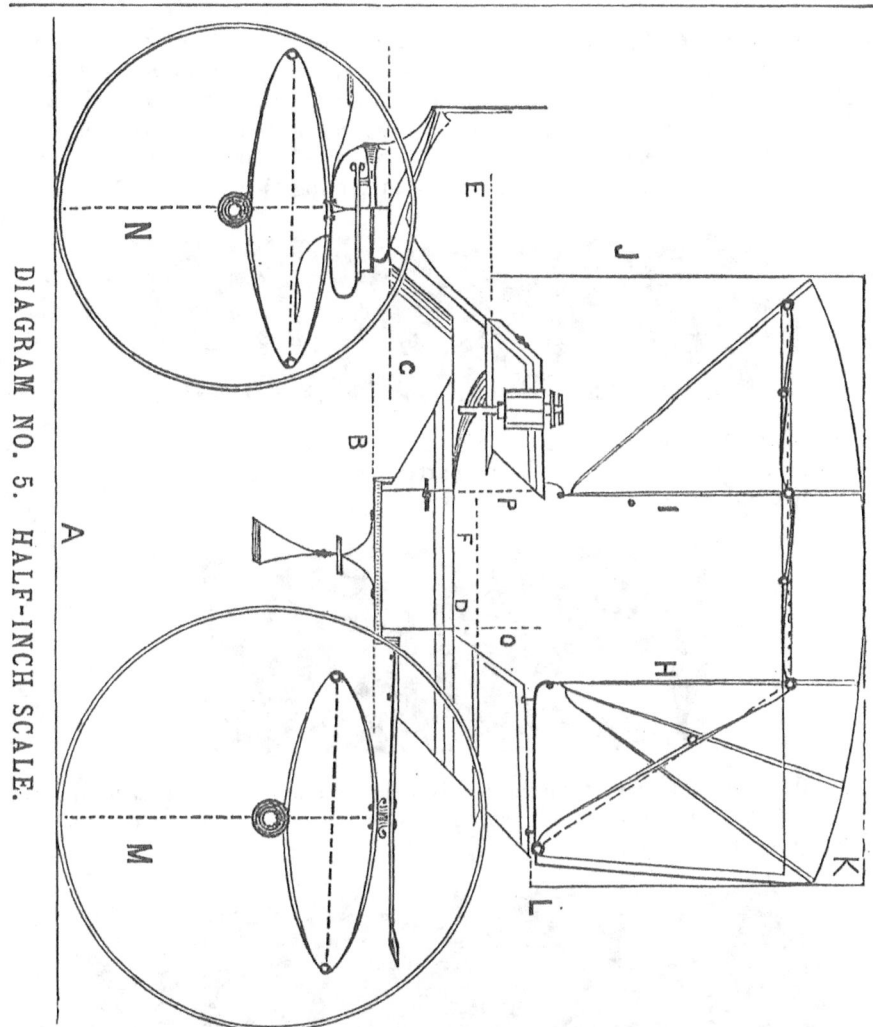

DIAGRAM NO. 5. HALF-INCH SCALE.

to make harmonize with the carriage and body. Next we establish height of front wheel, which is 3 feet 6 inches, and allows 4 inches clear on the line D when turned under the body. In fixing vertical line N, it depends on the width of track; this principle is fully illustrated in the French rule, that is, the space required for the wheel to turn was regulated by width of track. Having established vertical line N, strike the wheel. From center of wheel measure 12 inches, which gives the height of spring, also bottom plate, bottom bed, fifth wheels, top bed, which is curved, throwing the king bolt 8 inches in advance of center, leaving room for a turned collar for the body to rest upon.

Next lay out for the width of seats, the back one being 18 inches on seat frame, the bevels to be laid off according to the style in vogue. From the extreme point of back seat measure three inches, and draw vertical line L; this line showing the flare of the back of top. Lay out for the shifting rail, and draw vertical line H. From this measure off 21½ inches, and draw vertical line I; this giving the width of the two center bows. From front of front seat measure 6 inches, and draw vertical line J; also, from same point, measure 22 inches, and establish the dash line. Having now noted all the points from base line upward, and from vertical line N to the dash line, we will apply the sweep patterns, where the square and protractor have not been used.

We will now apply sweep pattern (Fig. 6) to the top H, being the back part; it has the back, center and front bows laid on it, the top having 4½ inches round.

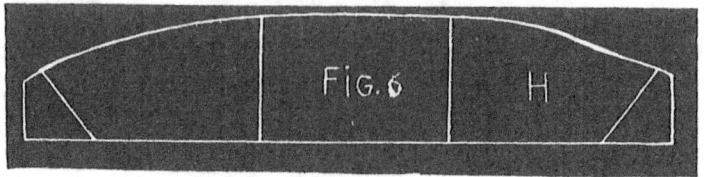

Fig. 7 is the spring pattern, dotted line No. 1 showing the horizontal center, and No 2 the vertical center. These we draw in quarter sections. This pattern will suit for the majority of elliptic springs, where side view is given.

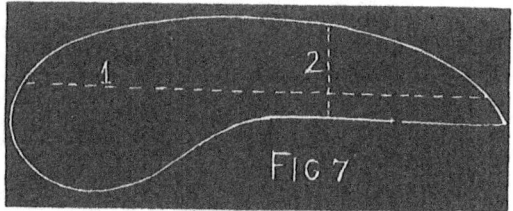

Fig. 8 will be found a very useful pattern in forming short curves, as each illustrates a different sweep. The other sweep patterns necessary for completing the body line have been previously given.

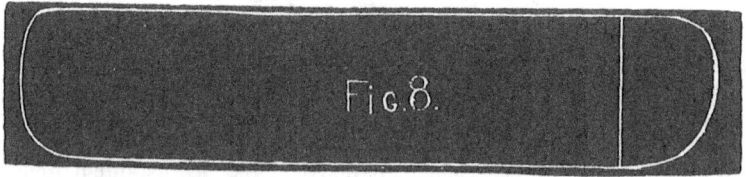

The joints, door handle, lamps, and all other points not mentioned, to be drawn as laid out in the diagram.

DIAGRAM NO. 6. HALF-INCH SCALE.

The manner of inking or tinting the draft having previously been given, will not need to be repeated. As the laying out and inking in of the outlines is necessarily the first and most important part of any draft, the tinting of the panels and proper gradation of the lighting and shading is a part of the work more artistic than mechanical, and much will depend on the natural taste of the draftsman.

While the majority of persons may be deficient in natural taste at this stage of the drawing, a very commendable degree of excellence may be acquired by careful study and practice.

In passing, we would say, that to become a skillful draftsman, one who can produce a draft which will bear to be constructed as shown, a practical knowledge of the construction of the body is absolutely necessary. An architect may produce a handsome drawing for a new edifice, and yet when the master-mechanic shall attempt to

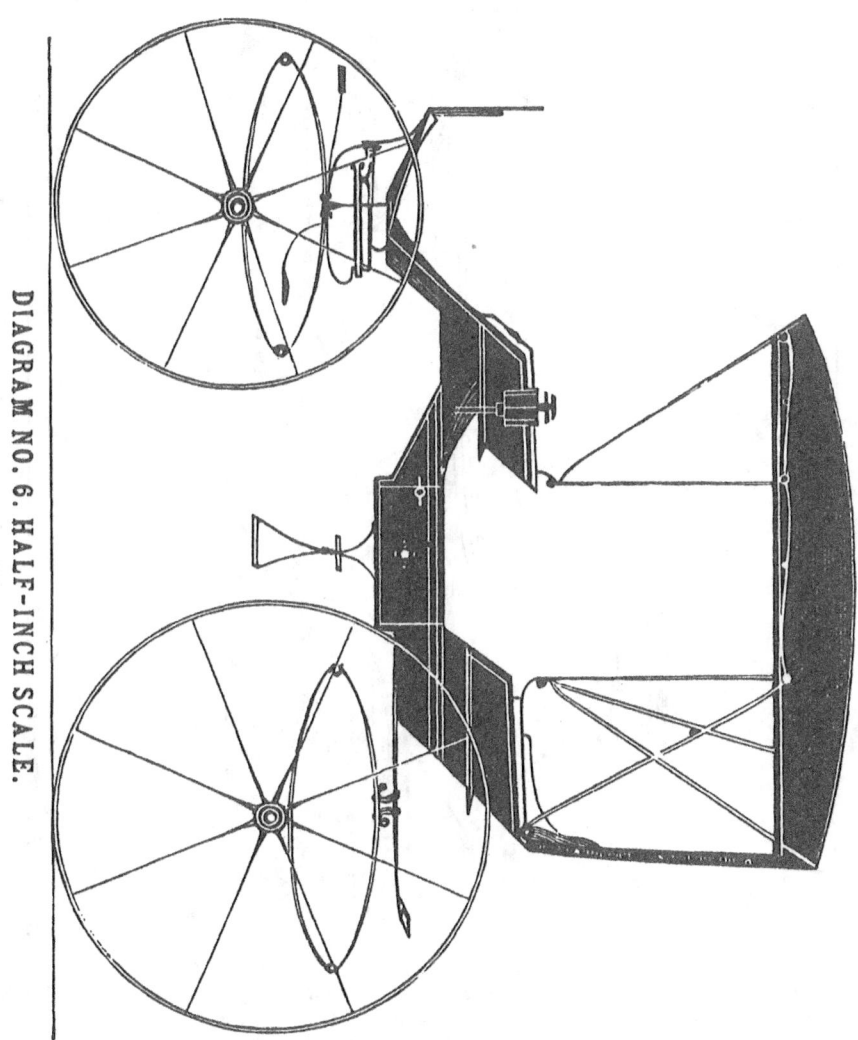

DIAGRAM NO. 6. HALF-INCH SCALE.

carry out the ideas, and construct the building, may find that certain fixed laws in mechanics will not admit of the house being built according to the design shown on paper. This is equally true with the carriage architect; he may possess an artistic eye, and make pretty drafts to look at, but wholly impracticable when an attempt is made to construct the carriage.

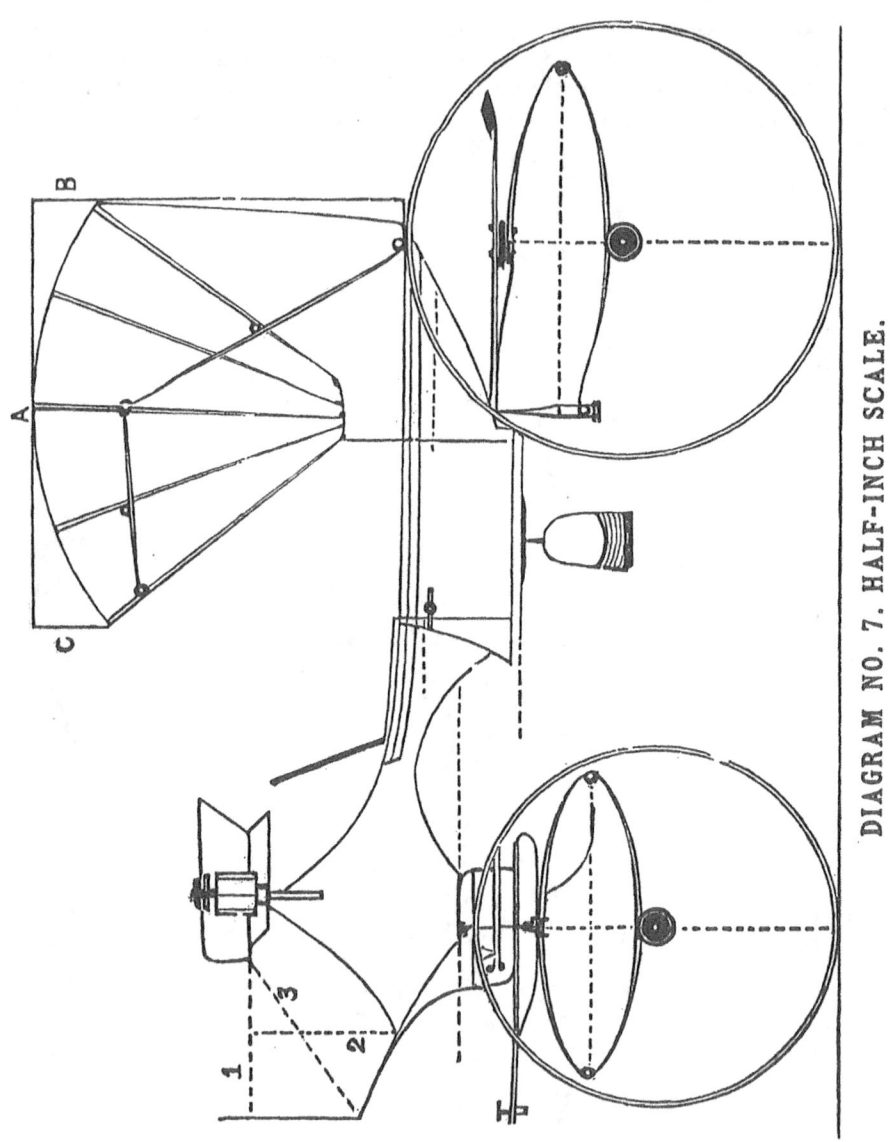

Diagram No. 7 shows a six-seat sociable. 1st. Draw the ground line, then the base line of body, depth of panel, and establish the height or pump-handle and bottom of boot. The height of seats are shown by the dotted lines. From the seat line, measure 47 inches, and draw top line A. Lay out the door and length of back quarter, and draw lines B and C. Nos. 1, 2 and 3 show the height and length necessary for the foot-room. The sweep patterns, which have been given, will furnish all the curves required on this draft. Any further items may be gathered from what has been already given.

As full instructions have been given in the use of India ink, for outline and shading, we shall hereafter give only the outlines of different classes of work, and, at the close of these exercises, give the full manner of applying the colors.

SCALE DRAFTING.   35

It is a good plan for the body-builder to make a draft to the scale of everybody possessing anything novel, preserving them for future reference. The habit once formed, the task will be a light and pleasant one. The time thus spent will bring its reward, if not in dollars, direct, it will come indirectly through the superior skill and fertility of imagination gained by continued practice, which will not fail to show itself on your own work, and be appreciated by the employer.

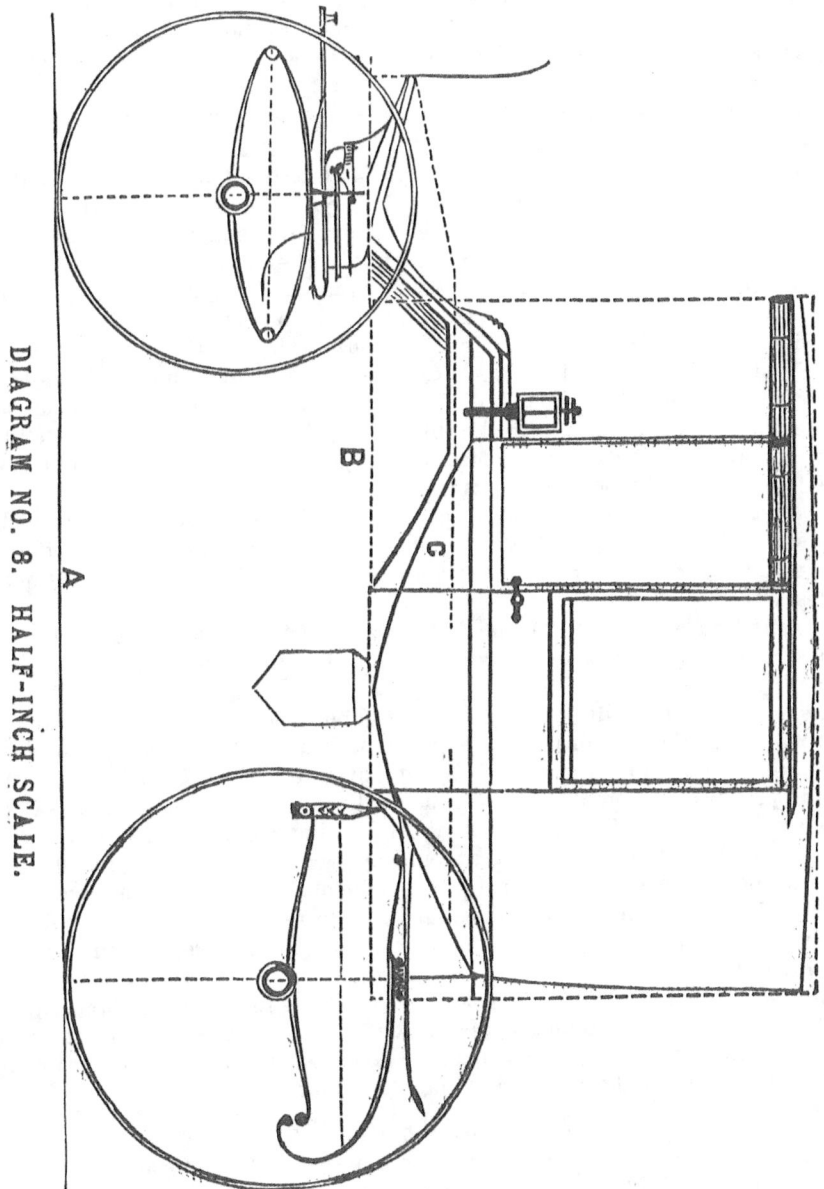

DIAGRAM NO. 8. HALF-INCH SCALE.

In the three concluding diagrams we give the method we employ in the drafting, ground coloring and finishing a colored draft. We will aim to make ourselves understood in this the concluding and most interesting part of our instructions, and trust that they who have followed us thus far will, at the close of these instructions, feel that they have been amply repaid.

This being the first standing-top introduced in this connection, we herewith present a six-seat rockaway, with sharp lines.

The body, to appear graceful, and reduce the rocker to an attractive size, the lines should be carried out front and back on full quarters. In shop parlance, we would say, that the sharp lines, in connection with the coupe pillar, give *too much belly*, and therefore *does not* produce the most pleasing exterior.

In laying out for this draft first draw base line A; then take in your dividers 36 inches, and from the base line produce dotted line B. Next, draw dotted line C, which is the height of seats, being 11 inches from dotted line B. This brings the seats on a line, the front seat forming the top panel of the arch. We secure, by this mode of construction, a lighter appearance on side elevation, for when the front seat is raised above the back one, the head room required above the front seat gives more than is needed on the inside, producing a deep side elevation. And, further, the seat line C stands 47 inches from base line A, allowing for a 42-inch front wheel, the hind wheel being 50 inches. The hind spring is clipped on top of axle, is 12 inches open; the depth of spring and back bar added to this will establish the line of bottom of pump-handle. The front spring should be clipped underneath the axle; the spring to be made 34 inches long, with 9-inch open. We make the front spring shorter to gain more stiffness, for it is necessary to calculate closely for the front platform carriage in the limited space allowed, where the front seat is low, as in this draft.

To establish the height of side elevation, measure 3 feet $8\frac{1}{2}$ inches from dotted line C, front seat, deducting the swell of the roof from this, and draw the top dotted line. Make the width of door 23 inches; back quarter, at roof rail, 25 inches; front quarter, 17 inches.

Both of the front seats should be made wide, as the trimmings are attached to a vertical back, which requires more seat room in order to add to the comfort of the occupants.

Having established the different points on the body, we next draw the horizontal and perpendicular lines. For this purpose we make use of the movable half of the T-head square, set so as to elevate the body one inch higher in front, the same bevel giving the proper direction of the perpendicular lines. The reason for this is that the draft, when completed, will appear to hang level, whereas, were the lines drawn perfectly horizontal and vertical, the body would appear to drop in front, a fault too frequently to be observed on drafts by professionals, who might improve in this particular. The swept lines are produced from patterns formed to suit the curves; the end of each line to be marked on the pattern as a guide in reproducing the pencil lines, which in coloring are nearly obliterated.

We will also direct attention to the fact that no ink is to be used in preparing for a colored draft; the horizontal and perpendicular lines are carried beyond the outlines of the body, thus giving the true direction when the coloring has so far proceeded as to require that they should be added to complete the draft.

DIAGRAM NO. 9. HALF-INCH SCALE.

The draft given on next page is that of a Clarence coach, and as the manner of laying out has been fully explained, we omit it here, as the principles are the same.

SCALE DRAFTING.

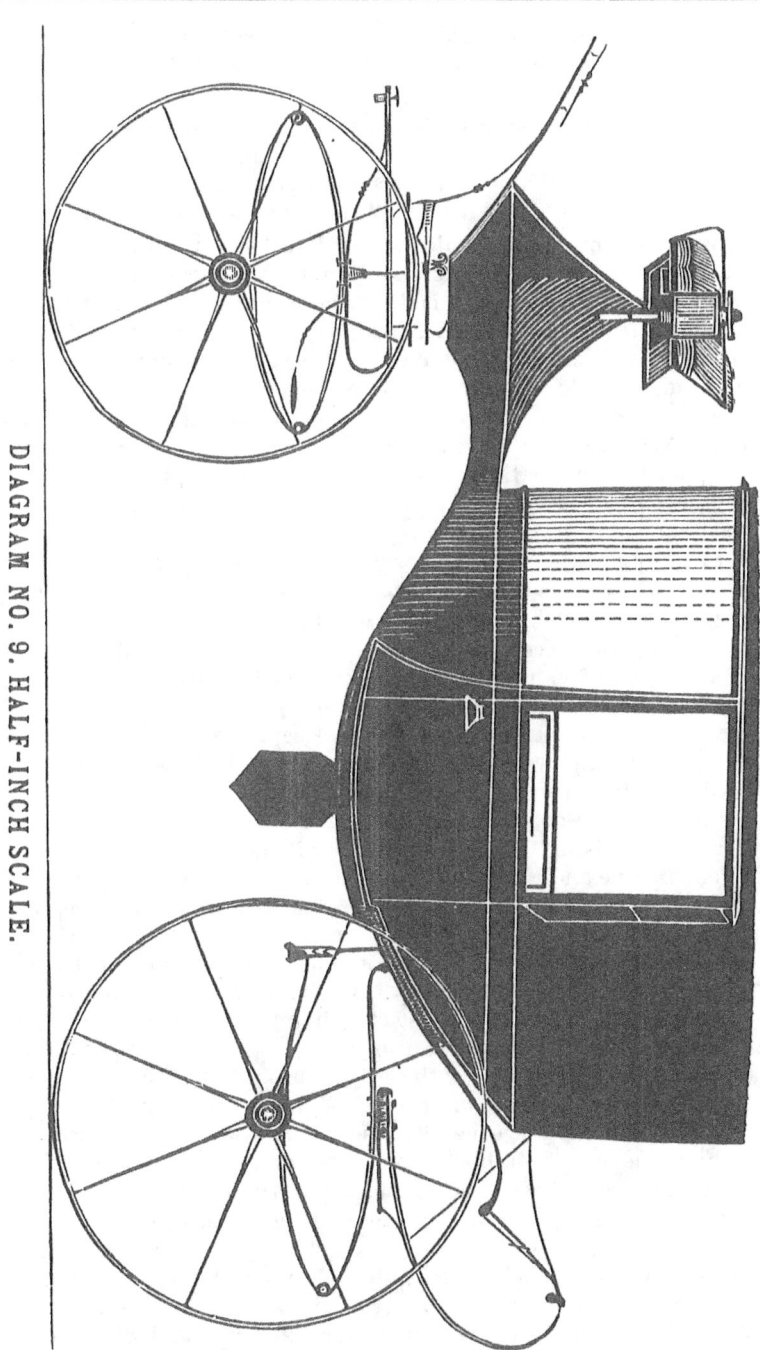

DIAGRAM NO. 9. HALF-INCH SCALE.

Having given the brushes and the colors necessary, and supposing that the student has secured a careful pencil drawing on Bristol board, and marked his patterns so as to be able to place them to their exact position when required, and extended all lines beyond the outline of the draft, which he will, of necessity, have to cover up on laying on the ground colors; having, we say, fortified himself against any confusion of ideas that might arise, he is ready to make his first essay.

Provide yourself with a nest of porcelain saucers, in which to mix the colors, and begin by rubbing up either lamp or ivory black. The colors should be rubbed up with soft water, and then thinned with prepared gum water. (Every article we may mention can be purchased of any respectable dealer in artists' materials.)

The novice or new beginner should not apply color to the draft until he has, by practice on a separate piece of Bristol board, brought the color to the proper consistency. It is better to have a space laid off the same size as the draft, and having coated this, then attempt the draft in hand; begin at the front of roof rail, and pass right on to the upper back quarter, coating it also; but in going over any surface, do not attempt to make a clean outline with the brush. Run as close as possible, and afterward, with the pen filled with the color being used, apply the patterns and true up the edges. Next coat the boot, which extends from the coupe pillar to the toe-board, paying no attention to the dividing line and lamp stem, where the latter projects on the boot. The first coating must cover, for in the attempt to apply the second the first will wash up and ruin the drawing.

Before proceeding with panel colors, we would state that lampblack is not as intense in color as ivory black, and although working and covering better than ivory black, we would recommend the use of the latter.

We will now prepare the color for the lower quarters. We decide to use a dark rich brown, which we mix of Vandyke brown, heightened with burnt sienna, and a rich cast given it by adding a small quantity of carmine or lake. Have the saucer and brush perfectly clean, and having tested the color on a separate pattern of the panels, and obtained the proper temper, coat the panels; start in at the back quarter, and carry the color forward, covering that portion of the hind wheel which crosses the back panel, also the molding lines on door, and the door handle.

Next in order is the belt. We propose to color it dark green. For this purpose we will take Prussian blue and raw sienna or yellow ochre. Begin as before at the back end, and work forward, covering any crossing lines that may interpose, stopping the color at the extreme front.

The body is now ground colored, and before the molding lines or anything else is reproduced the colors must remain two or three hours to dry. When dry, the molding lines may be attempted. We will select Indian red. Fill the pen, and with the T-square (using the swivel-head) strike the apparently horizontal and then the perpendicular lines, using *the patterns* for the *curved* lines. Next the door handle, brought out with liquid gold ink, representing Prince's Metal Mounting, which is to be used on other parts where mounting is generally placed. Follow now with the lamp, dickey-seat and cushion, the lamp with liquid gold, dickey-seat and cushion with black, delicately tinted; also, the toe-board, rocker and book-step black.

The circular front must be rounded by lines of black, as shown in the draft. On the body, where the appearance of a round is to be shown, use china white.

The carriage will next engage our attention. We select carmine. The pen should be used on the carriage in the same manner as directed for inking in, color being substituted for India ink. Strike the rims first, then the hubs, and where the hind rim runs on the back quarter, split the lines with china white; this carries the white ground of the paper over the dark color of the body. Complete the coloring in the following order:

Springs, front carriage, and lastly the spokes. The face of hubs liquid gold.

The manner of preparing to produce a colored draft, and the mere mechanical operation of applying the colors, are similar on whatever style of vehicle the draftsman

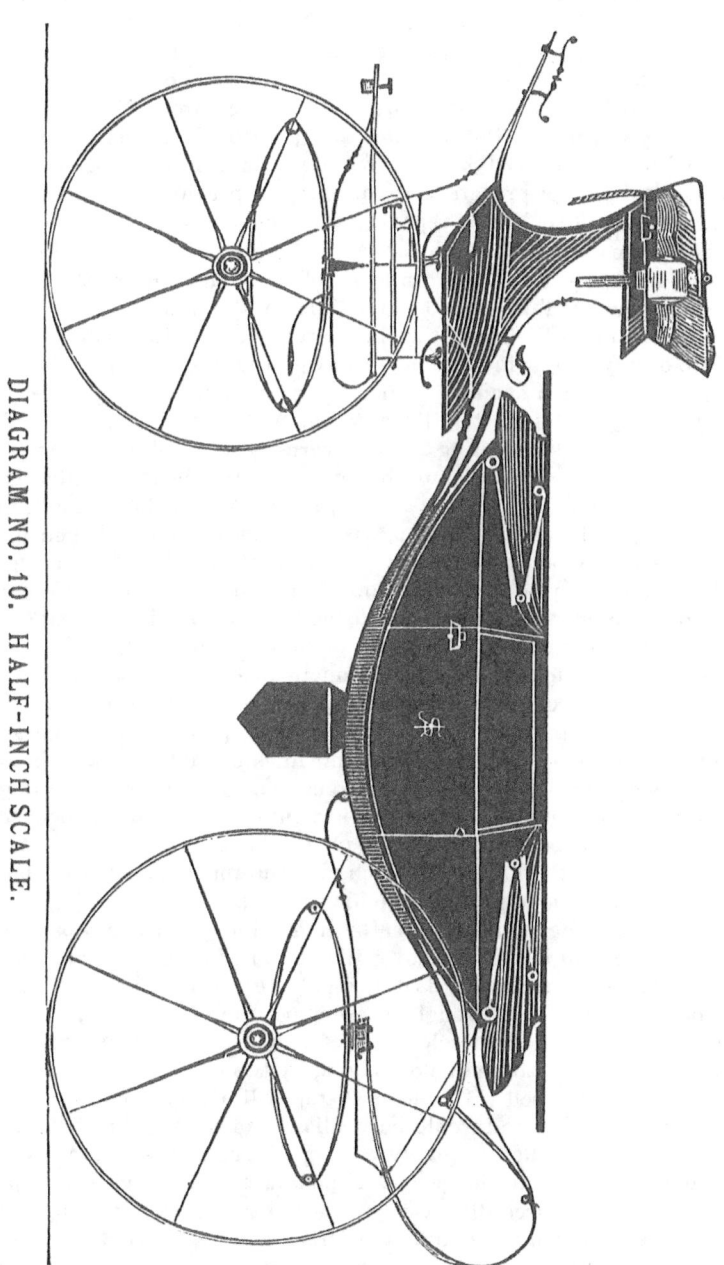

DIAGRAM NO. 10. HALF-INCH SCALE.

chooses to produce. But a variety of touch, and that delicate handling of the colors employed, which at once stamps a colored draft as first-class in every particular, depends not a little on artistic taste. On a standing-top coach, the colors require to be laid on evenly, and with clean edges, and as the greater part of the draft represents wooden panels, there is no effect to be given other than to show solid, unchangeable surface. A tassel or a festoon may be added, which would require some little knowledge of shading, so as to represent properly the folds made by the cord, which serves to loop the festoon up, or of the particular form of a tassel. But in the cut here shown (Diagram No. 10), with the top thrown down, it will be required of the student to represent a leather top, and that in the wrinkled appearance it has when lying down. This is a LANDAU, with the improved SAULSBURY BOOT. Having secured a correct draft with the lead pencil, and further prepared for coloring, as previously laid down, you will begin by addressing yourself to the representation of the top.

First strike the small circles, which represent the top props and the knuckles, with India ink or lampblack, and then ink or color in the joints; these are colored black, as are also the head pieces and the outlines of the bows.

We would here caution the student against carelessness in the matter of obtaining a correct outline of anything which he may essay to draw, and it is of the first importance, in the representation of an object in a folded or written form, that each portion should be represented by distinct outlines, which will at once present to the eye of a beholder the true form, the tinting to be governed by the outlines, or at least the tinting should not confuse or obliterate the outlines. In the representation, then, of this top, a distinct outline is first to be secured, and afterward the tinting is introduced with the pen, using India ink; or shading, using a fine-pointed sable pencil. The top having been completed, outline the dickey-stick, fall, lamp and boot, using black, as this gives the correct coloring of that portion of the finished carriage. When the outlines are finished, commence shading the cushion, then tinge in the seat skirt, which may be black at the back part and lightened off toward the front.

Next follow down on to the boot, laying on an even coat of black over its whole surface; also the toe-board. When the black is dry, bring out the division line, moldings and appearance of the round, with china-white, applied with the drafting pen. The dividing line on side of boot and moulding lines should be drawn heavier than those which are employed to give the appearance of a round. The moulding lines on boot to be drawn with white are the inner lines running from top of boot to the toe-board, and the front and bottom inner lines on the lower part of the boot. Now set the pen finer, and put in the fine lines, which give the appearance of the upper rounded portion of the boot, using a pattern to guide the pen.

The front loop should next engage the attention, which is also put on with china-white; that is, so much of it as crosses over the boot, including the step, and so much of the loop under it as is shown on this drawing. We will mention, in passing, that it is the better plan to use a color throughout every portion of the draft, where it is required, before taking up any other color; therefore, all iron work connected with the body and rocker should be painted before laying by the black.

Black having been dispensed with, the next step will be to decide on a panel color. Carmine is a very rich color, and would be brilliant and showy; but before we fully decide to employ it on this heavy, stately carriage, let us stop a moment and inquire whether it will accord well with its general appearance. Colors give the appearance of lightness and heaviness, according as they are light or dark; and as we wish the Landau to appear rather heavy because it is large, taken as a whole, we decide that carmine or any other light brilliant color would be out of keeping. Further, the trimmings should agree with the color used on the panels, and if carmine is suitable for the panels, it, or something approaching to it in color, is suitable for the trimming; thus carried out practically, we should have a bloody or a fiery-looking piece of work, offensive to persons of good taste.

Having set aside carmine, we will take quite a different color, one among several which may be employed, namely, PLUM COLOR. This color is produced by the use of LAKE and BLACK; we select PURPLE LAKE, and by saddening it with BLACK, produce the required shade of PLUM COLOR. It is to be applied to the panels of the body, and laid on in the manner heretofore given. When dry, proceed with the mouldings, using for the purpose, in this case, FRENCH BLUE.

The carriage part comes next. We will take for this draft a color a shade or two lighter than the panels, produced by a less proportion of BLACK being added to PURPLE LAKE.

Next in order is the representation of gold mounting. The parts to be so brought out are the lamp, dickey-seat handle, toe-board handle, door handle, face of hubs, and lastly the top prop nuts and the joints. The top nuts to be gilt in the center, leaving a black surrounding edge, and the joints to have a gold line through the center.

All that now remains to give a complete finish is to add an ornament or monogram. We propose to use a monogram as design on panel, and will employ FRENCH BLUE, lightened very delicately, if you please, with gold.

We have given illustrations of the principal drawing instruments and patterns, and ten were drafts, ranging from a buggy, through different styles of carriages, concluding with a Landau. The drawings and explanations have been attended with many hours of severe mechanical and mental labor, and we now feel a sense of relief as we pen the concluding sentences. From the careless and indifferent we do not expect to receive any thanks, but wherever there is an earnest wood-worker, desirous of improving himself in this particular, we feel assured that we shall receive our reward at his hands.

French or Square Rule.

—∞∘⋅⊗⋅∘∘—

THE object is to impart that practical knowledge of the square rule and the general system of drafting which is daily required in the workshop. This rule, as presented in the following pages, was selected as most important to the body-maker, being based on comprehensible principles, and becoming more and more necessary to be well understood.

By this rule we obtain the points by which we draw the correct side-sweep for the different pieces of the frame work, when the turn-under and side-swell is given by the operation of right lines drawn over the side elevation and cant of body, illustrated by diagrams in the drawing of a plain coach body. First. The design of the body or side elevation, with the ground plan for laying out the front carriage part, the height of wheels and turning under of same, establishing the arch or wheel house of the body.

Second. The manner of making the patterns for the body.

Third. The application of the square rule with the manner of construction fully shown in the diagrams.

DIAGRAM NO. 1

Illustrates the drafting of a coach, the laying out of the carriage part, the height of wheels and turning under of same.

You may now proceed to draw the ground line C. Measure 34 inches from ground line C, that being the height we wish to hang this coach, and draw the base line of body A; measure 3 inches from line A and draw line K, being the depth of rocker. From line K measure 12 inches, being the height of your seat, dotted lines. From this point measure 42 inches; draw the line E, being the height of body. Lay off for the width of the door, which is $22\frac{1}{2}$ inches; draw lines B and H. Measure from line B $26\frac{1}{4}$ inches, and draw line F, being the width of back quarter. From line H draw line N; measure 24 inches, this being the width of front quarter. Now you can sweep the body (with miniature patterns, scale drawing), and lay off for the guide rail of the door. Measure 26 inches from the hinge pillar, and draw perpendicular line—this point being the center of the hind wheel, which is 4 feet 2 inches. You will strike the wheel this size. Next proceed to lay out for the spring, which is 2 inches, 5 plates 42 inches long, $12\frac{1}{2}$ inches opening; this spring is clipped underneath the axle, which is $1\frac{1}{4}$ inches; measure from underneath the axle $12\frac{1}{2}$ inches, being the opening of the spring. Measure $1\frac{1}{4}$ inches for the five plates, being the depth of the top part of the spring, $1\frac{1}{4}$ inches for the depth of the back bar, $\frac{1}{2}$ inch for the brake or pump-handle plate. This point is the bottom of the brake; sweep the brake as represented in the diagram.

Now proceed to lay off for the front carriage, to ascertain the height of boot; the width of track being 5 feet, spring 2 inches, 5 plates 40 inches long, opening 14 inches, allowing $1\frac{1}{2}$ inches for the settling from the weight of the body. For the establishment of the point where we want the wheel to stand, when turned square under the

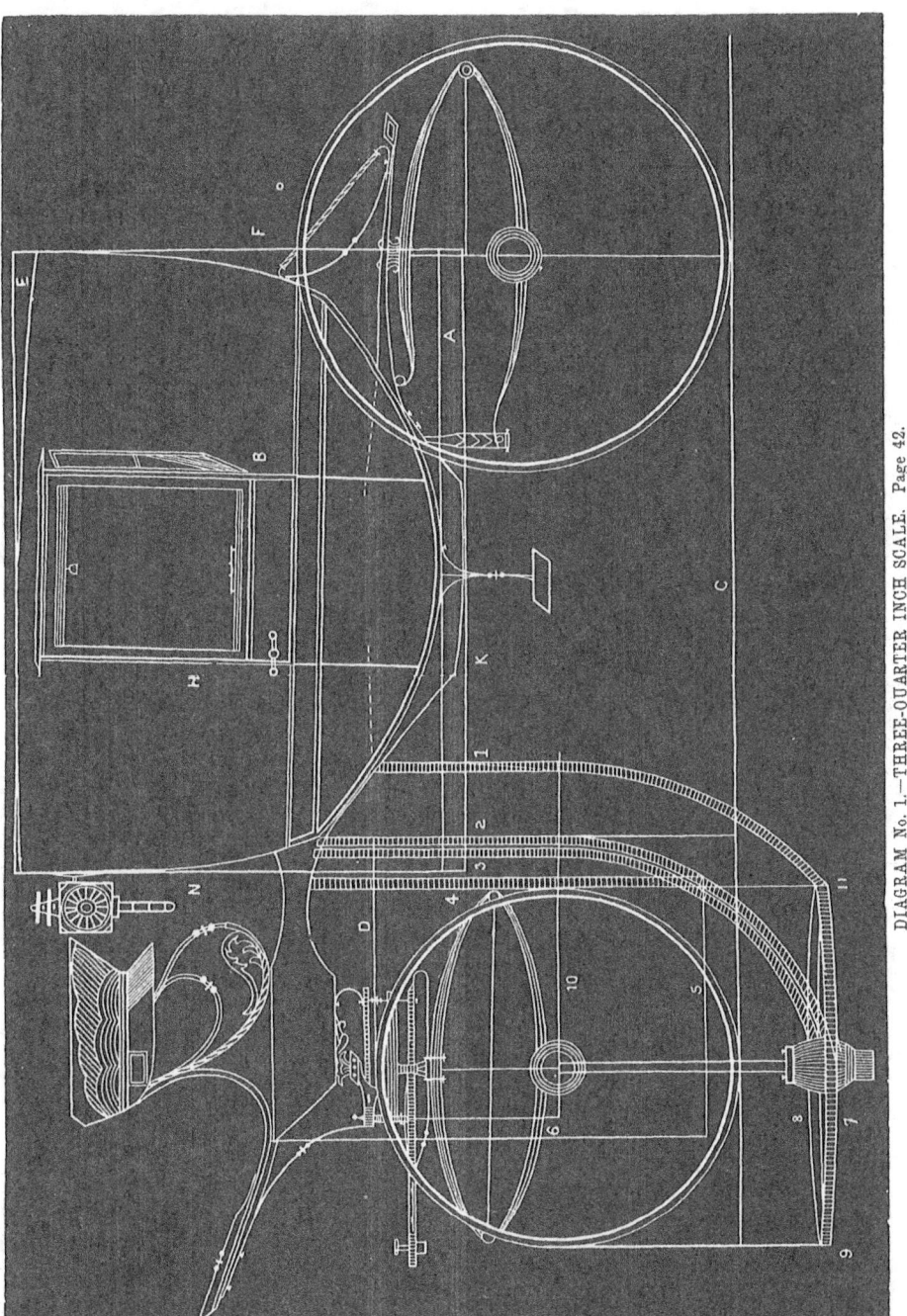

DIAGRAM No. 1.—THREE-QUARTER INCH SCALE. Page 42.

body, when the body is swept, extend the bottom sweep of the rocker to line N ; at the same time you can dot off the manner of paneling the bottom. This square corner is the establishment of line, figure 2, or the point we want the wheel to stand when turned square under the body. The object of this corner is to shorten the coupling as much as possible, which is an important point. Measure from this dotted corner in panel, on line No. 2, 7 inches, and draw parallel line D, this being the space required over the top of wheel for settling of the coach when loaded. Measure from line D, on line No. 2, to ground line C, 42 inches; this is the height of the front wheel. Measure from line No. 2, 27 inches; draw a perpendicular line 27 inches, this being the center of the wheel. Measure $5\frac{1}{2}$ inches from this line, the center of the wheel, and draw line, figure 6, this being the center of the king or body bolt. You can now strike the front wheel. Your front springs are clipped on top of the axle, which is $1\frac{1}{2}$ inches ; measure $1\frac{1}{4}$ inches, the depth of the bottom part of the spring, $12\frac{1}{2}$ inches opening, $1\frac{1}{4}$ inches for the top part of the spring, $\frac{1}{4}$ inch block top of spring, $\frac{1}{2}$ inch plate for the bottom of lower bed. This bed is 3 inches deep ; sweep up $1\frac{1}{4}$ inches, leaving $1\frac{7}{8}$ inches deep in center. Fifth wheels $\frac{1}{2}$ inch each. Top bed $1\frac{3}{4}$ inches deep ; $\frac{3}{8}$ inch plate for the bed. Lay off 3 inches for the depth of the body block; this point is the bottom of your boot ; proceed to sweep your boot as represented in diagram. It is necessary in hanging a coach to hang it 3 inches higher in front than back, calculating from a level line from the bottom of pump-handle to bottom of boot. We will now proceed to lay out and prove our calculation for turning of the wheel. We have given the top part as it passes under the body square on line figure 2. Next proceed to draw a parallel line, figure 10 being the center of front wheel, and also center of boot. Measure 17 inches from line, figure 10; draw line, figure 5, being half the width of boot; next draw perpendicular lines, figures 9 and 11, the size of the front wheel. Measure $32\frac{1}{4}$ inches from center of front wheel, and draw line 7, representing the top of the wheel. Measure $2\frac{1}{4}$ inches from line 7 and draw line 8, being the width of the wheels at the bottom. Take your dividers, place one point on line 6, where it intersects with line 10, this point being the center of king or body bolt, the other at the back part of the wheel, at point 11; you can strike the turning of the wheel to parallel line 10, and square from this point to the body as represented by line 1 ; this is the extreme back of the wheel as it turns under the body. You can proceed in like manner with line 2, starting at the top of the wheel. This is where the top of the wheel will stand when the carriage is turned square under the body, and also proves the establishment of this line. Line 3 represents the tread of the wheel upon the ground. You will notice in the diagram where the circle intersects line 5. This line being the outer edge of the boot at this intersecting point, you can square a perpendicular line to boot, which is 4. At this point is where the top of the wheel passes the outer edge of the boot, when turning under.

This manner of laying out work gives the builder an opportunity to provide the material for the wheels, to order the springs and axles, in fact everything pertaining to it, while the body is being built, so as there will be no delay in ironing or finishing, which often is the case, and causes a loss to the builder.

DIAGRAM NO. 2

shows the manner of making patterns. You will perceive we have taken the body from the carriage, and give the outlines of the side elevation of the body on the draft board, to make the patterns for the body so that they harmonize in shape with the lines drawn. Many different modes of accomplishing this important task have been devised, but we have never yet seen any rule so simple and complete.

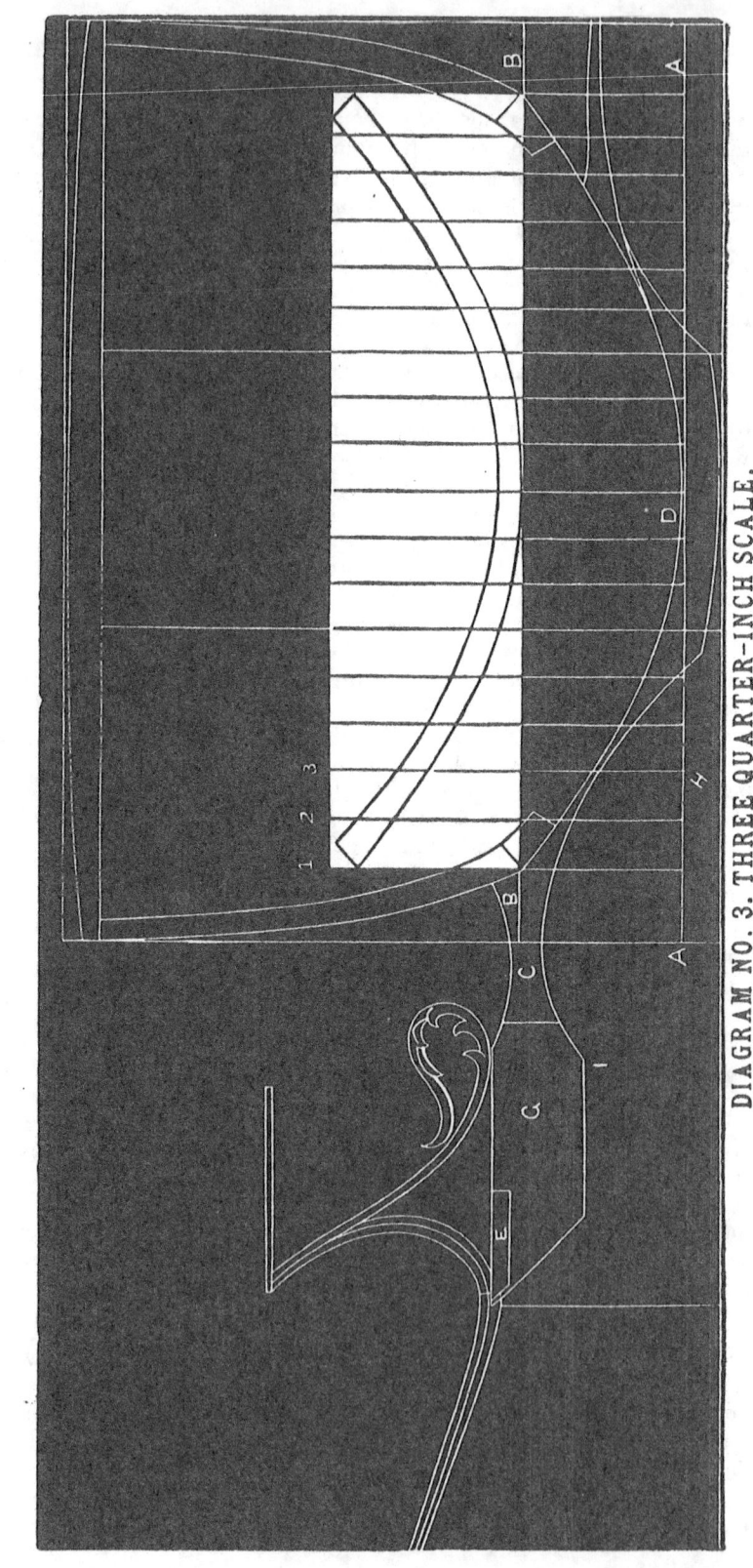

DIAGRAM NO. 3. THREE QUARTER-INCH SCALE.

You will proceed to draw the line B B, being the ends of the bottom side; straighten one edge of your panel; now lay it on the draft board, with the straightened edge on line B B, as the white space represents the panel in the diagram that the pattern is to be made from.

Having laid the panel in its proper place, square the perpendicular lines Nos. 1, 2 and 3 from the base line A A. Having drawn the perpendicular lines as described and illustrated, next take your large dividers, place one point on base line A, at perpendicular line No. 1. The other point you place on line B, where it crosses No. 1. Having this space with your dividers, take the point from A and prick it on the panel on line No 1. You will proceed in like manner with Nos. 2 and 3, and with all the perpendicular lines the full length of the bottom side.

You now proceed to make the sweep according to the prick marks. Then gauge 1¾ inches for the width of your pattern. All patterns are made in the same manner, excepting the standing pillar, roof rail and curve patterns, C, the neck pattern, to extend from I to H, to form the concave boot. G, boot pattern, extends back to inside of the front pillar. E bracket made as represented in the diagram.

DIAGRAM NO. 3.

You will now proceed to lay out for making the standing or door-pillar pattern. It will require a panel 5¼ inches wide, laid upon the draft board perpendicular. You will mark the shoulders at the roof rail, bottom side, and the extreme depth of the body on base line A. Now lay off the turn-under, which is 4½ inches, commencing the sweep about 4 inches above the arm rail. Leave the pillar ¾ of an inch deep at the bottom for the tenon, on base line A; at the guide rail, 2¼ inches; at the shoulders of roof rail, 1⅝ inches. Now you can space it off for the glass frame, and sweep the pattern as represented in the diagram. C represents the face of the bottom side, being 1¼ inches from the face of the pillar, to allow room for bolting the step after the door is cut through to the bottom.

Fig. 1 represents the roof rail pattern. You require a panel 7 inches wide for this pattern. We lay it underneath the draft board for convenience in this case. Square the perpendicular lines N H B and F across it. Now take the distance with your dividers from the perpendicular line C to line B. With this space you will place one point on line E, at the perpendicular line B, and prick it off, the same distance from line E on line H; these points being the outside of the roof rail at the standing pillars. Now place the back and front pillars in their proper places. Before we can decide on the sweep of the roof rail, mark off 1¼ inches, being the width of the pillar at the top, on lines N and F, from line E. Now you can sweep, as represented in the diagram, the width of it at the standing pillar, and at the doorway 1¼ inches, tapering inside front and back to 1½ inches at each end.

Having this pattern made, we will lay out for the width of the body, which is 53 inches over all; measure 26½ inches, being half the width from base line A, on lines H and B, and mark the two points. Now you can lay the roof-rail pattern at these two points, and mark off the sweep for the outside of body, which is the line G. You will take your dividers, place one point on perpendicular line C, being the face of bottom side at the foot of standing pillar, the other point on line B; with this space you can put one point on line G, where it intersects with line B, and mark it off on lines B and H; draw line D from these points, being the face of the bottom side.

Fig. 3 represents the back, half paneled and half framed, and also the sweep required for the curve pattern. You will swell the top 4 inches, the width of the body which is 53 inches, the same as laid out in side elevation, at the standing pillar. This

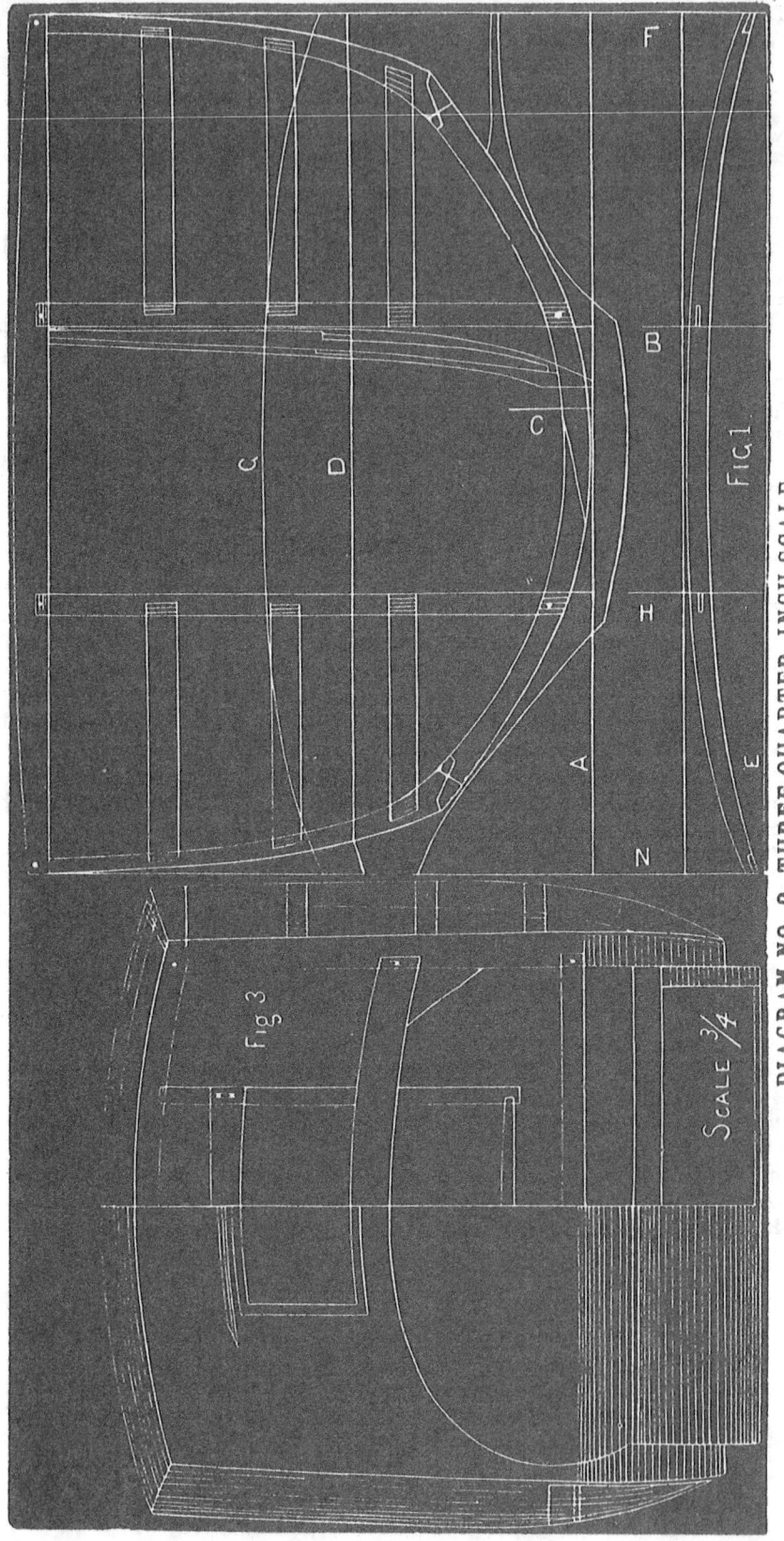

DIAGRAM NO. 2. THREE QUARTER-INCH SCALE.

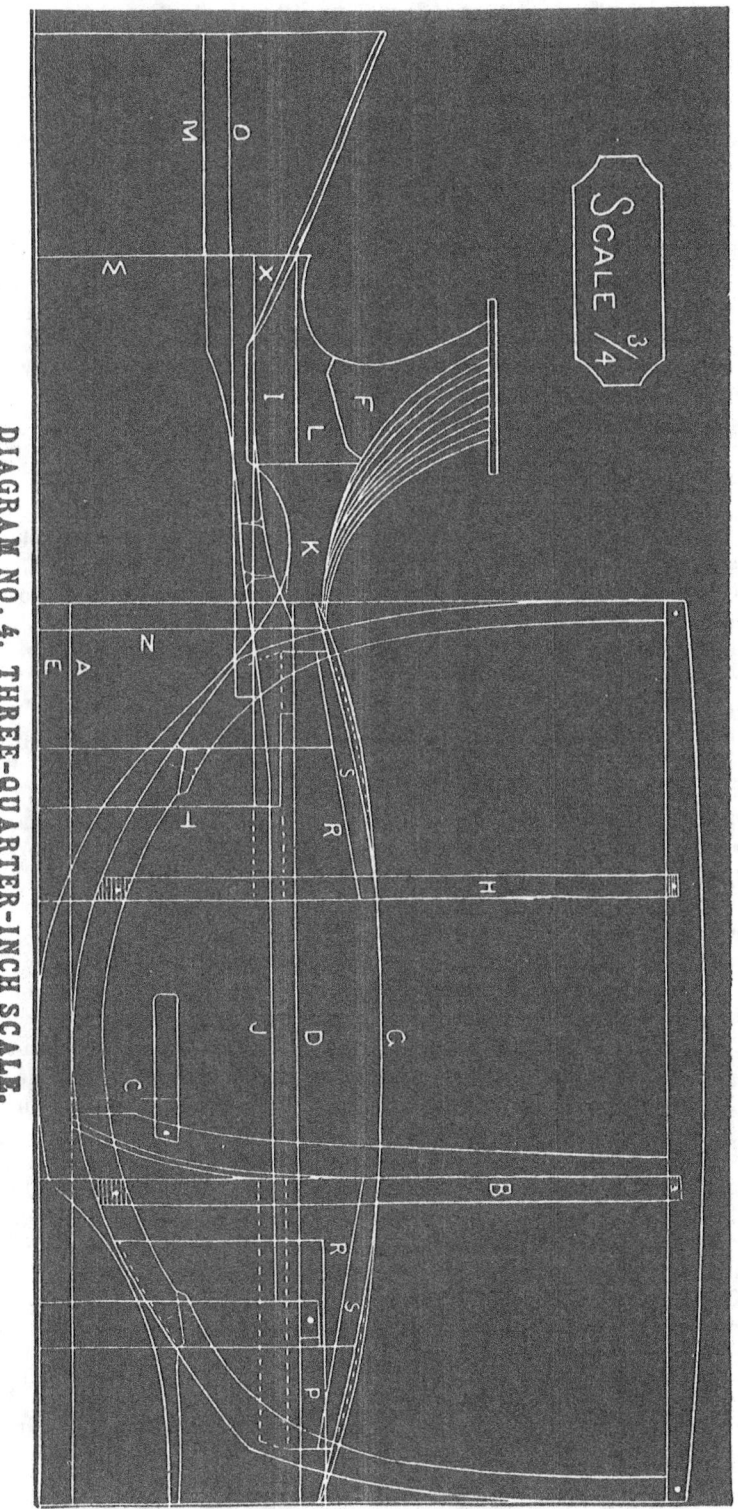

sweep makes a desirable swell, and gives a beautiful top when finished, although the present style is more flat. Back light, 18 by 11 inches; size of curve, $\frac{3}{4}$ inch deep, $1\frac{1}{4}$ inches wide; top end rails, 2 inches deep; back and front middle rails, $2\frac{3}{4}$ inches deep.

Having the patterns complete, you will lay them on the draft, and mark off all the shoulders. This is a saving of time in laying out the body, so that when the stuff is dressed, the shoulders are all struck off, and when swept are ready for gauging.

DIAGRAM NO. 4.

In laying down the cant we shall give it in the same system as practiced generally, the center of body being the lower edge of the draft board. The cant is the working draft of the body; all lines, except the side elevation, are called the cant, and laid out from half the width of the body. In laying it out we may introduce lines not known in this rule, but these points it is useless to discuss; by experience, we have found by the introduction of these lines, we can produce a much better swept body. The lines we allude to are the cheat lines S S, also the dotted lines between the roof rail and the cheat lines, which are the arm-rail sweeps. The object in these lines is to reduce the thickness of stuff of the bottom side at the foot of the back and front pillars, which carries an easy sweep up the pillars.

We now proceed and lay down the cant; width of body is 53 inches, line E being the bottom of draft board, or the center line. You can lay off half the width of body from line E, $26\frac{1}{2}$ inches on lines B and H; with the roof-rail pattern placed at these two points draw line G. This gives 3 feet 9 inches on the seat. Take your dividers, and from line C, being the face of the bottom side, and the point put to line B, being the hinge or standing pillar; having that distance, you can prick it on lines H and B, from line G, and draw D the full length of the body, being the face of the bottom side on the cant. P is the pump-handle space, the dotted lines the manner of letting into the bottom side. Measure from line D, $1\frac{3}{4}$ inches, and draw line J, being the thickness and inside of the rocker. Next you will draw the cheat lines S S, starting the sweeps on lines H and B, cheating it off $\frac{3}{4}$ inch at the front and back end of bottom side; by this line we will draw another line, which is the dotted line, being the sweep of the arm rails; measure 1 inch from the outside sweep, front and back, and draw lines R R, which is the inside of arm rail.

Now, lay off the concave boot, which is 34 inches wide; measure 17 inches from line E, and on lines W and T draw line X, this being the outside of the boot. Now draw the perpendicular line N; this line gives the point where the boot leaves the front pillar, and draw the concave line as represented in the diagram. This concave piece is the neck pattern K. Measure 2 inches from line X, and draw line O, which is the thickness of pattern L and the panel. Measure $1\frac{3}{4}$ inches from line O, and draw line M, which is the toe-board, bracket pattern. You can draw the line inside of boot (leveling for the rocker-plate). This plate is $2\frac{1}{2}$ inches by $\frac{1}{2}$ inch, corner turned on the back bar to receive two screws, front ending on the toe-board bracket.

We have now laid out the cant sufficient to ascertain the thickness of stuff required for the different parts of the body. You can take your dividers, and at the points of the draft, where the pieces are to be placed, get the thickness required. Bottom sides from 4-inch stuff, roof rail 4 inches, back and front pillars $3\frac{1}{2}$ inches, standing pillars 2 inches, rocker underneath the door $1\frac{1}{4}$ inches, pump-handle 2 inches; neck pattern K to form the concave, $3\frac{1}{4}$ inches. L $1\frac{1}{4}$ inches, F 2 inches, I $1\frac{3}{4}$ inches. You can mark the thickness on the patterns for convenience in getting out the stuff.

DIAGRAM NO. 5.

Taken for granted, the stuff, including the panels, roof boards, bottom lining and moldings, are ready for dressing, commence with the panels, and when dressed they

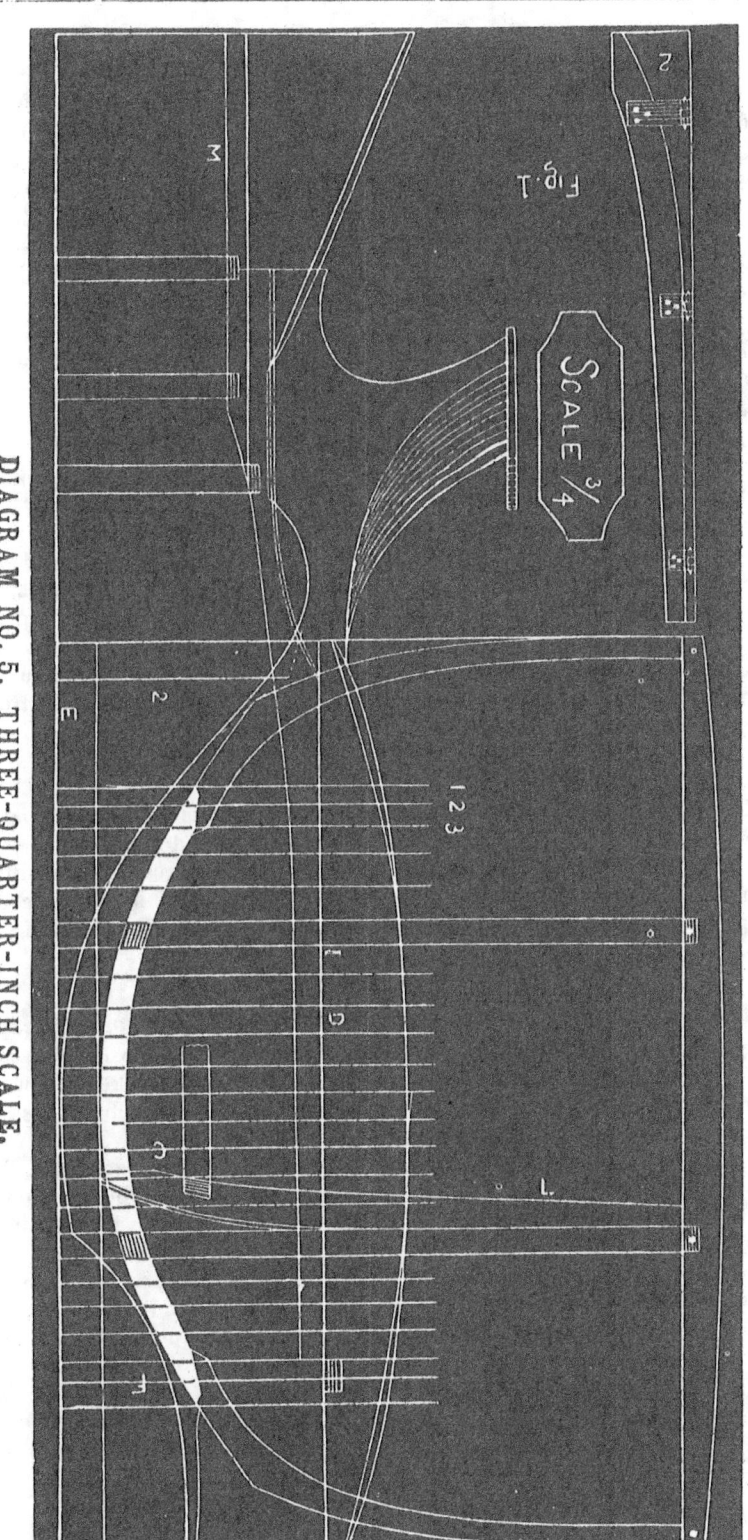

DIAGRAM NO. 5. THREE-QUARTER-INCH SCALE.

should be placed overhead on racks for this purpose, in a warm place, with sticks between. It being very important that they should be dry to prevent checking when used—if you wish to build a good body. Next dress your cross rails; when done, we will strike the shoulders, commencing with the back bottom bar; lay it on the draft as represented in the diagram, at space F; E being the bottom of the draft board, or the center of the body. D, the face of the bottom side on the cant. You will now mark your shoulder from line D, allowing 1¼ inches for the tenon, and also the center line from E on the bar. Now reverse your bar, keeping your center line at the bottom of the draft board, on line E, and mark the other shoulder in the same manner on line D.

Proceed in this same manner with the seat rails; with your dividers place one point on line C, the top of the seat rail, the other to inside of the standing pillar on line L, this space to be added to each end from the marks already taken. Bevel for the shoulders, from the top of the seat rail, inside of the pillar, on line L. Having this bevel strike your shoulders. In this same manner you mark your front toe-board rail, and all other cross rails, M being the face of the bracket on the draft.

You perceive that in this manner you can get the length of all your rails without the use of a rule, the draft being your guide. Next dress your bottom side, which when ready, place one bottom side on the draft, as shown by the white space in the diagram, ready to transfer the perpendicular lines on it; this to be done by commencing with the perpendicular line 1 at the foot of the front pillar, then marking it both top and bottom; also with 2 and 3, and so on the full length. You can at the same time make your cuts for the width of the standing pillars; square these marks across top and bottom; transfer these lines to the other bottom side in the same manner.

Fig. 1 represents the manner of letting in of hinges for hanging doors. No. 2 a piece of panel stuff swept inside to the standing pillar, the outside in width sufficient to allow the pins of the hinges to clear the moldings of the doors, dressed perpendicular; by placing this pattern to the pillar you can square across for your hinges, both standing and door pillar. We consider this a good idea for hanging calash doors, as they require to be exact to prevent the upper part drawing, when opening.

DIAGRAM NO. 6.

You will perceive in this diagram we have left out the belt or arm rails, and all other lines that are not needed for present consideration. We shall take up each piece separately, after being dressed, that is necessary to be pricked off, as this is the most difficult part for the beginner to understand. S S are the cheat lines to be used for pricking off the bottom sides. Underneath the draft board we have, for convenience, placed the two bottom sides, represented by the white space, with the bottom edge up, line F being their face, as laid together. These, you will perceive, are pricked off from the prick marks on the cant, which we proceed to explain the manner of doing.

You will notice, in the turn-under of the standing pillar, an outside line, which is the thickness of the molding to be left on the bottom side. You will take your dividers, place one point on perpendicular line C, where it intersects with the parallel line No. 1; the other point place on the line of the turn-under of the standing pillar. Having this space, with your dividers place one point on the cheat line S, where it intersects with the perpendicular line No. 1. You prick off this space from the cheat line S toward line D. From this prick to line D; the face of the bottom side is the thickness of the bottom side at the foot of the front pillar.

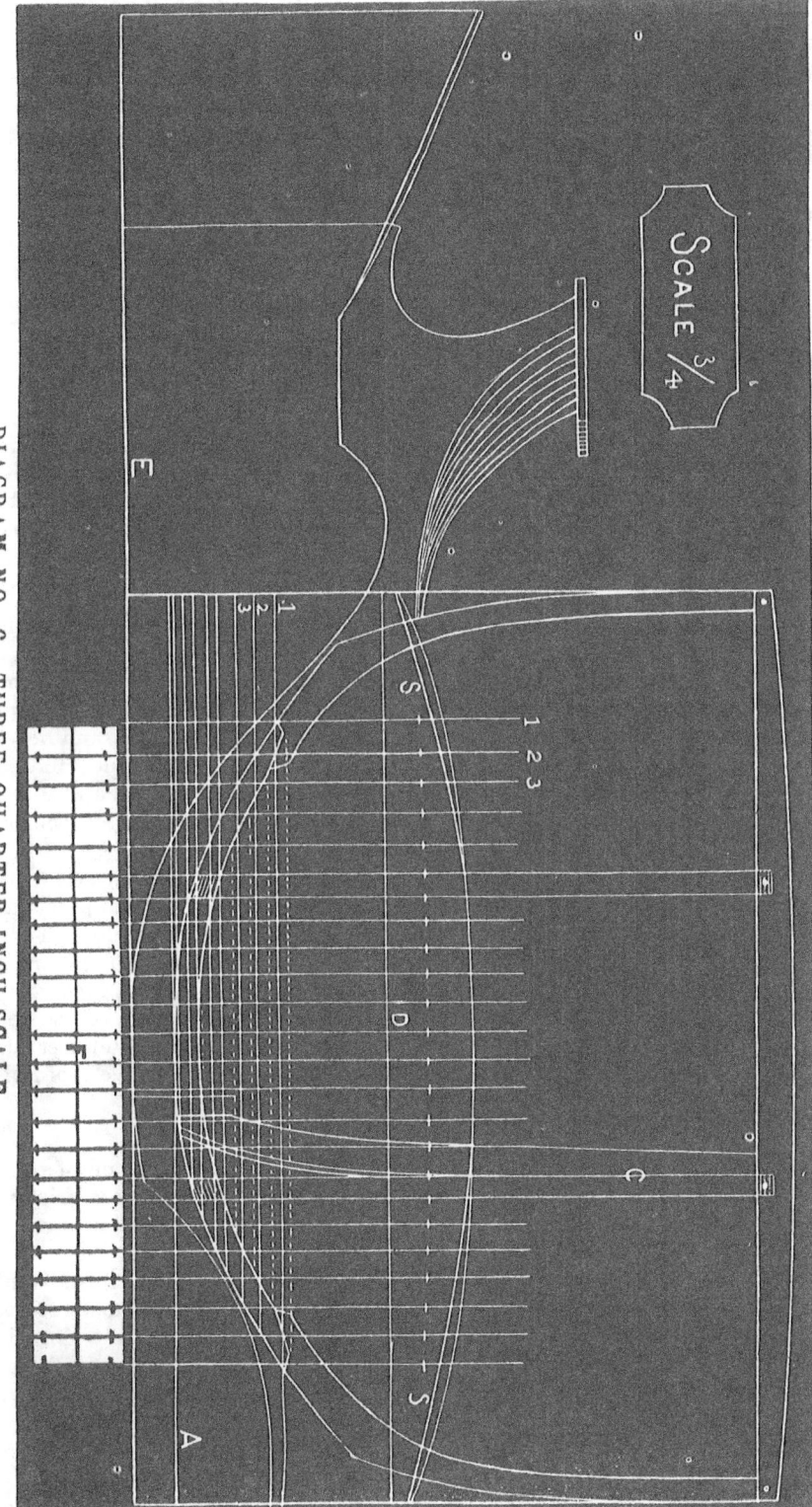

DIAGRAM NO. 6. THREE-QUARTER-INCH SCALE.

You will proceed in the same manner with the lines Nos. 2, 3, and all the horizontal lines, until you reach the base line A, which will be the center of the door. From this point, on the back part of the bottom side, as it sweeps up, you will raise on the horizontal lines, bearing in mind, where the horizontal and perpendicular lines intersect on the bottom edge of the bottom side, to have the turn-under deducted from line S on the perpendicular lines at the intersecting points, as represented in the diagram.

With your dividers you transfer the thickness, at the different points, to the bottom sides, as shown under the draft, commencing with the perpendicular line No. 1, from line D, being the face of the bottom side, the other point on the prick. Transfer this distance on the same line No. 1, one point of the dividers on the face line F; the other point you prick off, as shown in this diagram, and so on, transferring all the distances to the bottom sides, the full length.

Having the bottom sides now pricked off, on the lower edge, you turn them over and on the top side proceed in the same manner, taking the dotted lines where they intersect the perpendicular lines, top edge of the bottom sides, and deduct the turn-under from line S, and transfer the remainder of the space to line D, on to the bottom side. Now you sweep the bottom sides to these prick points, which gives both the sweep and bevel.

DIAGRAM NO. 7.

We will now take up the corner pillars; after being dressed, lay the back and front pillars in their proper places, and transfer the perpendicular lines marked 1, 2, 3 and 4; also, at the same time, strike off the shoulders at the foot of the pillars, arm rails and roof rail. You then square them across and transfer them to the opposite pillars. You will now proceed and prick, commencing with the perpendicular and horizontal lines No. 1, at the foot of the front and back pillars, deducting the turn-under space from the cheat lines S, on the perpendicular line No. 1, and so on with all the perpendicular lines, as illustrated and described in the manner of pricking off the bottom sides.

You will notice underneath the draft the four pillars, F being their faces with the outside edges up. Space from the cant to the pillars, commencing with perpendicular lines Nos. 1, 2, 3 and 4; take the space from line D to line S, on the lines F and N, being the width at the top of the pillars, which you can prick off on the pillars. With the straight-edge, from this point to the prick on cross line on No. 4, draw a straight line as shown in the diagram; then turn the pillars over; prick off in the same manner, keeping in mind where the perpendicular and horizontal lines intersect on the inside of the pillar. Set your bevel from line F, on line G, being the roof-rail line; sweep your pillar off, using this bevel for the top of the pillars.

DIAGRAM NO. 8.

In this diagram we have given all lines necessary for the completion of the body which have already been explained, excepting the sweep necessary for pricking off the lower edge of the belt rail; the object of this line is, when the rail is swept off for gauging from the outside, and when let in will lay level on the outer surface. We now explain how this line, with the other three, are established. G the roof-rail, and S S the cheat lines for pricking off the bottom sides before explained; the next line, for the top edge of the belt rail, that was introduced in diagram No. 4 as a dotted line. This line is drawn by using the roof-rail pattern, commencing it on perpendicular line No 1, where it intersects with the cheat line S, at the extreme front and back ends of the belt rail, and swelled up to line G, carrying an easy sweep across the door.

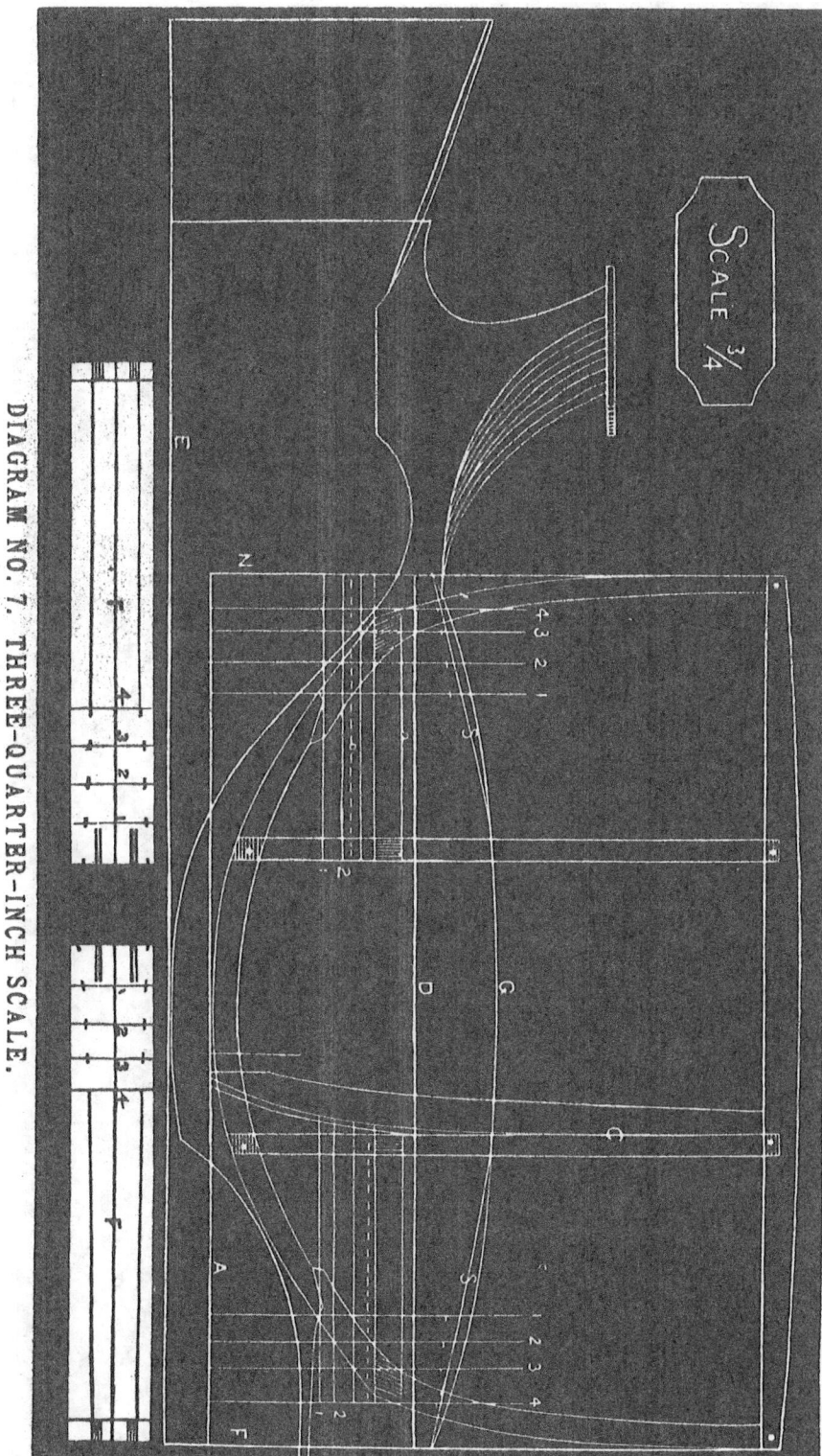

DIAGRAM NO. 7. THREE-QUARTER-INCH SCALE.

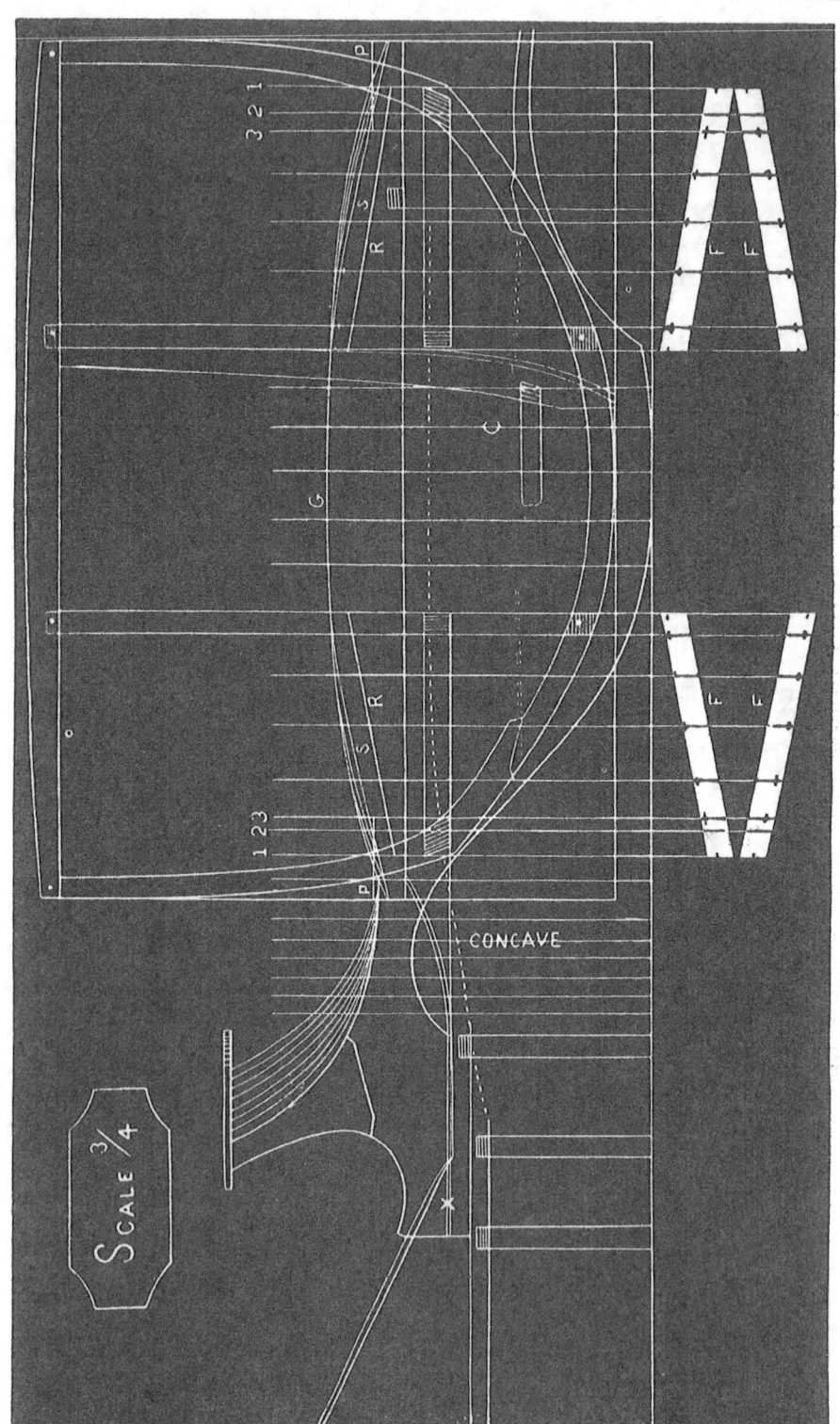

DIAGRAM NO. 8. THREE-QUARTER-INCH SCALE.

FRENCH OR SQUARE RULE.

The next is the establishment of the line for pricking off the lower edge of the belt rail. With your dividers place one point on parallel line P, where it intersects with perpendicular line 3, the other point on the cheat line S. This space is obtained by the horizontal line P, the distance from the top to the lower edge of the belt rail, as let into the front and back pillars; add to this space already taken the difference in the turn-under from the top to the lower edge of the belt, at the standing pillar. Proceed and prick it off from the cheat line S, on perpendicular line No. 1; from this prick sweep into line G. Next proceed and lay the belt rail in the proper place and transfer the perpendicular lines, setting your bevel from line R, being the face of the belt on the cant, on to any of the perpendicular lines. This you can bevel across top and bottom; also for your shoulders, as represented by the white space below the draft, showing the same position as on the cant, F being their faces.

Now proceed and prick off the top of belt, commencing with perpendicular line No. 1, deducting the turn-under from the line explained as the top of the belt, from this point to line R, being the face of the belt on a cant. This you can prick off, and so on the full length of the belt, as illustrated and described in the manner of pricking off the bottom sides and corner pillars. Having the top, turn over and prick the bottom; now they can be swept off, which gives you both the sweep and the bevel.

You will next prick off the neck piece for the concave, by laying it on the draft, in its proper place; transfer the perpendicular lines; after being squared across, can be pricked off, X line being the face. In dressing the standing pillars, mark off and dress the door pillars at the same time. Having all your stuff dressed, commence to lay out the whole body, including the doors, gauging from the outside; and when ready to frame, you will do it all at once, also in lightening out, with the same system, this being the quickest and most correct manner of constructing a body. Back and front paneled, canvased, lined up and cleaned off, ready to be set up. This closes the application of the French rule to a coach, being the plainest kind of a job.

In giving these explanations and drawings, we have endeavored to give it in a simple, clear manner, and we think our readers will bear us out in the assertion that it has never been given before where it can be so easily understood.

Working Drafts of Landaus, Landaulettes, Clarences, Coupes, Bretts, Phaetons, Rockaways, Etc.

LANDAU.--THREE-QUARTER-INCH SCALE.

THE accompanying diagram represents a landau containing the latest improvements. To indicate the different positions which the iron and wood work assume, when the top is up or thrown down, is a tedious operation, and is closely akin to that of drawing machinery. Examine every part of this outline, and you will find the most minute points, of any importance, carefully laid down.

After the side elevation has been correctly drawn, lay off for the height of door, which is established by the depth necessary to drop the glass frame level with the door, as shown in Fig. 1, line A being the hinging part of standing pillar, and also to have sufficient room to have the top lie nearly level (without coming in contact with the boot), as this is one of the beauties in the appearance of a landau when the top is down.

The glass frame is supported in its position, when up, by the patent iron glass frame supports, which are hinged at the points lettered B B. They are boxed into the standing pillars, leaving a rabbet inside, and also on roof rail. This allows the door to open when the glass frame is up. When the top is lowered, and the glass frame is dropped, the supports (before mentioned) are dropped down in the direction indicated by the dotted lines C C. The roof rail is hinged at the points D D, the opening of the top being at E. When thrown up perpendicular, the divisions are represented at F F.

We will now direct attention again to the point E, the joint of division in roof rail. At this point there is a curve framed each side level with the roof rail. To make the joint water-tight a groove is cut in the back curve. After the leather is put on, screw on an inch and a half band iron, covering the joint and groove, the ends turned over the roof rail. The edge to be chamfered and finished for painting.

The joint of roof rail is held even by *dowels*. For locking it together, the lever represented at Fig. 2 is the device made use of, two of which are required on each side of the top. The point G is a pivot inserted in a suitable piece of iron, which is screwed to the top; H is the knob for turning the lever. The slot I catches a pivot in the band iron before mentioned, which holds the top securely at that point. For hanging the doors, use concealed hinge at top, and at bottom a hinge swept to match it, as shown at Fig. 3. Bows to be plated on corners, and also corner plates from curve to roof rail.

There is a drop-light it front. The berth for receiving it, when up, is made of leather instead of styles to fall, giving a lighter appearance. A piece of wood is inserted between the leather and head lining for the purpose of securing the drip molding. The small black light slides sidewise in a wood frame inserted between the leather and lining.

The seats may be elevated when the top is down, giving the occupants a full view above the top. The mechanism required for the purpose is attached to the seats, and is operated by a lever under the falls.

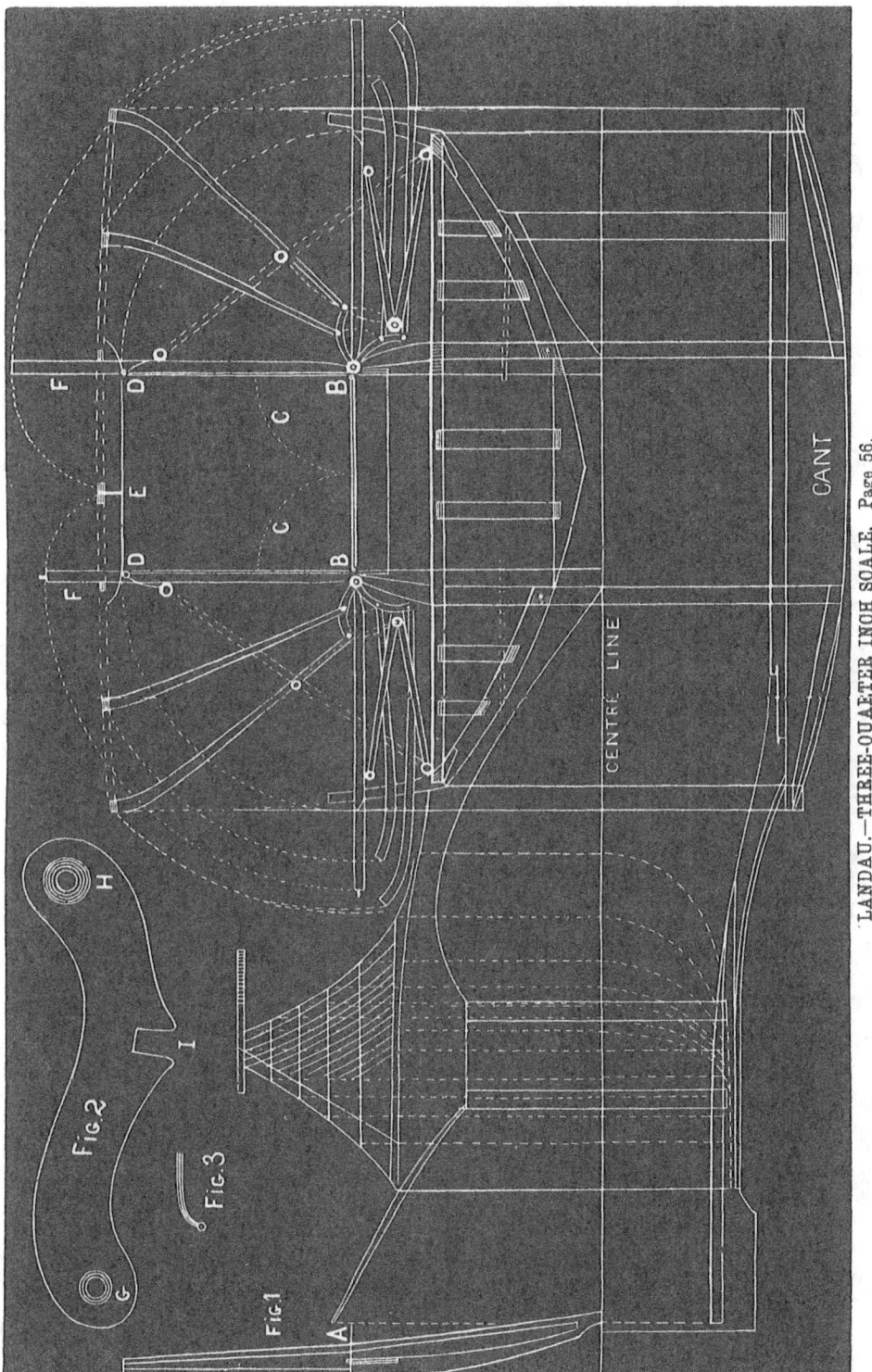

LANDAU.—THREE-QUARTER INCH SCALE. Page 56.

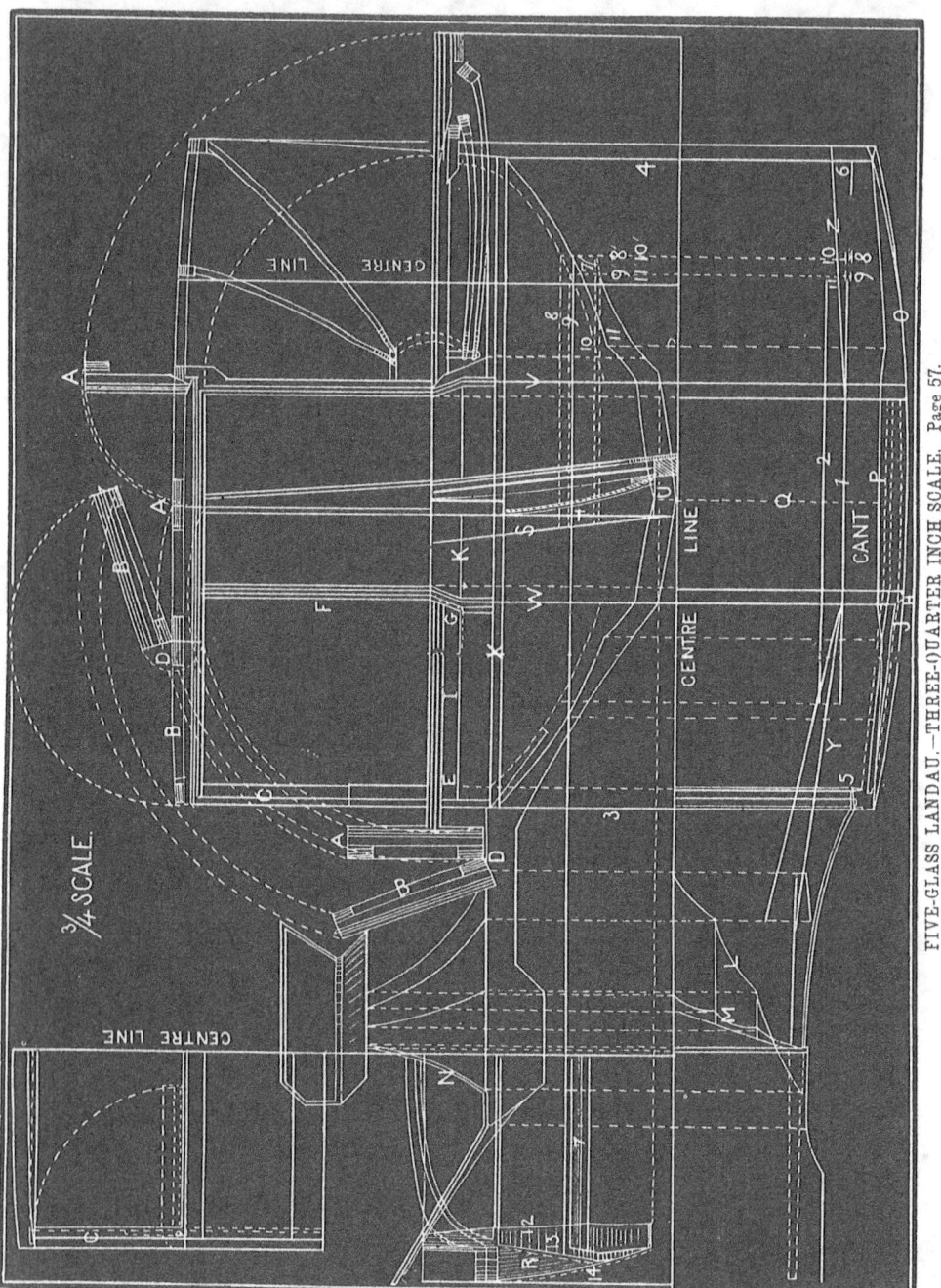

FIVE-GLASS LANDAU.—THREE-QUARTER INCH SCALE. Page 57.

Width of body laid out over all, 53 inches; 42 inches front and back; width of boot 30 inches. In building up for the round boot, the stuff requires to be well-seasoned poplar.

The boot to be prepared for the covering by first filling the nail holes with glue and sawdust, which, when dry, should be filed down level.

For covering use heavy linen; dip it in hot water, wring out dry, then saturate it in hot glue, and apply it, drawing very tight and smooth. When this is dry, sand-paper it off, and finish with a molding around the edge.

FIVE-GLASS LANDAU. THREE-QUARTER-INCH SCALE.

The five-glass Landau we consider the most complicated of any vehicle ever made in this country, and its description will be very much appreciated by those who are willing to learn, and to improve themselves in scientific and practical working. All the points of consequence are laid out in the most practical manner, and we wish our readers would study it through from point to point; the only way to learn it perfectly.

Our readers who may possess but a limited knowledge of this kind of work, by following the lines laid down will readily understand how the top is operated. The back of the body is just like the ordinary Landau.

The doors, which differ from the present manner of construction, have no flappers hinged to the door (better known as Wood's patent). The frame is operated by the door handle. When the frame is up, by turning the door handle the frame is thrown within the berth, and in falling strikes a spring, which prevents the door from being opened until the glass frame has reached the bottom of the door. We would mention that it is not necessary to employ the glass frame operator—which is not appreciated by every one on account of the heavy glass frame, which knocks the door to pieces—but we want the spring to prevent the door from opening when the glass frame is on the top. It would tear it to pieces if we did not prevent it through the spring.

To lower the top, drop the front light, run the front quarter glass frames backward, and drop them into the door berths, which are made of sufficient width to receive them and the door glass frame also.

We are now ready to unlock the top at division A, the top rail joint; having done so, unlock the top piece B, which is fastened with pillar C by a little spring, which prevents it from moving. Pillar C must now be dropped inward, which you see better in the top front view. Joint D has to be cut so as not to come in contact with the boot-neck panel, which is still prevented by a piece of round iron, covered with leather and fastened on E.

In lowering the remaining portion of the top, which travels in the direction of the double dotted lines, the back part is next thrown down in the usual manner, and is held in position by joints.

Pillar F is made of a solid iron piece stationary to the top rail, and the hinge fastened to it, which not only looks genteel, but is more solid than when made of wood. Dotted line G shows the hinge let in even with the wood. Letter H shows the pillar F, the top view in "cant." The panel marked I falls back the depth of the pillar F, in a straight line, which you can see in cant on line J; but only to line K. The boot is made similar to the "ribbed" or "shell boot."

We have but three pieces in our illustration. These should be nearly five inches thick, and concaved, as indicated by lines L and M.

The two middle pieces are glued together, forming one piece. This piece is then fitted and concaved, after which it is glued to the boot. The other four pieces are each fitted separately, then concaved and glued to their respective places. This done, glue the molding N at its place, projecting over the boot panel $\frac{3}{8}$ of an inch, and $\frac{3}{4}$ of an inch thickness.

We now lay out the width of the body: back of body, 43 inches; back door pillar, under arm rail, 51 inches; front lock pillar, $50\frac{1}{2}$ inches; front, 44 inches. This will give you the arm rail sweep.

The dotted lines in cant show the top rail piece, and the wood to be taken out where the front glass travels. Line P indicates the sweep of the body, without deducting the thickness of the molding. In taking the turn-under on line Q, and bottom of the door panel, you will find the door a little more rounding than the arm-rail sweep, because it gives a better sweep from the middle of the door with the whole back and front quarter. Line P from hinge pillar to the back is made in different ways; some that we have seen look very odd from the back; for that reason we drew the whole back view R, to show how odd it appears when you take the dotted line for the back pillar sweep; and the opposite, if you take too much wood away, for then it will have a very bad sweep from the middle of the door to the back quarter. The best method to obtain it is by making line P and pricking the whole length of the main sweep after it.

Draw line S to intersect with line T, on the bottom of the back pillar, which has to run in the same direction as the inside door pillar.

Place the compasses on dotted line U from inside door pillar line to line S, and prick the distance obtained on lines V and W, from line O, and from these points strike line 1, the outside line of the rocker, and line 2, the inside of the rocker, which is the thickness of a rocker of two inches. Next mark the thickness you require front and back pillars on lines 3 and 4, the thickness from the belt-rail line $2\frac{1}{8}$ inches; then take the distance between lines T and S on line X, and put it down on small lines 5 and 6, and lines 3 and 4, and from that mentioned point strike lines Y and Z. These are the two contracting lines of front and back pillars, and when the body is set up, the pillars on lines X, 3 and 4 will have the width measured from center line on lines 3 and 4, to *small* lines 5 and 6.

The center line of the back view is the same as the center line on ground view, for it passes directly through the center of body; and thus we have the plan for taking the width of the back bar. Dotted line 8 is the highest point of bar 7; follow line 8 to the turn-under line, and take the distance between lines T and S. Now follow line 8 back again to perpendicular line 8′, and pass down to line Z, and add the distance between lines T and S.

The same operation is necessary with line 9 and 9′, 10 and 10′, 11 and 11′. Small lines 8, 9, 10 and 11 will give you the length of the bar ready pricked off. Proceed in like manner in taking the width of the front and back bars above line X; also in taking the width of inside pillar 12, of the back. Main sweep 13, of back view, is taken from center line to line P; line 14 is taken from center line to arm-rail line P, deducting the turn-under.

To combine the door rocker with back and front pillars and the square-standing boot piece, we are obliged to make a separate draft, which we draw one inch scale, in the same manner as is commonly drawn in the shop, and shows the contracting lines, to mark out the wood for the door rocker, the front pillar and the inside rocker which combines the front quarter with the boot. The only change is with the center line, which is not in the center on this engraving; 21 inches is deducted from the whole width of the body, or $10\frac{1}{2}$ inches cant, or half width of the body. For top and bottom views we have to mention the same. The center line, which separates the two rockers, is taken from the width of the cant, and there is also deducted 10½ inches, the half width of the body; the same with the turn-under. The only reason for deducting some of the half width is to save room on the block; it would make our engraving much larger if we did not do so.

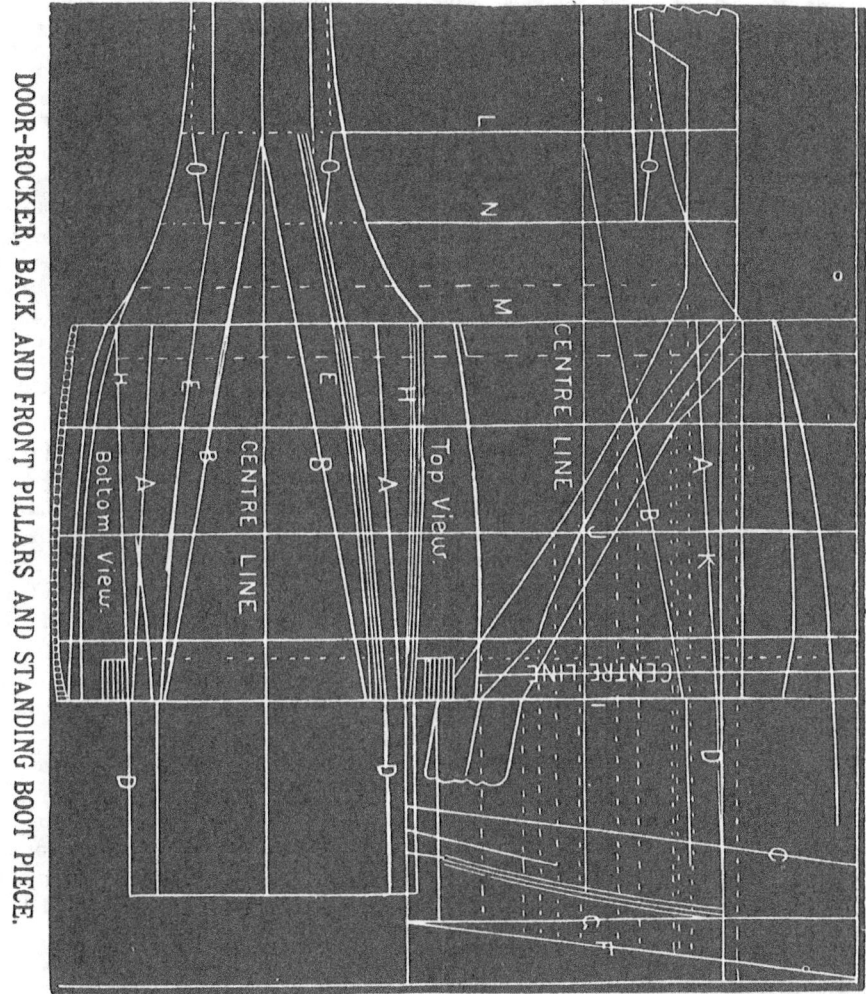

DOOR-ROCKER, BACK AND FRONT PILLARS AND STANDING BOOT PIECE.

In top and bottom views the contracting lines of the three pieces are drawn, as well as in cant; the others are the ready pricked off lines, given thus to lighten the labor of our students in reading the explanation.

To construct a body after that drawing, we have to mention first: The wood has to be sawed out larger, on account of the leaning out and contracted lines; and the more a piece of wood is turned out and contracted, the more wood it requires. Mark your back pillar and inside rocker after lines A, B and C, the inside back pillar; and the same in top and bottom view; the door rocker after contracted line D and line C; the student will see the three mentioned pieces are inclined after one line C, but are contracted in three different lines—A, B and D. Lines E in top and bottom views are the ready pricked off lines. These lines show the inside rocker piece when it is on its place, and is established from lines F and G, and contracted line B. We do not want that line in fact, but we only want to show the variation of the different pieces as produced through the inclination line F. If it was laid out square we would not want it at all. So it is with lines H, they are pricked off like the others from F and G, and contracted line A.

To joint the door rocker with back pillar is now very simple, when the draft board is well established. Suppose the joints are from I to J, door rocker D and back pillar leaning out in one direction, we have to cut off from the outside door rocker D the distance between line D and contracted line A from I to J; which, to make it plainer, we mark with letter K; this gives the joint for the door rocker and back pillar. We have still the inside rocker, which extends from line I to L. The distance between lines B and A, on I, is the desired thickness on that point; and between B and A, on line M, is the thickness on that place, and will joint to the inside back pillar. It is different with the joint on L and N, which combines the boot piece, because the boot piece is standing square. Letter O shows the joint—the best way to join it you can see in top view ready pricked off.

LANDAULETTE. THREE-QUARTER-INCH SCALE.

A Landaulette is a style akin to the Clarence coach, the main point of difference between the two being a leather back quarter and roof on the Landaulette, which is substituted for panel work, thus admitting of the front and top of doors being removed and the top thrown back, and when thus changed it presents the appearance of a brett.

Landaulettes vary in size, and are termed full, three-quarter and demi or half size, the latter being capable of such a reduction in weight as to require but one horse. In laying out the body, the first measurements taken should be the width of doors and front circle; next, the back quarter. There is one point just here we would call particular attention to, which is, that where the concealed hinge is used, and the joints dropped as shown, full allowance must be made for the space required for the concealed hinge to move in when the door is opened, so that it will not come in contact with the knuckle of the joint.

Next establish height of seats, as shown by the dotted lines, also height of roof. Space off for the guide rail for the glass frame to drop level, which is shown by hinging point B, and also in Fig. 1 at G. C C shows the boxes, which are hinged to guide rail B, which holds the glass frame when up, the advantage in this being, the door may be opened while the glass frame remains elevated. When the frame is dropped, a complete finish is secured by dropping the boxes, as shown by the dotted lines. D is the hinging point of the roof rail to the standing pillar E, the division in top. The standing pillar L L is taken from one piece, of sufficient size to allow for the bevel of door, and is framed up its full length without being cut at the hinge point. The hinge is screwed on, and the roof rail hinged at D, without being cut at E. The circular roof rail is lapped, as shown in the working draft, at K. The two bows at each side of division at E should be wide, to prevent springing; these and all other bows should be placed to their position. The stationary circle, which lies on top of deck panel, is fitted, dressed and glued to the deck panel. The next circle is fitted and framed to the pillar, but left unglued on line A, and has a molding worked on it to cover the joint. These circles are gotten out in two pieces, and lapped as shown at J. When all points of the work are ready, with a fine saw we cut the pillar at the division M, the roof rail at E, and the standing pillar at B.

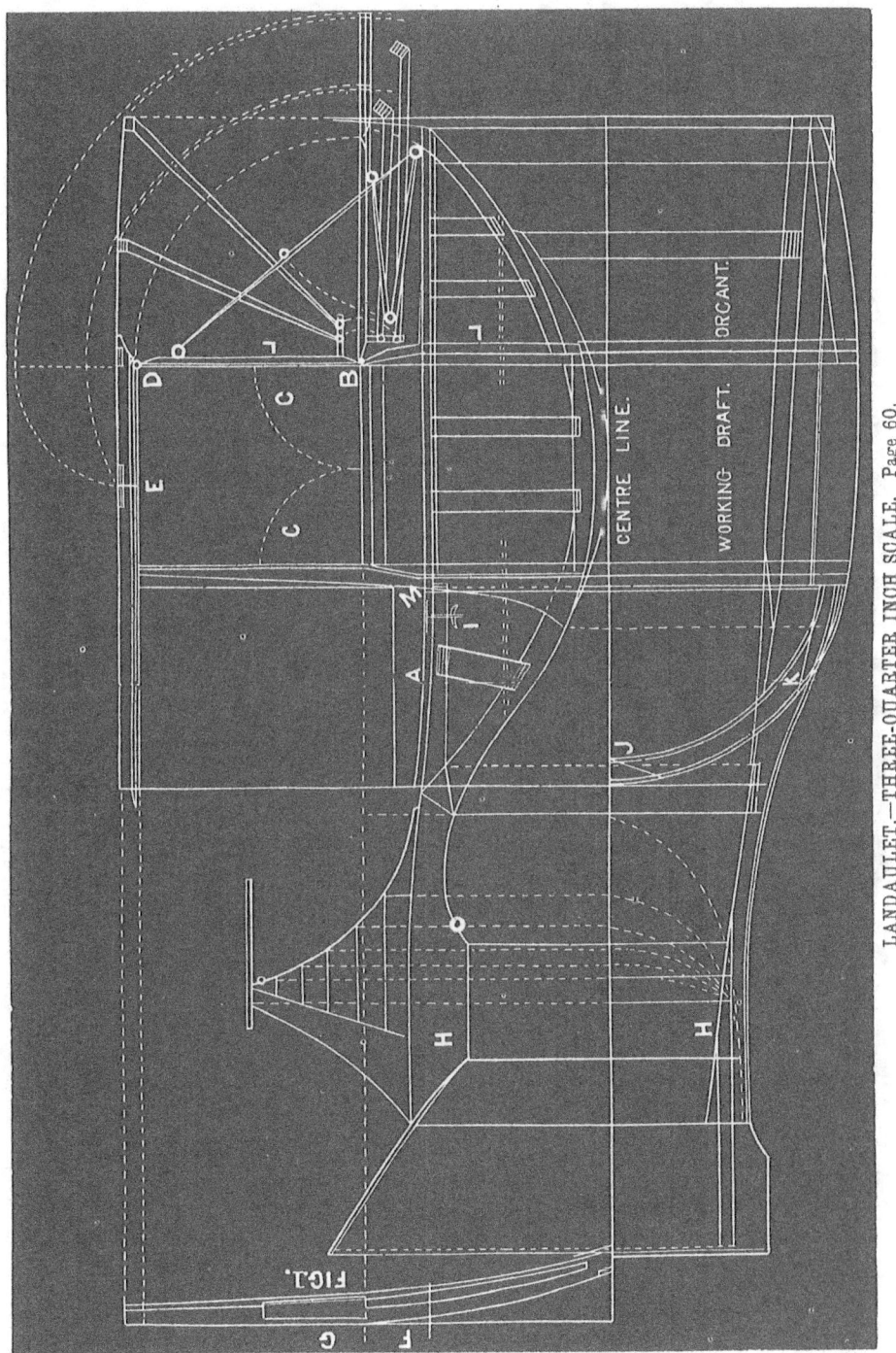

LANDAULET.—THREE-QUARTER INCH SCALE. Page 60.

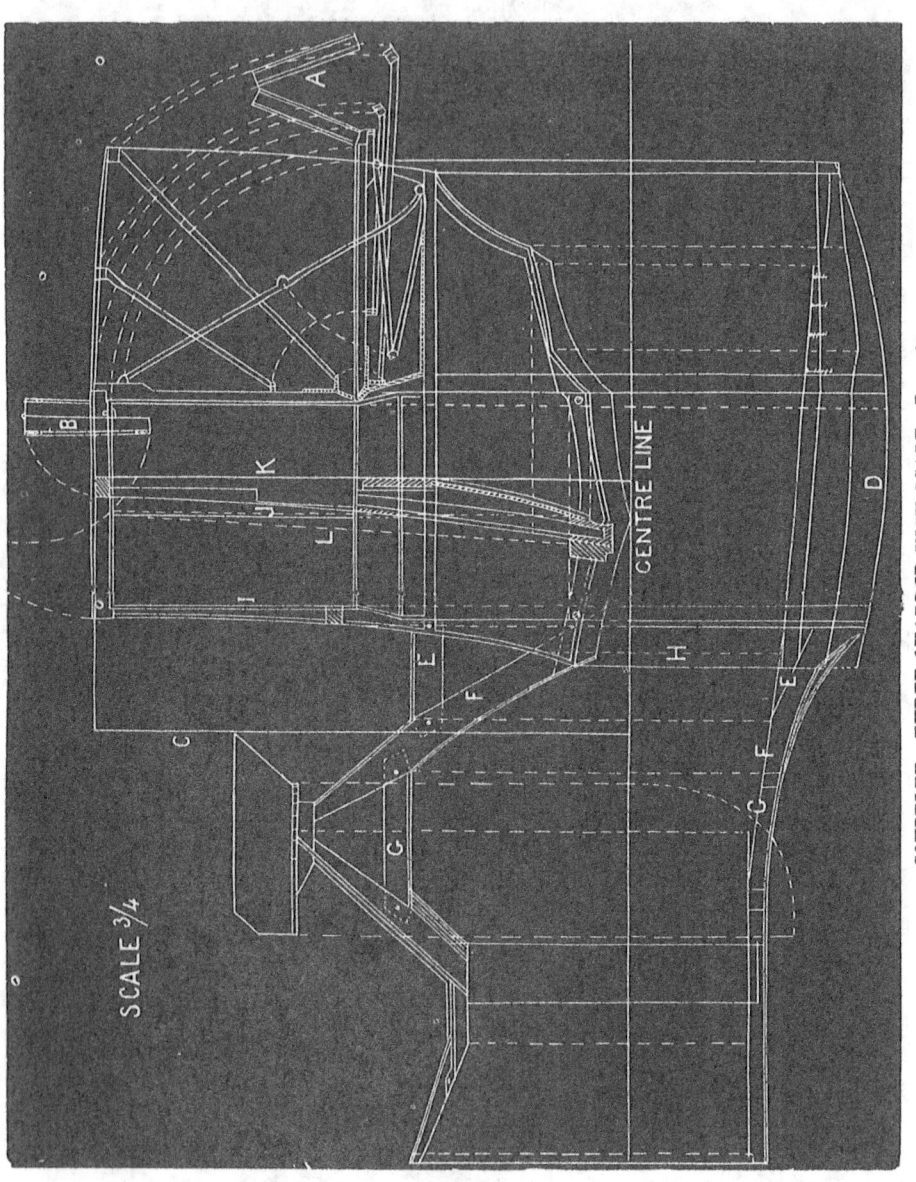

COUPELET.—THREE-QUARTER INCH SCALE.

In securing the front circle to the body, use on each side T-head bolts, with thumb screws, as shown at I. The heads of these bolts are let in and screwed from underneath the top circle; dowels are also required at intervals between the bolts. For convenience in removing the circular front, the trimming should be left loose, sufficiently to admit of the hand being inserted to operate the thumb screws. Having now given full directions as to the circular front and its attachments, it only remains for us to speak of the finish required when the front is removed.

There are two methods of forming a finish, one for mere appearance, and the other wherein a back is arranged for the front seat; each is secured by bolts and dowels as in that first mentioned.

For finishing purposes, one of them is hinged in the center; the other, forming the back, is made 10 inches high, and rounded down at each end. The division of the doors should be at line A, which is sawed open after being made as shown in Fig. 1, line F. This division is secured by dowels, thumb screws and plates on the inside, and plated to make a complete finish. When the upper part of doors are removed, finishing pieces are attached in the same manner.

In the working draft it is laid out for contraction, which requires the different pieces to be dressed on a bevel. Line H is a face of boot piece H, and shows the concave as framed into the front pillar.

COUPELETTE. THREE-QUARTER-INCH SCALE.

The advantages in this design are its convenience for riding as in an open carriage without the necessity of leaving the front at home, to be battered and scratched in lifting on or off.

The back part A, which forms the top door piece in its erect position, is given in two different manners. Some may not like the position of the door piece as given at A; we have, therefore, reversed the hinge, as at B, which shows it in its folded position. It is necessary to joint the door piece in center, to prevent it coming in contact with the C spring when folded.

The standing pillar we have drawn in the middle of the door; there being room enough, and much more convenient and familiar to experienced body-makers.

Line C is the half width of the "cant," or of the hanging pillar, which you can see by measuring from center line to arm-rail sweep, line D.

Each piece of the boot is laid out square, but three of them are contracted, which E, F and G indicate. We will mention that each piece contracted requires to be lengthened. Suppose we were to *not* lengthen piece E, which shortens nearly one inch, the consequence would be it would lift the front of the boot about three inches higher than the diagram.

Piece F starts from the door rocker up to the driver seat. It requires considerable thickness, nearly 5 inches on dotted line H. It can be made of 4-inch stuff or less, by gluing a piece of white wood on it to fill up the sweep of the boot. The front lock pillar will have to be jointed or hinged, so that the glass frame will not project above the joint.

The lock pillars I, from the jointing or hinging part to the top, must not be over half the width of the body, so they will not strike each other when folded over the glass frame. We will mention that the glass frame cannot fall farther than the rockers, and must run into the back pillar (see dotted line J). We have to do so on account of the thickness of pillar from K to J, which still has a thickness of four inches.

Dotted line L is the leaning-out line of the door lock pillar, which gives a thickness of $5\frac{3}{4}$ inches. Not being able to procure wood of that thickness, take 5 inch or less, and glue a piece on the bottom, but not on the top if you can help it.

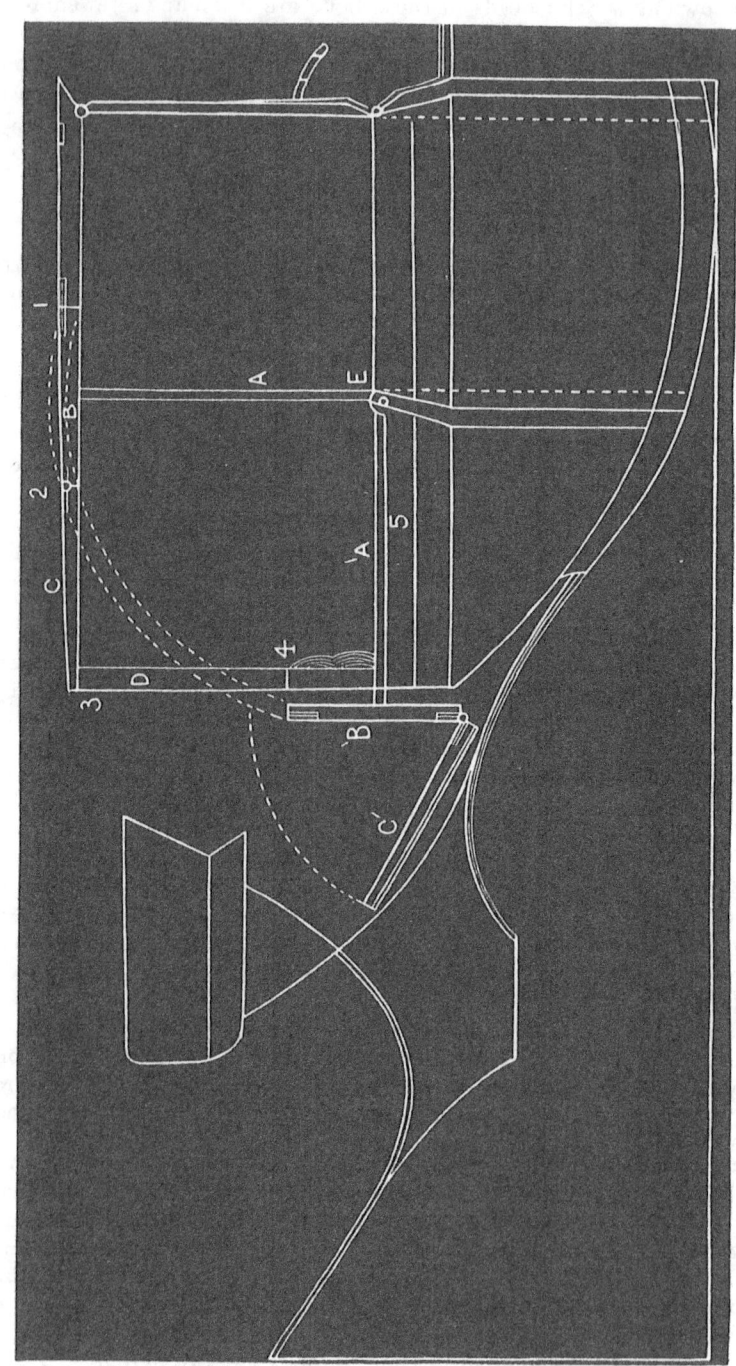

GLASS-FRONT LANDAU. THREE-QUARTER-INCH SCALE.

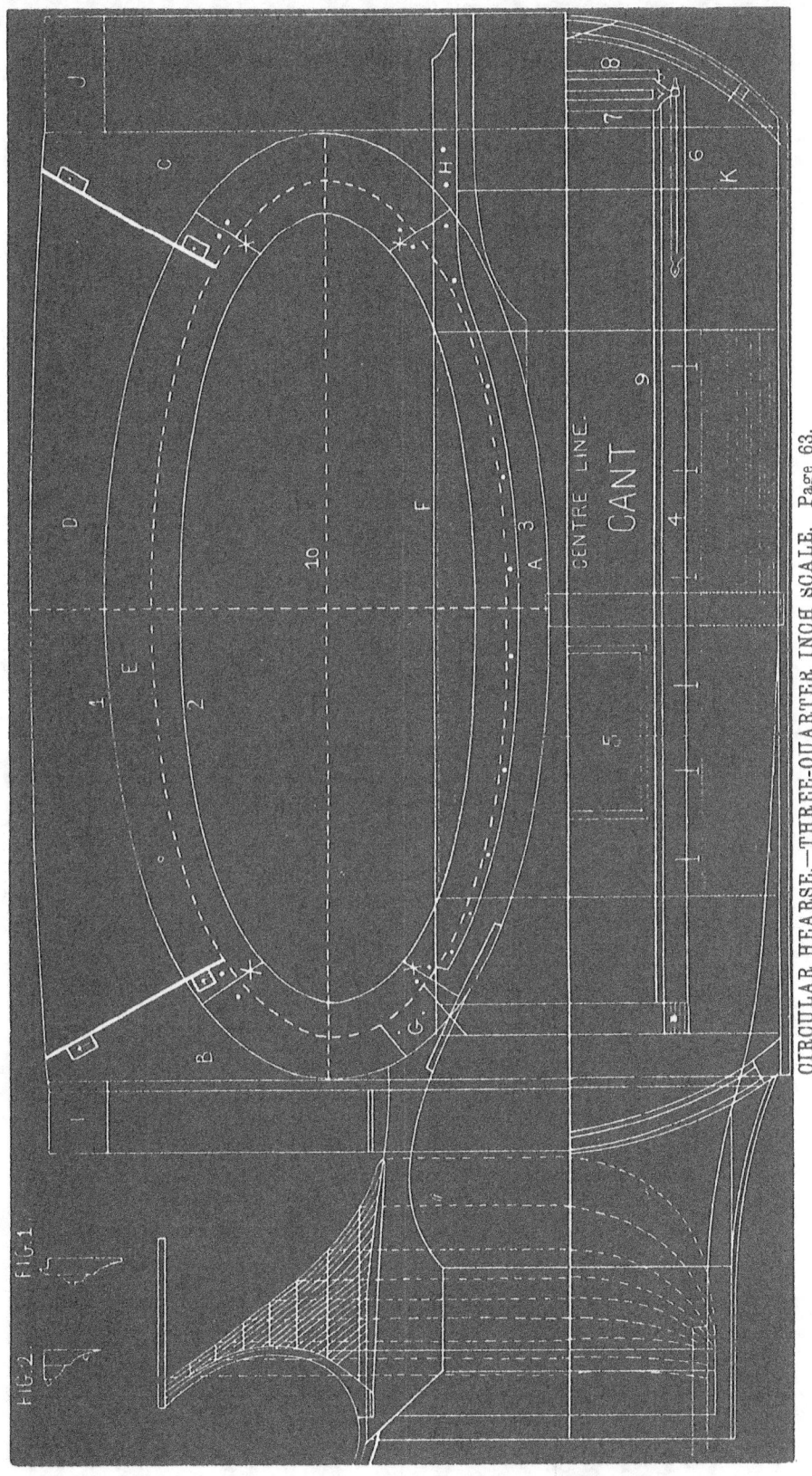

CIRCULAR HEARSE.—THREE-QUARTER INCH SCALE. Page 63.

Our diagram shows a section only of the body, namely, that portion from the hinge pillar forward, and including the skeleton boot and dickey seat. The front is glass as well as front quarters, and that there is no leather forward of hinge pillar save on the roof. We will now attempt to elucidate the method of arranging the several parts, and bringing them into harmonious working. We will first take up

THE DOOR,

which differs from the present manner of construction in the following particulars: 1st. The berth is left wide enough to receive the door frame and the glass quarter. 2d. Instead of using iron boxes that are hinged, the box is made in the hinge pillars. The frame is operated by the door handle. When the frame is up, by turning the door handle, it (the frame) is thrown within the berth, and in its descent strikes a spring, which prevents the door from being opened until the glass frame has reached it.

We will now direct attention to the sections of top frame. At A, B, C, D are the four divisions as seen in the erect position, and A′, B′, C′ show the positions of three parts when the top is down; D falls inward toward the center of the body, and, therefore, cannot be shown when down. The pillar A is made stationary to head-rail piece B, which extends from 1 to 2, this piece being hinged to C. D is hinged at 4 with a spring at the top, in order to form a connection with rail piece C. The panel marked 5 falls back the depth of the pillar A. At E is a hinging point plainly shown by the drawing.

TO LOWER THE TOP.

First drop the front light, next run the quarter lights back and drop them into the door berths, which are made of sufficient width to receive them and the door frames also. We are now prepared to unlock the top at division 1. Having done so, unlock head piece C, at the point marked 3, and throw it back in the direction of 1. Pillar D must now be dropped inward. The way is now clear for lowering the remaining portion of the top which travels in the direction of the double dotted lines. The back part is next thrown down in the usual manner, and is held in position by joints.

CIRCULAR HEARSE. THREE-QUARTER-INCH SCALE.

After drawing the side elevation, we will now proceed and dissect the different parts for framing, commencing with the bottom side, which is lettered A. This piece extends from the bottom line to the dotted line immediately above; follow the dotted line back to where the horizontal line F intersects it, and is lapped to the back pillar at H. Now, follow the dotted line to the front, where it forms a lap at point G with front pillar B. This is the bottom side, 4 inches wide by $1\frac{1}{2}$ inches thick, which is taken from ash. Back pillar C, which follows the dotted line on inside, is 1 inch thick, ash, and is lapped at the bottom side at H. You will perceive that there will be a half-inch projection on the inside; this will be made level by the shoulder coming on line F, which represents the top of the track. Our object is to use the stuff as thin as it will allow, so as to cut down any unnecessary weight. Front pillar B, extending into the dotted line, is $1\frac{1}{2}$ inches thick, ash, and is lapped at G. The top rail D, extending into the dotted line, is 1 inch thick, taken from poplar, with a stub tenon and two long tenons at each end.

Piece E, extending from line 1 to line 2, is $6\frac{3}{4}$ inches wide by $1\frac{1}{2}$ inches thick, poplar; the joints shown at points marked X are produced by slip tenons of ash. This piece receives the glass, and is shown by Fig. 1, with the rabbet taken out for the reception of the glass.

The next piece to be applied is 3 inches by ¾, poplar, the joints being at line 10; the finishing of the center is by the use of a half-round molding, ¾ by ½ inch. The concaves and rounds are all shown at Fig. 1. Fig. 2, the top moldings. The first piece in this molding is 5¼ by 1¾ inches; the second, 2⅛ by 1½ inches, finished with a half-round molding, as shown.

I is the front circular piece, width 6 inches, and 3 inches thick, which allows the glass frames to pass each other. The bottom circle the same thickness, and 2¼ inches deep. J, the top back circle, is 6½ inches deep by 2 inches thick, poplar, lapped in the center, as represented in cant. On the bottom back bar the outer circle is the same as J, extending into line K, is 2 inches thick, of ash, and is screwed to under part of bottom side, and receives the circular doors, which are 1½ inches thick. Next we take up the concave piece, which extends from line F to line 3; it is 7 inches square, poplar. It is rabbeted into the bottom side, and screwed from the outside, as represented by the dots, the back part resting on the back bar, the lower edge being the continuation of line 3.

We will now turn your attention to the top view of the piece last mentioned, as shown on the cant. Piece 4 is 3½ by 2 inches thick, poplar; this is glued and screwed to the concave piece, as shown in the cant, leaving the track 9 inches from the center line, 18 inches being the full width. The bottom boards ¾ thick, running cross-wise, grooved together and screwed from bottom of piece 4; 5 shows the lid, which covers the opening left to draw off the water when cleansing the inside; 6 is a slotted plate, in which plate 7 slides, which is screwed to the top of truck; 8 is the roller; 9 a half-round iron, which forms a track for the concave rollers of the truck to pass over.

From center line to outside of bottom side is 20½ inches, being half the width of body, the molding of Fig. 1 to be added. Half width of boot, 15 inches. The boot is shown free of hammer cloth, with the dickey-seat frame attached. The toe-board brackets are not shown full length for want of space.

ROCKAWAY LANDAULETTE. THREE-QUARTER-INCH SCALE.

The accompanying diagram represents a Rockaway Landaulette, and is presented with the latest improvements.

To indicate the back position which the iron and wood-work assumes when the top is up or thrown down, we would refer to explanation and working draft of Landau H. The joint which the molding J covers, is cut through to the hinging door pillar. Three fastenings are made and let into the circular pieces inside, represented by Fig. 3, also three dowels between the joint to hold it even. Line K, shows the joint of the two circle pieces. If the door wants to be cut, to bring it even with the front circle, these fastenings and dowels have to be used to fasten to top door piece L. The molding will cover the joint all around. If it is desired to ride with the front open, when the doors are cut they have to be made so as to be taken off (as indicated at L) with the whole front. When the front has been removed, the finish is made by pieces of white wood made the shape of the top door and circle, with a molding around which covers the joint. The same dowel holes are made use of to hold it in its position.

For hanging the doors, use concealed hinges at top, and at bottom a hinge swept to match it, so the pin will stand in a perpendicular position back and sidewise with the concealed hinge.

SIX-SEAT ROCKAWAY. THREE-QUARTER-INCH SCALE.

This drawing will show the manner of constructing this class of work—the side elevation being artistically drawn. By this we mean the outer, as well as the inner

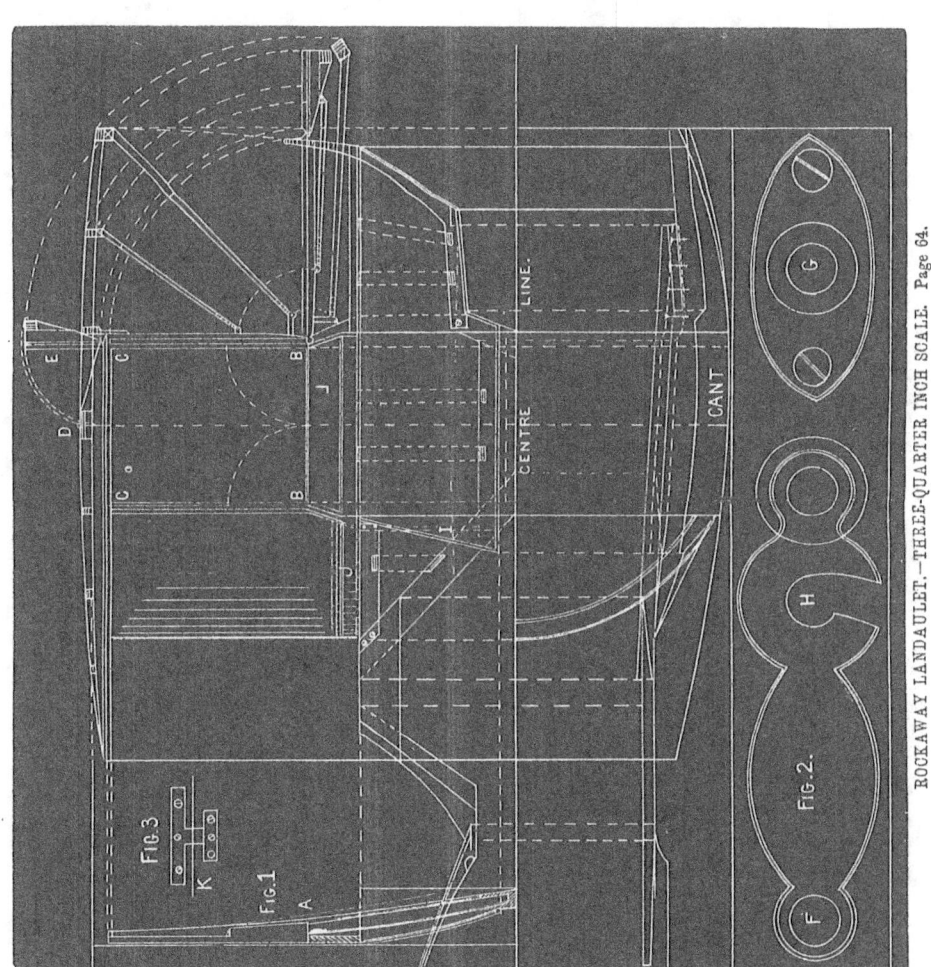

ROCKAWAY LANDAULET.—THREE-QUARTER INCH SCALE. Page 64.

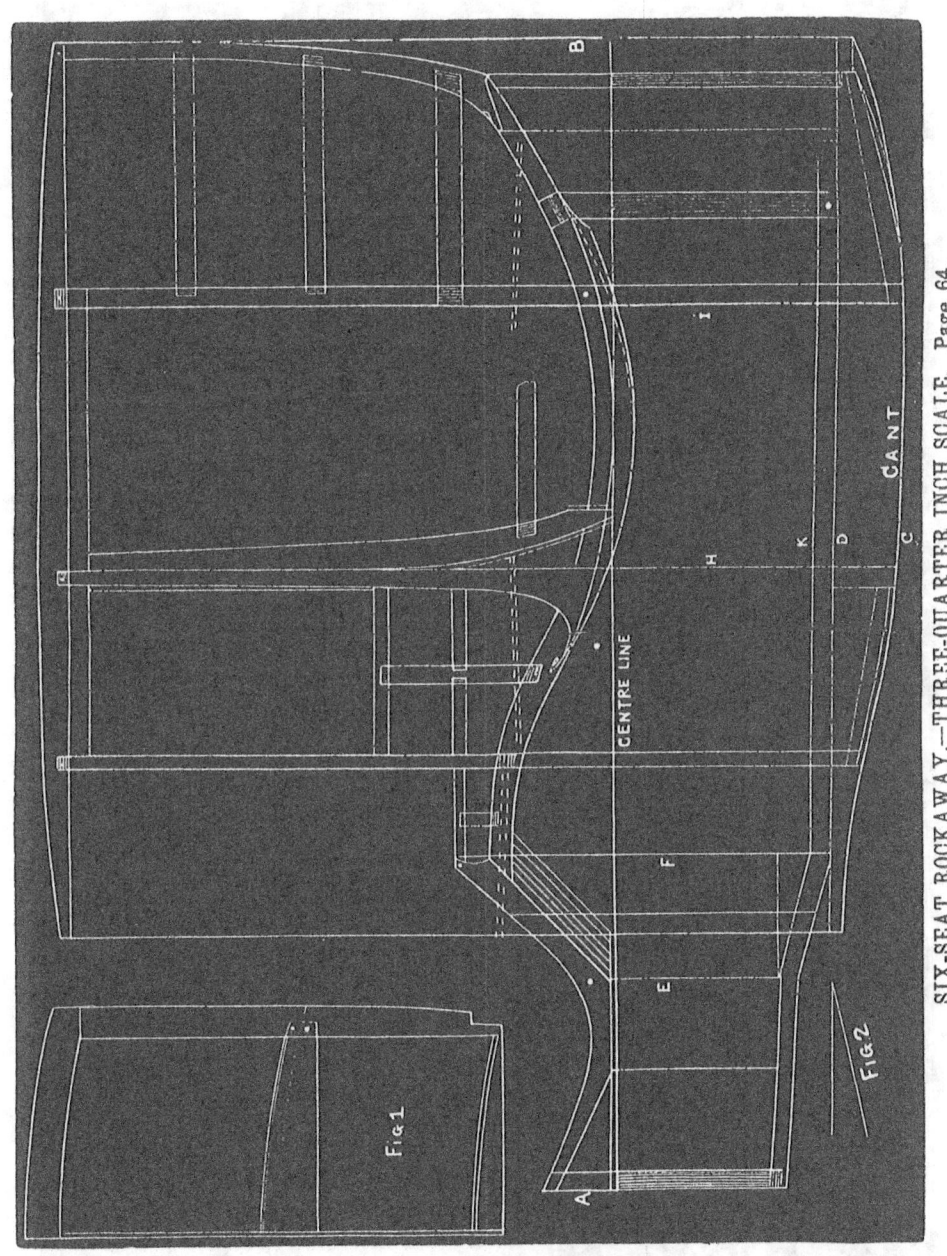

SIX-SEAT ROCKAWAY.—THREE-QUARTER INCH SCALE. Page 64.

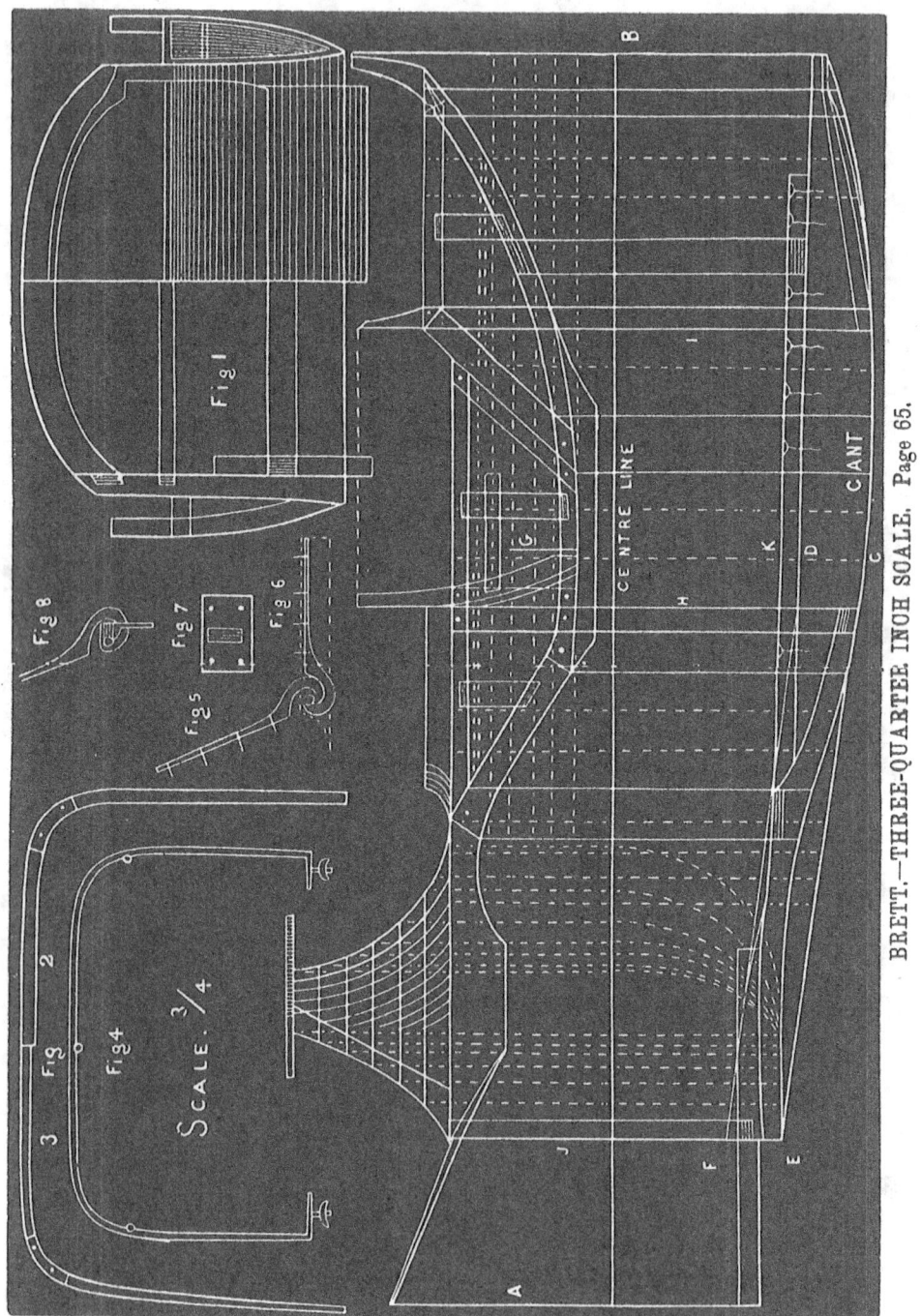

or molding lines, to produce a correct piece of carriage architecture. In this we give the latest and most improved manner of construction, doing away with the heavy thickness of timber at the inside of the coupe pillar, lightening out clear to the bottom side, thereby making, as the body-maker terms it, a "better job." We also give the manner of framing the front of door, doing away also with the deep rocker. This frame-work is all paneled over, coming into a groove in the coupe pillar, making all shoulders of the frame work to cross the grain of the panel. The scroll of pillar at the cross line O is glued on when the body is being finished.

This body has $4\frac{1}{2}$ inches turn-under, the doors shutting against the rocker. There should be a berth cut into the bottom door rail, allowing the glass frame to drop as near to the bottom of the rocker as possible; this gives a low door panel.

The short bottom-side from the hinge pillar back, coming down to the bottom of the rocker, the molding being boxed down, and the rocker being cut off, as represented by the dotted lines, is for the purpose of bolting the iron pump-handle, or break, too, giving it the appearance of growing out of the inside of the rocker. The rocker plates should be $2\frac{1}{2}$ by $\frac{1}{2}$-inch, with a flange to receive the cross stay for the cross spring, to take two bolts. It is also well to turn a corner on the back bar; this plate to be secured with No. 20 screws, 4 inches apart.

You will now proceed to lay out the cant, which is from center line to roof-rail line C, and from the top-rail line A to back-line B. Half the width of body on line B, from center line to line C is 21 inches; from the same points on line I, 26 inches; from same points on line H, $25\frac{1}{2}$ inches; same on line A, 15 inches. This produces a very narrow front, at the same time it gives a wide front inside seat. This is obtained in framing up; in the bevel of the piece between the lines E and F, and as laid down in the cant, and shown by the front view of it in Fig. 2, it gives the carriage, when completed, a graceful and very stylish appearance.

The swelled panel at this point is produced by the grain running the same as the lines drawn. We have given the width of the body at the different points, and also have described our improved manner of construction. For establishing different lines and points, we refer you to the French Rule.

Fig. 1 represents half view of the winter front. This is made with a wide center rail, and fitted between the front pillars, and sawed in two at the dotted line. The bottom part is made stationary, the top part dowled to it, and locked at the top, which can be removed for summer use.

BRETT. THREE-QUARTER-INCH SCALE.

You will perceive we have laid out the cant below the side elevation, for the purpose of its being clearly understood.

The side elevation being artistically drawn, you will determine the turn-under, which is three inches and a half, and also establish the depth of the pillar, and line G, being the face of the bottom side; this allows the doors to shut against the rocker. Complete the patterns. Proceed and draw perpendicular lines A and B, the extreme length of the body, and also lines H and I; then draw the horizontal center line.

Now lay out for the width of the body, which is 3 feet $11\frac{1}{2}$ inches on line I. Proceed and measure from center line on line I $23\frac{3}{4}$ inches (this being half the width of the body), and prick it off; this is one point of the establishment of line C. Measure off on line H $22\frac{3}{4}$ inches, being another point for the establishment of line C. This contracts the body two inches in the width of the door.

Take the dividers, place one point on line H, the other on line G; having this space, prick it off on lines H and I from the points last mentioned, which establishes line D.

5

From these two points, lay the straight edge and draw line D. Measure 1¾ inches from line D, and draw line K, which is the thickness of the rocker.

You will see by the diagram we have laid out the framing of the back pillar at the arm rail, and also the framing of the back bar in the rocker, the same passing back sufficient to take two screws, making a much more durable job. Measure 1¾ inches from line D on line B, and prick it off. Now proceed and lay out the width of boot, which is 30 inches; measure from the center line, on line J, 15 inches, being half the width of boot.

Draw line G from the different points established as represented by the diagram. The belt or arm rail as before explained. The space between the two lines E and F represents the thickness of stuff required to form the concave boot, F line being the face, laid on the draft for pricking off.

In forming the round boot, use 2-inch white wood or poplar, one glued over the other, until the desired height is obtained. When cleaned off cover with raw hide or linen.

Fig. 1 represents the back view, half paneled, half framed.

Fig. 2, half the front bow, the manner of framing. The two front bows should be the same width; the third, one inch narrow at the top, and so on, until the fifth bow is reached, which is Fig. 3, the back bow, which is the same width at the bottom, but three inches narrow at the top. This contracts the bows with the sweep of the body

Fig. 4 represents the storm cover bow. This is made from half-inch round iron; use stump joints at the three different points for folding. This is secured to the back of the dickey seat with thumb nuts.

Fig. 5 represents the concealed hinge, with the crank up.

Fig. 6, with it down.

Fig. 7, front view of the socket.

Fig. 8, side view, socket and crank attached.

In the construction of these hinges, there should be a wood model first made, for it is necessary they should be exact, to work well, so as to prevent alteration afterward. The sockets can be cast of brass, the crank made from the best wrought-iron.

FOUR-IN-HAND DRAG. THREE-QUARTER-INCH SCALE.

This drawing represents side elevation and cant. The side elevation is drawn to the ¾ inch scale, and the manner of framing shown, with short bottom sides framing into the lock and hinge pillar, the door shutting against the rocker. Panels put in a groove, with deep chamfered moldings. Proceed by laying out the cant or working draft, by drawing the center line, and also by squaring lines A and B.

Measure twenty inches from center line on lines A and B, and prick it off, which establishes two points on line C. Also half the width of the body, which is 40 inches wide. Square lines D and E. Also draw line F. This line is established from the points of line C by taking the space of the turn-under and depth of pillar, and prick it off from the points of line C on lines A and B.

Measure on line D, from line F two inches, and prick it off, and also on line E from the same point. This establishes the four points of line C. This line can be swept with an easy sweep, as shown in the diagram.

H and K shows the side flare of seats. This flare is obtained by the width required for the bows, which is here laid out for 48 inches. The bows decrease in width, following the variation of the seat line. The seat being narrowed behind, the bows should be arranged accordingly, thus producing a top in harmony with the swell of the seat, which are of the French pattern, with square corner. I and K shows the back and front

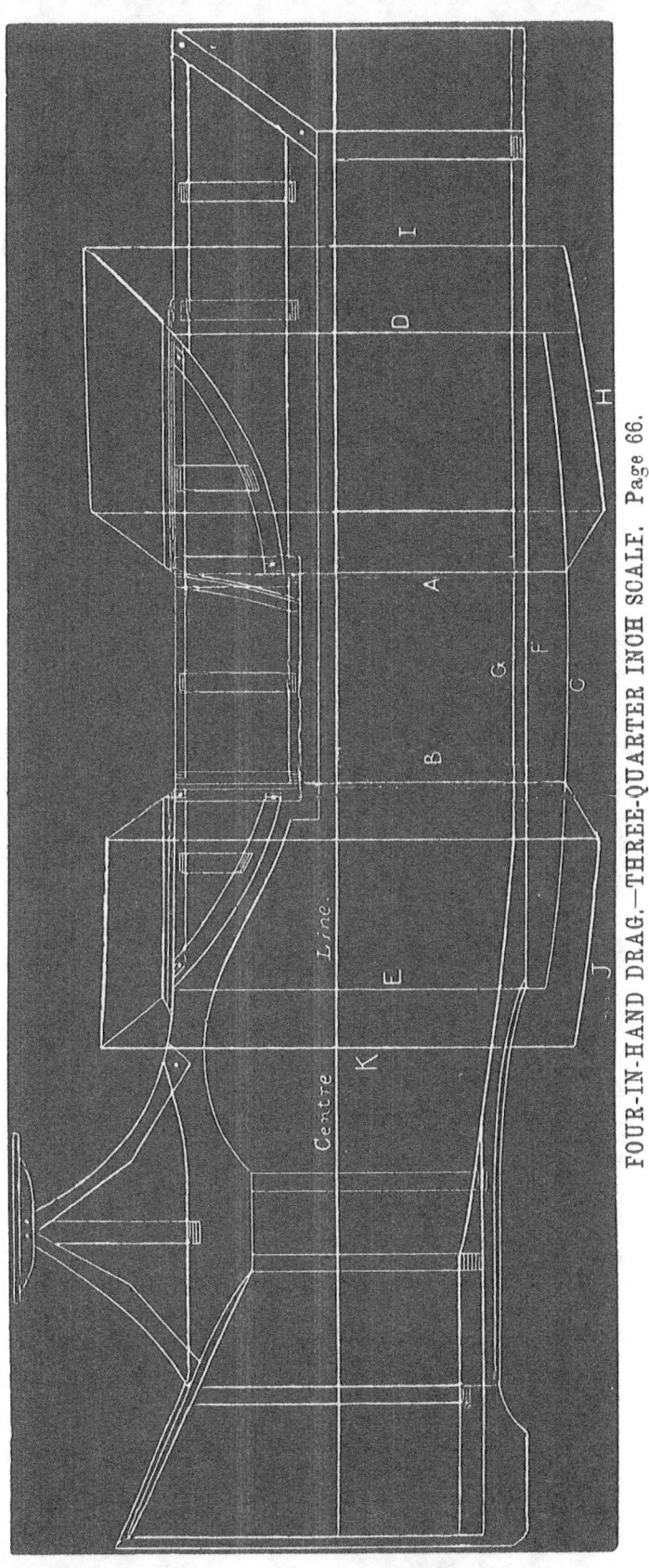

FOUR-IN-HAND DRAG.—THREE-QUARTER INCH SCALE. Page 66.

flare, and C the face of the rocker. The front is laid out for 28 inches, having a slight concave; also shows the manner of finishing the toe-board.

The top may be made either shifting or stationary. The stationary top has a high wooden back, secured to the seat in the ordinary way. It is 22 inches high from the top of the seat frame, and arched with wooden arm pieces, to which the trimming is attached.

The shifting top has the back and arm pieces secured to the rail by iron stays, which admits of the whole being removed when required; an extra back to be used for a no-top presentation of this vehicle, fastened to the ordinary eyes.

For rocker plates use $2\frac{1}{2}$ by $\frac{1}{2}$ inch, and $2\frac{1}{2}$ by $\frac{3}{8}$ Bessemer steel. Half-inch steel should be used in the door-way, running to the neck and any other weak points, but where panels help to stiffen the work, use $\frac{3}{8}$ inch steel. Turn a corner plate on to the back bar; this plate to be secured by No. 20 screws.

We have heard some objection to the use of Bessemer steel for the purposes named, but where it has been used of a proper temper and sufficient weight, it is superior to iron, in that it may be worked lighter than iron, and still contribute the requisite stiffness.

Rocker plates often fail to support the rocker because they are not properly bedded in white lead, and securely screwed. The holes should be drilled to the exact size of the body of the screws, and the heads sunk to the level of the rocker plate. The screws should be, say four or four and a half inches apart.

SIX-SEAT EXTENSION-TOP PHAETON. THREE-QUARTER-INCH SCALE.

On next page we show the skeleton frame-work of side elevation and cant. We will direct attention first to the boot. It will be observed that we give a sectional view showing the inside, whereas, in former drafts, we presented an outside view. The neck piece of boot is taken from $1\frac{1}{2}$-inch ash, and is paneled outside, giving a smooth surface for painting. A B are the two holes cut in the neck piece from the inside to within a quarter of an inch of outer surface, to receive bolts for the carriage. The bolts are of round iron, having square heads; they are inserted from beneath, the nuts being dropped into the holes A and B before mentioned, and the wrench applied to the head of bolts; C, in the side elevation, shows the inside view of toe-board bracket, and C in the working draft the top view, which is concaved on the inside to receive the rocker plate. D shows the inside view of boot bracket, which is 2 inches thick, and is lapped from the inside of neck piece, falling back one inch from outside of panel, which receives the circular pieces 1, 2, 3, 4, 5, shouldered and lapped on the inside, making a level surface on the outside with D. The joints are, of course, hidden by the canvas with which the entire boot is covered. Next observe the point marked Fig. 1 gives an inside view of the joint which supports the seat back, and shown in three positions. When turned down the joint rests on top of the cushion.

Fig. 2 is the front view, showing the vertical line the joint must work on. These joints can be laid out when the body is being drafted, thus furnishing the smith with a correct pattern to work from, as well as determining the width of end pieces of seat lid. E indicates the standing pillar, showing a front view with turn-under; short bottom sides are formed into E, doors shutting against the rocker. The brake or pump-handle is a continuation of the rocker, the rockers being spliced at center of door-way and at point marked F. In finishing the body there should be but one molding, which is placed at the top of panels and passes around the body. The back panel is mitered to the back quarter. Seats of the French pattern, square corners. In passing on to the working draft, the lines laid down are such as to require but a small share of attention compared with the side elevation.

The center line is plainly marked, and at G the sweep of body is given, and H H

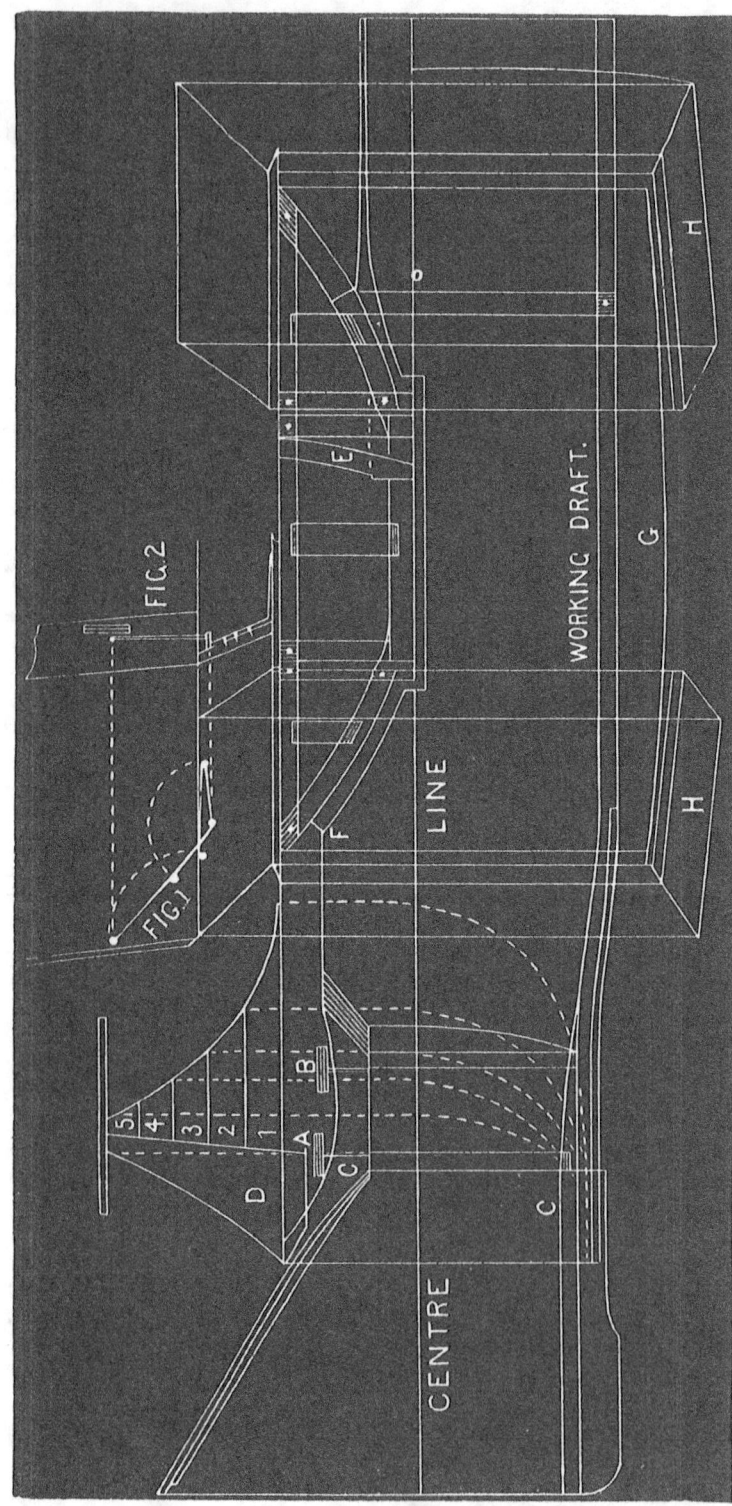

WORKING DRAFTS OF LANDAUS, ETC.

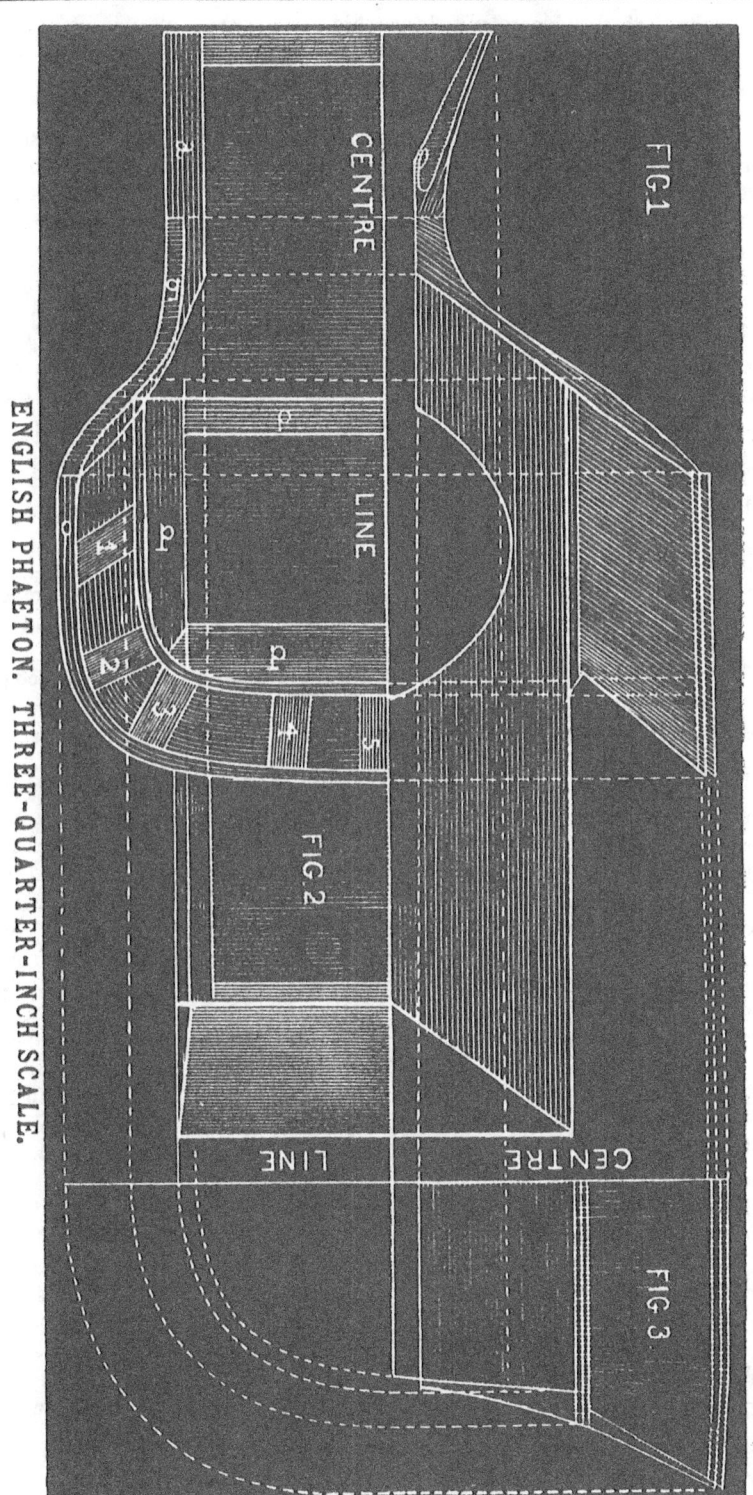

ENGLISH PHAETON. THREE-QUARTER-INCH SCALE.

indicates the full width of seats, which are laid out to take a 48-inch bow. Seat frames to project over body 1½ inches. Seat panels to fall back ¾-inch, with a beveled skirting.

ENGLISH PHAETON. THREE-QUARTER-INCH SCALE.

This drawing illustrates the method of laying out the English phaeton, showing the manner of constructing the panel seat, and, in connection with it, the front pillar, which is framed into the bottom side of body. The seat and front pillar being the parts of this body requiring special notice, we shall devote a large share of our explanations to them.

Fig. 1 furnishes a side view of the body as seen in its natural position when completed, and the dotted lines carried down on to Fig. 2 determine the points, which give a correct basis for showing a top view. At A is the bottom side; B the front pillar, and showing the manner of framing into the bottom side A; C the top arm rail, which is framed into the pillar B; D D D the seat frame; Nos. 1, 2, 3, 4, 5 show the strainers. The seat frame is grooved at the bottom to receive the panel and molding worked on the outer edge, which corresponds with the molding on arm rail C, which is also grooved for panels. To apply the seat panel properly, requires some skill in order to prevent the joints from appearing and marring the beauty of the finish. The most approved method to do this is to select wide panels, which must be cut quartering to allow them to bend around the corners. They are inserted at the pillars and seat frame and carried around, and the joints made on two of the strainers, a recess being left to receive a key panel, which forms keyed joints. When the wood is thoroughly seasoned, and well-painted, these joints will never show.

Fig. 3 gives a back view of the body, the dotted lines showing the degree of curvature of the seat corner. The end sought in this class of vehicles is to produce a massive appearance by deep panels, but not at the sacrifice of lightness.

FRENCH CABRIOLET AMERICANIZED. THREE-QUARTER-INCH SCALE.

Side elevation being correctly drawn and patterns complete, except the standing pillar pattern, we will determine the turn-under, which is 4½ inches, and complete this pattern.

Measure 1½ inches from the face of the standing pillar, and draw perpendicular line A, this being the face of the bottom side. Proceed and draw perpendicular lines B and C, the extreme length of the body; also lines E and D, and the horizontal center line. Now lay out for the width of the body, which is 3 feet 11½ inches on line D; this allows ¼ of an inch each side for the upper arm rail for fastening the top to, and will take a 4-feet front bow.

You will next measure from center line on line D 23¾ inches, this being half the width of the body, and prick it off; this is one pont of the establishment of line F. Take your dividers, place one point on base line I, where it intersects with perpendicular line D, the other on perpendicular line A. With this space place one point on perpendicular line D, at the first point of the establishment of line F; prick this off on line D, which is the establishment of one point of line G, and the face of the bottom side on the cant. Now proceed and lay out for the width of the boot, which is 31 inches. Measure from the center line on line E 15½ inches, and prick it off, this being half the width of the boot. Lay your straight-edge from this point, and to the last point explained, and draw line G; this is the face of the bottom side, and is contracted 2 inches. Measure 1¼ inches from line G, and draw line H, being the thickness of the rocker. Measure 1¾ inches from line G on C, and prick it off; this is the second point of the establishment of line F. Measure from line G on line E 3 inches, and prick it off, being the third point for the establishment of line F. Proceed and sweep line F

WORKING DRAFTS OF LANDAUS, ETC. 71

from these three points, as represented by the diagram, the cheat line for the bottom side, and also the belt rail, as explained and applied in French or square rule. The most difficult piece of this body is the front pillar, which should be pricked and swept

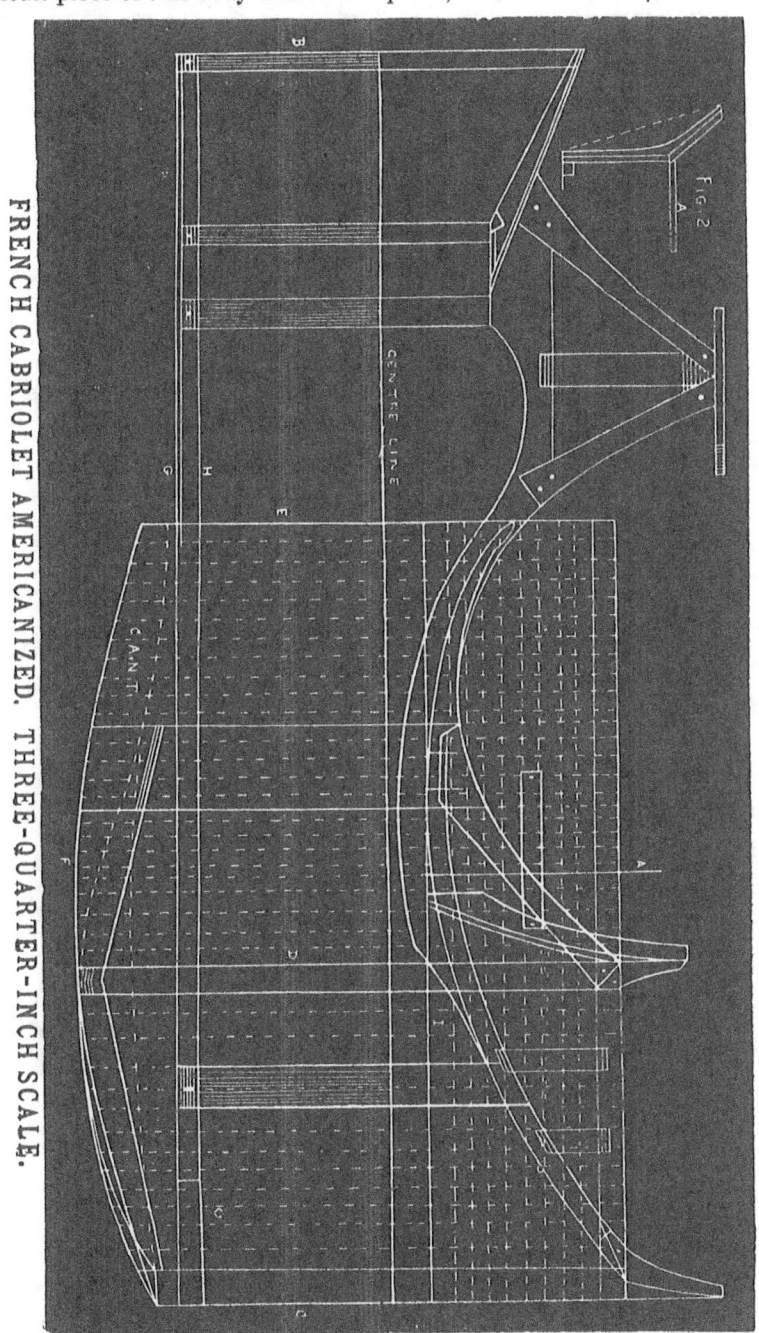

FRENCH CABRIOLET AMERICANIZED. THREE-QUARTER-INCH SCALE.

exact, both front and back. When this pillar, with the belt rail, are ready for framing, gauge from the outside.

You will notice the face lines for these pieces in the diagram, which are for pricking off, the outside of them being your guide. You will also notice the front of the front pillar is pricked off on the cant, and also the top of the bottom side, both sweeping in together.

The horizontal and perpendicular dotted lines represent the different points needed for pricking off.

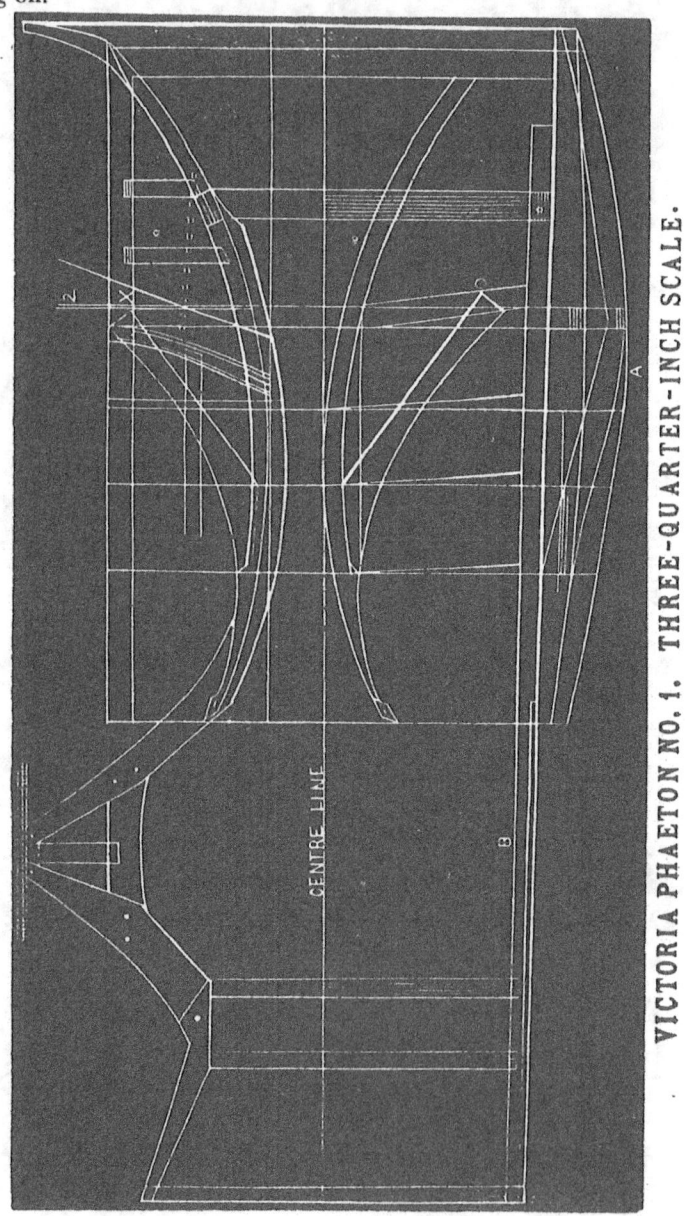

VICTORIA PHAETON NO. 1. THREE-QUARTER-INCH SCALE.

VICTORIA PHAETON, NO. 1.

This drawing illustrates the manner of constructing a light Victoria phaeton.

Having the outlines of side elevation gracefully drawn, and patterns complete, proceed and draw the center line.

You will notice, in the diagram, we have drawn the bottom side and front pillar in an opposite direction from side elevation, our object being to show the manner of lengthening the front pillar. This is done by first drawing line No. 1 parallel with the turnunder; then, with your dividers, take the distance from line No. 1 to line No. 2, and at point X, transfer this distance one point to line No. 2; the other point establishes line O. This will give you the required length of the pillar, when placed in its position.

Half the width of body is 23½ over all, and is contracted six inches, which throws all the different pieces on a bevel. This bevel is obtained from line B, being the face of the rocker, to any of the perpendicular lines. This bevel you use for marking the bottom sides and rockers, shoulders to cross bars, by beveling the ends, and marking with the pattern on both sides for dressing.

VICTORIA PHAETON, NO. 2.

MANNER OF ESTABLISHING THE DRAFT OF THE FRONT PILLAR.

It is necessary for the draftsman to make two patterns, one to draw the main outlines of the front pillar and the other to work by, which has to be longer than the first, on account of the inclination and contracting of the pillar.

In this illustration our front pillar is crooked, but is quite a compass sweep, and the change is not so perceptible as when a double sweep is lengthened.

If we would construct a Victoria Phaeton, the front pillar having a double sweep which had to run with a crooked arm rail, this would change the whole sweep of the front pillar and the arm rail considerably, and never would look like the original draft. For that purpose we have to establish a draft as we now give to get the right length and sweep.

We divide the illustration part I. and part II. to make it more distinct.

A is the front pillar in its original outlines; B is the turn-under of the body; C is the inside pillar line; D in part II. is the side sweep; E the contracted line of the front pillar; draw horizontal line F, in part I., from the top of the pillar to line J, and make the bottom line P to line J, and line G parallel with the two mentioned, which gives the bottom line of the body.

Line H has to intersect with line F, and turn-under line B; and in part II., H' has to connect with line D and perpendicular line K; line I has to intersect with line C the *inside* of the pillar, and line F the *top* of the pillar; line I', part II., intersects with line K and line M. Line J intersects with P and C; and line J', in part II., intersects with line K, and contracted line E.

We have to make first, perpendicular lines 1, 2, 3. There is no certain number for these lines. I always make as few lines as possible. No. 1 is the end of the pillar. To proceed any further we want first, lines L and M, which require to be pricked off. Line L represents the top pillar outside sweep, and line M is the inside pillar line. You will wonder that this line M is straight as shown on pillar C, part I., but a crooked piece contracted and inclined produces a curved line, as line M shows it. The intermediate space between lines L and M is the pillar with the outside sweep ready pricked off.

We are now prepared to prick off lines L and M. Line L has to be pricked off from perpendicular line H and turn-under line B. We will start from line F, the *top* of the pillar which intersects with perpendicular line H and turn-under line B. There is nothing to prick off, which you will see on perpendicular line K, and side-sweep line D, part II; line L, therefore, intersects with lines K and D. We go to part I. again, to horizontal line N, where it intersects with line 3, and go along that line to turn-under line B, take the intermediate space between lines H, and turn-under line B, and put it on line 3, part II., from side-sweep line D, to line L.

We proceed to line 2, part I., where it intersects with line O, and go along that line until we reach line B; take the intermediate space between lines H and B, and put it on line 2, part II., between side-sweep lines D and L, next to line 1, part I., where it intersects with line P; go along that line to B; take the space between lines H and B

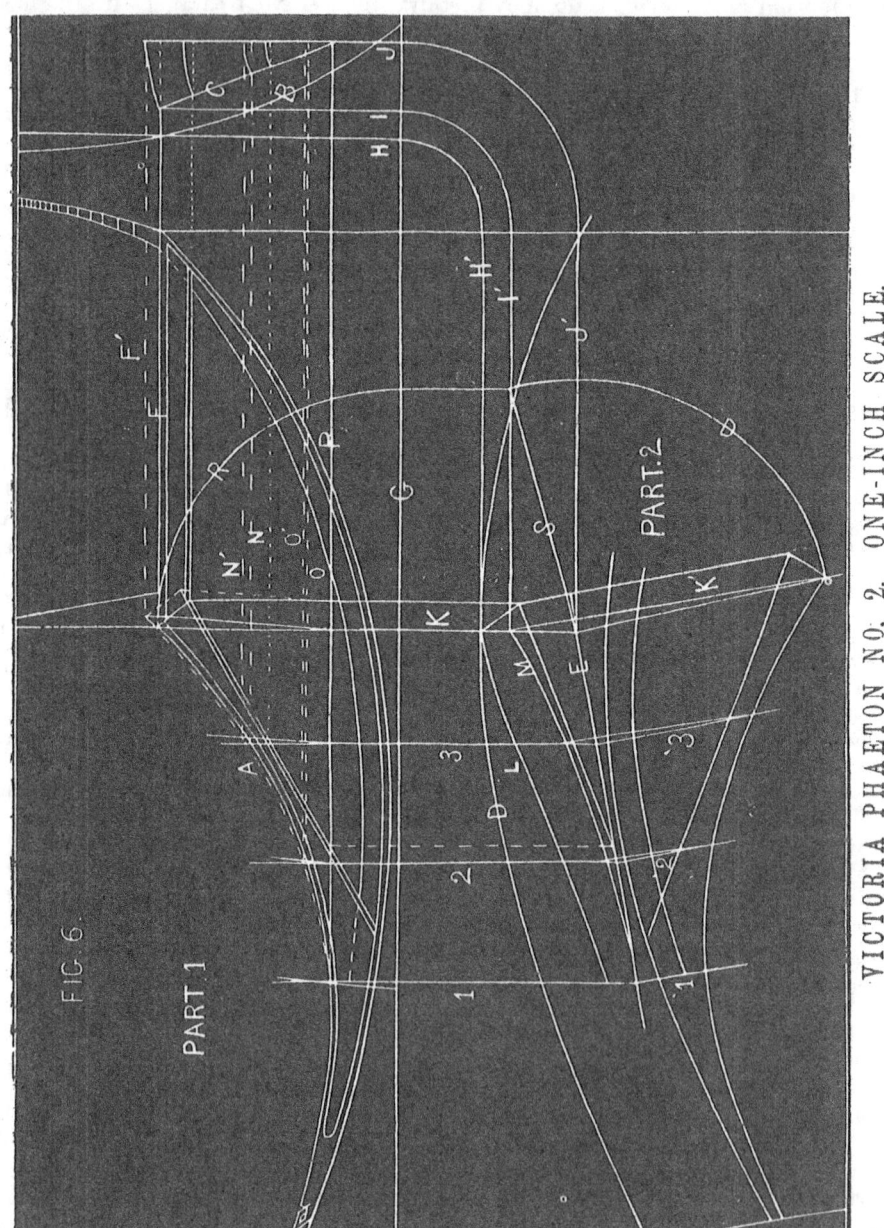

and put it on line 1, part II., from line D to L. So we have the full outside top sweep of the pillar which line L represents.

With line M we have the same operation to take the points, but from different lines. We commence again on line F, part I., take the intermediate space on line F, between lines C and J, and put it on line K, part II., from the contracted line E, to line M; proceed to line N, part I., take the intermediate space between lines C and J, and put it on line 3, part II., from the contracted line E to line M, and so you have to do with line 2. We proceed differently with line 1, because lines C and J intersect; we have, therefore, nothing to prick off, as we do in line 1, and contracted line E, part II.; line M intersects with E and 1. To complete second part, we have to put the square on contracted line E, and make line K′ start from inside pillar line M; and so on with 3′, 2′, 1′. All three square from the contracted line. This gives us the lengthening of the contracted line. Place the compass on line K, where it intersects with line G and make circular line R; square that line from line G to the inside line I′, which intersects with line M. Make line S, starting from the contracted line E, where it intersects with line K to line I′, where it intersects with R; take the compasses again, and put them where lines E, S, J′ and K intersect, and make line D; this is the lengthening of the inclination or of line C. You can take the lengthening just mentioned from lines J and G to F′, because line F′ is lengthened in a different way.

So you have to do with line N′; take the intermediate space from line G, on line J, to line N′, and put it on line 3′, from contracted line E, down to the pillar sweep; and so you proceed with line 2′. If you apply these methods as we direct, you will find your front pillar in its certain length and obtain the required sweep.

VICTORIA PHAETON, NO. 3.

MANNER OF ESTABLISHING THE DRAFT OF THE FRONT PILLAR AND BELT RAIL OUT OF ONE PIECE.

We must bear in mind that a straight piece of wood never changes its straight line through contracting or inclination, but changes the square into the bevel, and it has to be lengthened according to its degree of inclination and contraction. But a crooked piece of wood changes its lines from either a surface or side view after being inclined and contracted. We said, also, "Our front pillar is crooked, but is quite a compass sweep, and the change is not so perceptible as when a double sweep is lengthened."

In Fig. 8 we have a double sweep, and you will perceive the change at first sight. (See the dotted lines of piece P.) In this illustration our top piece P is lengthened only through its inclining line; in consequence, we do not want part 2, as in Fig. 6. To establish the cant put your desired width on line B from center line to D, which is 50 inches; take the width of back which is 36 inches—taking the width of back *always* on line A—where it intersects with the outside sweep of the back pillar, and draw perpendicular line C from back pillar and outside sweep. For the front width of the body, you have always to diminish the turn-under, from belt rail line D to the outside bottom sweep of the rocker, which we see on letter E. This gives you the points to establish the belt rail line D. The more sweep you have in your belt rail line, the more you have turn-under (in most cases), which give a nice appearance to the body.

In Victoria Phaetons and Cabriolets, the turn-under is arranged so as to give nearly a straight line to the outside sweep of the rocker F from the top of the arm rail to E, the front of the body. If the turn-under line G is made as we explained it, draw perpendicular line H to intersect with line A and turn-under line G. Here we would mention, the space between lines H and G, *under* line A, has to be diminished, and *over* line A between H and G, to be augmented. Most of the shops in America and

76   COACH-MAKERS' ILLUSTRATED HAND-BOOK.

England draw line H from the widest point of the body, and *diminish only;* but the practical way is to draw line H where it intersects with the horizontal line taken from line B and bottom of the arm rail, or in some cases, top of the arm rail as we do here; so then lines G H and A have to run in each other to one point. Our front pillar and belt rail P are taken out of one piece; for that purpose we have to draw line J. Take the thickness which you require for both ends of P and prick off from line G, on lines L and K, and strike lines J. Take the distance between lines H and M, and put it on

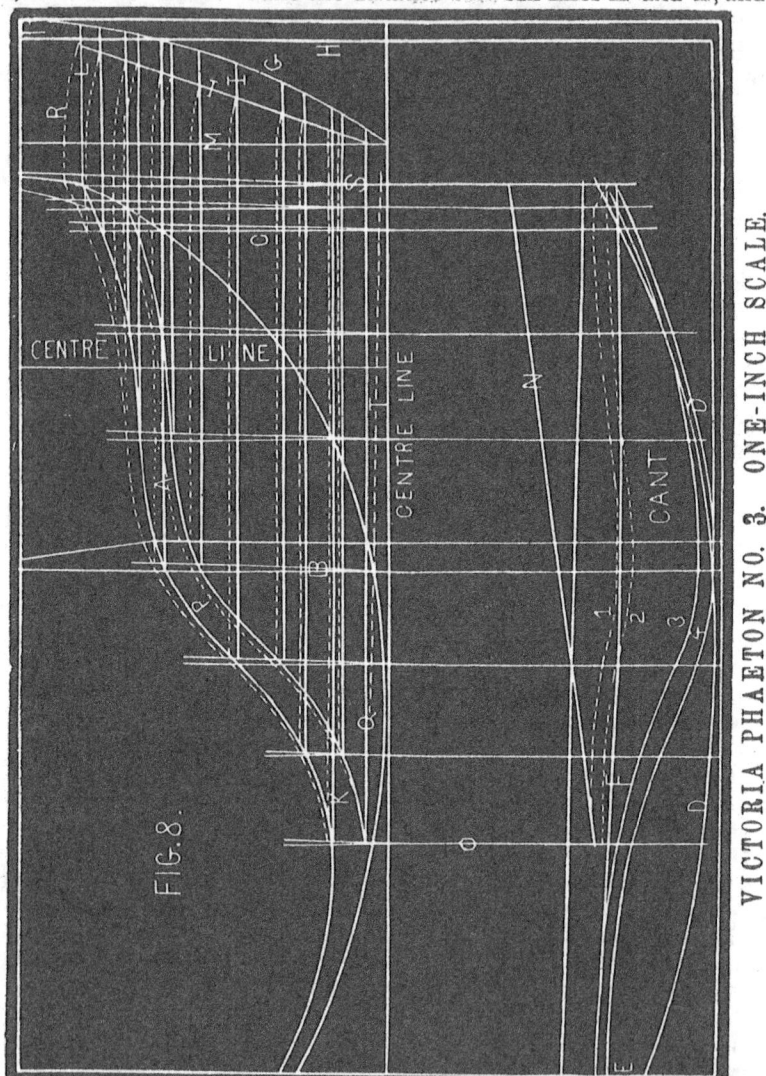

line C, from belt rail line to N, and put the same distance on line O from line D to N; this gives you the contracting line, but it must be remembered, having contracted line N, the piece P has consequently to be worked after that line. But when it is in its regular place it will not be contracted, because the piece P stands in its regular position on lines M O Q; then throw it outward of the body on line C and L, to top of piece P, toward line J. Suppose we should draw line N horizontally (or not contracted) starting from lines O and N, the consequence would be, to throw our piece out on lines C and A (on this drawing), eight inches, by counting the thickness of piece P; for that reason we have to take the distance between lines M and H and put it on line C, from line D to N, and this gives the line by which to work piece P. The pricking off of the piece P is proceeded with the same as explained in Fig. 6. The only difference in the piece P, above line A, has it to be augmented, which we will see in the four lines already pricked off for piece P. Dotted line I represents the inside bottom line of piece P, and is pricked off from lines M and J. Take the distance between these two lines, and always follow the horizontal lines which are made only for that purpose, which are, however, not necessary in practice, for a little stick will answer, by following from one point to the other.

Dotted line 2 represents the inside top line of piece P, taking the height of each point on the dividing or sectional lines. Take the distance between lines M and G and prick it off on the same dividing line you first take it from, and put it from line N to the dotted line 2. Line 3 represents the bottom outside line or piece P, and is pricked off from lines H and G, taking the distance between these two lines, and placing it on the sectional or dividing lines from line D. Line 4 represents the outside top sweep, and is pricked off as the above; you must not make any mistake in lines D, 3 and 4, because they run into each other. Mind also what we said with reference to the distance between G and H, above line A, for we have to add it to the outside of line D. Dotted line P shows the lengthening of the whole piece, and how to mark and prick off the sectional lines on the draft-board. In constructing a body we lay piece P either on the dotted lines, or on the main out-lines. If laid on the main out-lines you have to prick off the sectional lines that fall backward out of the square. This is a point needing particular attention, the difference being plainly observable in the joints back and front of the piece P To make these backward falling sectional lines take the distance between lines L and dotted line R, on line S, and put it on line S where Q intersects, and draw line T to where Q and O intersect. From dotted line T square the other backward leaning sectional lines. The above lines were not mentioned in the description of Fig. 6, and are of great importance to the workman.

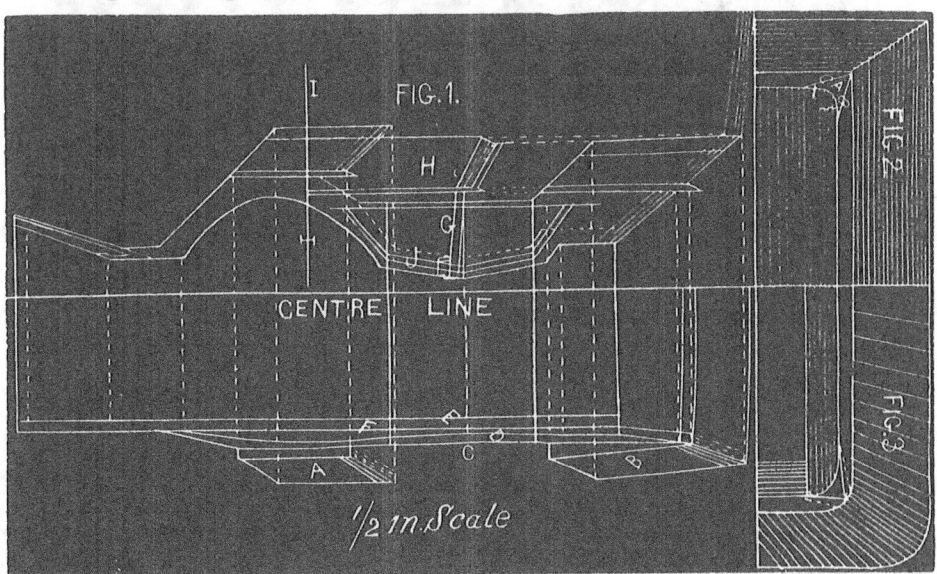

FOUR-SEAT PHAETON, NO. 1. HALF-INCH SCALE.

FOUR-SEAT PHAETON, NO. 1. HALF-INCH SCALE.

Fig. 1 represents side elevation and cant; drawn ½ inch scale, and is given in a manner to be readily understood.

Width of seats, 45 inches. Back, 42 inches. Lines A and B represent the flare of seats. Line C represents the top of the body and is drawn after the inclination of turn-under. Line D is drawn just the same. Lines E and F represent the rocker frame in and outside, which gives a thickness of 1⅜ inches. Line G is the turn-under and is 1¼ inches, and is nearly straight, because it has no side sweep. Letter H represents the front view of the back seat.

Line I is the half width of the body, and is taken from center line. Line J represents moldings which are nailed to the body when it is cleaned off, which gives to the body a very showy appearance. The back of the body and the two seats have rounded corners, which we will explain how to make.

Fig. 2 represents a square corner seat, and Fig. 3 a round corner seat.

In most cases when square corner seats are made the two boards are mitered together, and a piece of wood is glued inside to hold the corner together.

This practice is entirely wrong, as it leaves the wood too thick on the ends, and in consequence the wood inside is not strong enough to keep the joint from straining. To prevent this we adopt the theory, which is proven to be correct by practical application, by making a little pillar, as A, in Fig. 2; it is screwed to the frame of the seat, and the side and back pieces are planed off after lines B and C. This, you will see, takes the straining or force of the wood entirely away, and gives at the same time a long joint to the pillar, and side and back pieces. This will produce the best corner ever made.

In the round seat follow the same plan; make your side and back pieces near the joint as thin as possible in order to prevent the strain of the wood, shown at dotted lines in Fig. 3.

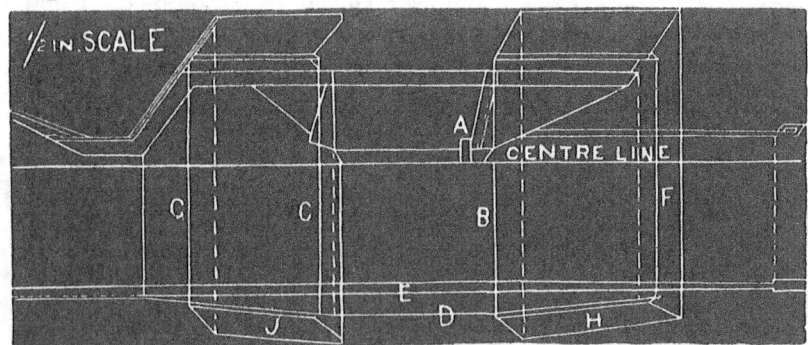

FOUR-SEAT PHAETON NO. 2. HALF-INCH SCALE.

This drawing represents side elevation and "cant" drawn to half-inch scale, and is given in a manner to be readily understood.

The manner of framing shown is with short bottom sides, framed into the lock and hinge pillar, the door shutting against the rocker.

To do it an easier way, frame the rockers together as usual, and screw your hinge and lock pillars against the rocker, as shown at A; put on the back and front whole quarters out of an entire piece, mark your side sweep, and turn under and plane it to the required thickness and sweep. Proceed by laying out the "cant" or working draft, by drawing the center line, and also by squaring lines B and C. Measure 18 inches from center line on lines C and B, and prick it off, which establishes two points

WORKING DRAFTS OF LANDAUS, ETC. 79

on line D; also half the width of the body, which is 36 inches wide. Draw turn-under line and the thickness of the pillar with the thickness of inside rocker, as line A shows it. Also draw line E. This line is established by taking the turn-under and depth of pillar, and prick it off from line D, on lines B and C. These lines have been often explained, and must now be well understood. Measure on line F from line E 1 inch, and prick it off, and also on line G from the same point. This establishes the four points of line D. This line in our diagram is new; it is swept in a straight line from F to B, from B to C and from C to G. H and J show the side flare of seats. This flare is obtained by the width required for the bows, which is here laid out 43 inches. The bows decrease and increase in width following the variations of the seat line.

For rocker plates use 2 by 7-16 inch steel. Half-inch steel should be used in the doorway running to the neck and in other weak points.

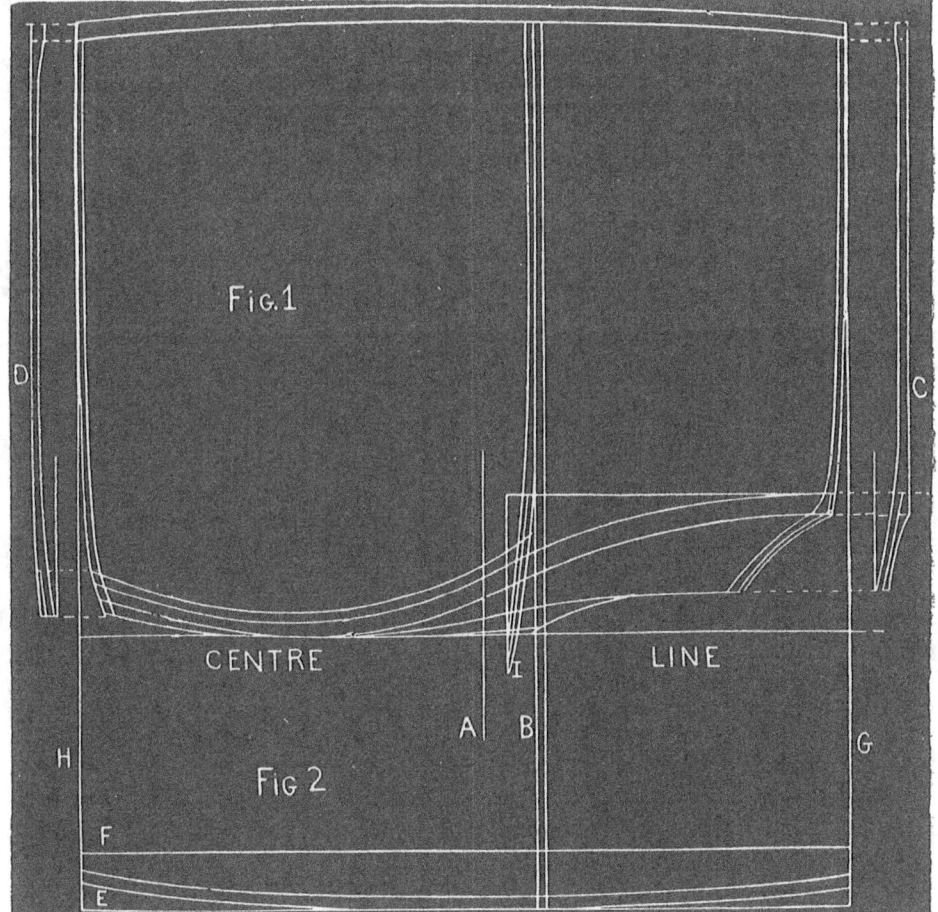

JUMP-SEAT ROCKAWAY. THREE-QUARTER-INCH SCALE.

Fig. 1 represents the skeleton side view. We will determine the turn-under, shown by line I; also the thickness of bottom-side required, shown by vertical line A. Next lay out for the back view of hind pillar, seen at C. D represents the front pillar. Having secured the three points named, produce the center line of body. Square the vertical lines H, G, B, and draw the horizontal line E, which is the outer line of roof rail. Then, with your dividers take the space from B to A, placing one leg on the line E, where it intersects with B, and prick it off in order to establish horizontal line F; this is the face of bottom side of the working draft. Next, take the turn-under and depth of back pillar C, placing one point on line F, at its intersection with G, and prick it off on line G. This establishes the outer sweep of roof rail at the top of the back pillar. Proceed with the front pillar D in the same manner, taking the depth F This gives the outer sweep of roof rail at the top of front pillar, and establishes the three points before mentioned for sweeping the roof rail. When the outer line has been swept, namely, line F, the pattern can be gauged the thickness required, and the inner surplus wood be removed. The advantages gained by the above mode of procedure are, that the pillars will stand vertical when placed in position without recourse to the old "fit and try" system. The body should be framed and put together, the side panels be fitted down on the bottom sides and pillars, aad swept off from the outer side, and the surplus wood be removed. We are now ready to glue on the panels and complete the job.

STANHOPE BUGGY. THREE-QUARTER-INCH SCALE.

After the side shall have been correctly drawn, proceed to lay out the width of the body, and establish the line at A, Fig. 1, which represents the shifting rail, and connected with it the slat iron of bow. Next produce the beveled seat line; also line B, which shows the mock pillar. After having established the bevel of seat and body

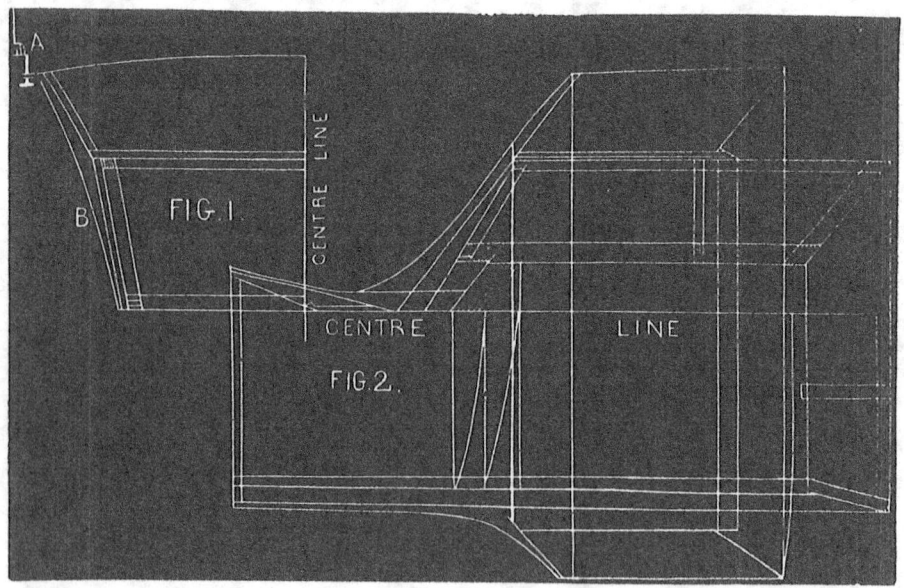

measure off 22 inches from point A, and draw the center line, being half width of body. Having first decided on the projection of the irons at the point A, we then measure half width of bow. This manner of working may be applied in obtaining the proper length of the cross bars without using the rule. Fig 2 shows the top view, the flare of seat and back panel.

PHYSICIAN'S CARRIAGE. THREE-QUARTER-INCH SCALE.

In this diagram we represent the cant of a Philadelphia physician's brougham. Side elevation being correctly drawn, care should be taken, as well as good judgment used, in laying out the cant, in order to produce a good turn body, as this is not so simple a job as the first appearance of it might seem. After determining the turn-under, and completing the standing pillar pattern, next lay out for the sweep of roof rail; measure from center line A, 23¼ inches on line E, being half the width of the body; proceed, and establish the different points and lines, as before explained. You will see by this diagram we have swept the front of the cant in, in order to produce a narrow

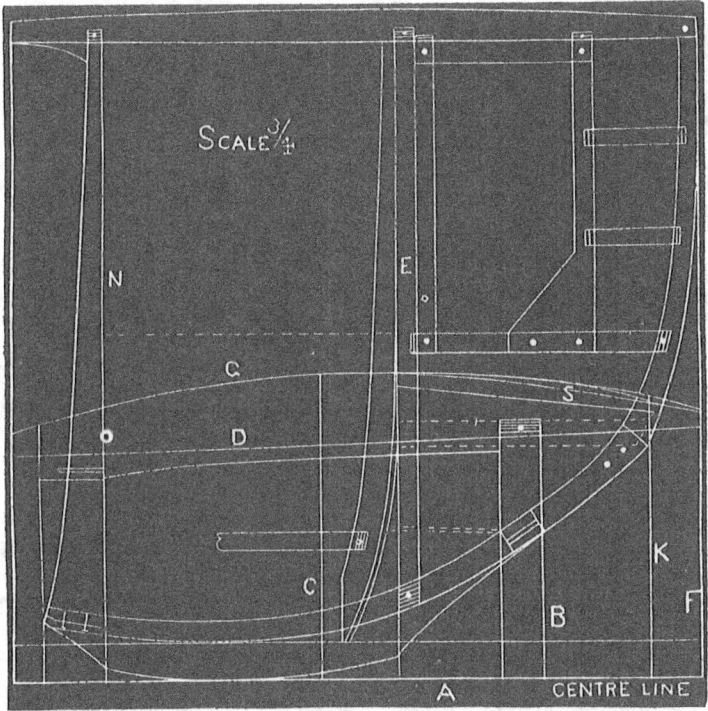

front, without giving too wide a back. In so doing we have laid out for the framing of the coupe pillar into the rocker instead of the bottom side. In this manner of construction you require a thick rocker. In sweeping off, the bottom side will be entirely cut through in the doorway. The rocker will substitute for the piece that is required by boxing down, and sweeping in and forming a bottom side out of the rocker, concaved on the inside. The object of this is to produce the required width on the seat, without too wide a back, with a narrow front; at the same time you will have a nice swell. To bring out these different points, contraction alone will not produce them

COAL-BOX BODY. HALF-INCH SCALE.

The diagram on next page will show the manner of laying out and making a concave coal-box body. There cannot be too much pains taken by the body-maker in building this kind of work, as we depend wholly upon glue to make a durable job. In preparing for gluing, you will level and rasp your frame work, giving it a woolly surface, and also your panels, so that the two parts, when glued, will adhere more firmly together. It is well to warm the parts to be glued, to prevent chilling of the glue, which should be hot. It requires the best quality of glue, properly prepared, right thickness, and a sufficient quantity applied to the parts to be glued, and griped as quick as possible. In short, all that is required is seasoned material, the different parts well fitted, glued with a sufficient quantity of the best quality of glue to make durable work.

Fig. 1 represents the side elevation, with the thickness required for the back panel, which is 1¼ inches. Fig. 2, end view, half of the width of body. You perceive by this manner of laying out this kind of body, you can obtain the thickness required for the side panels, which are 1 inch thick; it also gives you the flare, and half the length of all the cross rails, also half the width of the seat. Fig. 3 represents half the width

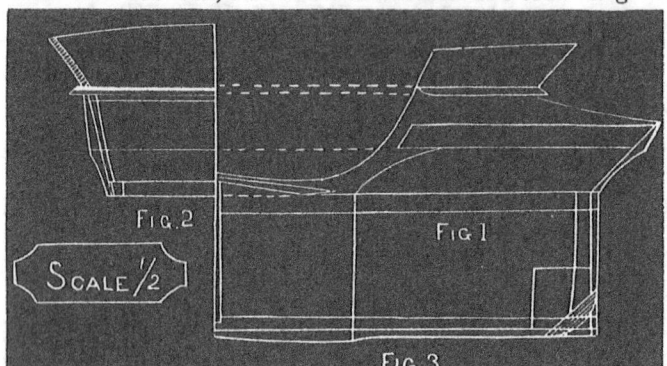

of the ground plan of bottom frame work. The bottom side 1½ inches thick; back bar 2½ inches by 1¼ inches, this being the depth of the bottom side at the point of framing this bar. To be lightened out in the center, leaving the full width at each end; to form the round corner it will require an edge plate on the bottom side, ⅛ inch thick, full length and width. Turn the corner on the back bar, to take two screws; this corner plate binds the corner together on the inside and adds strength to the corner. It is well in ironing this style of body to keep the loop or shackle well ahead from the corner. These corners are often started by the bearing of the shackles being too far back.

Having the sides glued on, you will glue in the corner blocks, 4½ inches square, resting on top of the bottom side. Next you will glue on your back panel; now you will proceed to cut the corner off, as represented in the diagram. Glue on the outside corner block. The grain running the same way as the side and back panels, you are ready to round off and lighten out the inside. This manner will allow you to give as round a corner as desirable. There is another way of forming round corners, where the corner cannot be cut off in the manner above: by rounding the corner into the pillar, and rasping the outside, forming a wool, and then canvasing over it. When dry, file the edge to a level surface. Corners well done in this manner will stand without the joints showing.

BUCK-BOARD WAGON.

Fig. 1 (on opposite page) represents the side elevation of body, and outside and center stay for axle. The dotted lines indicate the position of strap bolts, which fasten the body to bottom bars. These strap bolts are secured to the pillars which support the seat. There is no sill. Body is clear of the bottom ⅜ all around, except at point N.

Fig. 2 shows back view of axle riser and slats when on. Depth of bed and riser, from straight lines, 8½ inches. Bar F, ⅞ by ¼.

Fig. 3. Front axle, riser and bed. Depth, 8⅝.

Fig. 4. Bottom or slats are seven in number, and ½ inch apart, the two outside ones 2⅞ wide and ⅝ thick, the remaining slats are 5 inches wide. ⅝ thick back of center, and ½ in front of center. Length of bottom, 49 inches; width, 23 inches; N, bar for body to bolt to; M, bar for step and king-bolt guard and two stays from H H, Fig. 3.

WORKING DRAFTS OF LANDAUS, ETC.

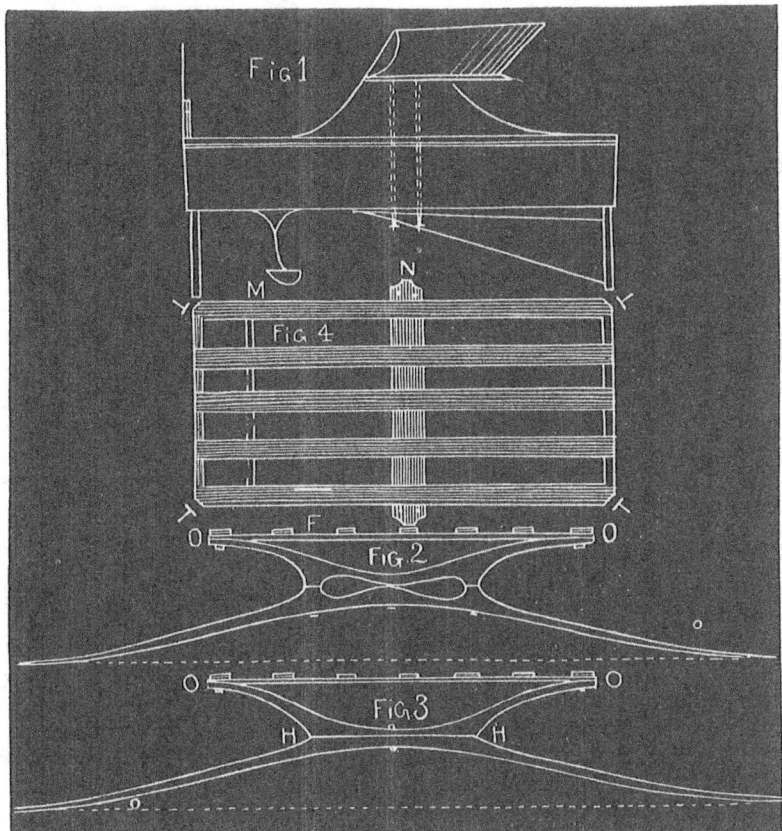

There are thin pieces of rubber inserted between riser O and bar F, the bolts passing through at slot T, and made fast to prevent rattling of body against the bottom. There should be a smooth plate of iron on the three-cornered block inside of body, for the rubber to work in.

FOLDING FRONT SEAT.

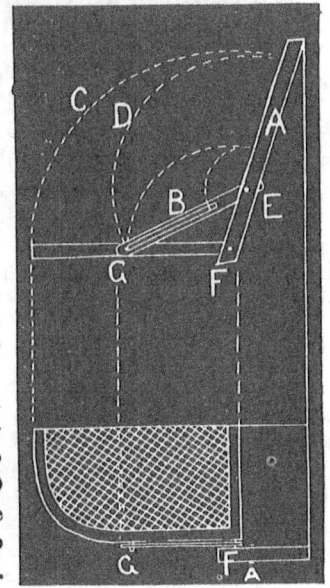

The accompanying cut gives the manner of constructing a folding front seat, mostly applied on small coupes, but can be applied to any other carriage intended for the same purpose.

Our illustration shows the seat in its open position, and every one will see the advantage of its simplicity, it being very strong in construction, and very convenient in putting it up and down.

To construct it make your seat any desired width and length; take two pillars as shown in A, plane them $1\frac{1}{4}$ inches square; the piece of iron B must not be made so long as to strike over the seat as dotted lines C and D indicate. E, F and G show the three pivots. The lower drawing furnishes a top view, and is lettered to correspond with the drawing showing the side view.

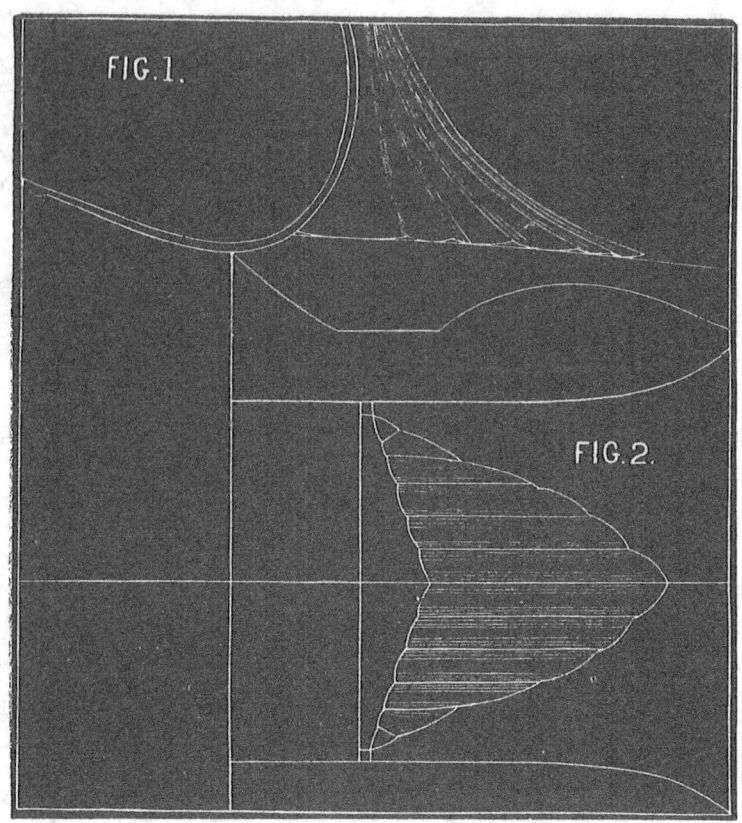

PHILADELPHIA RIBBED BOOT. THREE-QUARTER-INCH SCALE.

Fig. 1 represents the side elevation of boot; it will serve to convey an idea of the appearance of the ribbed portion.

Fig. 2 shows the variety and degree of curvature, and the pieces required in the construction, which are ten in number. Unlike the circular boot, which is constructed by building up with pieces laid horizontally, this has the sections placed in an inclined position, agreeing with the bevel of the boot. They are properly fitted, glued and screwed, so as to form a solid, substantial piece of work, and when the several pieces have been rounded up and finished, there is no canvasing required, thus presenting a foundation for the painting the same as the remainder of the body. The joints, when properly fitted, remain solid.

PANEL QUARTER.

The cut Fig. 3 (opposite page) represents the hind panel quarter of heavy-class work illustrating the method of obtaining the bevels of the four points of the wide belt rail, so that when it is swept off on the outside and tenoned it will lie level on the frame work. Probably some body-builders have experienced a difficulty in this particular, and where this has been the case, our effort to enlighten them will be fully appreciated. A shows the belt rail, H the standing pillar line, B the turn-under, E the sweep of roof rail on bed of body, F the face of belt, D the sweep of the bottom side. To establish line E, we

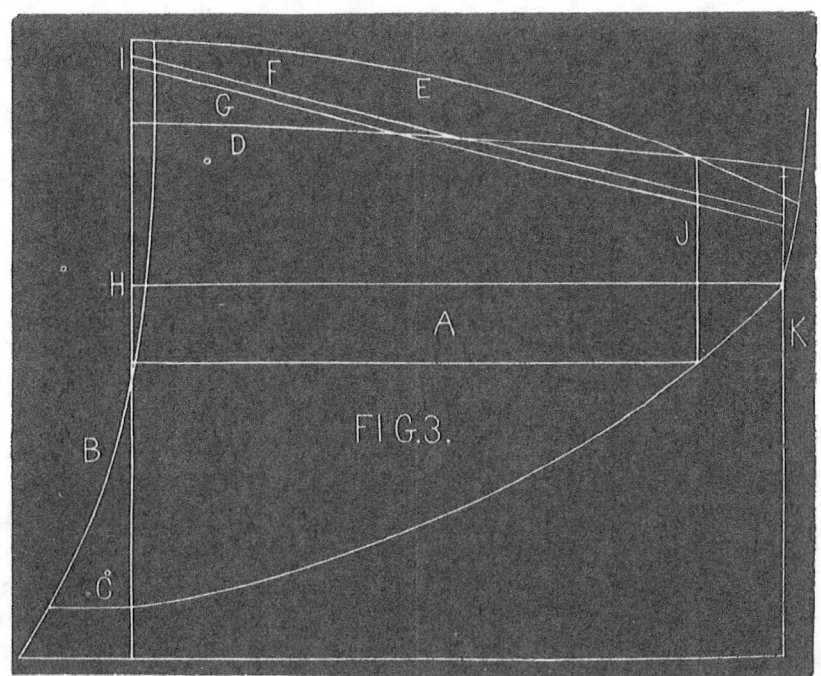

take in the dividers from line H to B on line C, and prick it off from line D on line H, and sweep it off where it intersects with line J. Line F is established by the thickness required for the belt. To obtain the bevel on standing pillar, we take in the dividers the space on line A, from B to H; then prick it off on line H from line F, which gives one point of line G; place one point of dividers on line J (being the vertical line of bottom of belt), where it intersects with line F, and the other point on line E at its intersection with D. Having obtained this space, put one point of dividers on line K, at its intersection with vertical line F, and prick it off toward line D, which will show by the dot on the line. From this dot on line K, take the remainder of the space to line D; we will now prick this off on line K from line F, giving us the other point of line G. This line may be now drawn, which is G, it being the top, and F the bottom line of belt.

TRAMMEL FOR OVALS.

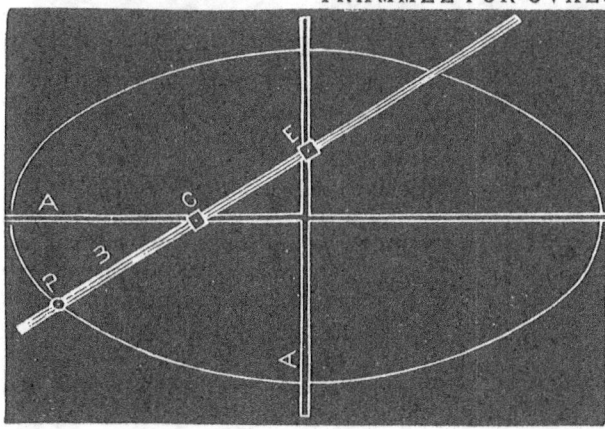

The adjoining is a diagram of a frame for striking an oval. This will strike a perfect oval of any size. As ovals are always used by coach-makers, it is necessary to know the quickest and most correct manner of obtaining them. The method of drawing by a string is defective, as the pencil varies in traversing the string. There is another mode of obtaining one, by squaring the length and width of the size required; then spacing the square into a given number of lines, and tracing to the intersecting points This mode requires too much time at the present day. The manner of making the trammel is as follows: A A, two pieces, one inch square, lapped together in the center with a one-fourth-inch groove on the top to receive the dowels of C E. It requires a pricker or pin near each end on the bottom of A A to prevent it slipping from its place. B is the shaft, half-inch square. C E are the sockets with the mortises cut large enough to slide on shaft B. Insert a dowel in the bottom of the two sockets C E, with a screw on top to prevent slipping on the shaft. P is the hole to receive a pencil in shaft B. Now it is ready for use. Slip sockets C and E on shaft B. Supposing your oval is to be 13 by 21, you set the dowel in socket C 6½ inches from pencil P, being half the width of the oval. Place E 10½ inches from pencil P, being half the length of the oval; tighten the screws to prevent the sockets from slipping on the shaft. Next take shaft B, and place dowels of C E in grooves A A. Now you are ready for turning the shaft B, keeping the dowels C E in grooves A A, they being your guide; which will strike an oval represented in diagram.

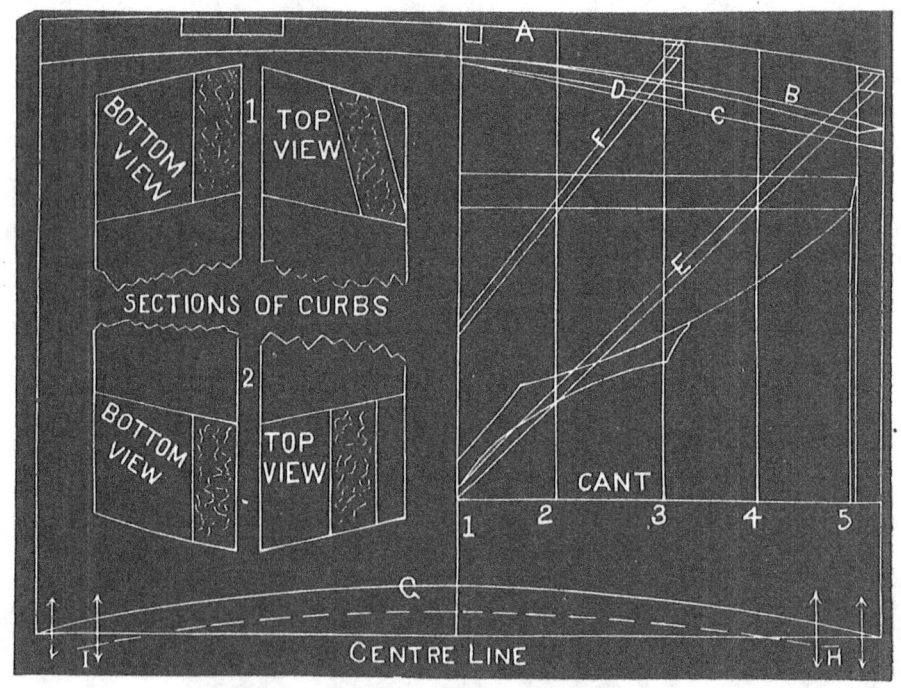

LANDAU BOWS. ONE-INCH SCALE.

The French excel all others in point of gaining time, as well as in durabilty, in the construction of Landau bows, and, if you demand our reason, we reply, the side pieces are gotten out straight, not curved, as the majority of our carriage-makers see fit to make them. The straight ones give to the tenon greater strength than the crooked or curved piece, which runs against the grain. The tenon also does not pass through the "curb" perpendicularly; but, on the contrary, in the same line with the bows, which will convince any one who gives it a thought that greater strength is secured thereby. The straight pieces are worked out in half the time that it takes to get out a set of the crooked ones.

THE DRAFT.

To lessen labor, it is the better plan to draw all the lines which are required for the top, within or on the body lines, as shown in the accompanying diagram. A represents the top sweep, and B the *side top* sweep. The two parallel lines, C D, are the contracted lines of side pieces, which we call E F. G shows the top cross sweep. The first thing the workman has to do is to work out the "curbs" after the line A, and perpendicular lines 1, 2, 3, 4 and 5, the pieces to be planed off one side, and the inside wants to be beveled, because it inclines outward and is contracted, therefore produces the same bevel as when leaning back and sideward.

If that bevel is put on you have to plane it to the required thickness, and take the points for the sweep; but before we take these we will be required to make our division lines. The number is not limited, but we always draw as few as possible. To prick the sweep off we take the thickness first on line 1, where line B intersects with line 1; you will take that little space between lines B and C, and prick it on both sides of your side pieces F, and so on with the other lines; it is the same operation at each point, but the result does not give the same thickness; for this reason we have to take each point apart from each bow.

THE LENGTH OF THE CURB.

We now proceed to mark the length of the curb. Take a pair of large compasses and put them on the cant and on line 3, and extend to line D. The width thus secured place on line G, on each side from the center line, because you took only the half width, that is the width inside the bow F; proceed in like manner with the bow E. Take the half width on line 5 from the cant to line C, and put them down as before stated, on each side from the center line.

We mark now the center of the curb and lay the curb on line G, and prick off its length. In pricking off the tenons we have two different methods:

In the first method we cut the curb parallel with the horizontal line, as shown at letter H, and when the bow is fitted together it should be as shown in sectional view, marked 1, top and bottom views.

The second method is as shown at No. 2, top and bottom views. In this it will be noticed the curb is not cut parallel with the horizontal line at letter I; it will be seen you can gauge off on the curb the same as on the side pieces E F.

PORTLAND CUTTER.

HOW TO MAKE IT.

The dash is paneled down to where the fender strikes the runner, and placed in a groove cut in the runner; the top rail at the goose-neck take from a piece of bent hickory rim; this sweeps the rail front, giving it style. The front should be made wide enough at the goose-neck so as to dispense with side wings. From the front beam to where the fender strikes the runner is another panel, which is rabbeted on the beam running up, crossing the fenders, and taking the lower rail of the dash panel. It is well to insert a bent piece for the sweep of the fender on each side, from the front beam to the runner, to stiffen the panel. The body, 30 inches on seat, with 2 inches flare on each side. Back panel, grain running lengthwise. The corners are formed by lapping the pillars on to the bottom side, sufficient in size for your corners. After your framing is complete, glue on back and front panels, afterward your sides. Your corners are then ready to round and to lighten out the inside, down to the seat. Screw and plug the panels. On the feather edges of the corners, nail with $\frac{3}{8}$ chair tacks, without sinking the heads; rasp the corners down, forming a woolly surface, then canvas over the corners; when dry, clean the edges down to a surface; this prevents the nails from showing. In molding off, use $\frac{3}{8}$ inch half diamond shape, 1 inch from the top edge, passing up the sweep of back. The belt moldings pass around the body;

the same on the bottom edge. These moldings are striped. On the top edge of the front corner of body, screw on a corner plate. Use light bolts instead of screws to fasten the body to the beams. The running part should be made light and strong— nothing but the best of timber being used throughout. The knees should be about $\frac{3}{4}$ inch thick, by $1\frac{1}{2}$ at the top, and the same thickness as the runner at bottom ; middle knees should not be less than 18 inches long between shoulders; runners 1 inch square.

CORNER BEVELS.

THE MANNER OF OBTAINING THE CORRECT CORNER BEVELS OR PIECES CONTRACTED OR INCLINED.

To take the bevel of a corner stick, or a seat corner without a stick, is an operation simple enough to those who are acquainted with geometrical lines. To render the explanation easy to be understood, we will endeavor at present to give it in the simplest manner, and follow with other lines, through which to obtain the bevel. We

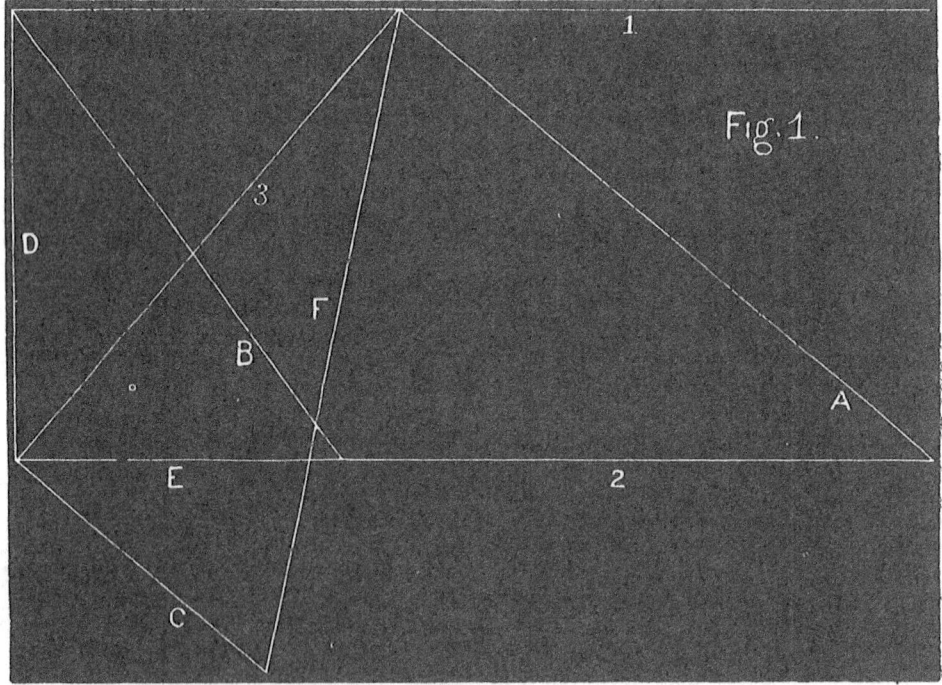

would remark that in the principle here given it will not matter how much bevel you put on from one side or the other, for through these lines you will always obtain the correct inclination. Referring to the accompanying diagram, Fig. 1, the lines A B indicate the bevel of back and side pieces. Now place the square on line A, and, at its intersection with top line 1 produce the line laid down. Again, we put the square on the square line 3, to obtain line C. For the next we have still the vertical line D, starting from top line 1, and line B, and running to bottom line 2. Now we have the intermediate space between lines B and D. On the bottom line at E, take that width and place it on line C. Next, take the straight-edge and make a line F, from that point to the top line; thus on the lines A and F you will obtain the correct bevel.

Suppose line No. 1, Fig. 2, to be the back piece bevel, and No. 2 the leaning out of the side pieces. Take your compasses and start from point O, and draw dotted line 4, from corners 1 and 3, and from the end of that dotted line make line 5 to perpendicular line c. Now draw line 7, from corners 5 and c to the bottom O. This is the bevel you require to cut the two ends of the back piece. With the other lines you have to go through the same operations to obtain the bevel of the side pieces. Take the compasses again, and draw dotted line a from line c and 2, and from corner of the bottom line O; draw a horizontal line b, to perpendicular line P; through these lines you obtain the other bevel of the side pieces. If line 6 be drawn from corners b and P to line O, through these lines we have only the bevels, by which to cut the ends, but we have not that bevel which is produced through the leaning out, which we

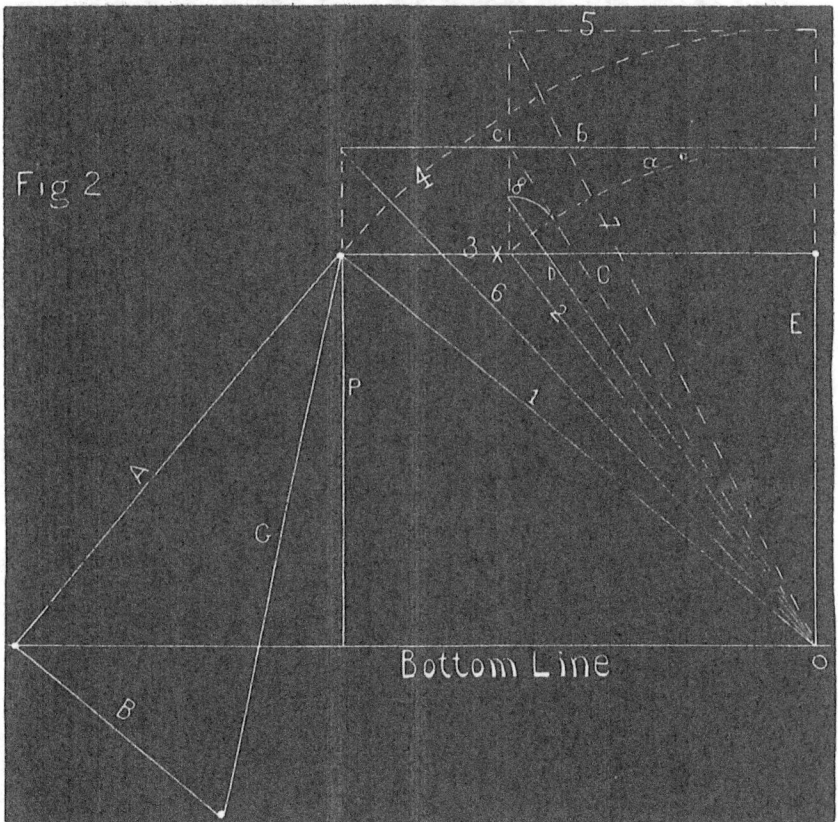

employ to get the bevel of the corner stick seat, or any other one which is leaning backward and sidewise. Take your square and place it on line 1, with the corner on the line 3, and draw line A to the bottom line. Again, take the square and put it on line A, and draw line B. Now draw line C from horizontal line b, and line c to corner O. With the compasses draw curved line 8 from point X. Draw the line which we call D from line c, and curved line 8 to the corner O. Now take the intersecting point between lines E and D on line 3, and put on line B, which will produce line G. To take the correct bevel, put your bevel on lines 1 and G, and you may rest assured of obtaining any bevel which is produced in leaning back and sideward. To prove this drawing, or any other kind, no matter what bevel it is, take two small pieces of wood, gotten out in the form as when you would construct a seat; put that bevel on the bottom of the back piece on line 7 and the bottom line, and on the side pieces to lines 2 and bottom line. Take line 7 and the bottom line, and cut the ends of back piece. Now take the bevel from line 6 and bottom line, and cut ends of the side pieces. By placing the square over D and 7 you will find the bevel of line 1 and G without taking it, because the square produced it through line 1 and bottom line, and line 2 and bottom line, and the same bevel lines 1 and G give you the miter.

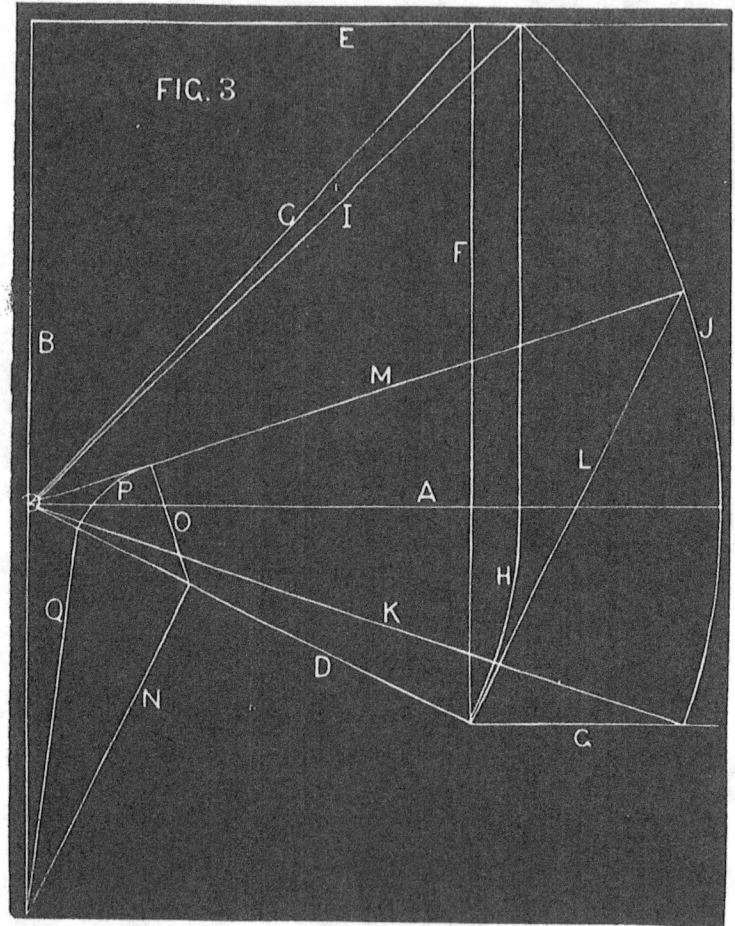

Fig. 3. First draw horizontal line A, and perpendicular line B. Put any desired bevel in the direction of line C, so it is with line D; any contracted or inclined line will answer. With these two lines we have the two inclinations.

Make the top line E to have a point to start from. Strike perpendicular line F, commencing at intersection of lines E and C, running to line D. To make line G, square line B from intersection of lines D and F. Place the compasses in the small circle where lines C and D intersect on line B, and strike line H from intersection of lines D, F and G to line A, and continue line H square, until you reach line E; from this point draw line I down to small circle. Place the compasses in small circle and sweep line J, commencing at intersection of lines E and H until you reach line G. From small circle to intersection of lines G and J, draw line K. Put the square on line D, at intersection of lines F and G, and draw line L to curved line J. From this point to small circle will give you line M.

Put the square again on line D, and make line N. This line can start from any point on line D you like : the further you go in the direction of line G, the more space you require, but the result is the same. Line O is squared from line M to intersect with N on line D. The small curved line P is made by placing compasses at intersection of lines O and N on D, commencing on M where O intersects, and running to D. From that point draw line Q, to intersect with lines B and N. Lines D and Q will be the desired bevel. Lines B and K the bevel of one shoulder, and lines A and J the bevel of the other shoulder.

If you wish to prove these bevels, take a piece of wood two or three inches square, cut bevels I and A. For the contracting lines take D and B, and plane the end of the piece to these bevels; after doing so, put the square on line I; mark and plane off all the wood outside the line. This will give you the result again.

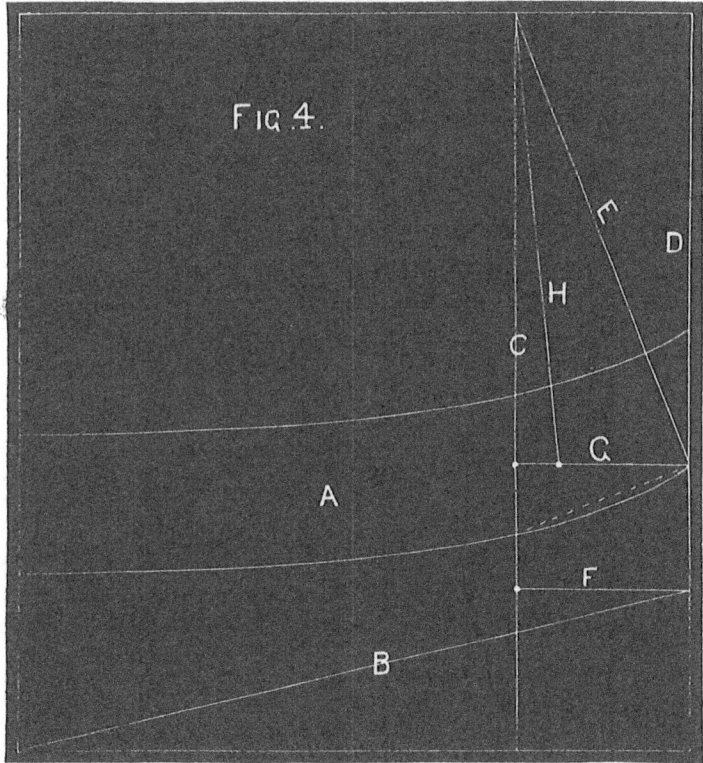

Fig. 4. As we remarked before, those who apply themselves will readily perceive how easy it is to obtain correct corner bevels without cutting the ends.

The corner-stick bevel we gave in three different methods. We now commence with the contracted pieces of wood.

Every body-maker of experience knows full well that each crooked piece turned out the square produces a certain bevel; the more crooked and contracted, the more bevel We will understand this better by referring to Fig. 5. Suppose A to be an arm rail—very much in fashion years ago—and B the contracted line. The first lines will be perpendicular lines, C and D; there is no certain distance for these lines to be drawn apart, but if drawn too far from each other will change the bevel. Now we take the square, put if on the bottom arm-rail line, where it intersects with perpendicular lines C and D, and draw line E. To make the matter clearer, put the square on line E, and the dotted line. Mark, *that an error in this line would fail to produce the desired bevel.* Now draw the two parallel lines F G. Take the width between lines B F on C, and put it on line G; starting from this point produce line H, running to the corner formed by the intersection of lines C and E.

Should you wish to prove the bevel, you will find that line H and horizontal line G will give the desired bevel.

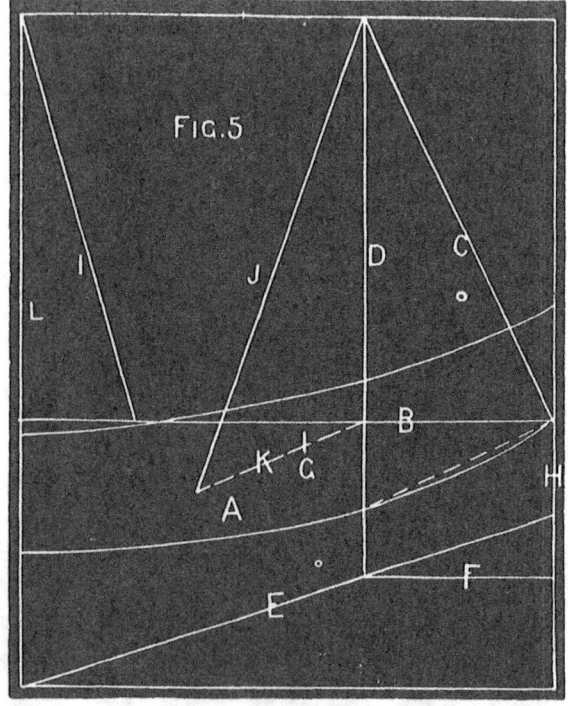

Fig. 5. These lines are nearly as those in Fig. 4, but showing a great difference in getting the bevel, because this one, although contracted like the other, still inclines outward, which will change the bevel a great deal, according to the inclination. A shows the rocker or any other piece of wood. Put the square on bottom rocker line where it intersects with lines D, B and H, and produce line C. E represents the contracted line. To draw line F, start from line E, where it intersects with D. Take the space between lines F and E. on line H, and put it on line K, which is marked G. Line K wants to be made square with line C, starting from perpendicular line D, and horizontal line B. If we made a line from letter G on line K, where lines C and D intersect, in taking the bevel from the horizontal line, this would be the bevel if it was not leaning out. Line I represents the leaning-out line. Take the space between lines I and

L on line B, and put it on line K, starting from G, and from that point make line I. If you put your bevel on line J, and horizontal line B, you will have the desired bevel of a piece of wood which inclines in three different ways.

THE USE OF GLUE.

It is the practice with body-makers, of late years, to omit the use of screws; but unless the panels are properly put on, and certain other conditions adhered too, which are noticed in what follows, it were better to cling to the old method.

To well understand the use of glue requires practice and considerable experience; it is surprising to see the want of this in country carriage shops.

First.—To do good gluing it requires thorough seasoned wood, and work well fitted.

Second.—In preparing for gluing use a scratch plane or rasp, to form a woolly surface upon the pieces to be joined together; the shop must be warm—and the pieces to be united must be heated well so as to allow the glue to flow freely. Having the glue properly prepared, spread it upon the parts so as to fill up all the pores and grain of the wood, and put the pieces together, then apply iron clamps which will force out all glue not required to hold the parts joined.

A great cause of bad gluing is caused by using an inferior quality of glue, and using it too thick. Before using a new quality of glue, the body-maker should always test it by taking a piece of poplar and ash and glue them together, and if, when dry, the joints give way under the force of the chisel, the glue is unfit for use. It must chip out into solid wood before giving way where glued together. This is a severe test, but by practicing it you will be amply repaid for your care in the stability of your work.

A WORD ON PLUGS AND BRADS.

In plugging screw holes, glue the edge of the plug; put no glue in the hole. By this means the surplus glue is left on the surface, and if the plug does not hit the screw it will seldom show.

Where brads are used the head should be well set in; then pass a sponge well saturated with hot water over them, filling the holes with water. This brings the wood more to its natural position, and closes by degrees over the brad heads. The brad must have a chance to expand when exposed to the heat of the sun, and not hit the putty; if it does it will force the putty out so as to show after finished.

IRON PLANES.

The introduction of iron planes into carriage wood shops is, so far as our knowledge extends, of recent date. Piano and furniture manufacturers, however, have been using them for some time, and from the letters of recommendation they have given, lead us to believe that the wooden plane has lost much of its former prestige. But we will answer you more directly, giving our own experience after using an iron plane six months. Our first impression was that it could not possibly excel the wooden plane, and as we grasped it, and attempted to put it into service, did not admire its working; and then it felt oddly in the hand, and appeared much more so in comparison with the familiar old wooden ones lying on the bench. So we hung up the iron plane, and did no more for a time than to cast odd glances at it. Finally, one day when not very busy, our attention was again directed to it, and after gazing for a few moments, decided to give it another trial; and, taking it from the rack, sharpened the bit, set it, oiled the face, and set to work in earnest to test its merits. We were pleased beyond all expectations, and since that have adopted it into our plane family. Curiosity led the hands in the shop to try the new comer, much to our annoyance; for at the moment when it was wanted for a certain kind of work, it was on a visit to a distant work-bench.

ITS ADVANTAGES OVER THE OLD WOODEN PLANE.

1st. It is self-adjustable in every respect.

2d. The parts are made interchangeable, and can be replaced at a trifling cost.

3d. The bits are of solid cast-steel of the finest quality and temper, rendering them vastly superior to the bits furnished with the old style wooden plane.

4th. For leveling panels, cutting miters, cutting end wood, and leveling surfaces for gluing, it, in our opinion, has no equal. Our candid opinion is, that the time is not far distant when they will wholly supercede the wooden planes.

SLATE FINISH.

A blackboard which will take the chalk pleasantly, and, when the draft is no longer needed, part with the chalk without leaving stains, is very desirable.

"Pearce's Slate Surface" is an article which will furnish a coating far superior to the ordinary method of preparing the draft board.

CANVASING OUTSIDE.

Heavy linen, and the best quality of scrims, are both used. The best shops use heavy linen. The threads of scrims being more open than linen, it is not so well adapted to painting over; and as surface is sought after nowadays fully as much as any other good quality in a carriage, we find that no pains are spared to secure that result. Before applying the linen, or scrims, rasp up the panel or corner to be covered. Cut the goods, dip it in hot water, and wring out dry, which shrinks it and removes the dressing. Then saturate in hot glue. Apply it, rubbing it down smoothly. When dry, file the edges to a surface and sandpaper canvas all over.

INSERTION OF SCREWS IN WOOD.

When screws are driven into soft wood and subjected to considerable strain they are very likely to work loose, and it is often difficult to make them hold. In such cases it is said that the use of glue is of service. A stick of about half the diameter of the screw to be used is to be first immersed in a thick glue, and then inserted in the hole prepared for the screw, which is then to be driven home as quickly as possible. When an article of furniture is to be hastily repaired and no glue is at hand, insert the stick, fill the rest of the cavity with pulverized resin, then heat the screw sufficiently to melt the resin as it is driven in. Chairs, tables, lounges, etc., are continually getting out of order in every house, and the proper time to repair them is when first noticed. The matter grows worse by neglect, and finally results in laying aside the article as worthless. If screws are driven into wood for a temporary purpose they can be removed more easily if dipped in oil before being inserted.

TO IMITATE MAHOGANY.

The surface of any close-grained wood is planed smooth, and then rubbed with a solution of nitrous acid. Next apply, with a soft brush, a mixture of one ounce of dragon's blood, dissolved in a pint of alcohol, and with the addition of one-third of an ounce of carbonate of soda. When the polish diminishes in brilliancy, it may be restored by the use of a little cold-drawn linseed oil.

HOW TO MAKE A CHEAP GLUE BRUSH.

Take a piece of linn bark, and soak the end well in hot water, then pound it lightly with a hammer. We have tried it, and can vouch for its answering a very good purpose.

BLACKBOARD PAINT.

For 1 quart of paint, use 1 quart of alcohol, 3 ounces pulverized pumicestone, 1½ ounces pulverized rotten stone, and 3 ounces lamp-black. Mix with some of the alcohol into a paste, grind, and add 3½ ounces dissolved shellac, and add balance of alcohol. Wipe out the old marks with a dry chamois.

CARRIAGE PARTS.

No part of a coach is more worthy of attention, and receives less, than that known as the "carriage" or "carriage part." It has always been the custom of bosses to give this work to apprentices and pretended journeymen, who could do nothing else, and the result was, instead of making carriages, they made what might with more propriety be called miscarriages, for the half-formed, ill-shaped things were worthy of no other name; and, we are sorry to say, that too many such workmen are still employed at this branch of our trade, not only in "cheap shops," but in shops which, in all other respects, turn out first-class work.

Good carriage-makers are scarce, and will be until employers are willing to pay a price that will induce competent men to devote themselves to this particular part; and then the supply will never be greater than the demand, for to be a *good* carriage-maker a man must first be a good mechanic. He must have some knowledge of body-making, ironing and carving, besides having a well-cultivated eye; and if he is something of a machinist so much the better, for the carriage is a machine made for the express purpose of carrying the body, and a little knowledge of the principles of machinery will often be a great benefit. But above all, he should have a "good eye," for patterns are of but little use except to assist in getting sweeps, so the greater part must be done with the eye only as a guide; then again, he needs it in giving the carriage a proportion to correspond with the body, for each part should be in harmony with the other, and not look, when it is done, as though it was made of remnants from several different kinds of jobs. This is one of the secrets of making a nice coach. No matter how well shaped the body, how tasty the carriage, how finely painted or elegantly trimmed the job may be, if there is no harmony in the different parts there will be no beauty in the whole.

Faults of this kind may be seen in every repository, and often prevent the sale of a job. The customer may not be able to tell where the fault lies, but he knows it is there, although he admits that is a well-shaped body, and well finished all through; but when he stands off and looks at it he don't like it, and nine times out of ten you will find the fault to be a *badly proportioned carriage.*

Although guess-work is not as common as it once was it is still practiced, and by men professing to be good mechanics, and who perhaps are, but practice it because they were so taught, and know of no other way in which certain things can be done. This way of working is a disgrace to any journeyman, and we hope soon to see it discarded by every respectable mechanic.

Guess-work is not confined to any one particular trade; all are affected by it, and none more so, perhaps, than our own. This is not because coach-makers are more ignorant or stupid than others, but because their calling requires more skill and knowledge of geometry than almost any other trade. Yet it must be admitted that, as a class, we are slow to use the means and opportunities which we have for our improvement, preferring to do everything in the old way (by main strength and stupidity), rather than adopt any "new-fangled notion." How long this state of things will exist we cannot tell, but we know that the publications of the present day are doing a good work in exciting the ambition of some to become better informed, more skillful and consequently more valuable workmen than their indifferent shopmates.

The point where the stay on a perch carriage part should connect with the perch, is one of the points which carriage-makers have been accustomed to guess at, but we will endeavor to explain the manner of doing it.

To ascertain that point, we will first take one of the front wheels, and find at what point it will strike the body in turning, which we do by measuring the height of the body from the ground on the wheel, as line A, Fig. 1; the top point of the line is the

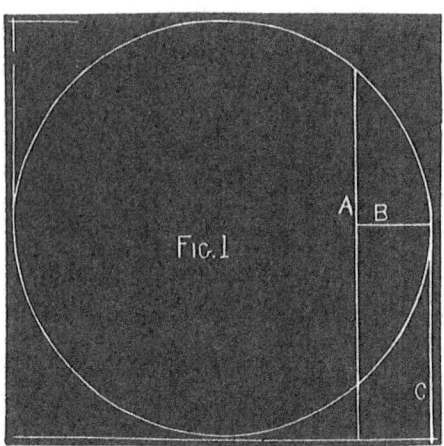
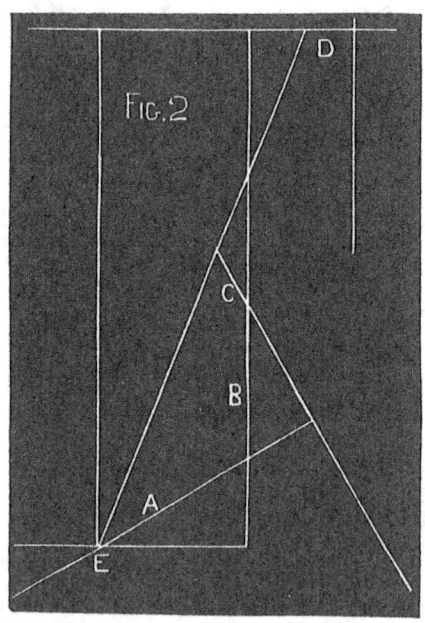

point where it will strike, and line B *is the distance which the wheel will reach under the body in the direction of the perch;* line C is the height of the perch from the ground. Now all that remains is to find *where*, and at what *angle*, the wheel will be when turned under, which we will endeavor to do by giving half section of carriage part, Fig. 2. Line A represents the front axle turned until the wheel strikes the body, which is represented by line B. Line C is the distance which the wheel reaches under the body, as we ascertained by Fig. 1, and will compare in length to line B in that figure, which is the important thing to remember. Now by drawing a line for the stay from the given point of attachment on the back axle D, so that it will just touch the wheel, we find that the front end will meet the perch at the king bolt E. This is the correct way of ascertaining where the stay should strike the perch.

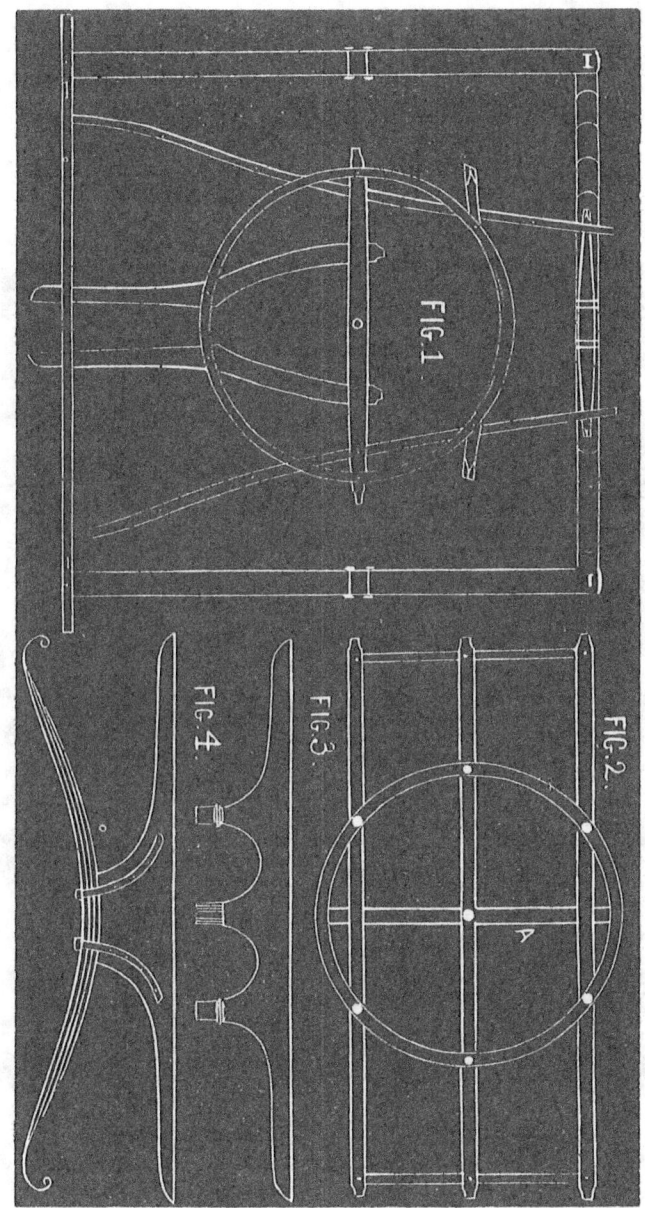

PLATFORM CARRIAGE FOR POLE AND SHAFTS.

Fig. 1 represents the lower part of front platform, with springs attached. The circle measures 24 inches in diameter. Short furchels are framed into the bed to receive the pole. Long furchels on the outside run back, and are clipped to the cross-spring bar.

The side spring measures $18\frac{1}{2}$ inches long from center, and $22\frac{1}{2}$ front from center to splinter bar. The single-tree for shafts is attached to the splinter bar, and when the pole is used the single-tree is removed, and the double-trees or bars are attached.

Fig. 2 shows the top part of carriage having three cross bars, which are 18½ inches from out to out at the ends. Bar A rests upon the fifth wheel, and supports the three cross bars before mentioned.

Fig. 3 shows a front view of the bars, with turned collars, which rest upon the fifth wheel, the center leg resting on bar A.

Fig. 4 shows the back view of hind cross spring and bar attached, which is bolted to body under the tail-board.

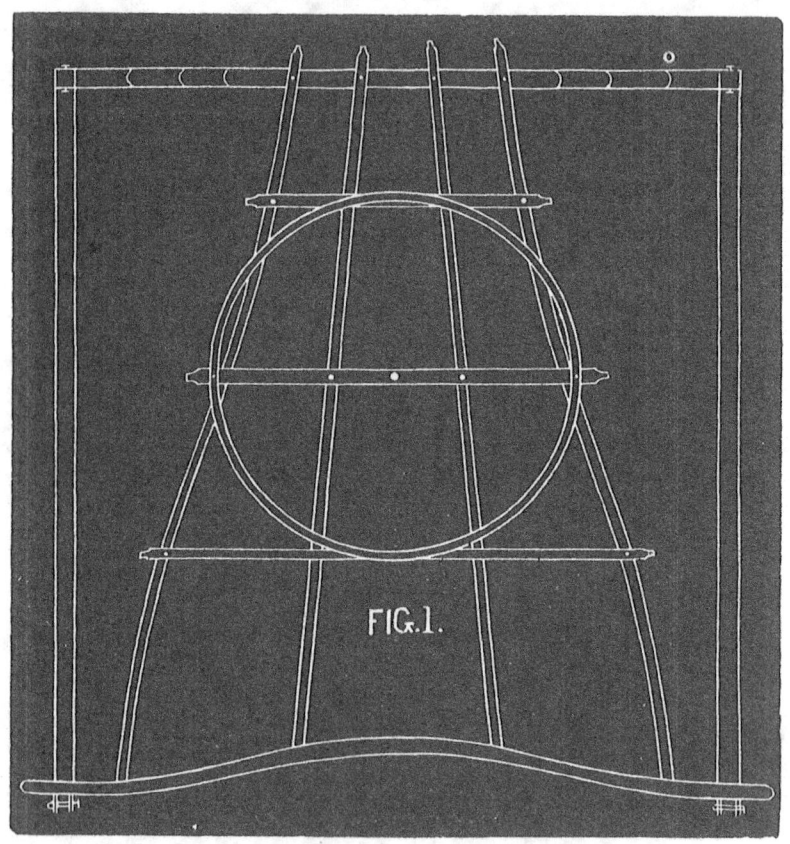

EXPRESS CARRIAGE PART.

The size of the bars depends altogether on the weight the wagon is designed to carry. For a wagon, say to carry 1,000 lbs., the two outside and the front bars should be 1¼ inches wide; inside bars, 1½ inches. The fifth wheel, 22 inches in diameter; springs, 5-plate 1½ inches, No. 3 steel, and 42 inches long. We attach the steps to the back bar, so that when the pole is turned either way a little they are thrown out beyond the body.

In Fig. 1 is shown the front carriage, which will give a correct idea of how it is constructed. It is drawn on a scale of one inch to the foot.

Fig. 2 is the back end view, showing tail-gate, spring and horn-stay.

WORKING DRAFTS OF LANDAUS, ETC. 99

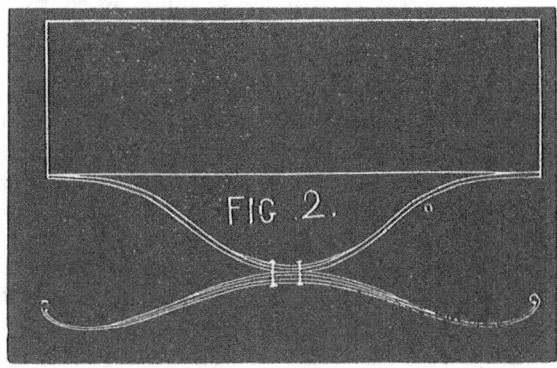

WOOD CARVING.

The taste for ornamentation in wood and other materials having been practiced from the earliest times, it is safe to conclude that it is a natural gift. The variety of forms presented by nature among plants and flowers was the free school to which man, even in the savage state, was welcome. In his first artistic attempts he may have felt the lack of proper materials for copying nature, but a stick would serve in the place of chalk, and mother earth would furnish a substitute for paper, for, in all probability, drawing on a flat surface preceded any attempt at reproducing objects in nature by molding them, or carving in either wood or stone. Plants and flowers appear to have been the favorite studies, and deemed the most appropriate for decorations of all kinds; nor have they, down to the present time, entirely lost their charm. The Egyptian is termed the primary style of art, as we possess no knowledge of any other anterior to it.

In nature's free school the artists drew their inspirations from the organic forms about them—the lotus, papyrus and palm tree furnishing them with studies in harmony with their symbolism, and sufficiently beautiful, when artistically arranged, to satisfy their notions of what was proper. We would not, however, care to copy the models they have left to posterity, if those we generally meet with are fair samples. The Greeks possessed great skill in seizing upon suggestions in nature, and producing therefrom ornamental carving. They seem to have been endowed by the Creator with the fullest measure of artistic knowledge, and, in the range of subjects which they treated, left succeeding generations to take the position of copyists. The art of embellishing wood by richly carved patterns has been confined principally to the adornment of private residences, public buildings and churches. Furniture at different periods has claimed a share of attention, and the indications are that we are again entering upon another term when carving will be in great demand, and if so, the manufacturers of carriages will be compelled to introduce it in order to satisfy the prevailing taste. The ornamentation of certain parts of the carriage, by means of carving, was quite fashionable fifteen or twenty years ago. We cannot, however, claim for it, in its palmiest days, any degree of skill worth naming. Occasionally, at that time, one in his travels would find a coach shop where a skilled carver was employed, but the large majority of shops depended alone on some one or more of their workmen who happened to have a little taste in that direction, but not sufficient energy to procure good models of carving, and, by close study and practice, catch the spirit they contained, and breathe it over their own productions. The consequence was, spring bars, axle beds, head blocks, pump-handles and body blocks were tortured into a resemblance of vines and leaves, possessing little or no variety, and certainly no beauty.

It was not uncommon to see leaves carved as if growing out of each other, the painter finding, when he attempted to "cut up" the carving, a leaf or more too many to carry out the main design. Instead of the beautiful flowing lines, the plumpness and variety of forms given by a skilled workman, carriage work was, and continues to be, so far as carving is used, afflicted with a gaunt, lean style of leafing, and great poverty in design. What the trade lacked was a knowledge of drawing and designing. There were but a few who could take the pencil and sketch the simplest form from a natural object, or copy a good model with fidelity. Too frequently a first-class piece of carving, taken as a pattern, lost some of its beauty at the hands of the first copyist, and those who copied from *this* departed still further, until, at length, there was left no resemblance to the original.

It is not required of the carriage-maker that he should be able to carve equal to England's first sculptor, Gibbons, of whom it has been said, " he exhibited a pot of flowers so exquisitely carved that the individual leaves quivered and shook with the motion of passing coaches," and under whose chisel "stone seemed touched with vegetable life, and wood became lilies of the valley and fruit from the tree."

The coach and carriage of smaller dimensions will not admit of great boldness in design, but every portion requiring decoration will admit of a well-conceived design and delicacy of workmanship. Should carving again become popular, we trust we may witness a change for the better. Our free drawing schools may have already done a good work in teaching our young wood-workers, and with a good eye, correct taste and free pencil, they may, when the necessity arises, be prepared to produce ornamental work worthy of the trade.

SCROLL PATTERNS FOR BAR ENDS.

Nos. 1 and 2 are designed for light carriage parts, and make a very neat finish.

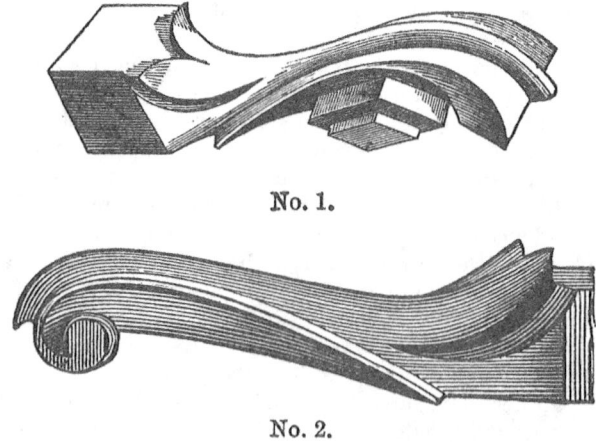

No. 1.

No. 2.

No. 3 is a new style of scroll end for front bar, or bed piece on platform work. When carved as shown, and tastily striped up by the painter, it will not fail to please the eye, and add much to the appearance of the carriage part.

WORKING DRAFTS OF LANDAUS, ETC. 101

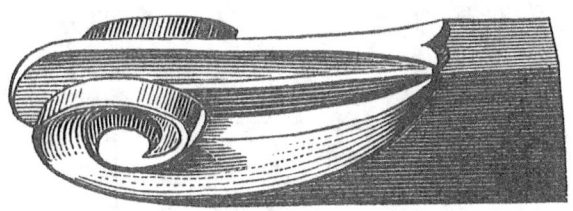

No. 3.

No. 4 represents a late design for the end finish of pump-handles for heavy work. Cast in brass or iron, a great deal of labor is saved in carving. The cut represents a side and top view, or rather a three-quarter view. The surfaces are worked out smoothly, presenting to the painter a chance to display his taste with the striping and cutting-up pencil.

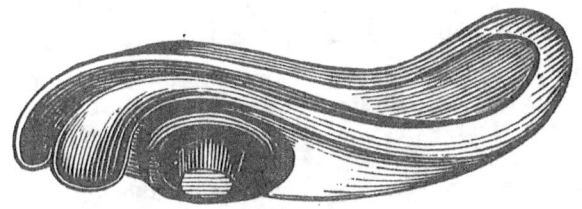

No. 4.

DESIGNS FOR CARVED BODY BLOCKS.

Nos. 1 and 2 represent two different designs of long body blocks, having but a single bearing on the front bed. The back iron stay is bolted to the boot, and the block is fitted over the L plate, producing a complete finish. Other designs may be suggested by these; for, after having decided on the proper size of the blocks, it is immaterial what form is chosen to ornament them.

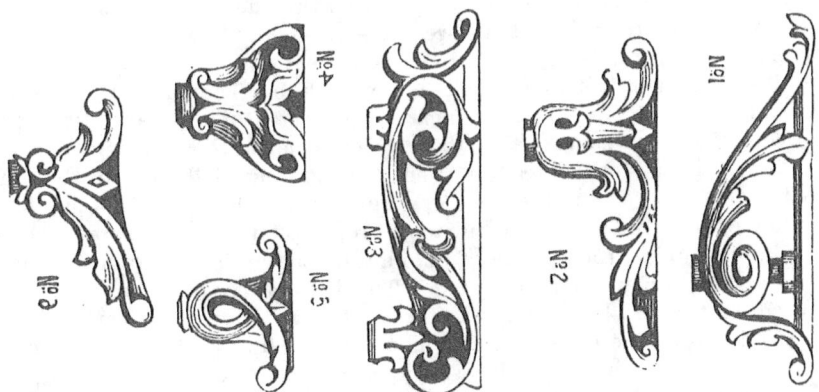

No. 3 is a more elaborate pattern, and will require a skillful carver to bring out its beauties. It will be noticed that it has two bearings, arranged for bolting through the front bed and back bar.

Nos. 4 and 5 show two short blocks of different designs, both of which are easily carved, and when finished and properly touched up by the painter, will look very neat. They are to be bolted through the front bed to the boot, and the back supported by an iron stay.

Fig. 6 is also a short block, but is designed for a swept body, as the top curve of the block will at once suggest.

SIZES FOR WHEELS.

Length	Diameter	Front End	Back	Size of Mortises	No. Mortises	Spoke	Rim	Tread
6	3¼	2¼	2⅝	¾x5-16	14	⅞	1	¾
6½	3½	2⅜	2¾	¾x5-16	14	⅞	1	¾
6½	3¾	2⅝	3	⅝x⅜	14	1 1-16	1	⅞
6½	4	2¾	3⅛	⅞x⅜	14	1⅛	1⅛	⅞
7	4¼	3	3⅜	11-16x⅜	14	1⅛ or 1 3-16	1⅛	1
7	4½	3¼	3⅝	1⅛x⅜	14-16	1¼	1¼	1⅛
7	4¾	3⅜	3¾	1¼x7-16	14-16	1⅜	1⅜	1⅛
7½	5	3½	3⅞	1¼x7-16	14	1⅜	1⅜	1¼
7½	5¼	3¾	4⅛	1¼x½	14	1⅜	1⅜	1¼
8½	5¾	4	4½	1½x9-16	14	1½	1½	1⅜
9	6	4¼	4¾	1⅝x9-16	14	1¾	1¾	1½

It will be borne in mind that the sizes taking in the one-eighths will be regulated by the size next the largest; to illustrate, we will name the four-inch hub. We would make the three seven-eighths the same on front and back, with the same size of mortise, same size spoke, same *number* of mortises and depth of rim, and the same on the tread. In the use of the different sizes of wheels, as above, the wheeler will be governed in his proportions by the heft of the carriage part and body. It is sometimes the case that patrons desire a heavier wheel in appearance; if, in such cases, the wheeler can use the same material and the wheel present a size heavier in appearance, or the same should it be desired, to be lighter. These are matters that strictly belong to the trade, and should alike be understood by the employer and employee. There is another very important matter in the use of wheels or hubs, and that is, the size of the axle or box. We should avoid getting less in the length of the tenon in the hub than what we have in the rim, if we would avoid the effects of the rim binding. If we have more tenon in the rim than we have in the hub, with the dampness that penetrates the tenon from under the tire, causing the same to swell, you can readily perceive the advantages it has over the tenon in the hub. There is yet another reason, when we consider that the hub, being nearer the center of the revolution, it is first attacked with the concussions or the labors of the wheel, if the wheel is the least rim-bound. When concussion strikes the wheel, this binding of the rim tends to lift the spoke out of the hub. When the lower or ground part of the wheel is doing the labor, it forces the tenon in rim and in the hub to its proper position; consequently, the upper portion is forced up just the amount of the binding on the rim; the result is, the spoke must work in the hub and not on the rim. This is the cause of so many wheels being pronounced worthless, when the real fault is not with the work as performed by the wheeler; and I would here say, that too many employers attach too little importance to this subject, by not watching, with a jealous eye their work when it first goes or is turned out, when possible so to do.

They should make themselves acquainted with these important facts for their own especial benefit, as protecting their own reputation.

Wheels should be made with a sufficient number of spokes to properly divide the space on the rim, and afford sufficient support to prevent sinking in between the spokes, and at the same time avoid too many to weaken the hub. The less number of spokes the stronger the hub, and weaker the felloe. Judgment should be used in dividing the difference, so as to make each part of the wheel strong in proportion.

For wheels from 3 feet 10 inches to 4 feet 2 inches fourteen spokes will be a proper number, but as the size of the wheel increases or decreases, add or diminish the number of spokes, being careful to reduce your tenon proportionately, in order to not weaken the hub.

CENTERING SQUARE.

This is a useful instrument for centering of all circular work, especially in shops where they mortise their own hubs. Its construction is very simple, being but a T-square whose stock is a portion of a circle. Let A C be the stock made of one piece of hard wood well seasoned (the extremities of which at A and C should have a small piece of steel affixed so that it should not be subject to wear by use), into which stock the blade B D is tenanted, so that A B is exactly equal to B C, and at the same time perpendicular or square to the cord A C. It is evident that if this instrument is applied to any circle, so that the parts A and C touch it, the blade B D will pass through the center of the circle, and by two applications the center will be found.

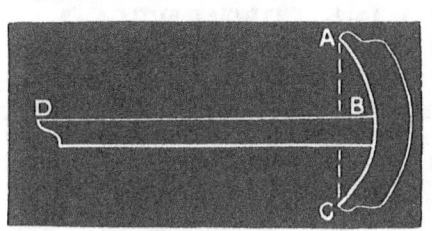

PART II.—BLACKSMITH DEPARTMENT.

THE BLACKSMITH.

> "Ho, ye who at the anvil toil, and strike the sounding blow,
> Where, from the burning iron's breast, the sparks fly to and fro,
> While answering to the hammer's ring, and fire's intenser glow!—
> Oh, while ye feel 'tis hard to toil and sweat the long day through,
> Remember, it is harder still *to have no work to do!*"

WE imagine we hear the sons of Vulcan throughout the land bemoaning their hard fate. "Oh, who," they say, "are doomed to lead such a fiery life as we. Although the sun may pour his sultry rays, causing the atmosphere to quiver above the parched earth, we may not cool the forge's heated breath. Enfeebled by the atmosphere from without, we must nevertheless keep up an intense heat in the shop, and for ten long hours stand between these two fires, while the violent exercise which our trade demands causes the blood to boil almost within our veins." Very true, my fellow shopmate, but while you may think your calling is one full of hard work, and rendered more unpleasant in summer by the ever-glowing forge, the flowing perspiration and the soot, nature comes to your aid, and repays you with glowing health and muscles well nigh as firm as the metal you work. The sounding blows delivered by the blacksmith make music wherever civilization extends, bringing into proper shape both iron and steel, adapting them to the numberless purposes for which they alone are fitted. The tools required by the various trades, the implements of husbandry, the shoes which give to the horse a sure footing, enabling him to apply his strength to better advantage, vehicles for pleasure or profit, vessels engaged in commerce, and their guardians, the "men-of-war," all have been dependent for strength, durability and efficiency on the workers in iron. Yes, there is music in the ringing anvil, and beauty in the sparks as they fly to and fro from beneath the forming hammers; and then the smithy has furnished artists with many fine subjects. "The Village Blacksmith" is a charming picture. The statuette by Rogers, "How the Fort was Taken;" the scene is laid in a smith shop, and, besides its life-like beauty, possesses a national interest. These, 'tis true, do not represent the glare of the forges, and the activity of the workmen, but the scenes are laid in the smith shop—the work shop above all others with which the masses are well acquainted. Wherever the pioneer pushes out into the wilderness the blacksmith must soon follow, for without him there could be no progress made. Artificers in metals other than iron and steel occupy honorable positions, and their products are of great value, but the metals they manipulate could be dispensed without entailing on the world such serious consequences as would ensue should iron be annihilated. The blacksmith may well be proud of his calling, for the eyes of all are upon him, paying him homage in their inner thoughts, if they are too proud to acknowledge it openly.

THE TIDY BLACKSMITH.

We do not mean tidiness in dress alone, but order and cleanliness in the shop while at work.

"Humph," say a dozen smiths at once, "it is foolish to talk of such a thing. The idea of men working in the midst of coal, smoke, and handling iron and grease continually, attempting to be tidy. A pretty figure we would cut to go mincing around our work, fearful of soiling our hands and clothing."

We reply, we are aware that the smith cannot be as tidy as the wood-worker or trimmer; neither do we require it of him. But is it not possible for him to improve considerably on the plan usually adopted by smiths? We believe it is. In fact, we know that it is possible for even a country blacksmith, who shoes horses, irons carriages, and takes a hand at the regular routine of work brought to a country smith shop, to keep himself, and the work he handles, free from greasy hand marks, and besides, keep his tools in perfect order.

We say we know just such an individual, and if there be *one* such, why may there not be as many more as the number who choose to follow the example set?

Some years ago we knew a smith (his name has now passed out of mind) who was a model of neatness. Everything connected with his forge and anvil had its appropriate place, and in his person he was scrupulously nice. His clothing was not costly, yet it was whole and clean; and, in order to appear on the street genteely, as well as to feel comfortable after the day's toil, it was his custom to change when he came to the shop in the morning, and place his clothes under cover, by this means securing to himself a dry, clean suit.

One visiting the smith shop, after the hands had all retired, would not fail to notice the contrast between this man's forge and its surroundings and those of the other hands. His tools were all in order, and his working suit and apron were hung up in a certain place, and there was no appearance as if he had pitched things to the right and left at the signal for quitting work.

Now, this habit of cleanliness manifested itself still further, for he was very particular about handling a piece of painted work. He managed, somehow, to train his helper, and between them they never clawed a job over with greasy fingers. An omnibus painted in the most delicate tints would come from his hands fully ironed, with scarcely a soiled spot on it, and a scratch, bruised or burned place was of such rare occurrence that the painter would rather invent an excuse for him than call his attention to an occasional mishap. He was, in fact, a model blacksmith.

Another, we would mention, who was not only scrupulously exact in matters pertaining to the shop, but in dress so tidy that he was styled foppish by his fellow-workmen. But it appeared as if it was a part of his nature to be so; and as he was a quick, excellent mechanic, high spirited and independent, his foppishness was overlooked by those who were well acquainted with him. As an evidence of his extreme nicety, we mention that he dressed in broadcloth and carried a cane every day in the week. To see him entering or retiring from the shop, one would be led to suppose that he was a customer of ample means rather than a horny-handed workman at the forge.

From the examples given we infer that the *majority* of the workmen at our branch might make great improvement in their shop habits without any extra expense or loss of time, and certainly it would add greatly to the interior appearance of the smith shop if each forge was kept in perfect trim.

In strong contrast to those named are the careless, sputtering, dash-about kind, who ever have their tools in confusion, and who throw things here and there with a recklessness that is marvelous. A body placed in the possession of one such becomes the receptacle for all the tools he may require, from the time he receives the body until it is ready to go back to the paint room. He handles the gearing with a filthy grip, and acts in this respect as though lubricating oil was the only proper foundation over which to lay paint. Tools are pitched into the body from any convenient distance, and should the inside of panel seats be cut by the edges of chisels, and bruised by other tools, he "cannot see that that amounts to much, for the scars won't show, or the painter can easily putty them." On quitting work his apron is tossed at the anvil, but more frequently at the body; and if it hangs to either, well and good; if not, there is no harm done.

From the latter class please deliver us.

We have aimed in the above examples to awaken careless workmen in **our branch** to a sense of their duty in regard to their personal appearance, and the carrying out of the same principles of neatness in the shop. It is not necessary for them to imitate the dandy with his cane, when they appear on the street, but that would be preferable to going to the other extreme. Let order and neatness prevail at the forge and throughout the smith shop, and when the day's work is done, be prepared to step out in clean apparel.

TOOLS.

Without paying any notice to the coarsest kind of a hammer, as sledge hammer and its use, we will give two patterns for hand hammers—a swadge hammer and iron and

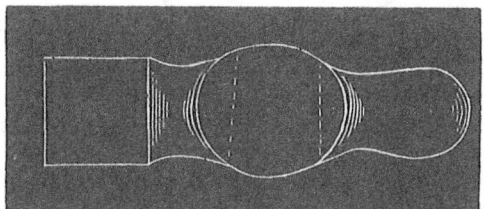

Fig. 1.

a riveting hammer. The hand hammer, Figs. 1 and 2, having a flat circular face and a globular pane, is a tool which a hand commonly owns, and a great deal of its effectiveness depends on the right proportion and size of that tool. The flat face is used

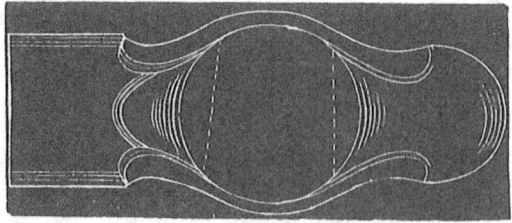

Fig. 2.

for pounding or stretching and finishing off the desired shape; the pane serves to forge out hollow places, tapering off collars, etc. The cavity for the reception of the handle is shown by dotted lines. Some prefer to widen out at the top and drive the wood out by wedging.

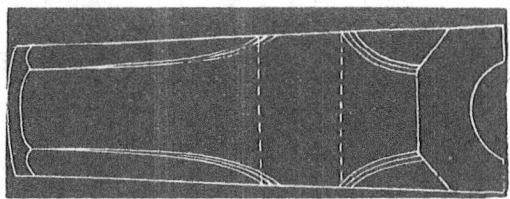

Fig. 3.

Fig. 3.—A swage hammer for 1-inch finish.

Fig. 4.—Swage iron, top view, corresponding with swage hammer, Fig. 3; the moulds vary from ½ inch to 2 inches in width, and widen a trifle at either end of the

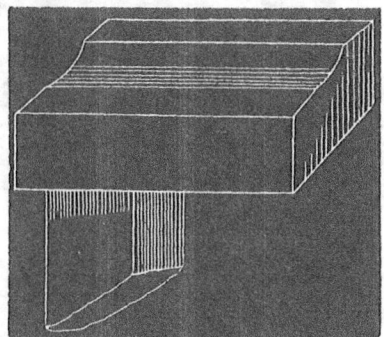

Fig. 4.

concave mould. This prepares the piece to be swaged, to get true after passing the middle of the swaging block.

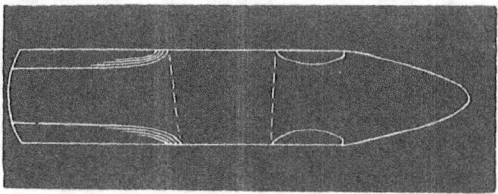

Fig. 5.

Fig. 5.—Riveting hammer, which only needs notice as to its proportion and shape.

FULLER, Fig. 6, and FULLING BLOCK, Fig. 7.—Fullers are reversed swages, and are used for setting off collars, carriage stays, steps, or any part requiring a scolloped or concave finish. The width of face varies between 3-16 up to 1½ inch. Small sectional cut A shows the shape of face. The largest ones to finish slight bends, as for fifth wheel sockets or brackets on heavy work; a glance at our illustrations shows us that they are indispensable for making tools.

SQUARE PUNCHEON, Fig. 8.—Their size varies as to the more or less complicated work a smith is called to perform; a corresponding square hole receives the piece cut out, to keep the edges of the cut true. The utmost care must be taken to punch the hole square, as the striping will afterward show the deficiency only too plain.

GAUGE CHISEL, Fig. 9.—It only deserves notice as to its use, either for trimming step sockets or plates, or shaping the ends of rounded narrow plates; the object of its illustration is to show compact proportion, which prevents jarring.

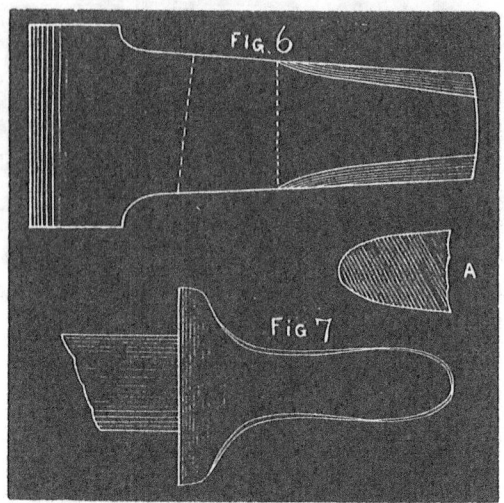

Figs. 6 and 7.

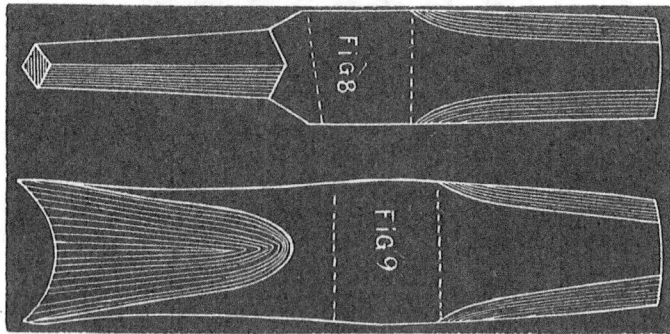

Figs. 8 and 9.

FLATTER FOR CLIPS, Fig. 10, especially for fifth wheel clips, spring or bracket clips. A tool for this purpose has to be made very exact, as it saves time and finishes a piece as in a drop.

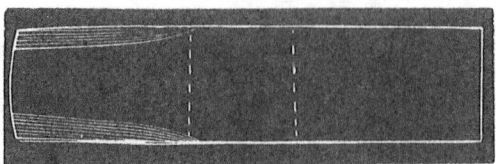

Fig. 10.

FLATTER, Fig. 11, stands, in regard to utility, in the same line with the plane of the wood-worker, for a 2½ inch square-faced flatter will issue an exact counterfeit of the inside of rocker to the rocker plate fitted thereon. Rocker plates for light work are beveled on the inside top edge of parts exposed to view, also in the space of doors on heavy work, which depends in point of neatness solely on the skillful handling of the flatter.

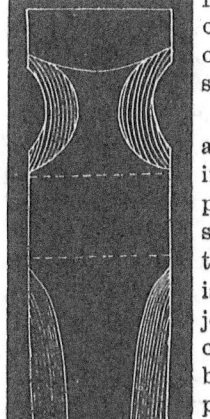

Fig. 11.

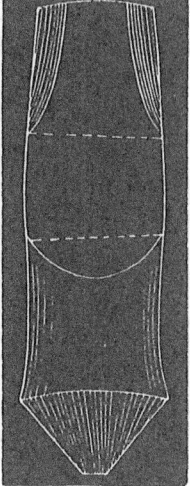

Fig. 12.

BOB PUNCH, Fig. 12, has a round face similar to a countersink, to press a cavity in a flat piece of iron, intended to have an upright, round or square piece welded on to it; as, for instance, a square socket to a swedge block, or a round lamp socket to the bearing plate. The piece to be welded on is fitted to the impression of the bob, to give the job a firm bearing. Rapid heating and shutting out of the air are essential at welding. The pieces to be welded, if heavy, are coated with sand, for the purpose of floating with the dross on the surface of the heated piece, shutting off the air, and agitating the heat necessary for welding; before hammering together, remove the dross. If you take green, grassy coal, you can dispense with the sand, only cut in the surface with your chisel, which will aid the hot air as well to penetrate as to escape, just the same as the Bessemer process of melting steel. Use borax for covering for light dash rails, which require a thick heat. Soft iron and hard iron will weld difficult; the former is apt to shift off. You will have to heat again and strike carefully. Some weld steel to iron, giving the steel a wedge-like shape, and the iron a corresponding opening. To prevent the steel from sliding out, cut burrs on the surface of the steel and strike light at first. Iron containing less carbon requires a high degree of heat for welding; therefore, the better quality of iron is to be brought to a white heat, almost commencing to burn under lively sparkling. A good weld joint should defy detection, and only show a slightly darker margin.

Small cast-steel tools, as punches, etc., when burned, might be restored by the following mixture: two ounces of bichromate of potassa, one ounce of pure nitre, one ounce of gum aloes, one ounce of gum arabic and two ounces of resin. After powdering and mixing well, heat the steel to a low red heat, and put the powder on. It is then heated again, as before, and cooled, which will make the steel very hard.

TOOLS FOR MAKING LAMP SOCKETS, AND HOW SAID TOOLS ARE MADE. ALSO, THE MODE OF MAKING SAID SOCKETS, ETC.

To make the matter plain, should any one see fit to make a set of tools, and try their hand in making the sockets, it would be proper and perhaps necessary to give a word of caution in regard to the weight and the process of making, for the reason that they are subjected to such powerful hammering.

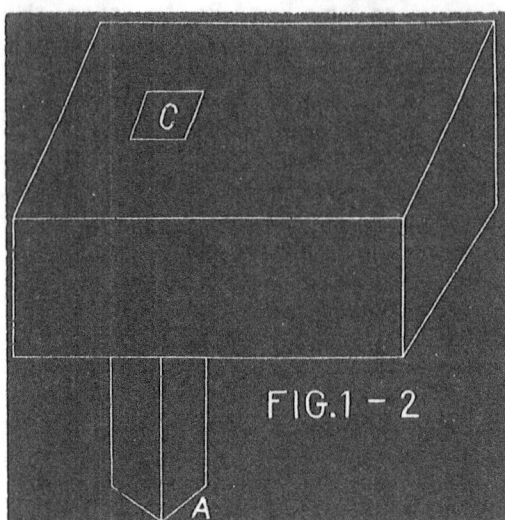

FIG. 1 – 2

In view of this fact, we would recommend them to be laid with good cast-steel to that of half the thickness. Use cast-steel for the reason that steel of a softer character would be liable to settle around the holes. Some might ask the question, why not harden or temper them? We would answer, for the simple reason that it would be almost impossible to keep the temper in them; yet, with extra care, it could be done. You will observe that it is necessary to use two blocks or heading tools—one cut representing both, as is shown. Fig. 1 represents both first and second of said block or heading tools.

Dimensions of Fig. 1 and 2, 3¾ of an inch long, 2¼ inches wide, 1½ inches in thickness; shank A formed to fit the hole in the anvil, and C a ⅜-inch square hole in the face, extending through the shank, diverging so as to leave the hole ¾ of an inch at A. The formation of said hole is rather difficult, and some may be benefited by a few hints concerning it: First, then, drill from the face C half way, then from shank A to meet the same with a 5-16 inch drill. Drill again at C, using a half-inch drill, drilling two-thirds the way through the block. Shift again, and use a ⅜-inch drill, beginning at A, and drill to meet said half-inch drill. Drilling with a ⅜-inch drill through shank A will obviate in a measure splitting the shank, while straining it with a square punch sufficient to meet all demands.

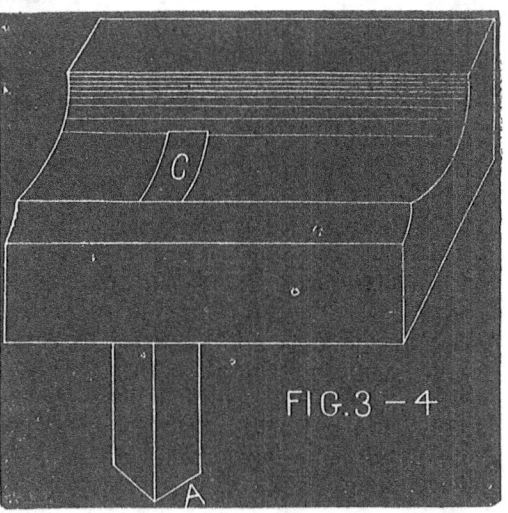

FIG. 3 – 4

Second, to prevent forming a cold shut in the shank near the surface of the anvil, where the most strength is desired, after the hole is all squared up complete in the face at C, drive in Fig. 9, with the straight side of the pin from, and the shoulder toward the smith (which represents the pin or former), up to the horizontal line E, and the tool is complete.

For the second use Fig. 10, driving in as before to horizontal line D. Fig. 11 represents the size of Figs. 9 and 10 the other way. Fig. 4 represents the swage or swages.

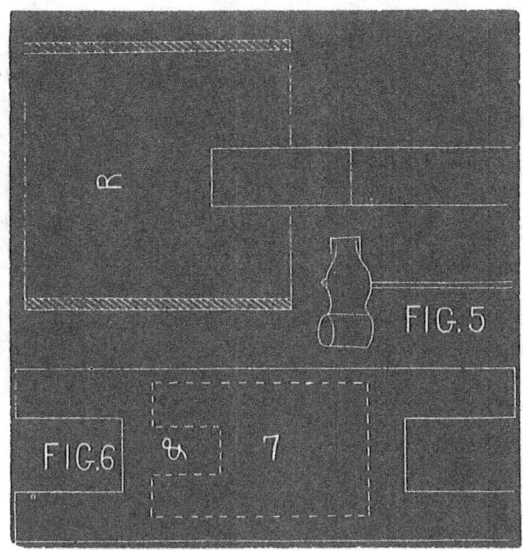

It is necessary to have two of these swages, made in the same proportion as the blocks ot heading tools, as above described. Make the swage part 1¾ inches wide by ⅝ of an inch deep.

These tools should be made to fit the anvil so that the shank will be only ¼ of an inch from the edge or left hand side of said swages. These shanks are to be drilled and the same process performed as the blocks referred to, with the exception of driving in Figs. 9 and 10 a mere trifle past the horizontal lines E and D, so that the shanks of the sockets can be slipped in and out easily. These tools are used as formers, in connection with Fig. 5, as will hereafter be shown.

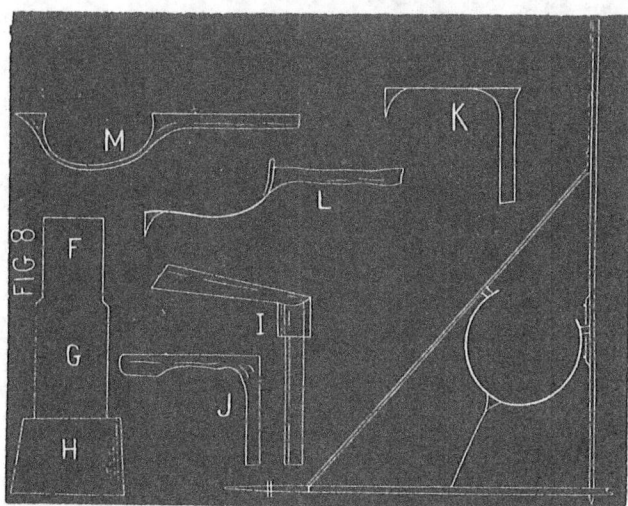

Fig. 8 represents a mandrel to bend or to form the sockets to or over. There are two sizes represented here: F represents the small size, 1⅝ inches; G the large size 1¾ inches in diameter; H represents the base or flat part, 2 inches by 1⅛.

Fig. 6 represents the large-sized gauge, made of 1¾ inches band iron, with an opening filed in each end to slip over the shank of said socket to mark by for trimming. Fig. 7 represents the small-sized gauge, made of 1⅜ inches band iron, and used as Fig. 6 (we give a half-view in dotted lines). These are all the tools necessary, except a pair of half-collar swages, in which to form the shoulder to the thumb screws, together with the tools the smith necessarily has on his bench. Taking it for granted that the tools are all in readiness, we will proceed briefly to describe the mode of making the sockets.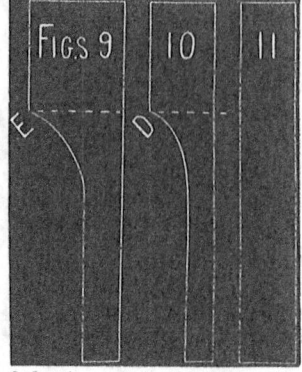

First, the iron should be of the first quality of Norway, 1 inch by ½-inch ; place Fig. 1 in the anvil, draw the shank properly to fit hole C; then place, while hot, the full size of the iron 1 inch from the shoulder on the small part of the beekhorn, say one or two inches from the point, working lively with both hammer and sledge. Flatten it only at the place where it rests on the beekhorn, then bend as letter I represents; quickly placing it in the heading tool, keeping up the same lively strokes until it reaches the point J represents; having the dividers set 4½ inches, placing one point on the outside of shank J, and at the other point mark with a center punch, cutting off ¼ of an inch longer to draw out the small end K represents. After this done, turn the other end and go through the same process as represented in I and J, then both ends will be formed.

The socket is now blocked out, which brings us to the point where, it may be possible, we have too much stock or not enough. This we must take into consideration before we commence plating. If we see we are going to lack stock we must draw to

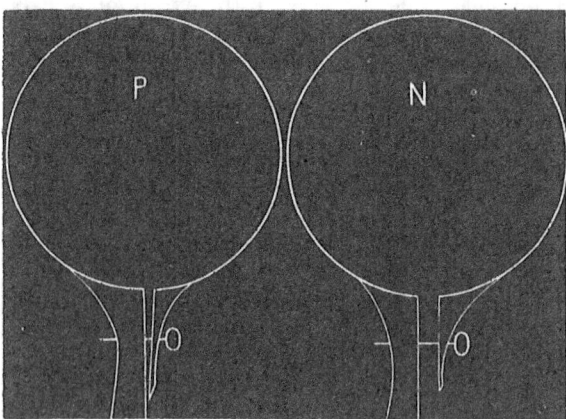

some extent over the beekhorn; we do this because, when drawing over the horn, it draws in length and saves stock, whereas, if drawn over the face of the anvil it would plate out in width and lose that which we need to procure the length; thus we should use both beekhorn and face, frequently trying with the dividers to ascertain what course to pursue, all the time watching the thickness, and also the evenness, judging how much to draw on both horn and face until we bring it to its desired length. Then, again, if we find we have too much stock, we must do all the drawing and plating on the face, keeping a strict watch as in the former case; if necessary, occasionally piece the other way to throw the stock out in width, which we can trim off, and thus, in both cases, secure a good job.

K represents the socket plated complete ; L represents the shank of socket shown in I bent in swage, Fig. 3, using the fuller, Fig. 5. After this shank has been set down by the fuller, it should be trimmed off as M represents, with a sharp cold chisel, close to the shank, squaring them nicely with a file before bending them over the mandrel. Bend the small end in Fig. 4, represented in Figs. 3 and 4 ; trim and file as before.

M represents a socket ready for bending over the mandrel, Fig. 8. N represents a socket complete edgewise; P represents a socket when the lamp is in; R represents the body of socket in full.

When bending these sockets over the mandrel, or, in other words, the former, if we should find one rather long, we hold it together with a pair of tongs near the place for the thumb screw, placing the edge of said socket ¼ of an inch above the shoulder G on the mandrel, and upset or close in the edges with a light riveting hammer until small enough. This can be done successfully. The same can be performed on the small size in the same way by holding edge over the top F represents.

The length of the large socket, when ready for bending, should measure 6¼ or 6⅜ after plating. The length of the small size should be 5 inches when hot. Observe these rules and you will be sure of success. We will add, that to make said tools out of clear cast steel would be an improvement. On cut Fig. 8 will be found a very pretty style of stay irons for perch carriage.

TOOLS FOR WELDING PINS IN SHIFTING RAILS.

LOWER TOOLS.

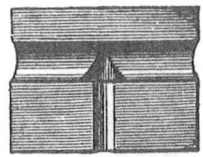

Form of Pin before welding.

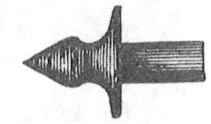

Form of Pin before welding, to be made in half collar tool.

When you have the tools made, make your pins something larger than is required when finished; then take iron, ⅜ by ⅝, and use a flat punch, with rounded edges, to make the hole for the pin; have the pin cold and the iron hot; drive the pin through, and rivet slightly, to prevent its falling out while getting your heat; then place it in the bottom tool and put the top tool on, and let the helper strike two blows, and you have it welded; then clean the scales from both iron and tool, and replace them; two blows more will finish very clean.

DIE PLATE AND SCREW.

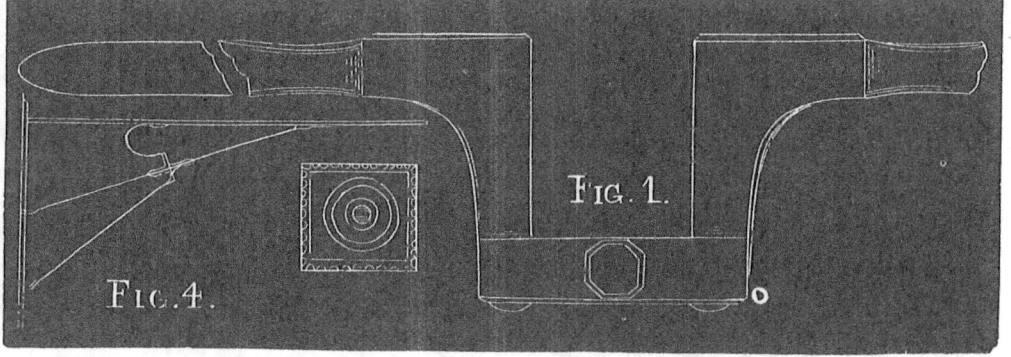

Fig. 1 represents a die plate for cutting the thread of clips, where a common or straight die plate would interfere with the opposite shank of the clip, to that intended to cut; bend both shanks of the clip sidewise, so as to bring your die plate in a diagonal position, and you can operate it at pleasure.

Fig. 2. Top view to Fig. 1. The dies are mitred, and held firm in their corresponding cavity, by plate A, Fig. 1, screwed to the bottom of the frame. Carriage bolts are got up cheaper, more uniform and superior to those made by hand in manufactories for that purpose; all that remains to be done in the smith shop is to replace the nuts, which always seem to be, more or less, wasted or lost.

Common nuts are cut out of medium flat iron, of the size of the nut; then, at the second heat, punched, upset and finished over the puncheon. Larger and more solid nuts for axle-trees, etc., are rolled out of a piece of flat iron of the appropriate size and proportion of the nut, in connection with a square piece of the same material, partly cut off, turned over the end of the cylindrical piece, and then the whole welded. This last named piece forms the shoulder of the nut. This proceeding causes the grain of the iron to run parallel with the thread, giving it more strength. It is then driven out in a block, with a plain round cavity, and formed by degrees into a square or six-cornered nut.

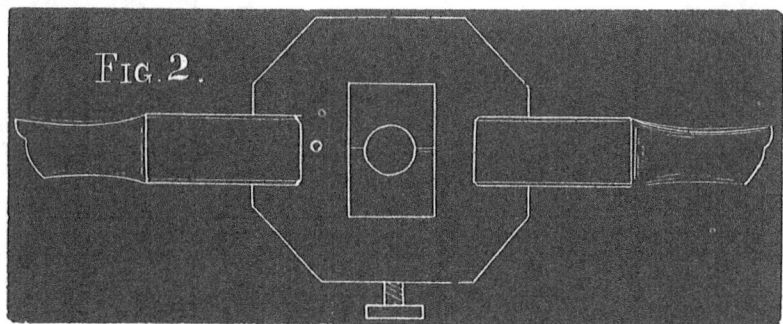

To thread a bolt, screw the same lower and upward, in the vise; press your dies gradually around the bolt; begin to turn cautious and slow until you have obtained the imprint of the thread; turn the die plate back, screw up closer, and repeat the operation until you obtain a perfect thread. The cutting of the thread in the nut is effected in a similar way. The nut is held in the vise, and the tap turned by the tap wrench performs the cutting. The head of the tap must be of the same size as the body of the screw, to allow the tap, after using the full length of it, to drop through the nut. This saves the turning back, which injures the thread in most cases.

For a mechanic of a thoroughly investigating turn of mind it is but a matter of consequence to know how screws are delineated. A screw is a medium to exert great lever power through small spaces. We distinguish two classes, male or external, and

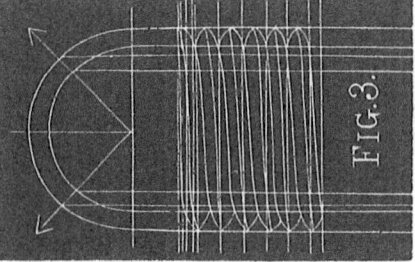

female or internal screws. A tap, either plug or tapering, consists of a cylinder, having on its surface a projecting spiral viz.: the thread. This might be either square or V-shaped, and running right or left. Our drawing, Fig 3, shows a right-hand V-thread, which, when reversed, becomes a left-hand thread.

The internal screw for the nut performs upon the same principle of a cylindrical curve,

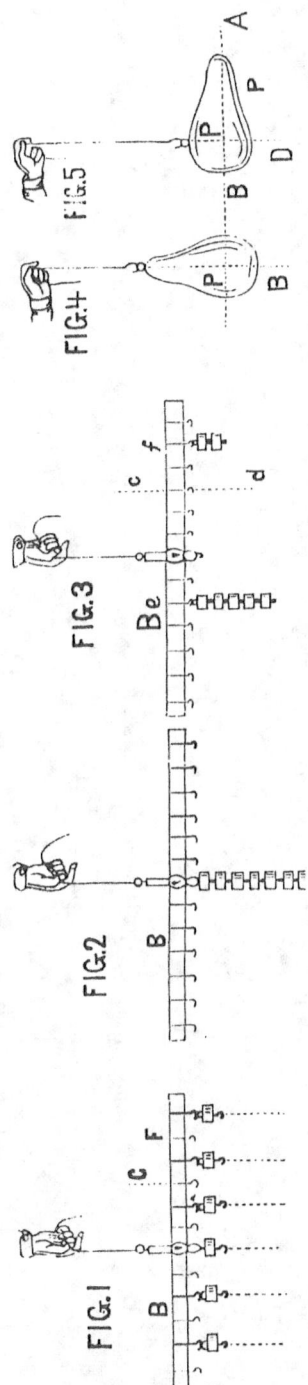

TO HANG OFF A BRETT OR EIGHT-SPRING LANDAU.—THREE-QUARTER INCH SCALE. Page 115.

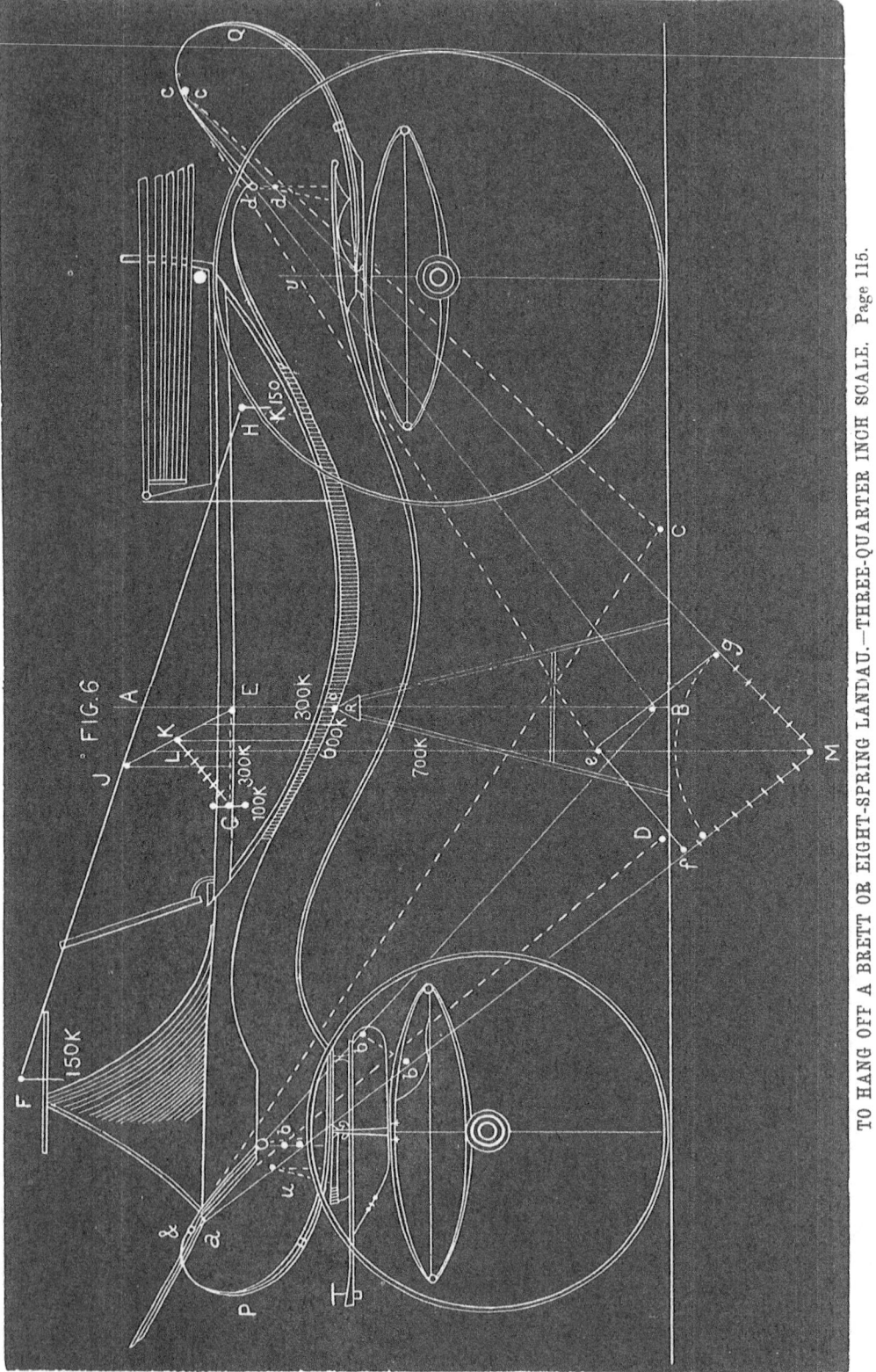

TO HANG OFF A BRETT OR EIGHT-SPRING LANDAU.—THREE-QUARTER INCH SCALE. Page 115.

revolving around a certain point, and parallel to the axis, reaching that point after a full turn. The larger semi-circle, Fig. 3, is equal to the diameter; the smaller semi-circle measures the depth of the thread. We then divide both circles into four equal spaces, which might be increased if the curve of the thread requires a peculiar delineation. Vertical lines are drawn from those five points, at the intersecting of both semi-circles, as long as required. Next set off the space between the thread, and divide into six equal parts, marked at our diagram by seven horizontal lines. These lines will aid to find the true curve of the screw, by connecting the different points of intersection through diagonals, which, if repeated at each interval, will delineate the correct outline for the whole length of the screw. To find the bottom curve of the thread, draw a diagonal line from the outside of thread to the horizontal line, marking the bottom of thread, where it intersects the vertical line, projecting from smaller semi-circle.

A practical eye will enable the student to work out the longer curves by drawing them parallel with the first one, by the aid of horizontal lines crossing top and center of the thread.

Fig. 4 represents patterns of straight stay and step. The smith may display considerable taste on this stay, and produce a piece of work which will be very pleasing to the eye. The *step*, worked out as shown, makes a very genteel finish.

TO HANG OFF A BRETT OR EIGHT-SPRING LANDAU. THREE-QUARTER-INCH SCALE.

In the accompanying diagram is illustrated the method of hanging an eight-spring Brett, which is applicable to all vehicles having C springs connected with platform and elliptic springs. This is one of the most important subjects to be considered in the construction of this class of carriages. The body may be faultless in outline, and perfect in its individual adaptation to the comfort of the occupants. But if the draftsman be ignorant of the proper method of hanging the body, and does not calculate understandingly as to the bearing of the body on the supports, or springs, with and without its weight when fully loaded, the carriage will prove to be worthless, until it be rehung by one seeing the faults, and is competent to apply the proper remedy. The difficulties besetting the pathway of every builder ignorant of the principles laid down in the following treatise, will be in equalizing the bearings so that the springs will remain level when the weight is added. For, when the bearings are unequal, the springs will turn in, and cause the opening of the leaves.

In our experience we have seen instances of this kind, the mechanics in charge being unable to apply the proper remedy at once, but sought to search it out by altering and re-fitting certain parts, every change serving only to lead further from the right way, thus making the matter worse instead of better; and not until a skilled mechanic was consulted could the defective part be rectified.

With these preliminary remarks we pass on to consider the direction and length of the suspension straps, and the strength of C springs in vehicles having eight springs, and others with supporting straps in front and behind.

EXPLANATION.

1. To aid the reader in comprehending the rule herein laid down, it will be necessary to give some explanations in regard to the technicalities used in the mechanical treatise. These explanations will be found very useful in order to prove the proportion of resistance to the tension of the supporting straps. We will number all the paragraphs, so that the student will more readily understand the whole treatise.

2. *Force or power.*—We apply the term force to all causes capable of producing motion, or moderating the same. There are different kinds of forces. The force which is created or formed by gravity, and is measured by the aid of the pressure, or that of tension, for the purpose of determining weight, is termed the molecular and magnetic forces. We shall, however, confine ourselves in the explanation to that kind of force which is produced by gravity.

3. *Direction of a given force.*—The direction of a force, and the direction of a motion which this force has communicated to a body in a case where this force was the only agent. If a leaden ball were suspended by a thread, and the thread were broken at the extreme end, the ball would fall to the ground in a vertical or straight downward line. This vertical force determines the direction of the leaden bullet in its fall.

4. *Parallel forces.*—We apply the term parallel to such forces as move alongside of one another on the same body. For instance, if we take the case of the hook, which seems to suspend the seven weights, according to square B, Fig. 1, were missing (to give way), the weights would all fall to the ground at the same time, and would describe respectively the vertical lines 1, 2, 3, 4, 5, 6, 7. These lines are all parallel, and we conclude that weights or forces move in parallel directions.

5. *Simultaneous forces* are such as act at the same time and point, where we had intended to produce a given effect. The bearing straps a, b, c, d (Fig. 6), are simultaneous forces. The forces which exert themselves on b and d, have the same action on the bearing straps, if the weight were intended to be concentrated on point b, which is the focus or center of power of the direction of the bearing straps.

6. *Origin and Composition.*—All forces which produce the same effect, as several forces combined, is called their origin; the other antagonistic forces are called composite parts. When we lift the body (Fig. 6) by means of the moving forces in the direction of the bearing straps a, b, c, d, it is evident that the body cannot be lifted in both directions at one and the same time, but will move in the line A B, which is the origin of the forces a, b, c, d. We could, therefore, replace the forces a, b, c, d, by a single force A B, applied in the same manner as to weight in b and d. This single force produces by itself the same effect as those applied in the direction of the bearing straps when the body is in motion. We will now give a few examples of the union and separation of parallel forces by means of the apparatus marked 1, 2 and 3. A *beam* is suspended horizontally by means of a hook, and tapers off like a knife-blade. Commencing at the suspended axis, the beam is divided into six equal divisions; each division has little hooks attached to it on which to suspend the weights, which are also provided with rings to hang them on any hook we please.

If we now suspend seven weights along the whole length of the beam, each weighing one pound, one at the point where the beam is suspended, and the others at equal distances apart, commencing at the center, you will find that it will always hang in a horizontal line. Both sides are therefore subject to the six even forces, which are arranged parallel and symmetrically on both ends of the beam.

If we now take the weights off which were suspended on both sides of the center, and hang them all one under the other on the middle hook, as shown in Fig. 2, the beam will still be horizontal. Therefore, in both cases, we find that the pressure is concentrated on the axis.

Again, let us take the beam (Fig. 1) and divide the weights which it carries into two groups by means of the lines c and d. Now we can proceed as before, and take the weights off each group, and concentrate their force on the points e and f, which are themselves at equal distances from the middle. We have here the result as shown in

Fig. 3, without changing the horizontal position of the beam. By accompanying Figs. 2 and 3 we will find only two weights, one weighing five pounds, the other two pounds, which, being suspended at *e* and *f*, produce the same effect as the weight designated at O. It will be seen that the distance from O *e*, the point of suspension to the point where five weights are suspended, contains two divisions, which means that as many divisions as there are, that there are a like number of weights. Now *f*, and the distance O *f*, contains five divisions, as many as there are hooks at *e*. These terms (words) are given in a mechanical treatise through the following presentation :

1st. Two parallel forces attached to a solid body, having a parallel origin in both are alike in weight.

2d. This attached point of this origin divides the distance of the attached points of these ingredients into two parts, which are in inverse proportion to the forces of this ingredient.

3d. Center of gravity.

The phenomena which are produced by these weights in the mechanical system are often the same when these weights were united at one single point in the interior of the body. This is the point called the center of gravity. This center of gravity of a solid body will always be found to be in the same position as the direction of the origin of the weights of all the parts of the body. For instance, if we place a suspended body P, consecutively in two different positions, as shown in Figs. 4 and 5, the center of gravity will be found at *g*, the position of the thread C D; and in the second position intersects the direction which the thread A B has in the first.

The knowledge of the center of gravity facilitates the operations and divisions which we will have to make in the direction of the suspension straps. We might also concentrate the following weights of the carriage and the persons upon one point, through which the whole origin of the weights and all the parts have to pass.

We will here apply the foregoing origin to a Landau. The weight of the persons in this carriage changes considerably the direction of the bearing straps. All that may be said in regard to this carriage can be applied to that which relates to the direction of the suspension straps.

For the purpose of fulfilling all the conditions in hanging a carriage well in the direction of the suspension straps, it is necessary to calculate exactly the length and strength of the springs, so that all points of the body fall vertically. For instance, if the point *d*, Fig. 6, the shackle head or eye should descend three inches down to *d*, caused by the weight of the persons, it must also fall three inches in front, so that it may remain in the vertical line. In short it is necessary that, being loaded or empty, the body must rest squarely on the wheels.

The direction of the suspension straps and the strength and length of the springs can be determined without resort to the weights and the suspended object, and without knowledge of the style of hanging the weights. Let us suppose that the whole body, together with all appendages, will weigh about 600 pounds. For the purpose of placing it in proper balance we will find the point R at the point O, through *which the origin A B of the weights of all the parts of the body has to go*. This origin passes through the center of gravity (3), and should, above all, be vertical. The direction *a* B and *c* B of the bearing straps fall on the origin *a* B at any point above or below B, just according to the inclination of different views.

The tension, which is exerted on the bearing straps, by no means depends upon their length, but upon the sum of their inclination from their origin. So, also, the forces of all the parts of the body produce the same effect, as shown at *b* and *d*.

The whole weight being stated at 600 lbs. (shown at B), on the lengthened bearing straps, the body will be held where it was first placed. The bearing straps should converge to the point of origin of the weights, and all the parts of the body attracted. For, were it otherwise, if, for instance, it should tend toward a C, c C, or a D, c D, the origin would have to go through one of the points C or D, which are just the contrary of those we mentioned before.

Direction of the bearing straps under the weight of body and the persons.—Taking the weight of the body to be 600 lbs., as indicated at E, and the two persons sitting on the driver's seat at 300 lbs., as shown at F, and the two persons sitting on the hind seat at 300 lbs., seen at H, and furthermore, the two persons on the front seats at 200 lbs., indicated at G, all the forces and weights tend in the same parallel direction. Now, we must ascertain the origin of all these forces, and in order to do this, we must first find the origin of two of them, and from this follows the origin of a third force, and so on until the whole origin is found.

The two forces, F and H, each at 300 lbs. make an aggregate of 600 lbs. This force has for its origin the point J, midway between their distances. This new force of 600 lbs. can be added to the same sum shown at E, by which we get a force of 1,200 lbs., as shown at K, the middle between the points J and E. There yet remains two forces to be added together, one of 1,200 lbs., and the other of 200 lbs. The greater of these is six times that of the smaller one; we may therefore divide the aggregate into seven divisions. If G, K, the point of application, be divided into seven equal parts, it will be found that from the point of application to the whole equal origin will make 1,400 lbs. at L, upon the first division from the point K. The change, which is in the position of the origin, which afterward acts upon the new force, necessarily causes a change in the direction of the suspension straps. In the beginning the point of intersection should be at the new origin, but afterward the increase of the forces causes the springs to bend, and consequently the body descends. Suppose we take the points A and C the heads of the spring would have reached A and C, and suppose the case that the middle of the hindermost shackle-head had fallen down to D, a distance of 3 inches. Now, we will direct or lead C, D, through the direction of the suspension straps to the opposite origin, M, and unite the points at A, M; in this way we have the new direction of the suspension straps. Now we compare these new directions of the suspension straps with the old ones, which will remain different as to length. In the addition of the new forces the bearing straps have altogether changed their direction, but their distance from the heads of the springs to the middle of the shackle-head, will be seen to be the same in the same position. For the purpose of satisfying our conditions, those which we mentioned in paragraph 10, the middle of the shackle-head must be in a vertical direction, b, b, and by means of this we get the length between the connections, of the length and direction of the bearing straps, which we have given at 3 inches; otherwise the length a, b, and c, d, of the bearing straps in the first direction should be exactly like the direction of the length of a, b, c, d, in the second. The length of the front suspension straps is 11 inches. If made of any other length, say, for instance, 30 inches from a, b, to b, and carry the distance over from a to b, in this manner it will be seen that the shackle-head could never go in the vertical direction; for in such conditions the coach would be very badly hung—not only would the body be altogether inclined forward, but it would lean more to one side than our vertical position would allow.

Powers or strength of the C springs.—In order that the two C springs, P and Q, can properly play under the total weight of the body, they must possess the requisite strength, proportionate to the tension of the bearing straps, upon which the body rests.

Now suppose we take a point—e, on the line of origin, L, M, and draw two lines, e, f, and e, g, running parallel to the bearing straps, and two others m, e, and m a. We have a parallelogram e, f, and m, g, of which the sides of f, m, and m, g are proportioned to the tension of the bearing straps, a, b, c, d. In other words, let us suppose f, m, $7\frac{1}{2}$ inches long, and m, g, is $6\frac{1}{2}$ inches, the weight of the body 1,400 lbs. The side m, f, is subjected to a pressure of 750 lbs., and the opposite side, m, g, to a pressure of 650 lbs., which taken together makes the aggregate 1,400 lbs. It follows, therefore, that the spring P has to bear 750 lbs., and the spring Q 650 lbs.

PLATFORM SPRING CARRIAGE.

Before entering in detail upon the description and measurement of the different parts constituting a platform carriage, we will define in a few words what would be expected of a mechanic, of either experience or gifted with a well cultivated eye, to fit him to work out his job in proportion to the body, from a heavy coach to a light calash or six-seat phaeton. A graceful, yet strongly ironed, carriage part, will make a job saleable, for the customer, who is every day becoming more familiar with the requisite qualities of a stylish turnout, will bestow a large share of attention to the finish and proportion of the carriage part. Why is it that we find only a limited number of mechanics of general good judgment capable of forging a carriage in perfect harmony with the body, and tasty in its different parts? Because few have a knowledge of body-making—or the action of a carriage as a machine, designed to carry the body—as some understanding of the principles of machinery will materially assist to combine lightness and strength in the different parts of a carriage—and, lastly, for want of a schooled eye, for patterns are of but little use, except to assist in altering sweeps, as necessity demands.

GEARING NO. 1.

The annexed drawing (Gearing No. 1) illustrates a light fore carriage, with drop pole and shafts, suitable for light phaetons, coupes, Victorias and other light carriages. It was taken from an original French drawing, which is very clumsy in appearance, and remodeled, or, in other words, Americanized to suit the wants of our builders.

Fig. 1 represents the lower part or under section. The new mode of constructing this carriage is the doing away with the bent furchells, and using puncheons in their places; the inside front stay is worked in one piece; in the center is formed the socket to receive the king or body bolt. This stay rests on top of the two puncheons. There is a T plate formed solid with these stays running back to the bed, and at A forms the inner part of the socket for receiving the shafts. The back stay passes around under the puncheons, crossing the bed to the front, and bolted where the front stay crosses the puncheons; the other end extending to the front, forming the outside of the socket for receiving the shafts. A A represents a piece of hickory bolted between the two stays.

Springs $1\frac{1}{2}$ inches, four plates, 37 inches long, $11\frac{1}{2}$ opening, or vary to suit the body the carriage is intended for. Lower bed $1\frac{1}{4}$ by $1\frac{1}{8}$; $\frac{3}{8}$ plate on the bottom. Wheel or stay iron $\frac{1}{2}$-inch, round, increasing in size to the puncheons. The box clips over the bottom bed, with clip bars, are worked solid. The clips are put from underneath the springs, and nutted on top. The size of half fifth wheel 1 inch by $\frac{1}{2}$-inch flat, with brass scrolls riveted on.

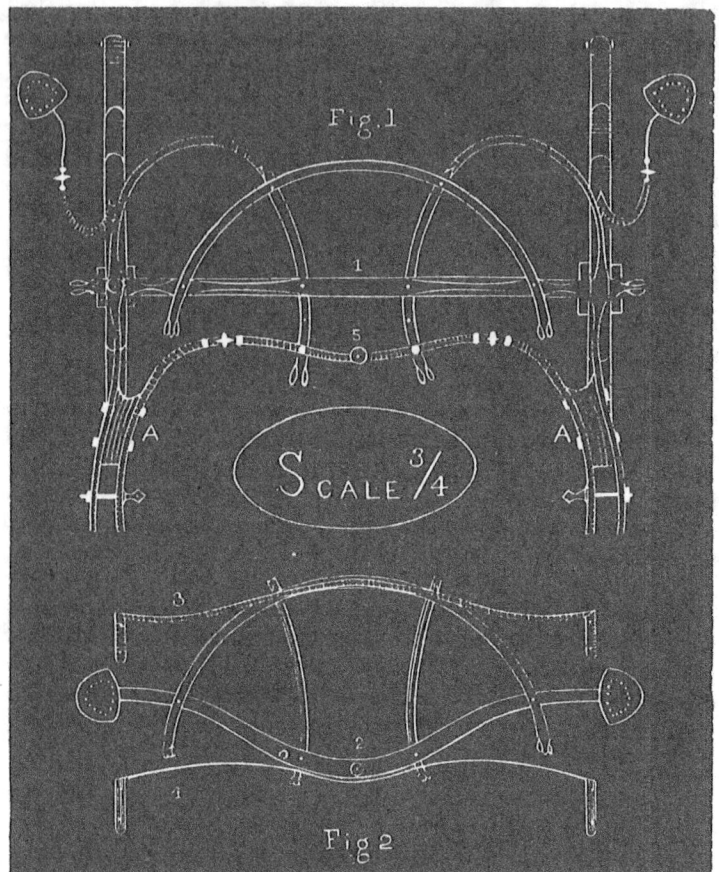

PLATFORM GEARING NO. 1.

Fig. 2 represents the top part of carriage; No. 2 the bent bed $1\frac{1}{8}$ by $1\frac{1}{4}$ in center; $\frac{1}{4}$ plate on top. Step at each end, as shown in diagram, or finish with scroll. No. 3, iron back stay, or back bar. By making this of iron it gives the carriage a lighter appearance. This iron is turned L shape, and bolted to the body. No. 4, the horn strap. The top half of fifth wheel half round iron, with brass scrolls lapped and riveted as shown. Care should be taken and judgment used in proportioning both wood and iron for the body intended for.

GEARING NO. 2

Illustrates a light French carriage, with bent furchells, suitable for Victoria phaetons, coupes, and other light carriages, and is superior to many other carriage parts. It can be used with stiff pole, draw-bar shafts or drop pole. For dimensions we refer you to gearing No. 1—the different parts proportioned to the size of the body intended for.

Fig. 1 represents the lower part; the cross lines at A A is where the stiff bar detaches to receive the shaft or the drop poles—the furchells extending to the cross lines A A. The wheel iron or stay, on the outside of the furchell, extending ahead $5\frac{1}{2}$ inches, and also the plate on the inside of the furchells the same distance. These irons require to be a good thickness, and tapering to the end. The blocks are fitted on to this space, and scrolled at the end, front of bar, which is put on top, as represented

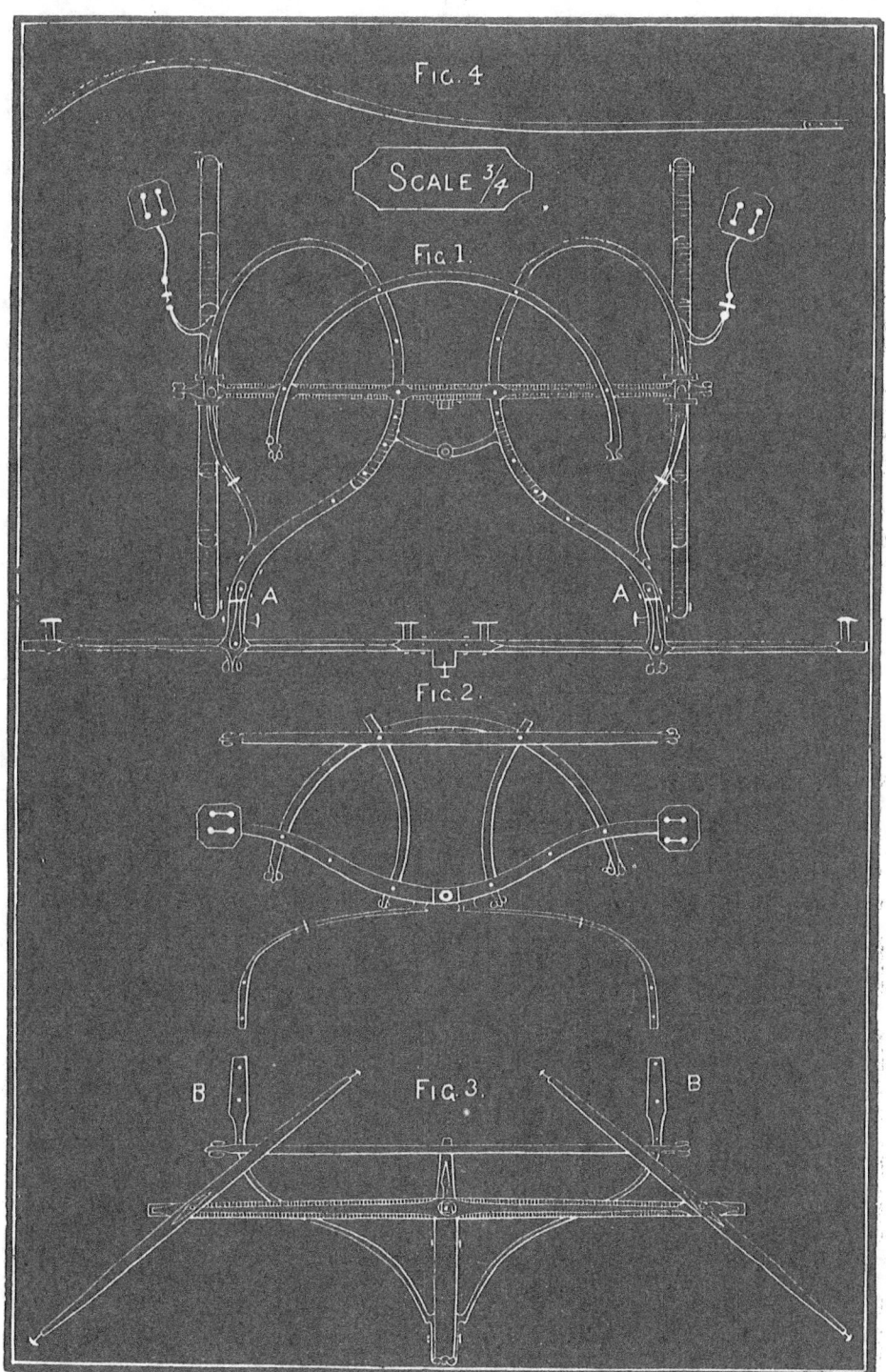

PLATFORM GEARING NO. 2.

in diagram. This bolt should be made with an eye on the inside, to receive the trace of the harness when the shafts are used. The stiff bar is bolted on top of these blocks with a stay running back and bolting through the furchells. The splinter bar is represented in diagram as showing a front view, to more fully show the socket for the pole underneath, and the trace blocks or top of bar. By removing these two bolts, the blocks and stiff bar are detached. It often becomes necessary for the builder to make carriages as low as possible, without lessening the opening of the springs. This can be done by cutting off the bottom bed inside, and butting against the spring, and setting the scroll outside in the same manner. These are secured by a T plate on top from the wheel iron, and also one on the bottom, made solid with the bottom plate of bed. This manner of construction was first used in Paris, and since introduced in this country for its utility.

Fig. 2. The top part of the carriage made as before explained.

Fig. 3. The drop pole. The side stays pass underneath the back bar, clip from the top, and nutted underneath. This stay at the B points is made $\frac{1}{8}$ inch thick top and bottom—the sizes to fit the spaces A A when the stiff bar is detached. It will require a block, between these two irons, and bolted through, as represented in the diagram.

Fig. 4 represents the Tillbury shaft. These shafts are made without a crossbar, ends to fit in the spaces A A, and two side plates, screwed on at the back end, 6 inches long, and also 1 inch oval plate, screwed on the bottom, running about 2 feet from the back end, tapering in front. This plate prevents breaking of the shaft from the side strain.

GEARING NO. 3.

Figs. 1, 2, 3, 4 represent a carriage of a medium size, with a top bed of $4\frac{3}{4}$ curve, which throws the wheels forward accordingly, thereby allowing the boot to be placed nearer the body, and affording more freedom, to give the former a stylish outline.

Fig. 3. Bottom bed. Distance from outside to outside of spring, 3 feet 4 inches; supposing the track to be 4 feet 5 inches, $6\frac{1}{2}$ inches must be added to either side, to correspond with the width of the track and length of the axle. The wood-work of

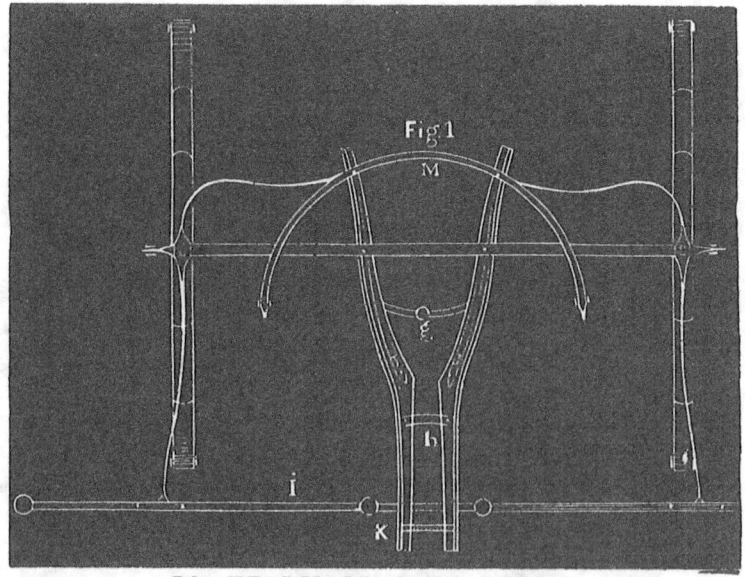

PLATFORM GEARING NO. 3.

the bed has the following dimensions: height in center, 2¼ inches; the same at point A, or the shoulder for the fifth wheel, mounted by a cast-iron socket 1⅝ inches high; height at scroll end, 1⅝ inches; total height from base line *b* to top of socket, including the bend and ½ inch bottom plate, 4⅛ inches; width of the bottom bed, 1¾ inches.

Figs. 2 and 4. Top bed—top and front view. Fig. 4 illustrates the length and height of the bed. The height in the center is 1¾ inches; at point *d* 2⅝ inches from base line *c*. The boot is generally 32 inches wide; the length of the bed is 44 inches, of which 5 inches at either end project outside of the boot and bear the step.

This bed has a plate bolted to the top, terminating ½ inch inside of the straight bearing at *d*, forming in the center the socket *e* for the king bolt, at a thickness of 1 inch, tapering down to ¼ inch at both ends, and rounded nicely at the top.

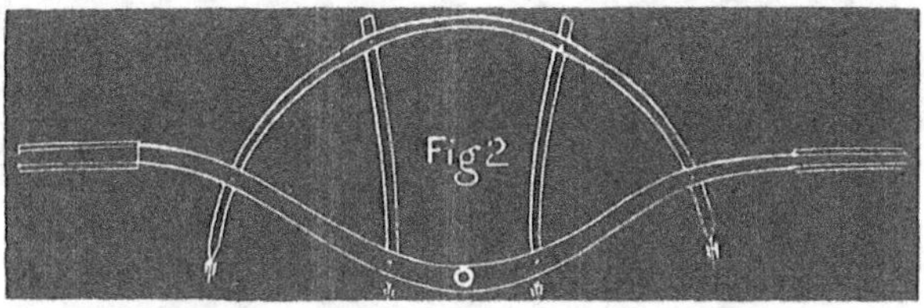

Fig. 1 is a full top view of lower carriage. The connection and dimensions of the iron work of the furchells demand our attention principally in this view. Pole plate *k*, 2 by ¼ inch, and heel plate *h*, 9-16 inch thick, secure the top of the pole. The draw bar *i* has a half-inch plate flush with the outside of the furchells at the intersection with the bar, and forming the rest for the pole. Upper and under plates *l l*, ¾ inch thick, back of the lower bed, and 1¼ inch wide where the fifth wheel rests on, tapering at the top to ¼ inch toward the heel, and wide enough to clear the bead molding;

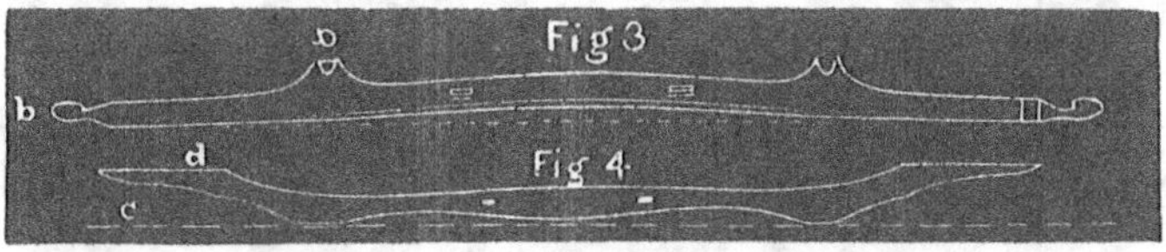

the under plate, similar to the upper one, runs the full length of the lower side of the furchell, tapering down to ⅜ inch in front of the same. The bridge *g*, 1 3-16 by ⅞ inch, welded to the upper plate, completes the king bolt socket. The stays are of a novel plan, and although simplified, combine neatness and strength, two points which the leading blacksmiths of our city make their study, and respond to the qualities set forth in our introductory remarks.

GEARING NO. 4.

Fig. 1 represents the lower part, with springs attached. A is the bottom bed, 1½ inch by 1⅞ deep, arched 1⅛ inch. This bed requires a ½-inch plate on the bottom. B is the draw bar, 1¼ by 2⅛ inch wide in the center, tapered to 1¾ wide at the end. This bar requires a plate ⅛ of an inch thick, screwed the full length of the bar on the bottom. The furchells, 1⅜ thick, front of the bed. These are plated underneath, running the full length. The half fifth wheel, 1¼ wide by ½ inch thick. The C C C C represent the wheel or stay irons. The box clip over the bottom bed, with clip bars, are worked solid. The clips are put from underneath the spring and nutted on top. The size of this iron is ¾ round at the C points, increasing in size and flattening on coming to the

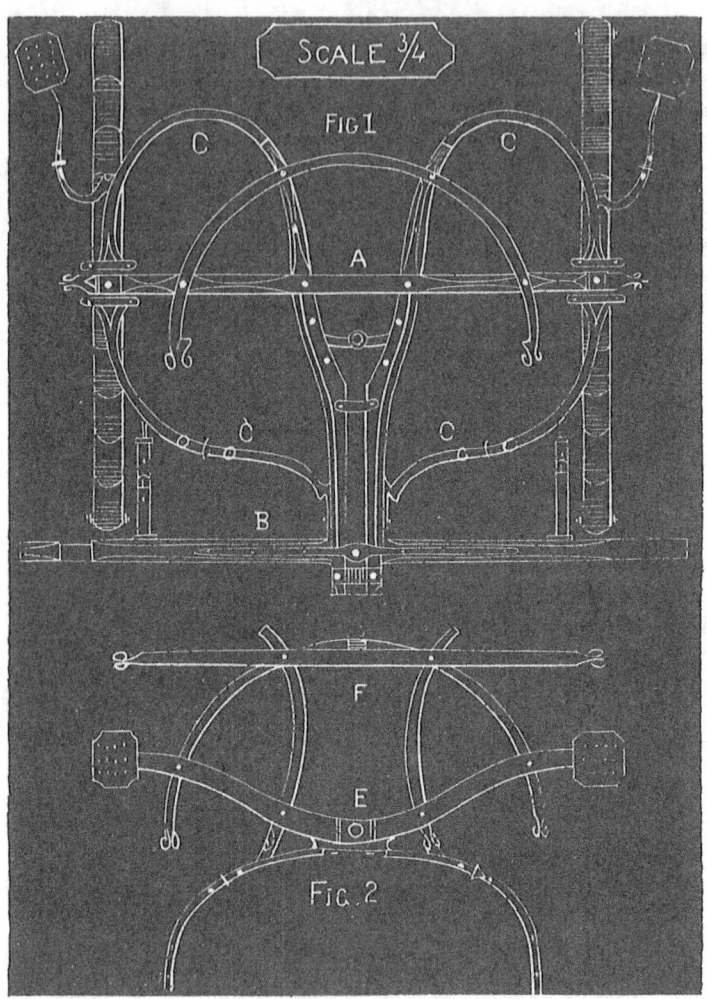

PLATFORM GEARING NO. 4.

bed. It is necessary that this iron should be of a good size at the back part, to prevent the weight of the pole bending it, causing the front of the springs and pole to drop.

Fig. 2 represents the top part of the carriage. E, the front or bent bed, $1\frac{3}{4}$ deep by 1 $1\frac{1}{2}$ in the center, with a $\frac{3}{8}$ plate on the top, steps at each end, as represented in the diagram. Fig. F is the back bar, $1\frac{1}{4}$ by $1\frac{1}{2}$ deep, $1\frac{1}{4}$ square at the ends. This bar will require plating on the front side. Size of the plate, $\frac{1}{4}$ by the full depth of the bar, screwed on by one-inch No. 12 screws. This edge plate prevents the bar from springing in the center, which causes the drooping of the pole and springs. This half fifth wheel is made $1\frac{1}{4}$ half round, with brass scrolls lapped and riveted, as represented in the diagram. Whiffletrees, 34 inches long. Size of pole at the draw bar, $3\frac{1}{8}$ deep by $2\frac{3}{8}$ wide; length, 9 feet from the same point.

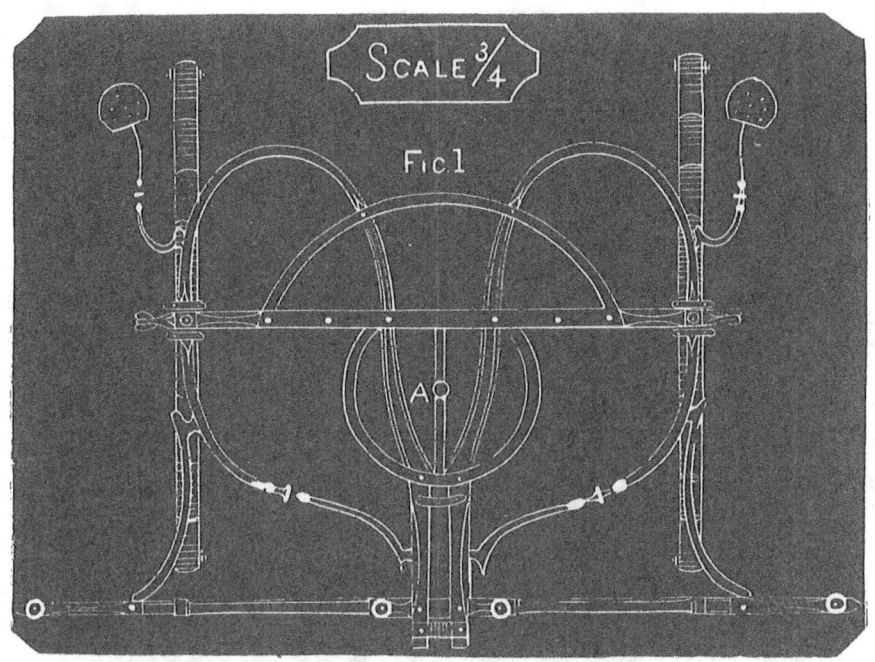

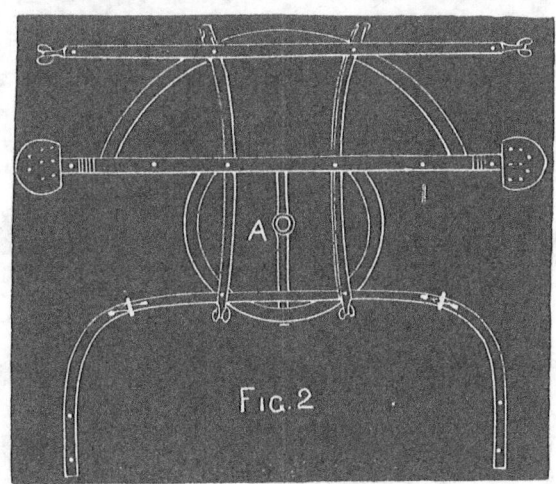

PLATFORM GEARING NO. 5

This carriage is intended for hard service, although it does not have the style of other carriages. With this kind of fifth wheel we get a good bearing when it is turned under the body. These are made solid with a plate on the two beds. Fig. 1 represents the lower part. A A the king or body bolt; for dimensions we refer you to Gearing No. 4, half fifth wheel coach carriage. Fig. 2, the top part of carriage.

GEARING NO. 6.

Illustrates a full wheel coach carriage with bent beds; this carriage-part for **style and** durability cannot be excelled; in whatever position it is turned the bearing **is the** same. Fig. 1 represents the lower part with king or body bolt, 4 inches in advance of center, of ends, of bed and axle. B the body bolt.

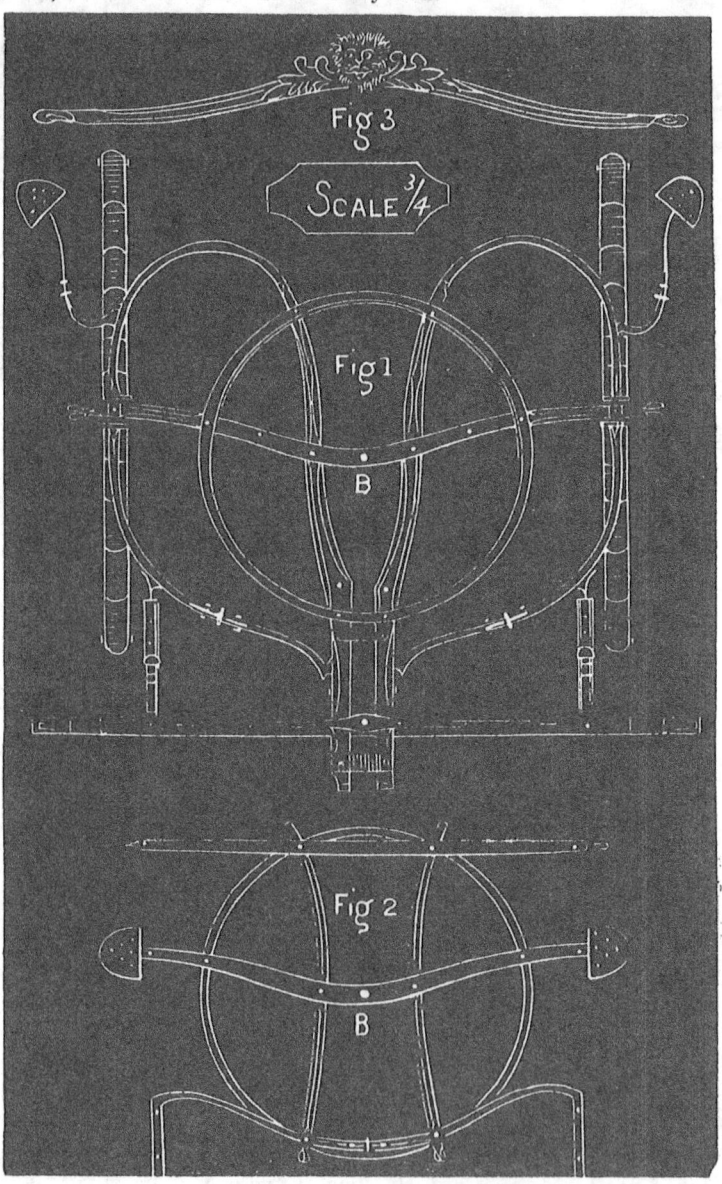

PLATFORM GEARING NO. 6.

Fig. 2 the top part; B the king or body bolt, which passes through both **beds. It is** well to insert in the top fifth wheel oil tubes at intervals, for oiling when **necessary.**

This is done by drilling ¼ holes, and inserting brass tubes with a thread cut on each and screwed in. For dimensions, we refer you to half fifth wheel coach carriage, plate No. 4.

Fig. 3. The back view of a back bar, with a lion's head in the center, the leaves branching from it. The top part of the head should be layed well over the bar, and the chin to the lower edge. The design should be small, and cut deep to give a bold look.

GEARING NO. 7.

is intended for a drop pole and shafts. This can be made as light as desired. The dimensions we give are for a four passenger phaeton.

Fig. 1 is the back bar, 1 inch thick, swept up 1½ inches ie the center.

Fig. 2 represents the lower part, with the springs attached, springs 1½ inches, three plates 37 inches long, 11½ inches opening, or varied to suit the body the carriage is intended for. No. 1, the lower bed 1¼x1⅛ in., ⅜ in. plate on the bottom. A is the wheel or stay iron. The box clips over the bottom bed, with clip bars, are worked solid.

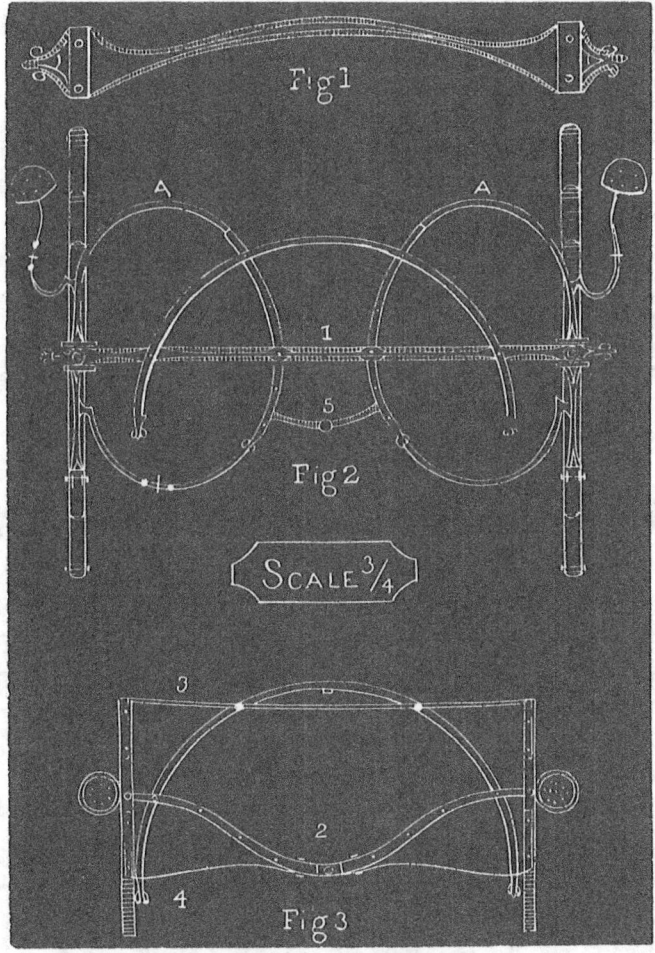

PLATFORM GEARING NO. 7.

The clips are put from underneath the spring, and nutted on the top. This iron passes round under the nutting bars, and lapped in the center of the bed; size ½ inch at **A** point, increasing in size on coming to the bed. No. 5 is the cross stay that receives the

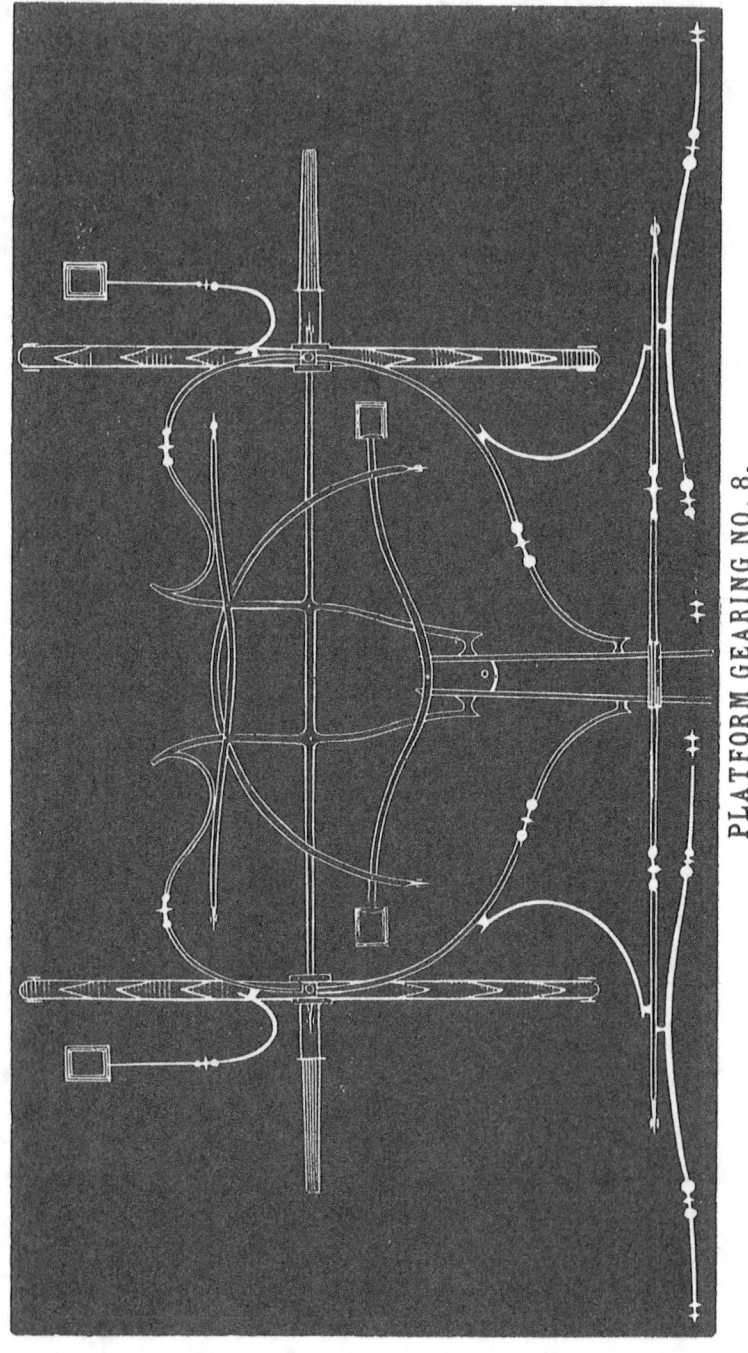

PLATFORM GEARING NO. 8.

BLACKSMITH DEPARTMENT.   129

king or body bolt. This stay is put top of nutting bars, bolting through the wheel iron, as shown in the diagram.

Fig. 3 represents the top part of the carriage, No. 2, bent bed ⅞ inch deep by 1¼ inches in the center, ¼ plate on top step at each end, as shown in the diagram No. 3 iron back stay or back bar. By making this of iron, it gives the carriage a lighter appearance; this iron is turned L shape, and bolted to the body. With this kind of finish we insert a collar between the body and front bed, in place of a body block, which gives a much lighter finish.

Fig. 4. Horn stay. Half fifth wheel, 1 inch iron, half round, with brass scrolls lapped and riveted, as shown in diagram.

GEARING NO. 8.

The carriage part is all iron, and for originality and beauty is hard to be excelled. It is well worthy the study of manufacturers and workmen. It is all iron, with but one bolt in the carriage part, that is the king bolt; all other parts are forged solid or welded together, so that it makes but two pieces in the whole front.

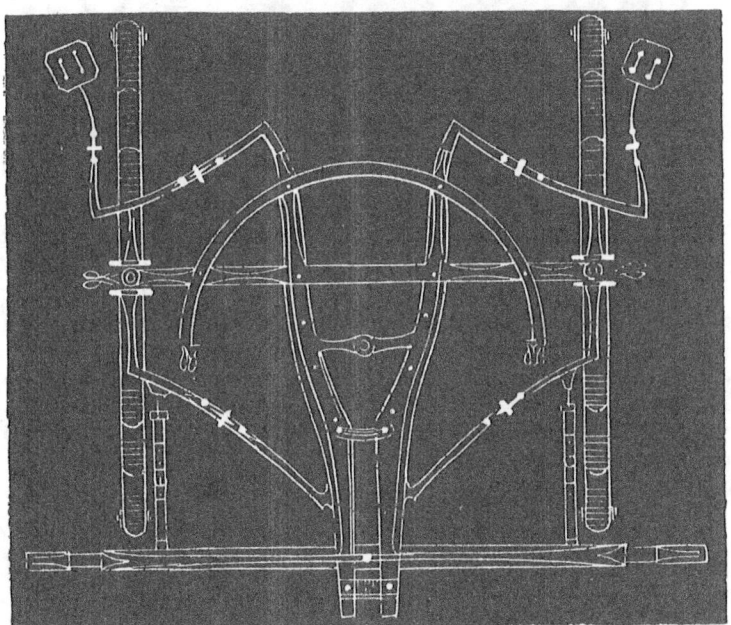

PLATFORM GEARING NO. 9.

At the first glance, it may be said it has a stiff appearance, by the sweeps, which are nearly straight; but the same opinion prevailed at the first introduction of straight joints for carriage tops. Now it is difficult to find a good builder who uses anything else. This carriage to be proportioned to the body intended for.

GEARING NO. 10.

FRONT RUNNING PART ON LOOPS FOR A PARK PHAETON.

As a novelty in carriage parts, we have selected one that has only one piece made of wood, viz.: the spring bar; all the rest is made by the blacksmith, and therefore of interest for all that have a call for iron architecture. It is light but strong, and if properly executed looks stylish. It is designed for a light phaeton, or cabriolet, for two seats.

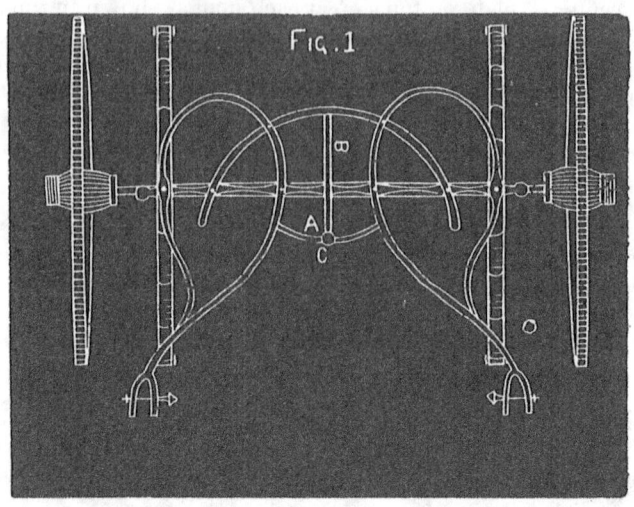

PLATFORM GEARING NO. 10.

Fig. 1.—*Top view.* In the place of a bent furchell we have a similar stay, as represented in plate No. 1, although the manner of connecting these stays with the spring bar is different. We also do away with the puncheons. The head block A is forged of the best quality of iron, ⅞ by 1 inch, and has ¼ of an inch added at the socket for the king bolt. It is finished with a flat surface at the top and bottom of scant ½ inch, and the sides rounded to. A bridge B is welded to it, starting at the socket, and resting on the top fifth wheel, supporting the middle loop, which steadies the bearing at the back end of the carriage. The stays for the reception of the shaft have a cross piece at C welded between, forming the lower socket for the king bolt. The dimensions of this cross piece are 1¼ inch at the start from the stays, by ⅞ thick.

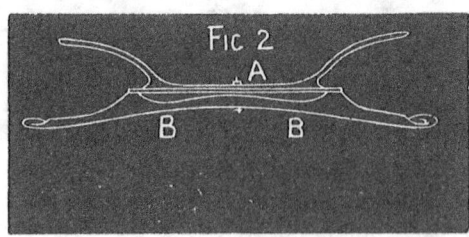

Fig. 2.—*Front elevation.* Showing the shape of spring bar, ¾ by ⅞ inch, being plated at the bottom ½-inch. B B shows where the stay iron is bolted to the wooden bar. The way the iron or horn bar A is raised at each side is also shown. For measuring use the ½ inch scale.

Fig. 3.—*Side elevation.* Represents wheel, springs and loops, the bend of the shaft stay, and the manner in which the center loop and bridge B is connected; it forms one solid piece of the head block, or horn bar, and the center loop, playing square on the spring.

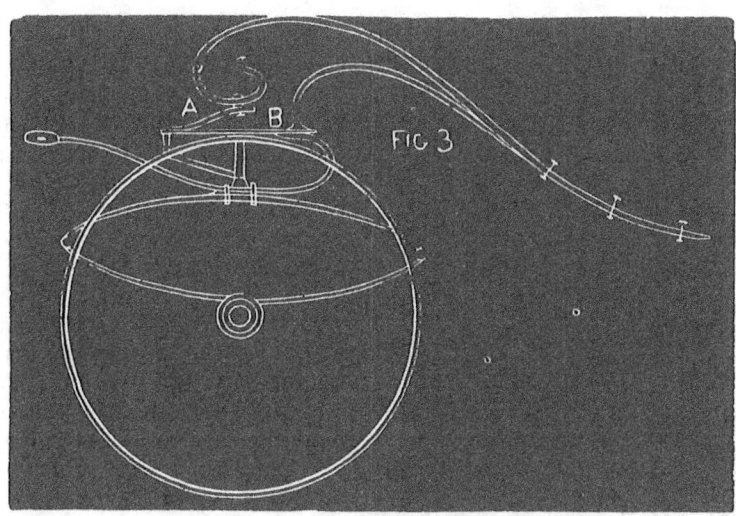

Fig. 4.—*Top view.*

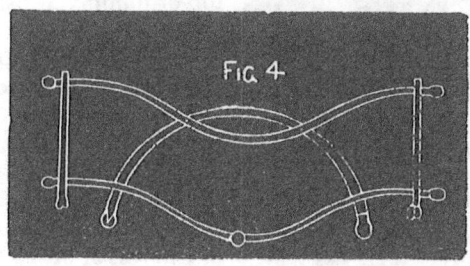

GEARING NO. 11. STAYS FOR FOUR-SEAT ROCKAWAY DROP POLE.

This design for stays is intended to match the present style of straight lines, English quarter, etc.

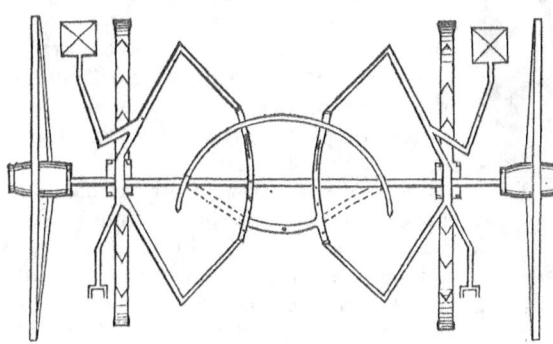

PLATFORM GEARING NO. 11.

As novelty is now sought eagerly by carriage-makers, the pattern here given will doubtless attract attention, and find those who will copy it. The principle of carrying out a design on every part of the carriage is as correct an one in regard to lines as in that of colors; and those builders who pay greatest attention to the matter of harmony between the colors of the painting and trimming, and of the distribution of similar lines over the entire carriage, have taken hold of the secret of producing pleasing work. The angular lines on the drawing here given fall into their places very naturally. To have them perfect in all their parts when completed will require considerable labor in forging and filing, aside from the mental skill of deciding on the proportions the parts should bear to each other.

GEARING NO. 12. CARRIAGE PART ON C AND PLATFORM SPRINGS WITH IRON PERCH.

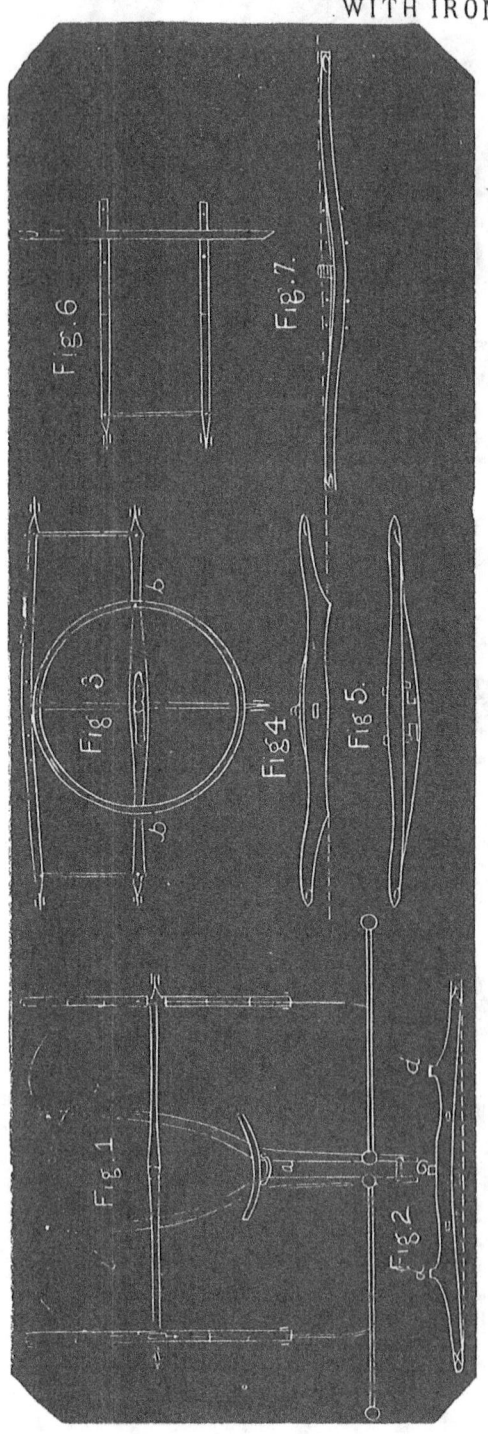

A carriage part on eight springs is the most complicated job of all similar or plainer jobs, and it would take up a great deal of space and time to describe it minutely; we have shortened the task without neglecting to give a full description of the principal parts of both wood and iron work.

The length of the perch is generally 7 feet 10 inches between the axles. The thickness $1\frac{3}{4}$ inches, and depth 2 inches, are nearly maintained all along the perch, with the exception at the end, where it tapers down to $\frac{1}{4}$ inch. The front springs are 20 inches, and the hind ones 21 inches apart; the platform springs are with a solid top or dummy, finished oval.

Fig. 1.—FRONT CARRIAGE PART (LOWER) TOP VIEW.—The fifth wheel $1\frac{1}{2}$ inches wide, 7-16 inch thick, is bolted to the furchells, $30\frac{1}{4}$ inches from front end to the middle of the bed, and 13 inches from there to back end. The lower bed is 2 3-16 inches wide in the center, $1\frac{7}{8}$ inches at the bottom and $1\frac{1}{2}$ at the top. The top plate 5-16 inch, ends at the rest of fifth wheel; the bottom plate $\frac{1}{4}$ inch thick, runs the whole length of the furchell. Stop a arrests the turning of the fifth wheel at nearly a quarter turn by the closing in of shoulder bb, on upper fifth wheel, Fig. 3. As then the act of turning commences, the carriage describes a circle, twice the length of reach from center to center of axle, the width of track counted in.

Fig. 2.—FRONT VIEW OF LOWER BED.—The curve of the bed is $2\frac{1}{4}$ inches from straight line to bottom plate. The depth of bed is $2\frac{7}{8}$-inch in center, socket or rest dd for fifth wheel $\frac{5}{8}$ inch high, and in line with the middle of top plate $\frac{1}{2}$ inch thick, with a thimble e 1 inch by 3-16 inch thick, fitting tight into a like one, let in the lower plate of top bar.

Fig. 3.—FRONT CARRIAGE PART (TOP) TOP VIEW.—The width of top bed is $2\frac{1}{4}$ inches in the center, and $1\frac{1}{2}$ inches at the end. The back bar is $1\frac{5}{8}$ inches wide in the center, diminishing slightly toward the end;

plates bolted to top and bottom. This bar rests on the top of the perch, and has a strong stay 1 inch thick and ⅝ inch wide (see Fig. 5, front view,) bolted to the bottom of the perch, while the bottom plate of the bar is secured to the top of perch; for this end the perch has two short cross plates welded to top and bottom for the reception of the bolts, fastening the perch to back bar and stay. At the end of the bar, and right under the middle of the C springs, passes a straight stay to top bed, keeping them firm and receiving the bolts for the lower end of the C spring. The fifth wheel, 24 inches in diameter, is fastened at the back by a clip fitted around the perch; the front part of the wheel connects by a bolt passing through a roller with the puncheon. Top bed, front view, Fig. 4, is 2¾ inches deep in center, 2⅝ inches where the fifth wheel passes, and is bent up 3½ inches from a straight line to the end of top plate. The back bar, Fig. 5, is 2 inches deep in the middle, and 1¼ inches at the end.

Fig. 6.—HIND CARRIAGE PART, TOP VIEW.—This consists of two straight cross bars for the support of the back C springs. The straight stays, as on the front carriage, are bolted underneath the ends of the bars, keeping them square. These bars are 1⅝ inches wide, with ½ inch plates. Depth of bars 1⅞ inch.

Fig. 7, front view, shows the curve of 1¾ inch from a straight line at the top. The perch rests on the top of the bars, and is bolted to the top plate by short cross plates welded to the bottom of the perch, which latter is perfectly straight to within 8 inches inside of front cross bar, and thence swept to follow the bottom side.

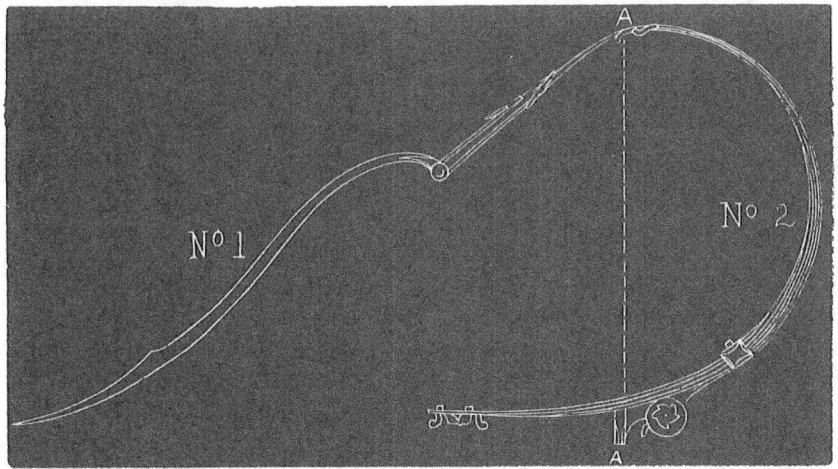

The main feature of this job lays in the application of the body loops. Shape and position are of an endless variety; and through the lack of a certain rule, the issue depends wholly on the skill of the workman. Practice in scale and free-hand drawing, and a sharp eye, facilitate the work to a great extent. The loops are swept after the draft, and fitted to the body, which, for this purpose, is turned upside down. The loops might then be easily laid on, instead of holding and pressing them under in the reverse position. Care must be taken to have the body level; then measure from any point at the back of the body to the head of the loops crosswise, to bring the loops square. Then ascertain whether the body frame is square, by measuring the same cornerwise; a slight difference might be straightened through the fitting of the loops. The space between (across) the loop-heads must correspond with that between the hook of the C spring; and further, the first must have the same width as the latter, to avoid the play of the strap. The front loops are generally longer than the hind ones, and bent in such a manner as to shorten the strap, and to do away with any unnecessary swing of the body between the front C springs. If the body hangs too low in front, bend the front loops down; and in the opposite case, if raised too much, bend the loop upward, without altering the length of the strap. In any case, try to bring the body as far back as possible between the hind wheels without interfering with the opening of the doors, as it always crowds toward the front.

Dimensions of the loop: at the butt, 2 inches wide, 1⅞ inches thick, the head 1⅝ by 1¼ inches. (See No. 1.)

A detached or dickey seat, on stays, rests with the back stay in a square socket, fastened to the front pillar. The front or supporting stay of the heel board is screwed to an arm, extending from the front loop-head, and kept firm by a square shoulder fitted against a corresponding projection on the inside of the loop-head. The rocker plates should be ⅜ inch thick, and 3 inches deep, but might vary according to the depth of the rocker. They are made of Ulster or common English iron, and fitted to the inner side of the rocker.

A careful workman will avoid bringing a red-hot plate too close to the wood, and do the rest of the fitting in a state of heating, just bearable to the hand. The holes should fit No. 20 screws, and be well countersunk. The ends of the plate, front as well as back, are bent in an angle to fit against the front and tail bar, 5 inches long, tapering down at the end, and well rounded. A rocker plate should run all along the bottom side, and should be as solid as possible at the bearing point above the top bed and cross bar. The C spring, No. 2, has six plates, 2 inches wide and ¼ inch thick. Parts of this spring are the band, 2½ by 3-16 inch; the jack, whose back or upright part is bolted to the back bar; the top part follows the sweep of the spring, and slips between the bottom of the spring and the band; a bolt passes also through here. Line A A, plumb from the tip of the spring to the back bar, aids to obtain a graceful and correct curve for the spring. Lastly, the body is hung upon the carriage part, finished previously, the springs tried again, whether they bear square, and the job finished.

GEARING NO. 13.

LIGHT PLATFORM SUITABLE FOR CUT-UNDER BUGGY.

Fig. 4 represents the lower part, with springs attached, the bottom bed of iron (making the entire carriage of iron), on account of lightness of appearance. It can be taken 1 inch square, lightened down, except the bearing of the fifth wheel, which is shown by Fig. 6, which is the front view.

The scrolls for the end finish can be of brass, lapped and riveted under the box-clip.

The Manner of Producing these Scrolls.—Take two pieces of whitewood, thickness required, doweled together (scroll to be cut to taste), with a running lap, lapping to the iron. When the pattern is complete, can be taken apart in center for the molder to produce his castings. By this manner the builder can retain his own pattern of scroll in brass as well as in wood, which is generally practiced by good builders, and their work is known by its finish.

Half-fifth wheel and iron pump-handle scrolls are produced in the same manner, which is represented by Fig. 7, made in the same manner.

Fig. 5 represents the top part of carriage, with an iron bent bed; can be swept to suit the turning of the wheel under the body.

In Fig. 4 the side wings are made whole, 9-16 round at two points, inceasing in size and flattening at top to receive (3) the bridge, which is bolted at the top and finished off even with the side wing. Half-fifth wheel, ⅞ by ⅜.

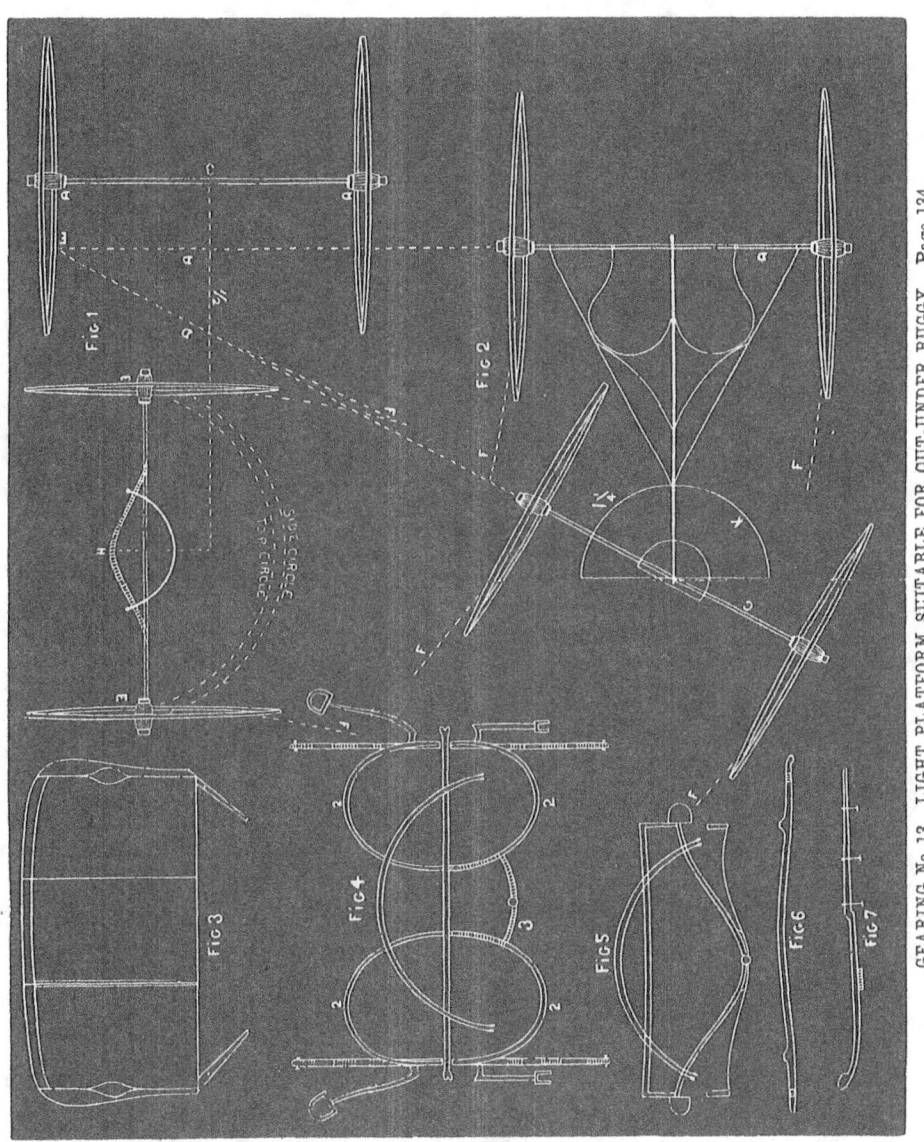

GEARING No. 13. LIGHT PLATFORM, SUITABLE FOR OUT-UNDER BUGGY. Page 134.
SHORT AND EASY TURNING. Page 135.

We have given dimensions in former diagrams, but it requires a good smith with good judgment to proportion—to give strength where needed, and to lighten down at other points to produce a light and durable job.

SHORT AND EASY TURNING.

To bring a carriage into a different course from a straight one, requires a circular motion at a half turn a carriage has established itself in a right angle to its position at rest.

A two-wheeled vehicle turns on one wheel, which forms the center at the place where it touches the ground; the opposite wheel forms the circle of said center. The body in this instance follows the circular motion exactly as the axle, and consequently retains a steady position above the wheels.

A four-wheeled vehicle remains in a straight line, when first the front pair of wheels are turned under, then by the effect of the draught as the hind pair of wheels follow in a wider circle. To effect a turning, we bring the front axle first in a corresponding direction with the desired turn.

We make distinction between the moment of turning, or the angular position of the axles, previous to the turning itself, and the effected turning of the vehicle around a center or king bolt, according to the construction of the carriage part. The wheels have to be brought in a position corresponding with the direction of the turning. The body must be fully supported after the turning, and the front or dickey of a carriage to stand in a right angle to fore-axle.

We have to consider a few points relative to the height of the front wheel, and the elevation of the body above the ground, which averages generally 30 inches. To give a front wheel its proper height (between 3 feet 4 inches and 3 feet 6 inches), and have it turn a full circle, we sweep the body at the required place, viz.: put in the wheel-house of a proportioned length and a depth between 3 and 4½ inches.

The front carriage part is fastened around the king bolt, turning that part horizontal. This action causes the front wheel to describe a circle, whose diameter is the width of the track; but as the wheel leans over at the top through the dish, we have a larger circle in the middle and top of the wheel. We therefore find first a top circle, having a diameter equal to the width between the highest point of the wheel, and a side circle following the termination of the cross diameter of the wheel, having as a center the king bolt.

We refer now to the diagram, Figs. 1 and 2, showing the difference between a full turn at a four-wheeled carriage, and a broken turn at a two-spring, reach gearing.

Front wheels E E turned half way, or in a right angle with the dickey; points D D or top width of the hind wheels will, after the application of the draught, follow the turn of the front wheel parallel with circle F F, having its center in G. Top and side circle center in H.

Looking next to Fig. 2, we find by comparing the longer circle F F to the corresponding one in the foregoing illustration, that it takes considerable twisting and jarring to effect a short turn. Lines C D and B D, at their intersection E, the center of the circle, require such a space to turn fairly. This is a reach gear causing the hind wheels to turn around the same center, shown in lines F F. We have applied a segment K to show how near we can get to a quarter turn. This applies to most of the light bodies without a demicut or wheel-house. Both figures drawn of the half-inch scale.

Fig. 3, a neat pattern of a dash, 15 inches high, having a handle formed by the inside rail, welded to the outside rail. This dash might be covered entirely, leaving the handle open, or from the second rail.

THE IRONING OF A CARRIAGE PART.

The carriage part of a rockaway is a fit specimen for plain ironing, inasmuch as it is closely connected with the higher order of four-spring and eight-spring carriages.

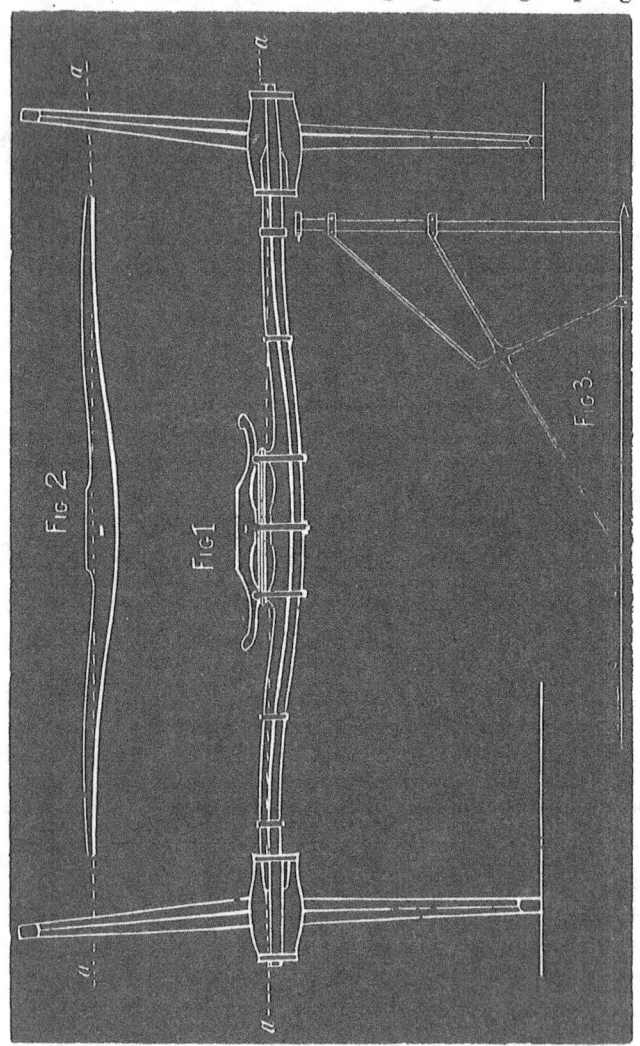

Our job, drawn out, has bent axle beds, wheels 3 feet 4 inches by 4 feet, hub 7 inches by 4⅜ inches, 1¼ by 1⅜ inch felloes, spokes 1⅜ by 1 inch.

The axle bed serves as a pattern, not only for the shape of the axle, but for the width of track, which is calculated to be 4 feet 8 inches. Supposing a smith would have to iron a carriage different from such as he has already patterned out, then he must set himself to work out just such a drawing as the one in Fig. 1. It is a medium bent axle, but just enough to give it a graceful curve at the shoulder of the axle, which would degenerate into an ugly kink if bent shorter.

The axle arm is 7 inches long, even with the back of the hub and the flange of the nut let into the front of the hub. It has scant $\frac{3}{8}$ inch taper, while the wheel has $\frac{1}{2}$ inch dish. This gives 4 inches more width at the top than the bottom of the wheel, and brings the lowest spoke plumb. As the welding of an axle shank is familiar to every smith, we need only say that the bottom side of the axle is slightly rounded and left square at the places designed for the clip. The reach generally runs parallel with the bottom of the body, at a distance between 10 and 11 inches from the latter. The hind wheel is 4 inches higher than the front one. Bodies of a straight or slightly curved bottom side have these 2 inches, or half the difference in the height of wheels made up by the lessening of the bend of the hind axle $\frac{3}{8}$ inch, shown by dotted lines a, a (Figs. 1 and 2). The bearing of the fifth wheel clip $1\frac{1}{4}$ inch, the thickness of upper and lower fifth wheel, made of $\frac{3}{4}$-inch half round iron $\frac{7}{8}$ inch, and the mortise in head block $\frac{3}{4}$ inch above fifth wheel, making $2\frac{3}{4}$ inches in all; from which deduct $\frac{3}{8}$ inch less bend of hind axle bed, further the mortise in hind axle bed, being $\frac{3}{8}$ inch higher than that of head block, measuring from bottom of head block and axle bed, which leaves the two ends of the perch at the same distance from your base line. This answers for our special case. Very irregular bottom sides demand crooked reaches; one tenon as much as 5 inches above the other, then the shape of the box loop regulates the space for the spring.

The iron plate, lining the under side of the perch is made of 1-inch half-oval iron, left flat 12 inches from the head block, where it is welded to the plate of the head block, and gathered up to 5-16 inch in the middle, forming the socket for the king bolt. Measure angling from head block to end of hind axle bed to get it perfectly square. The flat space of the reach plate, mentioned before, forms the surface for the fitting of the fifth wheel clip, bolted to it; also the king bolt stay. The reach plate also forms a clamp at the back end, fitted tight around the hind axle, running up to the bottom of axle bed, rounded and finished half oval.

Place a strong straight edge under the wheels, and measure up to the axle, at equal distances, to get your curve true. The springs are made to range, and to stand square, by measuring to end or bolt from a straight edge placed on the floor. This done, measure crosswise, from bolt to bolt, to get them square. The springs are set drooping $\frac{1}{2}$ inch to have them come plumb after the body rests on them. The spring bars are handled in the same manner, and in line with the face of spring. The two bolt holes weaken a spring bar; and in good shops narrow clips are used in place of the bolts. Fig. 3 shows the manner in which the stays, made of half-round iron, are applied. The cross rods are steadied by small scolloped flanges at their intersection, of the same thickness as the stays, and round edged. The long stay strikes the reach 26 inches from the hind axle.

SKELETON BOOTS.

This sketch represents an angular and straight lined Landau. We will give a brief outline only, with the chief points of government, which are about as follows:

W represents the front quarter (as per sketch) of the Landau; the obliquely dotted line represents the main front bow (let down); the continuation of the same O is that part which forms the upper portion of the doorway, when the top is up; the vertical, obliquely dotted lines P is the position occupied by O, when letting the top down. The dotted lines B and C are the two remaining bows. X is the front post or pillar to which the hinge is adjusted, Y is the case for the front glass frame, when the top is down; the circular dotted line R denotes the circle described by the lower bow C, in raising or lowering the top; the dotted line H H the seat valance or skirt, N the seat rail, L the front of boot or bracket, E front seat leg, D back seat leg, E and M, where the boot is attached to the suspender or loop G, by means of bolts K K—bolts passing through the loop and securing the same to the front gear T, scroll finish on extreme front end of loop, U U U bolts securing loop to body.

Having explained the outlines of our sketch, we will now proceed with the construction. The body must first be suspended from behind in its natural position. We then place the stool under the front of the body, and have the front at the proper elevation. The next step is to place the front gear in such a position, so that when it is turned that there will be space enough between the wheel and loop to prevent collision when the carriage is employed.

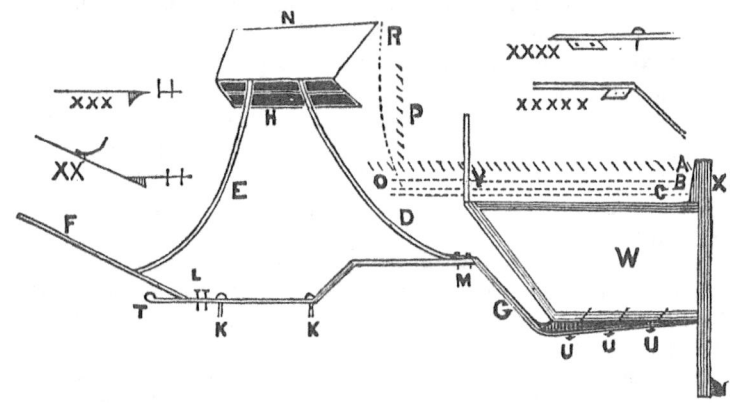

This much having been done we proceed to construct our loop, which, for a job as shown us, would require 1¾ inch square iron (best American iron), swaged oval, leaving or presenting an edge ¾ of an inch wide, forming the base of this size and gradually tapering the same to the center of the arch. Make the front end of 1½ inch square, and swage oval, welding both, when properly fitted, at the center of the arch. For the support of the boot or seat there is a flange forged upon the inside of the loop, at or near the back end of the arch, as shown by the sub-sketch X X X X X. The support for the front has, also, a flange forged on the inside in front of the bolt K, as shown by sub-sketch X X X X, and also concaved on extreme end, to allow of fitting scroll finish; sub-sketch X X shows the method of getting out the front leg and bracket; sub-sketch X X X the back seat leg. Before putting up the seat legs they should be nicely fitted to the flanges, so as to procure precision. E and F are in one piece, and secured to the seat frames by means of bolts; D is a separate piece, and also secured to the seat by means of bolts. The distance from the upper point of F to the front of the seat frame ought not to be greater than 22 inches, the seat to set enough above the body to secure harmony. The pitch of the bracket is regulated by its connection with the loop, for by extending the front flange forward we elevate the upper point of the bracket; by shortening, we depress. The position to be occupied by the back seat leg depends upon the top, as does, also, the position of the seat. The seat leg must occupy such a position as to allow of that portion of the top marked O to become horizontal when the top is down, without, in any manner, interfering with the seat leg. The seat must occupy a position that will not allow of the bows B and C coming in contact or colliding with the seat skirt.

Make the back seat leg 1¼ inches by ⅞ inch oval. Make the front on the bracket part, that takes the foot board, 1½ inches by ¾ half oval; the part E 1¼ inches by ⅞ inch oval, and the lower portion, which connects with the loop, 1½ inches by 1 inch oval. Before securing the foot board, the seat legs should be finally finished and thoroughly secured to the loops. The foot board may be then fitted and secured with bolts to the bracket, being careful, however, to have a strip of band iron, one-eighth of an inch in thickness on the upper or inside of the foot board, to prevent the bolts from turning and the foot board from warping and splitting.

The sketch upon which we have operated, as before mentioned, was an angular Landau; hence, our presenting the angular-shaped loop, etc., as shown in our sketch.

IRONING PLATFORM EXPRESS, MEDIUM WEIGHT.

In the accompanying cuts we give sectional views of the most important parts of the iron work requiring to be forged. The sizes of iron to be used may be found in what follows:

Fig. 1 furnishes a front view of splinter bar arranged for both pole and shafts. Fig. 2, splinter bar designed for shafts only. The first one mentioned is very ingeniously arranged, and is of great value in point of convenience. Fig. 3, furchell stay and splinter

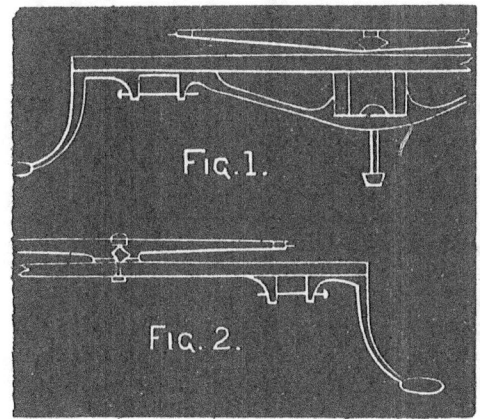

bar and iron, showing the manner of lapping and bolting the irons together. Furchell stays, 1⅜ by ⅞, swaged in center one inch oval. Splinter-bar iron 1½ by ½, to run back under furchell stay as shown. Fig. 4, top view of splinter-bar iron, which is 1½ by ½ inch. Fig. 5 shows the step and connecting irons. Steps, 15 inches long, to be tapered from bend, which will be 1½ by ½ to ¾ round at tread. The tread, 6 inches, round. Transom plates, socket center, one to work in the other bearing about 2½ inches. Bars bolted to body with ⅜ or 7-16 bolts. Fig. 6, section of body with double side stays and strap loop. Inside stay, ½ inch. Outside stay, ⅝, to be well tapered from the bar. Shank through bar, 7-16 under shoulder, and ⅜ for the nut. Inside straps, ¾ by ½ inch round, with shank to go through sill and nut on bottom. Fig. 7, front view of stay. Fig. 8, crab for springs at middle bar. It is made by welding across from the lugs and bent down similar to a shaft box. Fig. 9, crab for ends of springs. Take 1 by ⅜ iron weld across and punch the holes 2½ inches from center, all around, and bend two ends each way. Fig. 10, handle T-shaped at one end, and an eye at the other, swelled in center to ⅝, taper to ⅜. Fig. 11, toe rail ⅞, half oval, placed 2 inches above toe-board at center, with slight arch to ends of toe-board. Fig. 12, shaft iron, the iron at heel 3 inches from shaft cross bar. Having noticed the cuts as here given we would again call attention to the body. The tail gate should have No. 6 sheet-iron instead of wood panel hinges, 1½ by ¼. Rod, 7-16. The top rail of body requires 1¼ by ⅛ band iron all around, the

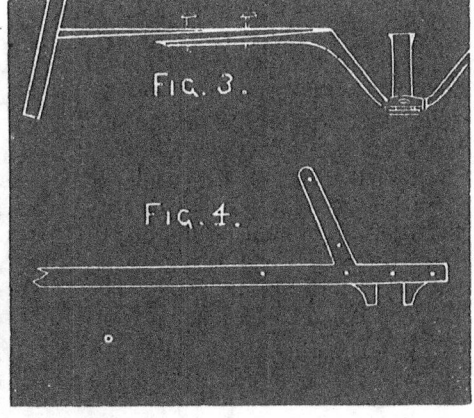

same to be screwed down. Back stays of body and hinges of tail gate supplied with

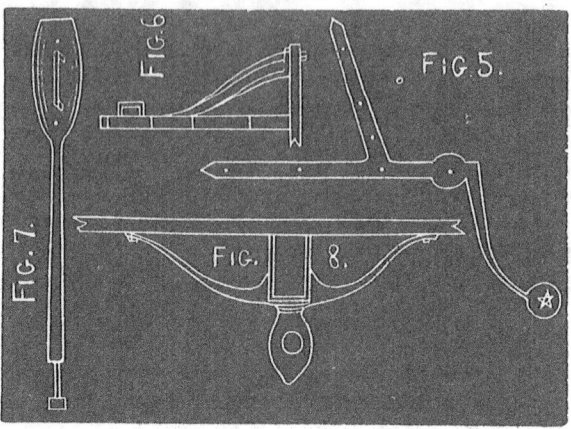

staples for strap, 2 inches in clear, ¾ wide and ¼ thick. Body irons should be fastened

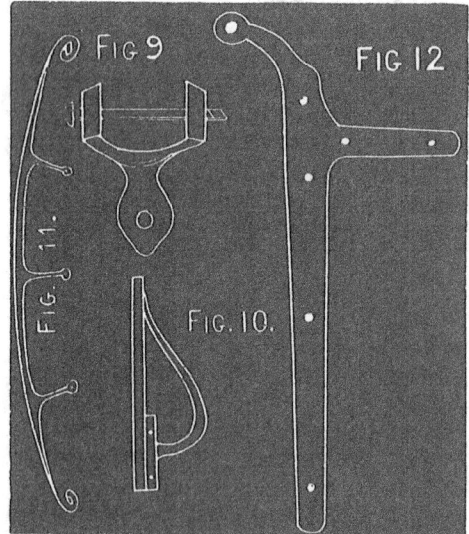

with 3-16 button-head rivets. Body bolt ⅞ round. Hub bands, front 2¼ by 3-16, back do. With sand bands, 1 by ⅛ inch; without sand bands, 1½ by ⅛. Fifth wheel, 1⅜ by ½ inch; diameter, 24 inches. Block from fifth wheel bed to furchell stay, 3 inches high, 1¼ inches thick. Side springs, 39 inches long, 2 inches wide; five plates, with hole 3 inches out of center. Cross springs, 40 inches long, hole in center. Spring, clips ½ inch square iron. Spring blocks should be 3 inches deep, and hug the axle about ¼ inch. Tire to project fully 1-16 of an inch over tread of felloe; or rather the felloe should be tapered a full sixteenth. The body to be hung an inch lower behind than front. Axles, 1⅝ inches. Axle arms 8⅛ inches. Tire, 1½ by ½ inches. Wheels 3 feet 3 inches and 4 feet 2 or 3 inches.

IRON BACK BARS FOR A LANDAULETTE.

For a carriage where lightness and a decidedly stylish effect is sought after, this method will be found far superior to those taken from wood.

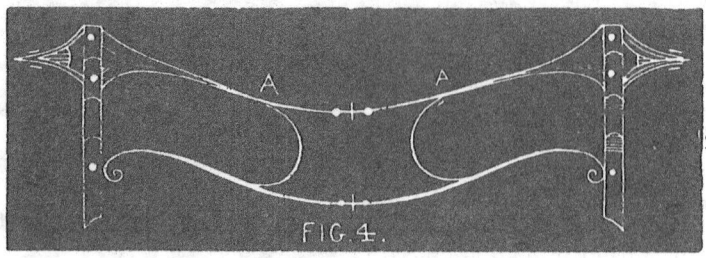

The front or main bar is ⅜ thick at the bearing of C springs. The end carving of bar is of wood, the top part of iron bar being shouldered down a fourth of an inch from the outer edge of C spring; this allows it to cover the joint, and is tapered down to a running lap. When the wood is inserted it is screwed from underneath with wood screws, carved out and finished. When finished, the bar has the appearance of having been made throughout with iron. The nutting stays are welded to the back bar, and bolted through the main bar at points A A. The ends of the back bar are finished with a scroll, the ends running under the C springs, and bolting through.

This style of back bars, when used in connection with an iron carriage, produces a style of finish not far from perfection.

BACK QUARTER WINGS FOR LANDAUS.

To place a wing upon the back quarter of a Landau, and to have the same in perfect harmony with the rest of the structure, is what might be termed one of the almost impossible things. We have seen many attempts made, nearly all of which were complicated or complete failures.

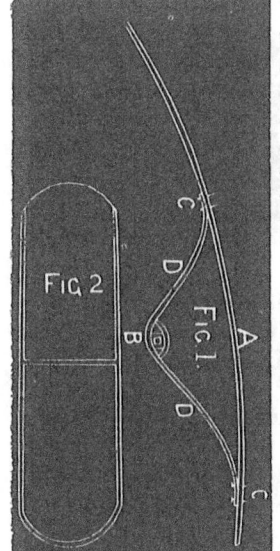

The accompanying figures will serve to give an idea of the most approved method. The wing is made as shown in Fig. 2, without any projections, of the requisite sized oval iron, having the inner bar of heavier material than the outside one.

Fig. 1 is an outline of the wing stay, and inside bar of the wing; A is the main or inside bar of the wing; B is an eye, or block with square eye, which fits to the back or lower back prop of the Landau outside of the joint; the arms D D are the supporters of the wing, and are secured to the wing at the inside bar by means of bolts, as represented by C C. These arms, where they connect with the wing, should be concaved so as to fit it nicely without cutting the leather. The bar A and eye B and arms D D should be made from the better quality of iron, as they are subject to much strain.

That section of the prop which takes the eye B ought not to be less than three-quarters or five-eighths of an inch square.

The wing to extend back of the center of the axle at least six inches, and in front enough to allow of about two inches space "clear" between the door when opened, and the lower or front end of the wing. Four and one-half inches or five inches will be ample width for the wing; wider than this adds weight to the whole and necessitates the making of the works, as described, heavier and unpleasant in appearance.

If the doors of Landaus were like the doors of Cabriolet Caleches, made to open or swing to the front in opening, the putting on of wings would be a matter of little trouble.

"MONITOR DASHES."

Fig. 1 is intended for a round corner, and makes a very neat pattern; the rail, including handle, is swept with the corner; the collar and handle to be plated. It shows a straight dash, that is, there is no curve at top of dash.

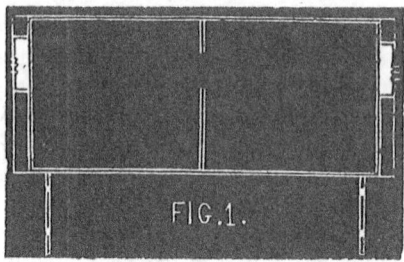 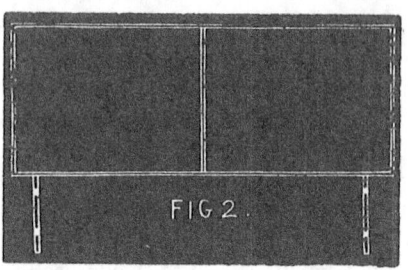

Fig. 2 is a plain square dash having no curve. We give this, inasmuch as it serves to show the slight variation between the other two, and, as customers differ widely in their estimates of what is about right, the plain dash will be found to suit a large number.

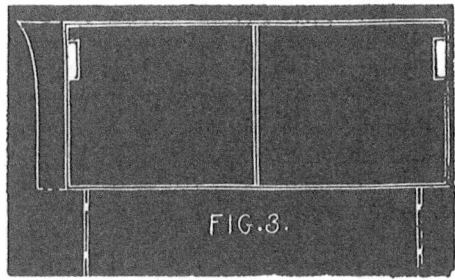

Fig. 3 shows a curved dash with side handles. This one and No. 2 are intended for square corners. No. 3 is at present the most fashionable, and we agree with those who pronounce it the prettiest pattern. The line to the left of the dash shows the curve or end view.

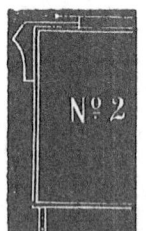

Cut No. 2 shows a section of dash, with the latest style of French dash rail. The lines are all straight, except the part where the hand would naturally grasp it. This is curved, and the iron forged full in the middle, tapering off at either end, so as to conform more naturally to the hand when closed around it.

The rail is not full plated. The ends or handles are plated throughout their extent; the middle portion of the top rail is plated to about the distance shown in the cut, or to speak more distinctly, so much of the the top rail at each end as is here shown should be painted black, the remainder plated. By opposing the black on the ends of top rail to the gilt handles, and the dash being black, the plating has a more brilliant appearance.

The hand hold may be covered with fair leather and wrapped with gold wire, if a different effect is desired, or the black and gold may be distributed according to the taste, being careful always to place them in direct contrast.

Fig. 4 is suitable for coupe or other work. It is made perfectly plain, and has the handles formed within the corners of the top, leaving, when covered, triangular openings. By this means the square corners of the dash are preserved, making a clean finish.

BLACKSMITH DEPARTMENT 143

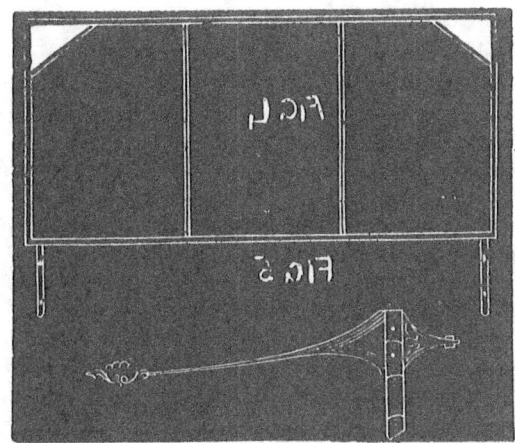

Fig. 5 is an iron back bar. It has a carved center in place of a collar. The bar is flattened where the center carving is to be placed, and the center ornament is of metal, riveted to the bar. Among the many improvements of late, perhaps none have shown more artistic taste than that displayed on back bars. The heavy carved wooden back bar has given place to those of iron of such a slender and delicate form as to awaken a feeling of distrust as to their safety in the minds of those unacquainted with their strength. The cut shows but half of the bar.

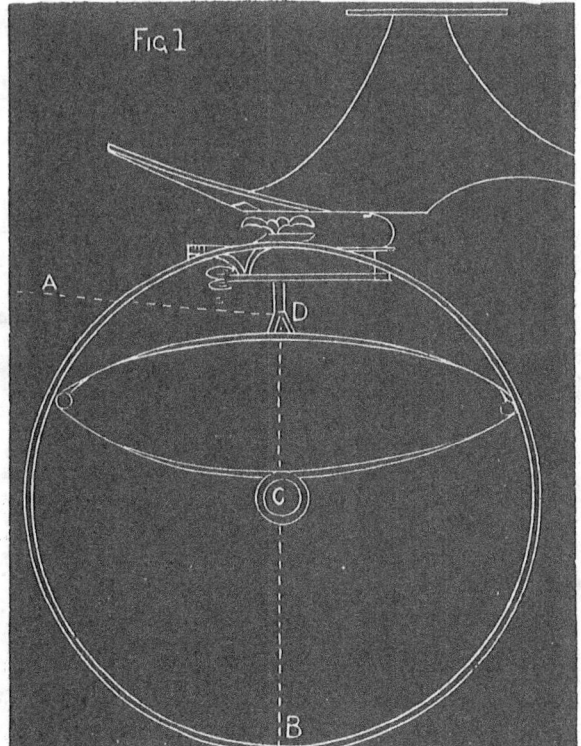

THE HEIGHT OF WHEELS AND LINE OF DRAUGHT.

The greatest power of draught is obtained by a horizontal, or a line in the same height with the breast of the horse; said line operating in a square angle of the diameter of the wheel. The height of the wheel depends, consequently, on the height of the line of draught.

It would require a seven foot wheel to bring the line of draught in a corresponding height, or parallel with the ground. Suppose a horse, measuring 3½ feet to the proper elevation, were to start the traces, then we could calculate how high, possibly, we could raise our splinter bar, shaft or pole, to fetch the line of draught as near horizontal as possible; if a platform spring carriage, the height of spring and head block would give us 3 feet to the point A, where, as a rule, the line of draught A D should be placed. The proportion between front and hind wheel is generally 3½ to 4. The action of draught causes the axle arm to press against the face of the box and compels the wheel to turn; the lower or supporting spoke B acts as a simple lever, resting on the ground and acting at C in the center of the wheel.

The higher a wheel and smaller the axle arm, the less power required to turn a carriage. A wrong proportion in the hanging of a body above the axle hastens the wear of a carriage. If we should, as many do, suppose that the nearer the load is to the wheel the lighter the bearing. A body on platform springs, at a rapid motion, will swing forward, causing more or less vehement shocks, which in heavy built bodies tend to suffer the front spring to lean over and would soon wear the gearing down. Therefore the center (C) of the wheel should be placed in as near a perpendicular line with the extremity of the boot as the fastening bolt through the rocker at E allows, leaving the king bolt 4½ inches forward of the axle.

Bodies on C springs have a still greater forward swing, especially on plastered roads, and are balanced, as a weight, by the length of the loops and spring traps; the front C springs to contain one extra leaf for steadying the body. Two-wheeled carts, dennies, etc., have the body, fully loaded, hung balancing over the axle.

SHIFTING-SEAT RAIL. (ALL IRON.)

Fig. 2.—*Top View.* This is easily understood, bearing in mind that the lower and upper rails, A and B, Fig. 3, cover or correspond. Take the two irons, C and D, ⅞ and ½ inch, and fit them exactly to the inside of the seat, then weld them to the seat rail and drill a hole in the middle between top of seat and rail to receive the thumb bolt to screw the top rail down. The top rail covers the seat rail, and is made of oval iron,

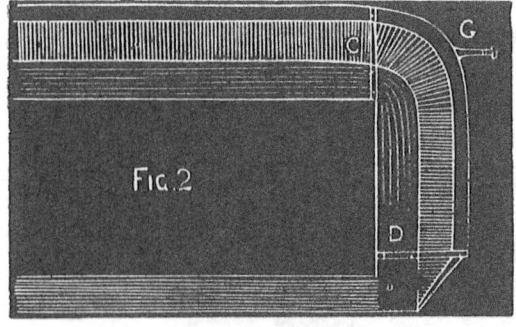
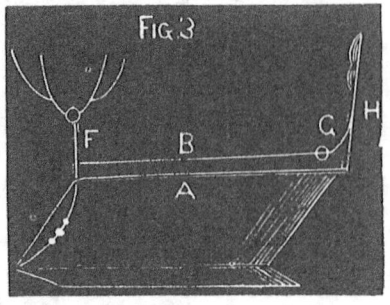

9-16 by ¼ inch, welded to the pivot F, then to the prop G scant ⅝ inch square, having the lateral part, where it joins the rail, widening to give it firmness. Bend the ends of pivot F and lazy back H square to cover the slats C and D, and have a thread cut into

BLACKSMITH DEPARTMENT.

receive the four thumb bolts. From the prop G runs a short stay up to the lazy back to strengthen it. Knobs are riveted to the rail to hold the squabs and back curtain.

SELF-SHUTTING DOOR STEP.

The present design, we think, will be found superior to anything before given to the

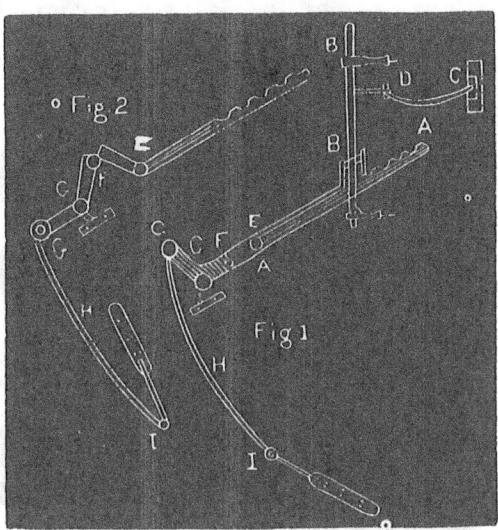

public. The straight gear A A is secured to the inside of the rocker, sliding in a corresponding tin case, and by the turning of spindle-gear, B B, slides forward, as in Fig. 1, opening the door; this throws short lever D C, connecting the door pillar with the spindle B, in the position, as shown in Fig. 3.

The other end of straight gear, A A, connects with joint E F, which again moves the lever G G, being at its middle joint secured to the inside of rocker. The extreme end of said lever G G raises or lowers, as required, the step bar H at G being broken at I, and in this way, revolving in the different sockets, opens the step.

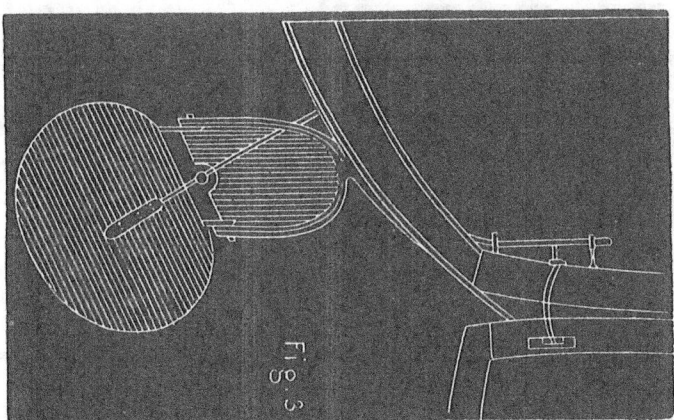

Fig. 2 illustrates the position of the lever G G and joint E F, when shut, spindle B B being shifted to the extreme end of straight gear A A, from the left to the right. The end of step bar H is drawn up, and the step closed. The spindle B B is to be placed behind the door pillar and covered by the trimming. For the explanation of Fig. 3 apply the diagram Fig. 1, which, as shown, is hidden entirely by the rocker, and although this *modus operandi* has a slight pendular motion in bar H at joint G, it is a very effective and neat way to work a self-shutting step.

DOUBLE-FOLDING COACH STEP.

Fig. 1.—*Front view*. Showing the thickness of the slats, their sockets, and the three turning points, A B C. This view is only drawn to the dotted center line, the points A B C being corresponding.

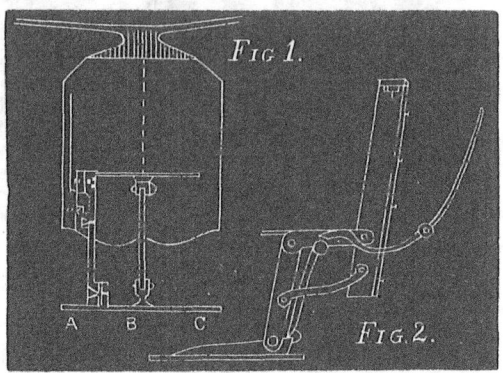

Fig. 2.—*Side view*. Representing the slanting position of the frame, the width of and length of the step bars, and the lever operated by the opening or shutting of the door. Scale, 1 inch to the foot.

A RULE APPLICABLE TO FIFTH WHEELS.

There are no doubt a goodly number of smiths who labor under disadvantage in forging circles, or fifth wheels, simply because they understand no rule by which they can obtain the length of the wheel before bending. For the benefit of those who need such instruction we would call special attention the following

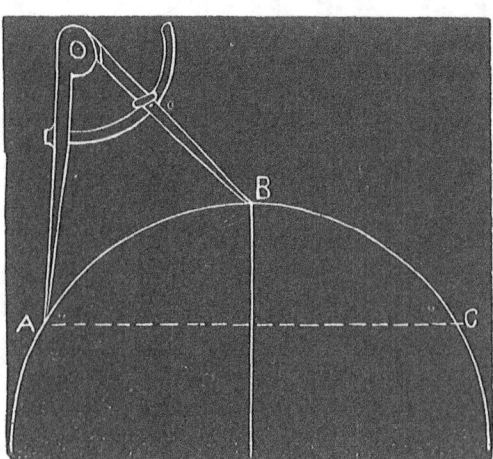

RULE OR METHOD.

In order then to obtain the length of a circle, or fifth wheel, before bending, we will set the compasses by the scribe, or mark made on the carriage part or upper bed, made to represent the circumference or the diameter of the wheel. If we find the wheel to be twenty-eight inches in diameter, the compasses will be set fourteen inches, then strike the circumference of the wheel on a "former" or draft board, as above represented; then place the bed on the circle, struck in a manner so that the marks or scribes will fit or correspond with the outer edge of the wheel, which will bring the back part of the bed against the inner side of the back lip to the point of A and C. This accomplished, we will next find the length of the wheel before it is bent, by dividing the distance between A and C, to the point B. Bear in mind that twice the distance of the compasses, as above represented, will give the length of the wheel when straight; care should be taken to place the points of the compasses on the outer edge of the wheel, and the inner edge of the back lip; remember, *inner edge of the back lip*, and not the inside of the wheel. Keep the points of compasses on the outer edge in both cases, straight or curved. The varying sizes of iron from one inch up to two inches will not affect the result. The inner edge of the iron will upset just as much more in proportion as the iron is wider, and will stretch on the outer edge the same, governed by the same principle.

WEAR IRONS FOR CONCAVE BODIES.

The accompanying sketch for wear iron is particularly adapted to bodies having concaves; also a sectional side view of a body with concave rockers and wear iron applied.

Fig. 1 is an outline of the wear iron in question. A is that portion which fits im-

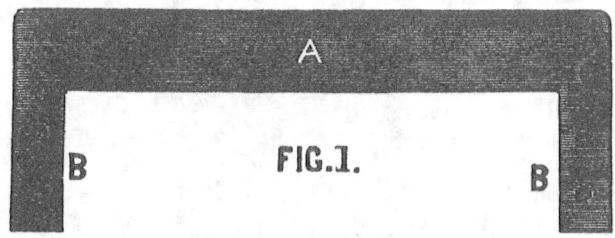

mediately under the panel at that point where the concave commences, and which also projects beyond the side of the body to guard off the wheel. The angles B B are the portion which fits or applies to the concave, and being made narrow are more easily fitted than if the intervening space were solid metal.

The part A is made of tire steel, 7-8x5-32, being about the size best adapted for general work. The angles B B are of Norway iron, and welded on in clip shape, are drawn quite thin, the whole being secured to the body by the screws, as indicated in sketch. The ends or angle being of Norway iron, are easily fitted.

Fig. 2 is a sectional view of a side of a body. A the panel, B the concave, C the flat

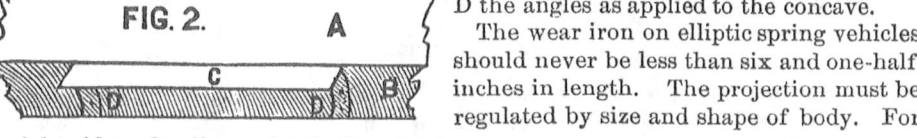

or projecting portion of the wear iron, D D the angles as applied to the concave.

The wear iron on elliptic spring vehicles should never be less than six and one-half inches in length. The projection must be regulated by size and shape of body. For straight sides of ordinary depth, ⅜ projection is ample; if the body bevels much, the iron should project enough to prevent any possibility of the wheel coming in contact with the body. For half elliptic vehicles, of ordinary dimensions, four and one-half inches will be sufficiently long for the wear iron, the matter of projection to be governed as above.

It is well to harden the outer edges of the wear iron, and in cases where steel is not available, and the iron is made of *iron*, it may be case-hardened, for which purpose *Ferro cyanide of potasium* is the cheapest and best, and they will perform nearly as much service as will steel.

DRAFTING JOINTS.

The plan here laid out is for the purpose of making the joints, and ascertaining the correct point for the knuckle without laying the top down to see whether correct. In finishing fine falling-top carriages a great deal depends on a smooth top to make a perfect job. Nothing looks so bad in a show-room as a wrinkled top (particularly a close top), often caused by raising and lowering in making joints.

In explanation, you will prepare a board, length and width not particular, say 3 feet 6 inches long, and from 10 to 12 inches wide. Straighten one edge, and from the left end of the board, at the lower edge, mark the point figure 1. Next ascertain the distance from figure 1 to figure 2, and see if figure 2 is higher or lower than figure 1; on a level line mark the difference, if any, the bottom edge of the board being your guide. Now, proceed to lay off for center of bows, as presented when the top is down; measure two inches from figure 2 on a perpendicular line for the center of back bow, the same distance to the second bow, and so on, until the center of bows are all complete. Now, lay your straight edge at the center of figure 1, and on the point marked for the bow, and draw a straight line through. Measure 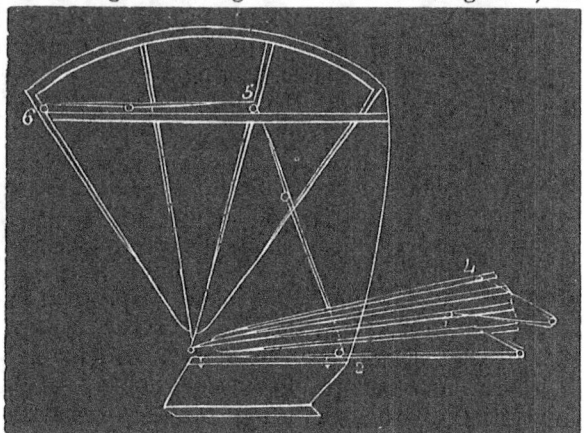 from No. 1 to No. 5, top prop. Mark this on the line of the third bow, and it will give you No. 3. Measure from No. 1 to No. 6, and you have the distance to No. 4. Now you will need 4 pins at Nos. 1, 2, 3 and 4. You will now prepare a cord with a leather ring at each end. Take the distance from No. 2 to No. 5 with cord; with this distance ascertained, you will take the ring from No. 5, and place it on pin No. 3; the ring on No. 3, the other on No. 2. Draw out your cord, as shown in diagram by figures of back joint. This point is for the center of your knuckle. Proceed in like manner with the front joint. You may place the joints, when forged, on the pins of the board, and you will see if correct. This rule can be applied to extension tops in the same manner, by adding the extra bows, and changing the pins on the board to suit the top that the joints are required for.

ANOTHER MODE.

Take an inch board, say 1 foot wide and 5 feet long, and 2 feet from one end of the board, and four inches from the side, insert an iron pin just the diameter you wish the eye of the joint, and 2 inches long. On one end of the board, and 32 inches from this pin, and 4 inches from the edge of the board, cut a slot ⅜ of an inch wide, and 6 inches long, and at the other end of the board cut a similar slot 22 inches from the pin. In each slot put a pin of the same diameter as the first, with a shoulder on it; top to rest on the side of board, and a thumbscrew for the bottom side. These pins should be tapered at the point. Now, take a straight edge and square, and obtain the length of the long joint on the right side of the top, adding say ¼ or ⅜ of an inch to the length so obtained; then place the end of straight edge against the stationary pin, and move the adjustable pin up to the square; screw fast with the thumbscrew, and you now have an exact and permanent draft of the long joint. On the opposite end of the board you can draft the short joint. Upon the opposite of the board you can arrange similar pins and slots, and thus have a draft for the whole set of joints at once, and the ironer will know when he has the joints just according to order without further trial. The knuckles can also be located according to the mode referred to, and the shape of the joints chalked out.

DEVICE FOR ELEVATING LANDAU SEAT.

In the annexed cut is shown a view of Landau seat, standing in position as when elevated. There are different methods of making the attachments, but the one here given is about as simple in construction as any we have noticed. When the top of a Landau is laid down, it cannot, of course, lie flat with the body, and the elevation of the back seat, or the addition of an extra cushion, is required, in order to add to the comfort as well as pleasure of the occupants. An extra cushion would answer the purpose, but it would by no means be so convenient as the device here shown. By

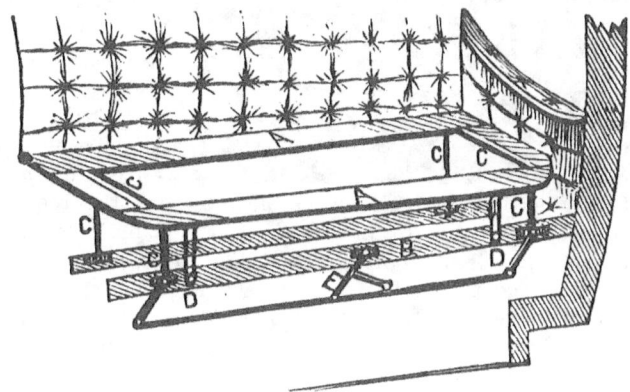

referring to the engraving, A A shows the seat frame, and B the rail connected with the body on which the seat rests when lowered down, and where the pivoting points of the levers are attached. C C C C C mark the position of the irons, which are shown passing from the front back under the seat frame, the legs having attachments to the back-seat rail. The irons immediately above D D are slitted and play up and down, serving to guide and steady the seat. E is the handle by which the device is operated. The arms of the levers which raise the seat describe a quarter circle. Two straps, one on each side, are fastened to the back of the seat to assist in steadying it, and the seat bottom is caned. When a Landau is finished, the "fall" covers up the machinery, and to the unitiated there would be no thought of such a convenience being attached.

IRONING SULKIES.

For a sulky to weigh seventy pounds, more or less, and to track about 4 feet 4 inches, or 4 feet 6 inches, we would say set your shafts about $4\frac{1}{2}$ inches above the axle, and make the short stay from the axle to the shafts of 5-8x5-16 oval, and the long stays of 9-16x5-16 oval. Set the seat, at the front part, $9\frac{1}{2}$ inches above the shafts, with the back part of the seat 1-2 or 3-4 inch more elevated. Make the two side stays of 9-16x5-16 oval, and all the other seat stays 1-2x5-16 inch oval. Have the stirrups 22 inches from front corner of seat, and make them of 9-16x5-16 oval. Make side rails only of 5-16 round, $1\frac{1}{2}$ inches high, and 1 inch flare. Use 3-4x6 inch axles, well drawn on shank. and let into the bed, and set wheels under a plumb spoke ¼ inch. Use 3-4x3-32-inch tires.

HOW TO FIND THE RIGHT SWEEP AND PROPER WIDTH FOR SIDE AS WELL AS DASH FENDERS.

In many yet simple cases of drawing out a sweep for a plain fender, a blacksmith, especially one not used to drafting, will feel puzzled how to proceed. Patterns to match the side sweep cannot be steadied by only holding it against the spot where it is designed to place the fender, and if it is one running across the front of a body, it is

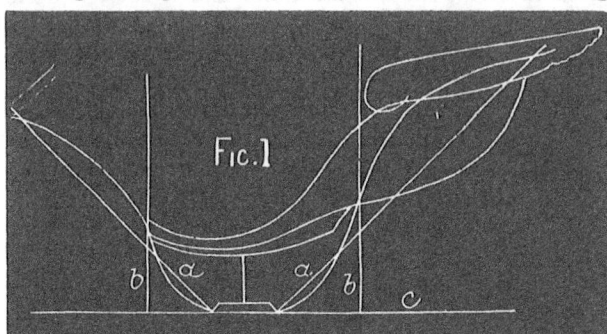

nothing but the "cut and try" rule before a blacksmith will succeed in making a decent as well as properly faced fender. Beside the many projecting parts, as for instance, on cabriolets, with or without a dickey, many places have to be looked after before a man could forge out a job that would add to the symmetrical appearance, as well as the practical utility, of a carriage. But after getting the side sweep, how to give the required width, the rounding top, the clearing of a lamp; how to get the right place for the stays rising from the loops to fall in, nicely supporting the weight of a fender, or, if intended for a reach job or a platform gearing, how to clear the wheel. Such points we will make the subject for a series of lessons in another department, and will begin by showing the difference of a fender for a body with a (Fig. 1) platform spring carriage part, and the same with a reach (Fig. 2).

In regard to sweeping off a double fender, we first draw line C parallel with base line of the body, touching the end of fender and step, which is from 10 to 12 inches below the body. From line C we draw line A A in an angle of 45 degrees, or half a

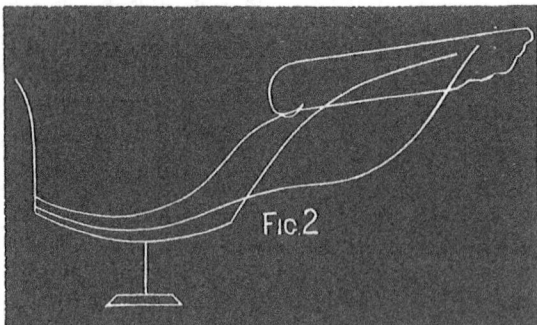

right angle, which sets off both sweeps symmetrically. Lines B B are perpendicular from base line C, at equal distance from center of step, and show the place where the fender sweeps touch the outline of the body. After gaining these lines, it is only a matter of taste to draw the curves above and below the intersecting of lines B B and A A.

In this part we have drawn a full side view, Fig. 3, showing the position of wheels, springs and lamps, where we have tried, as every mechanic should, to solve the different questions, as how to put the front spring to clear the step, and the rim of the front wheel to turn between the back fender and back wheel. The fender, as shown in Fig. 2, is between ten and eleven inches in width, and will cover in accordance of the width of the body and track of the wheels. For a carriage hung on a two-spring or reach gearing the width of the fender is six inches; because then a fender is only an ornament, and is cheaper to make as on a gig or dog carts. The springs have generally a play of four inches and a half, consequently your bolt in your front spring must be so much below the step, and is fastened below the axle. The king bolt is three inches in front of the axle. The lamp is situated in a cut of the extreme end of the front fender.

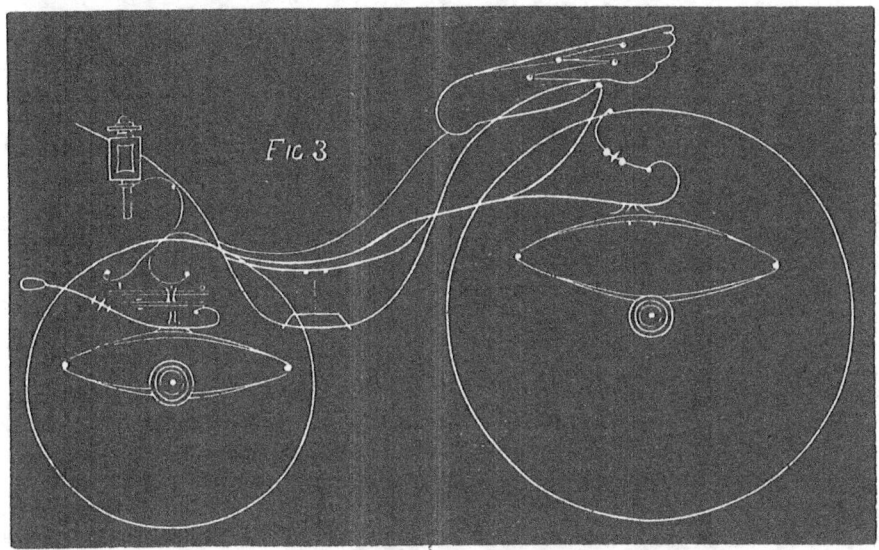

Fig. 4 illustrates the shape of the fender seen from the top; it is a half, or a view from a center line. It shows the width of the body and track. The lamp A is placed

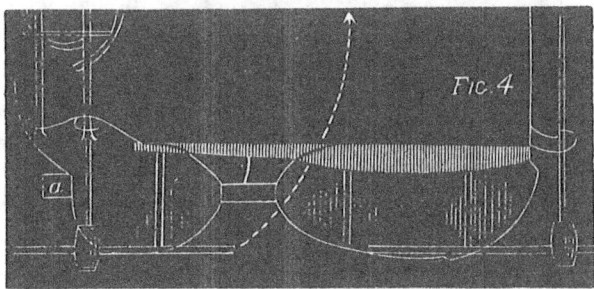

on a stay running up from the loop, supporting the fender, as shown in Fig. 3. These illustrations will suffice in any case to construct a fender properly for a pony phaeton.

Fig. 5 illustrates the Prince Albert width, three springs and reach, suitable for a pony phaeton. The front fender is simply an ornamental dash-fender and 6 inches wide, where it strikes the body; it gives more protection against the splash, but the mode of constructing the carriage part prevents the covering of the front wheel: this has to be turned forward, as in all cases on reach jobs. The back fender is likewise 6 inches wide, but covers the wheel, and is brought off the body the required distance by having the two supporting stays made of the proper length, according to the width of the track. The stay screwed to the prop rises outside the joints $4\frac{1}{2}$ inches, to give play for the back spring.

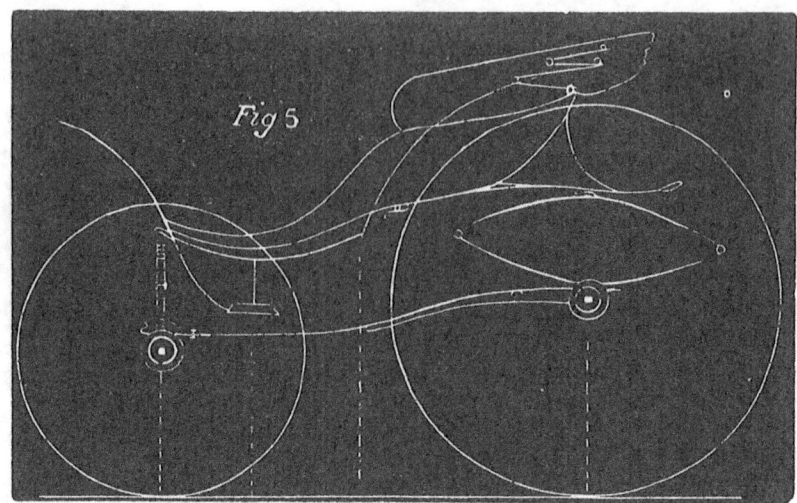

Fig. 6 shows the top view of the carriage with fenders. The dotted lines transfer the axles, the step and lower end of the back fender to the side view, Fig. 1. The back fender is disconnected from the step to clear the passage of the front wheel. In order

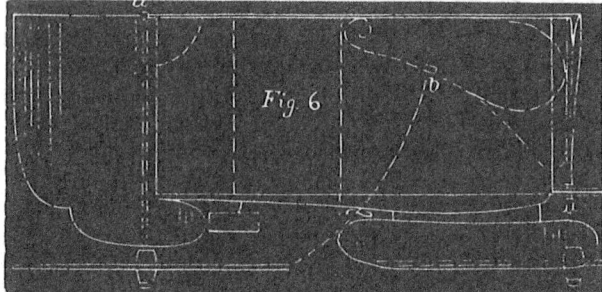

to find out these requirements, you set your compasses with one point at letter A, or king bolt, and take the half width of track, which we suppose to b 2 feet 5 inches, and strike from the hub, or the extreme heigh of the rim of the front wheel toward B, and you will find in this way that one part of your rim will strike the body at the rub iron, while the lower part of the rim will touch point B, at the stay of the reach, having a small rub iron at that point. This view gives you the length of the stays and the exact position of the back fender, and, in fact, a full description of all points to b considered for this class of fenders.

STEEL.—WELDING STEEL AXLES, ETC.

All steel has in it more or less carbon, and the higher the steel the more carbon it contains, rendering it sonorous and brittle, and too high heating will force the carbon from the steel, and the more carbon that is extracted by heating, the nearer it will approach to iron. And again, as the steel increases in heat, its affinity to oxygen is increased, while the lower the heat, the less affinity it has to oxygen. Steel, once being overheated, can never be restored to its original state. Many axles used are made of what is called Bessemer steel, which is, in our judgment, nothing more or less than iron filled with carbon. It has been said that by using iron filings, to assist in welding steel axles, the pores of the steel would open, so that the borax would carry particles of the iron into said openings, in fact would become iron again, which is a mistaken idea; for if the steel is not overheated, it retains the same amount of carbon, and its life is not impaired. Axles may be so highly carbonized as to require a great deal of skill and attention on the part of the smith in order to weld them properly, while some may fail entirely.

The higher the grade of steel the more difficult it is to weld. Iron filings are not objectionable, but they will not turn steel to iron. To be successful in welding steel axles, everything preparatory to the task must be attended to.

First, then, after having well calcined the borax, add a small quantity of sal ammoniac, say one part to twelve or fifteen of borax, which renders it easier to melt or fuse, and will aid in keeping the steel clean. The borax, opening the pores of the steel, will cause it to melt or fuse before it gets so hot as to burn it. Have the fire clean and free from new coal, to prevent sulphur getting on the steel. Of course, all coal has more or less sulphur in it, and iron or steel cannot be united when there is much sulphur in the fire; for this reason new coal should be excluded from the fire, especially on such extra occasions.

Now, then, having all things ready, compose yourself in the best possible manner, be calm and confident of success. The reason some fail in the attempt to weld steel axles is because they are timid. Some may laugh and scout the idea, but it is none the less true. Self-confidence and self-reliance, together with due caution, will surmount innumerable difficulties; can even astonish kings and queens. Then be calm in the undertaking. Place the ends of the axles in a clean, bright fire, heat nearly to a borax welding heat, then remove, upset and scaff, and near the heel of each scaff put them about half-way off with a sharp chisel; then lock them together and set them down with a blow or two from the sledge, after which cover well with powdered borax. Again place them carefully in the fire, gently covering them with coked coal, letting the ends reach just through the hottest part of the fire to prevent their burning off. Let them lie perfectly still, giving a strong but even blast, carefully watching the appearance of the steel as the heat penetrates it, and if the heat does not appear to be even, turn them gently over for a moment or two; and in turning back, if you discover the heat is uniform, let them lie until the heat is raised so high that it is on the point of fusion; or, in other words, as hot as the steel will bear, having, if help is plenty, two men standing ready to use the sledges, and one to take it out. When taken out upon the anvil, place the scaff in the notches, prepared in a manner to keep them from slipping, holding one sledge upon the axle, near the weld, to keep it from any jar to prevent it from uniting, while the smith gives a couple of sharp blows on the extreme end of the lap, and if it adheres, come on with both sledges, and hammer with all power. Let the blows be rapid as possible, until about the size and length desired, which should be left fullest in the center. These directions followed, one need never be afraid of failure in consequence of a bad weld.

In welding axles you will sometimes find that, when you think you have a good heat, in placing the laps together, a blow, however heavy or light, will only jar them apart, and the lap will possess the appearance of small round particles, like that of meal, which is conclusive evidence of their having been overheated, and when the heat is lost in consequence of overheating, the more difficult it will be to weld at all. This being the case, the easiest and surest way to finally secure a weld will be to place it in the vise, and screw it up tight, after having taken as good a heat as possible. It will adhere in this way when it will not in any other. Then replace it in the fire, and proceed to weld as usual. Still, the best method to remedy a failure, save time and insure success, is to cut off four inches from each stale, and weld on a couple of short pieces of iron, taking a borax heat on the steel, and a good sand heat on the iron, and they will weld with perfect ease; this done, upset and weld in center, as any common iron axle. In this way you can get a perfect job.

There has been more trouble among smiths to get good welds, in consequence of using borax too freely, than in any other way. It is very necessary, when welding steel axles, to be cautious, and not use borax so as to allow it to drip in the fire, for the borax and the dirt will immediately settle upon the tweer iron, and form a crust over the mouth, and shut off a portion of the blast, which will often render it absolutely impossible to make a good weld. Borax possesses the property of iron, and when it has thus settled upon the tweer iron, it resembles that of iron. We have heard smiths attribute their bad luck in welding steel axles to there being melted iron covering up the mouth of the tweer iron, when it was nothing but melted borax and dirt, and rather difficult to remove. To get rid of it, all the coal and dirt must be removed from the fire, for the reason, live coals will retain the borax, and just the idea of removing the dirt and cinders from the fire will not remove all the difficulty; and, to a certain extent, it will be only a repetition of the former. To sum up the whole thing in reference to steel axles, they are not worth one straw when iron can be had; they are not reliable, neither durable, and never ought to be used. One breakdown may cost the manufacturer two or three hundred dollars, and the buyer much trouble. Steel cannot be case hardened, for the reason that it will break, while iron can be case hardened, and be made more durable.

SETTING AXLES.

The wheels should be boxed and ready before the axles are welded up. If a platform job, get the length of the axle. Say you wish to track your axles to suit the city railroad, 5 feet 2½ inches, which is the given or starting point; then take the dish of your wheels, by laying a straight edge on the face of the felloe, and measure from the straight edge to the back end of hub, which we will say is 5 inches in each wheel; 10 inches is then to be taken from the 5 feet 2½ inches, which gives you the length of your wheels standing plumb; add half your swing, 2 inches (if your wheels are high; if low you will not want so much swing), you will then have the length properly. Your wheels should first be hooped to insure the proper length.

EXAMPLE.

	Inches.
Width of track, 5 feet 2½ inches	62½
Dish of wheels deducted	10
	—52½
Swing, 4 inches, add 2 inches	2
	54½

Reduced to feet, makes 4 feet 6½ inches.

The annexed cut represents the gauge or axle set. A and B represents the side we use to give the pitch or set in the axle. This gauge should be set so as to give the arm ⅜ pitch; that is, on seven inches; or in other words, if the arm is 7 inches long it must be dropped ⅜ of an inch. It makes no difference if the arm is five, six or eight inches long, the proportion of ⅜ of an inch will answer for all. As you have them set by this gauge while warm slip the wheels on the axle, keeping the wheels snug to the shoulders; roll the axle bottom side up; that will throw the track on the upper side or top of the wheels; then measure to see how near right you have the track. If you find the track half an inch or more too wide, measure with a small rod of iron, say from the off shoulder of the axle to the top edge of the tire on the near wheel; then reverse

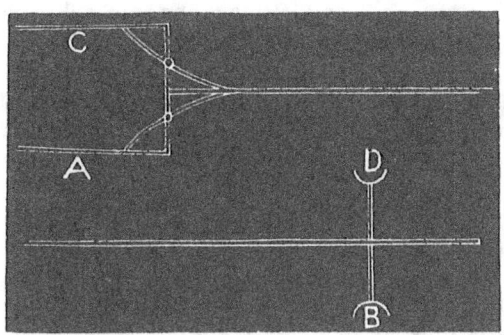

from the near shoulder to the tire on the off wheel, keeping your thumb on the spot where the edge of the tire strikes the rod on both wheels, and if you find one wheel measures one-half inch farther off than the other, then that wheel has one-half inch more dish than the other; the arm, then, that that wheel belongs to, wants to be pitched a trifle, enough to bring it the same as the other wheel; but if the wheels should both dish alike, and you find each wheel stands ¼ of an inch off, pitch both arms a little more, enough to have them track the width desired. If one wheel is dished more than another, you will find when the axle is rolled the right side up that the wheels will not measure alike on the top. Thus you see it is necessary to measure on the track side of the axle. The axles for a platform carriage want no gather in them, for the reason that the front end of the springs are raised from 1½ to 2 inches higher than the back ends, which, of course, rolls the axle up, and by rolling the axle up you roll in a portion of the pitch, which forms a sufficient gather.

The great object to be obtained is to give the arm the right pitch every way, to make the carriage run easy and light as possible, even in the absence of a plumb spoke. All carriages do not look best running on a plumb spoke, especially where the wheels are very dishing, and should not be sought when it can be avoided. Some coaches or carriages, made after the coach order, we are obliged to give a heavy pitch, for the purpose of carrying the wheels away from the body, so as to bring them some specified width of track, to suit some particular customer, or run in some place where a certain width of track is demanded. In such cases, of course, we vary from the general rule, as we must be governed by circumstances.

The patent *axle-set*, which is used at the present day, amounts to just nothing in most of cases; not that there is any fault attached to the same, but for the reason that not one-half of the smiths know how to adjust them; but even if they did understand it perfectly well, they would not, as a general thing, take the pains to alter the gauge more than one time out of ten.

Supposing the wheels are not dished uniformly; of course, to be exact about it as we should, the axle-set or gauge must be adjusted to suit the convenience of each and every wheel. Again, through the neglect of the boss or something unavoidable, the wheels are not always ready, and thus the new thing becomes an old one, and the whole shop use the patent axle-set indiscriminately on all wheels, rendering it no better than the old-fashioned axle-set. In all cases, and under all circumstances, the wheels should be ready for the smith before they weld up their axles.

In setting axles for buggies, and, in fact, all kinds of perch work, be governed by the same rule previously mentioned on platform work, only instead of giving ⅜ inch pitch on 7 inches from a straight line, give only 5-16 of an inch on 7 inches.

A gives the set in the axle, C is intended to give the gather; both sides are furnished with circle slides, fastened by means of thumbscrews, so that the operator can adjust it as he thinks proper. If we are going to depend upon the axle-set alone to give the pitch and also the proper gather, we should first weld up the axles and give them the proper pitch, that is, 5-16 of an inch on 7 inches from a straight line; then adjust C to the taper of the arm, try it on both back and front side of the arm to determine whether it is straight or not. When made straight, gauge C should be adjusted so as to fit both sides of the arm, while D rests on the other arm. We now have the axle straight and without any gather. Move gauge C so it will stand off the point of the arm; if 7 inches from the shoulder 1-16 of an inch, or, in other words, give the arm 1-16 of an inch gather, which is enough to give any carriage that is built at the present day; and, in regard to the taper of the axle, there is not that variation in the axles now in use which will make any material difference as to the result.

In consequence of the wheels not being dished alike, it would be a matter of taste whether we move the arm from where the gauge placed it or not; the wheels will range better to let them remain as they are, and also run better. But in order to suit the notions of some, that is, to square the wheels up, it would be necessary to use the small rod formerly spoken of, and measure from the shoulder on the front side of the axle to the edge of the tire from the opposite shoulder of said wheel, parallel with the axle both ways from the front, and in case one stands off more than the other, start the arm enough to bring them equal. After this is done, measure again with the rod across the front from one wheel to the other. After measuring the back in the same manner, if we find the wheels stand ¼ of an inch narrower front than back, the gather then is sufficient. This measurement only occupies a moment's time when a little accustomed to it, and should never be neglected. Wheels running with one-half inch gather would cross each other's track often, if they could run the way they are tending.

Any one who is quick of apprehension can readily see the impropriety of so much gather, when taking into consideration the unnatural strain felt throughout the whole carriage, harness, man and beast. How can a carriage run light and free when the wheels are all struggling for the center of the highway; acting almost upon the same principle as a snow-scraper, rolling up furrows of mud instead of snow, tenaciously hugging and wearing the shoulders, cutting the washers, heating and absorbing the oil, etc. The tire will sooner wear out in consequence of so much slipping; the felloes are liable to protrude from the tire, while the mud is flying all over the carriage, and the horse chafing and foaming with sweat. All this talk about a plumb spoke, how much swing to give a high wheel, how much to give a low wheel, and also how to obtain a proper pitch and gather by a rule in geometry, or by geometrical mathematics—throw them all to the winds; they are not worth a thought.

While we are spending our time trying to solve mathematical problems, we lose sight of the one great principle that underlies the whole, that is, how to set axles and give the proper gather, with as little friction as possible. To the observing mind it is easy to determine by watching the result of every day's experience. Carriages are continually coming to the shop to be remodeled, which makes a grand school for experience.

Supposing we go to work and set axles by mathematical calculations, it will then all be ambiguous; practical observations must determine the result. Mechanics are few who understand geometry; and in view of this fact, we readily see that it must render such calculations impracticable. while the most unlearned can take either the axle-set, or the simple rod previously represented, and set his axles perfect. There are no axle-sets in use by which axles can be set with accuracy, giving both pitch and gather, bu what imperfections can be detected by using the rod.

SETTING AXLES WHEN COLD.

The top iron D is a bar 2 feet 1 inch long, 1¾ inch square at the fulcrum A, and tapered each way. A hole is punched in to allow the screw B to go through, said hole to be oval or long, so as to allow the screw to move either way; the screw has an eye

BLACKSMITH DEPARTMENT. 157

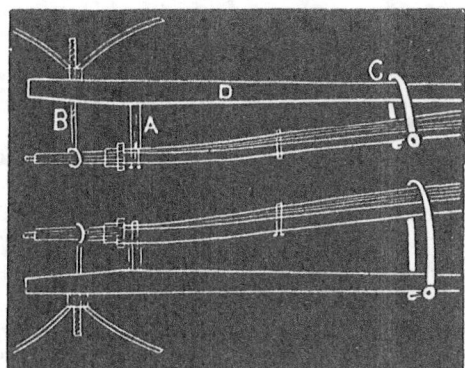

large enough to go on the axle arm; slip it on to about the center of arm, then place the clevis C on the back end of the bar, or lever D; place the fulcrum A either on top of the clip at the shoulder, or underneath, as you may want to set the axle in or out When you place the fulcrum on top, place a strip of harness leather on the axle bed and on top of that an iron, say six inches long, hollow to fit the shape of the axle bed; on that set the fulcrum, and you have power enough with the screw to draw the arm where you want it. When you want to reverse it, put the leather and hollow iron under the clevis. With this machine you can either put in, gather, or take it out the same way.

CARRIAGE SPRINGS.

We know of no rule whereby one can be governed in making springs sufficient to carry any certain load or number of pounds. Neither do we believe there is or can be any such rule, although we do not profess to be better versed in the mechanical line than thousands of others. But if there is any such rule in existence, or even could be, it might be changed into something near as many shapes and forms as the letters of the English alphabet. For instance, we want a set of springs for a market wagon, where the load will be mostly over the back spring. The back spring must, of course be the heaviest. The question is, how much? This we determine by an actual test. If we find the spring or springs too limber, we change them for heavier ones, and in the country add new leaves until they are stiff enough for the use to which they are intended. The test in country places is often made by putting them into actual service, and if we find they meet the required demands we make a record of it, recording length of spring, width of steel, the proportion of said springs from ears up, on both forward and back springs, also the thickness and quality of steel. There is a quality of steel that will not stand up under a load into one-third as much as a first-rate quality; and even if we had no more to contend with than the mere quality of steel, it would be almost impossible to be governed by any rule that would be reliable. Experience and good judgment, together with the record above mentioned, will make us successful manufacturers, as far as the springs are concerned, which is one of the most essential points in the construction of a wagon or a nice carriage. All carriage manufacturers gain their knowledge of what is requisite in a set of springs, to be used in the particular class of work they are in the habit of building, from the lightest to the heaviest work. The buggy, phaeton, rockaway, landaulet, landau and clarence, after they are hung up, and while in the smith's shop, are tested by loading them down with from two to five and eight men, and then springing them up and down, at first carefully, while the foreman or superintendent watches the result, and if everything appears right, increase the force of the test; if so far satisfactory, cramp the forward wheels under the body and spring as before, to see if the body will spring clear of the wheels; but if we find by this test that the springs are too limber, we remove them and put under stiffer ones. Perhaps the springs will differ only in length, say one or two inches shorter, as the case may be, or it may be in the proportion of said springs, each leaf running down a trifle lower on its own bed, which will render the springs stiffer, or the failure may be in the tempering or quality of steel, and by putting under another set of the same dimensions, but of first-rate quality, would obviate the wrong.

If all these things were taken into consideration, and the results recorded, it would afford practical experience, such as every mechanic, as well as manufacturer of carriages, should give heed to and lay up—a knowledge which at some future day may be of use to them. But supposing the above-mentioned springs, after changing, proved to be all right, and while riding them down again we found that there appeared another difficulty, such as the rolling of the hind axle, what would be the cause of such motion?

The carriages are not few that are turned out yearly with this difficulty, and notice when we will, if we see such a carriage, the C springs are limber, perhaps made too light. The springs should be made in this proportion: if the back springs should measure three feet in length, that is, from ear to ear, with five leaves, one inch and a half wide, No. 2 steel; that is, a very plump quarter of an inch thick, the C spring should be made of one inch and three-quarters steel, No. 2, with six leaves, and there will be no danger of rocking. The axle will never rock or roll under a stiff C spring; no matter if too stiff, it will not roll, but will always roll with a C spring too light. Yet the carriage manufacturers learn by actual tests, and order their springs, giving the dimensions and proportions.

And now, as we know of no rule to be governed by in making springs, although there may be, we will just mention what we think would be adequate to hold up 1,500 lbs.; built with three springs; such a set of springs ought to weigh some 80 lbs., the front spring be made of five leaves, and 38 inches long, the two back springs 36 inches long, four leaves on each spring, all made of $1\frac{3}{4}$ No. 2 steel, and of the first-rate quality. It will not answer to bolt on the springs; they should be clipped on.

CROSS SPRINGS.

Upon the length of the cross spring, and the amount of weight which it will have to sustain, depends in a great measure the number of plates which it should contain.

If your side springs contain four plates, and your cross spring be not more than thirty-six inches long, four plates would be sufficient for all ordinary purposes. If you increase the length of your spring, it becomes absolutely necessary to increase the number of plates, or the width of the steel entering into its construction. We would use—if the cross spring were thirty-eight inches long, and there were four plates in the side spring—a cross spring with five plates.

The foregoing would apply to carriages of the heavier grades. For a pony phaeton, victorias, or a light platform buggy—or in cases where the amount of direct weight is not great—with side springs with four plates, and the cross spring thirty-six inches or less in length, we would have but three plates in the cross spring.

The cross spring ought to be set high enough, so that when the full weight is upon it, it is above the horizontal. Three and one-half inches, or four inches for a thirty-six inch spring, increasing the height as the spring increases in length.

TEMPERING SPRINGS.

A knowledge of the art of tempering springs is of some importance to the mechanic. There is, perhaps, no kind of tempering that requires so much care in manipulation as getting a good spring temper. It is necessary that the spring be carefully forged; not over-heated, and not hammered too cold. The one is as detrimental as the other. To insure a spring that will not warp in tempering, it is requisite also that both sides of the forging be equally wrought upon with the hammer; if not by the compression of the metal on one side more than another, it will be sure to warp and twist. We will suppose that the article has been carefully forged, finished up, and is ready for tempering. Clean out the forge, and make a brisk fire with good clean charcoal ; or, if bituminous coal must be used, see that it is well burned to a coke, in order to free it from the sulphur that it contains, as sulphur will destroy the "life" of the metal ; then carefully insert the steel in the fire, and slowly heat it evenly throughout its entire length. Give it time to heat through its thickness, and when the color shows a light red, plunge it evenly into lukewarm water, or water from which the cold chill has been taken off, so as not to chill the surface of the metal too quick before the inside can also harden, and let it lie in the water until it is of the same temperature as the water. A much better substitute for water is a good quality of animal oil—whale oil or lard oil is best ; as a substitute, we have used lard, by melting it before we inserted the heated steel in it. The advantage of using oil is, that it does not chill the steel so suddenly as water, and there is less liability to crack it. Remove the hardened spring from the water after it is sufficiently cooled, and prepare to temper it. To do this make a brisk fire with plenty of live coals, and then smear the hardened spring with tallow and hold it over the coals, but do not urge the draught of the fire with the bellows while so doing ; let the fire heat the steel very gradually and evenly ; if the spring is long move it slowly over the fire so as to receive the heat equally. In a few moments the tallow will melt ; then take fire and blaze for some time ; while the blaze continues, incline the spring, or carefully elevate either end, so that the blaze will freely circulate from end to end, and completely envelop it. The blaze will soon die out ; then smear it again with tallow, and blaze it off as before. If the spring is to be subjected to a great strain, or it will be required to perform much labor, it may be lightly blazed off a third time ; and if it is to be exposed to the vicissitudes of heat and cold, it must be left to cool off itself upon a corner of the forge, and not cooled by putting it in water or throwing it on the ground.

TEMPERING SPIRAL SPRINGS.

Spiral springs of steel wire are tempered by heating them in a close vessel with animal charcoal, or with bone-dust packed around them, similar to the process of case-hardening ; and when thoroughly heated, cool them in a bath of oil, and proceed to temper them by putting a handful of them in a sheet-iron pan, with tallow or oil, and agitate them over a brisk fire. The tallow will soon blaze, and the agitation will cause them to heat very evenly. The steel springs for fire-arms are tempered in this manner, and may be said to be literally "fried in oil." If a long slender spring is needed that requires a low temper, it can be made by simply beating the soft forging on a smooth anvil with a smooth-faced hammer. By this means the metal will be sufficiently compressed to form a very good spring without further tempering. Use a light hammer in the process and "many blows," and a spring will be made that will last for a long time, where it has to bear no great portion of labor in its action.

SETTING AND TEMPERING OLD SPRINGS.

In setting up old springs where they are inclined to settle, first take the bed leaf and bring it into shape ; then heat it about two feet in the center, plump to a cherry red ; then cool it off in cold water, as quick as possible. This will give the steel such a degree of hardness as to be liable to break, if let fall on the floor. To draw the temper, hold it over the blaze, carrying back and forward through the fire, until it becomes so hot that it will sparkle when the hammer handle is drawn across the edge ; then cool off, or not, just as you please.

Another mode is to harden the steel, as before stated, and draw the temper with oil or tallow ; tallow is the best. Say, take a candle, carry the spring as before through the fire, and occasionally draw the tallow the length hardened, until the tallow will burn off in a blaze, then cool. Every leaf is served alike.

HOOPING WHEELS.

What we call straight wheels are those which, by laying a straight edge across the face of the wheel, having both the felloes and spokes touch the straight edge.

First, then, we will examine them, and see what condition they are in for the tire, so that we can determine what draft to give them. See if the felloes are down snug on the shoulder of the spokes, and how much open there is in the rim. For instance, one set we will suppose to be 1½ inch felloe, open 3-16 inch, give good ¼ inch draft, 1⅜ inch felloe, 3-16 inch open, just ¼ inch draft; 1¼ inch felloe, 3-16 inch open, 3-16 inch draft; 1⅛ and 1 inch felloe, ⅛ inch open, ⅛ inch draft.

Now, in determining what draft to give the above wheels, we supposed them to be all good, sound, hard hickory felloes. If the felloes are rather soft timber, give just a mere trifle more draft. If the wheels should be above ¼ inch dish, the felloes would want only one-half of the opening, but give the same draft as the above. In running the tire, lay all the above tire in sets on the floor, roll the wheels on them, and allow 1 inch for taking up in bending; then mark the end of hub with chalk, 1, 2, 3, 4, etc., and the tire with a sharp cold chisel, mark thus, I, II, III, IV, and so on; then straighten on a block set up endwise, about 2 feet high, little concave or lowest in center of block; letting the helper strike while the smith manages the tires, until the kinks are all taken out of them; bend one end a little, so that it can be got in the machine, and take pains to get them round as possible.

In running the wheels with the "traveler," a wedge must be driven in one of the joints of the felloe, for the purpose of tightening the other joint or joints in the rim; then get the length of the felloe, and in running the tire, cut it ⅛ inch shorter than the rim measures. In this explanation we are supposed to have steel tire, and we have a kind of steel tire now that is very high, and difficult to weld, and there are many smiths that will profit by this lesson, if they will use the precaution we give. This tire steel will not stand as heavy heat as even cast-steel, and if it is overheated in the least, it will crack or break in two while hot.

There is one peculiar fact connected with it we find in no other kind of steel, and that is this: It is apt to slip, however good the heat, and to obviate this, after scaffing the ends down to a sharp edge, make a rather short lap, and while hot, take a sharp-pointed punch, and punch a hole nearly through both laps, and drive in a sharp pin, made of 3-16 inch steel wire, ½ inch long. This will not show on the outside of the tire when on the wheel; neither does it weaken the tire like a rivet. We have often seen tires broken where the rivet went through.

In welding, first have your fire perfectly clean, your coal pretty well charred, and the fire hot but rather small, for the smaller the fire, if hot, the less it will waste your tire each side of the weld; have the borax charred; put a little on the weld while hot; pull the fire open carefully with the poker, and place the lap in the hottest part; roll a few pieces of coked coal on the weld; blow steadily, carrying your tire back and forward through the fire, or stop the blast a moment, until the lap is heated alike all through; take it out and weld with hammer and sledge. With this precaution you will never fail getting a good welding heat, and need not upset your tire before welding. If the tire is upset before welding, it makes the lap so much thicker, that before it can be be heated through alike there is liability to overheat, and waste away the tire each side of the weld.

Some practice the **rule of cutting off the tire** the proper length, upsetting, scaffing, and punching the hole for the rivet before bending, getting the length by three times the thickness of the tire, etc.; but this rule is not reliable for three reasons: First, in the thickness it makes the lap. Second, the exact measurement we have to be subjected to in determining the length of tire, in consequence of the opening in the felloe, etc. Third, a limber or soft tire will take up more in bending than a stiff or hard tire; and the consequence is some of the laps or welds will have to be left thick and bungling, while others will have to be drawn thinner than the rest of the tire to make them reach the desired length.

In laying the tires down, the heaviest should be laid at the bottom, and leveled with brick, so that the tire will rest permanently on every brick or bearing, and the rest laid on top the way they will fit the best, to prevent warping the tire in the fire. A level stone should be used to lay the wheel on when the tire is put on. If the tires don't get warped in the fire, don't hammer them at all, without there are some kinks left in the tires in fitting them; avoid hammering, if possible, for it marks the tire; cool off gradually, pouring the water on out of the spout of a tea-kettle, until it shrinks enough so it can be taken up; then roll it in soap water to prevent it from hardening, until it is so cool it will not burn the felloes, truing up while the helper is rolling it in the water, with a mallet covered with thick leather, at both ends; let the third person take the wheel and finish truing the tire with a leather-covered mallet as above; while it is so hot that you can't bear your hand on it, the felloes move easily then under the tire, and should not be moved after they get cooled off, if it can be otherwise avoided; for this reason, when the tire gets cold, all its roughness and imperfections become imbedded in the felloe.

The tire once moved will move the easier next time. After the tires are all on, examine the wheels, and see if there are any crooked spots in the tire that do not set down to the rim; should there be any, heat a short piece of iron, and lay on the tire; it will soon heat it, so as to burn the felloe, but take it off before that time, and rap it down with a hammer. It is a bad practice to heat the tires on a forge as some do, for, in truing them in fitting, we have to bend them cold, and if heated on the forge, and one place red hot, you will often find there a short crook edge-wise. If some of the wheels are dished more than others, put them on the off-side of the carriage. Never take a tire off, if it can be avoided, without it is so loose or tight; as to spoil the wheel when run.

MATERIAL.

The quality of wrought-iron can hardly be ascertained on its surface. The best way to judge the texture is to have it cut apart; when cold, brittle iron will break after a few blows, and show a coarse grained, short cut; while a superior quality of iron bends, resists separation, and shows a fibrous cut, having either a light color, with a dull luster, or a dark color with a bright luster. Inferior iron will appear either bright and glistening, or dull and grayish at the cutting.

Iron heated bright red will also show breaks or flaws when bent, punched or upset. Steel is considerably harder than iron, but can be filed or drilled. If cooled off in water when red hot, it attains the hardness of glass, and is tempered by heating the polished surface to a bright yellow, and cooled in water, and has then the right temper for cold chisels, hammers, swages, etc. Steel for springs must have a degree of hardness between that just described and wrought-iron. The manner by which articles of wrought-iron become a steel-like surface, consists in heating the iron in a fire-proof box, filled with pulverized charcoal and wood ashes, creating carbon, which unites the more readily with the iron the longer it is exposed to the heat. Steel manufactured on this plan appears inferior after this first issue, and is refined by welding and hammering out, and by repeating the same process of heating as before, gives the refined steel. Steel moltened in an air-tight vessel, consolidating breaks or unsound spots, and being thoroughly saturated with carbon, makes the cast-steel.

11

AXLE CENTERS.
WHAT IRON TO USE IN THEM.

"The center of an axle of the lighter class of vehicle—more commonly known as a "buggy"—is a subject that will not bear much tampering with. If too *rigid* it is a failure; if too *elastic* or *weak*, it is also a failure; therefore it becomes necessary that we should employ an agent that is *elastic, tenacious*, and, withal, *flexible*—the latter in order that in cases of collision, etc., the axle may bend rather than break. Were we to construct a thousand vehicles every year, we should never employ any other agent for axle centers than "Low Moor" iron. This comprises all the qualities required, and is far superior to steel in every sense, and if given a fair trial will prove satisfactory to those trying it."

TEMPERING DRILLS.

If the drill is a small one, hold it in a jet of gas till a cherry-red heat appears; then dip the point, or as far up as you wish it tempered, into water or oil immediately. Then try if a smooth file will bite or file it; if it does, it must be done over again, but if it does not bite, but slips over the face without making any impression on it, it will do. Next clean the point or side of the face carefully on an oil-stone; then twist a piece of wire round the end of the drill, or hold the end of the drill in pillars in a candle or gas light till you see it (on the clean piece of the face) gradually turn from a light to a dark straw color; withdraw from the heat and allow it to cool gradually. If it goes past the dark straw color to a blue, it will be rather soft, and must be dipped in water or oil at once, or else the process gone over again. The above dark straw color will bore brass beautifully. If it is to bore steel or iron, have it just as hard as when it comes out of the water or oil in the first process, but care needs to be taken with it if very small in that condition, as it is very brittle

TEMPERING THIN TOOLS.

To prevent thin steel tools from warping requires no small amount of skill and practice on the part of the smiths who may attempt to give them the proper temper. The tempering of thin blades in a satisfactory manner being so delicate an operation, even with the edge-tool maker, it is not surprising that so many carriage-smiths, with but limited experience in the art, are unable to give the required temper, so as to prevent warping. Our manner of tempering such tools is to coat the blade with chalk. After heating, cool it in linseed oil. Next take a block of hot iron, and by applying the tool to the iron draw out the temper of the blade, until it attains a heat known as "bird's-eye blue." Cool it off again in the oil, and the process is completed. There is a solution for tempering thin steel tools, but not having had occasion to use it, we cannot speak from our own experience as to its merits.

ROCKER PLATES.

"Bessemer" steel is the best material for rocker plates for any carriage, and more especially for such as Landaus, where the points of bearing are so wide apart, and hung upon the platform gearings or carriage. The size used being generally 4 inches by $\frac{3}{8}$, although in some bodies $3\frac{1}{2}$ by $\frac{3}{8}$ inch would do as well as 4 inches by $\frac{3}{8}$ would in others. It depends greatly upon the style of body in one case, and the construction of the rockers in another. Some body-makers, either from a want of knowledge or care, make the splices too near the door, so that all the strain is on the plates, instead of the rocker bearing its fair share of the work.

Then, again, it is an easy matter to have a good pair of plates spoiled in the drilling.

The greatest care should be taken to have every screw and bolt fit to the holes in which they are placed. For, if the holes are too large, the consequence will be that, as soon as the carriage gets into wear, it will naturally drop until the screws get a bearing, and by that time the doors will not shut; and purchaser as well as builder, if they have not studied the subject, think the plates are not strong enough.

A CHEAP WAY TO MAKE A COLD CHISEL.

Take a flat file, break off one end, so that a piece will be left about eight inches long; heat it in a charcoal fire to redness, and bury it in wood ashes or dry-slaked lime, so that it may be a very long time in cooling. Grind one end of the file until the teeth have all disappeared; for wherever the scratch of a tooth is left the metal is apt to crack, the cause of which it is not necessary here to explain. Now take the file to any place where you can have the use of a hammer and an anvil, or any other heavy block of metal, and if you are too poor to pay for it, we have no doubt that any smith will let you heat your file and hammer it yourself without charging you much. Hammer out your file until it is sufficiently thin, and keep on hammering it with light blows until it cools to what is called a black heat—that is, a heat just visibly red in perfect darkness—after this, thrust the cutting end in a charcoal fire, until one inch is red hot. Now cool half an inch of the edge in cold water, which will render the edge quite too hard. Watch the color of the steel as the different shades appear near and at the cutting edge, and as soon as you see a light straw color on the surface plunge the chisel into cold water. But on this point, practice alone is a safe guide. The great points which require attention are to avoid burning or overheating, and to pack or hammer-harden the steel well.

VALUE OF IRON PER GROSS TON--2,240 LBS.

FROM TWO CENTS PER LB. UP.

Per Pound.		Per Pound.		Per Pound,		Per Pound,	
2 cents	$44 80	3⅛ cents	$70 00	4¼ cents	$95 20	5⅜ cents	$120 40
2⅛ "	47 60	3¼ "	72 80	4⅜ "	98 00	5½ "	123 20
2¼ "	50 40	3⅜ "	75 60	4½ "	100 80	5⅝ "	126 00
2⅜ "	53 20	3½ "	78 40	4⅝ "	103 00	5¾ "	128 80
2½ "	56 00	3⅝ "	81 20	4¾ "	106 40	5⅞ "	131 60
2⅝ "	58 80	3¾ "	84 00	4⅞ "	109 20	6 "	134 40
2¾ "	61 60	3⅞ "	86 80	5 "	112 00	6⅛ "	137 20
2⅞ "	64 40	4 "	89 60	5⅛ "	114 80		
3 "	67 20	4⅛ "	92 40	5¼ "	117 60		

WEIGHT OF ONE FOOT OF BAR IRON.

THESE CALCULATIONS ARE MADE FOR EXACT SIZES. ROLLED IRON IS USUALLY FULL. ALLOWANCE MUST BE MADE FOR THIS.

SMALL FLAT IRON.

					WIDTH.					
	Inch.	1	1⅛	1¼	1⅜	1½	1¾	2	2¼	2½	2¾
THICKNESS.		LBS.	LBS.	LBS.	LBS.	LBS.	LBS.	LBS.	LBS.	LBS.	LBS.
	¼	.830	.930	1.04	1.14	1.25	1.45	1.66	1.87	2.08	2.29
	⅜	1.25	1.40	1.56	1.71	1.87	2.18	2.50	2.81	3.12	3.43
	½	1.66	1.87	2.08	2.29	2.50	2.91	3.33	3.75	4.16	4.58
	⅝	2.08	2.34	2.60	2.86	3.12	3.64	4.16	4.68	5.20	5.72
	¾	2.50	2.81	3.12	3.40	3.75	4.37	5.00	5.63	6.25	6.87
	⅞	2.91	3.28	3.64	4.01	4.37	5.10	5.83	6.56	7.29	8.02
	1	3.33	3.75	4.16	4.58	5.00	5.83	6.66	7.50	8.33	9.16
	1¼	4.16	4.68	5.20	5.72	6.25	7.28	8.32	9.37	10.4	11.4

WEIGHT OF ROUND IRON, PER FOOT.

INCH.	LBS.	INCH.	LBS.
1/4	.163	2 3/8	14.7
3/8	.368	2 1/2	16.3
1/2	.654	2 5/8	18.0
5/8	1.02	2 3/4	19.7
3/4	1.47	2 7/8	21.6
7/8	2.00	3	23.5
1	2.61	3 1/8	25.5
1 1/8	3.31	3 1/4	27.6
1 1/4	4.09	3 3/8	29.8
1 3/8	4.94	3 1/2	32.0
1 1/2	5.89	3 5/8	34.4
1 5/8	6.91	3 3/4	36.8
1 3/4	8.01	4	41.8
1 7/8	9.20	4 1/4	47.2
2	10.4	4 1/2	53.0
2 1/8	11.8	5	65.4
2 1/4	13.2		

WEIGHT OF SQUARE IRON, PER FOOT.

INCH.	LBS.	INCH.	LBS.
1/4	.208	2 3/8	18.8
3/8	.468	2 1/2	20.8
1/2	.833	2 5/8	22.9
5/8	1.30	2 3/4	25.2
3/4	1.87	2 7/8	27.5
7/8	2.55	3	30.0
1	3.33	3 1/8	32.5
1 1/8	4.21	3 1/4	35.2
1 1/4	5.20	3 3/8	37.9
1 3/8	6.30	3 1/2	40.3
1 1/2	7.50	3 5/8	43.8
1 5/8	8.80	3 3/4	46.8
1 3/4	10.2	4	53.3
1 7/8	11.7	4 1/4	60.2
2	13.3	4 1/2	67.5
2 1/8	15.0	5	83.3
2 1/4	16.8		

WEIGHT OF ELLIPTIC SPRINGS.

1 1/8 × 3 × 36 inch, weigh about	28 lbs. per pair
1 1/8 × 4 × 36 " " "	34 " "
1 1/8 × 4 × 38 " " "	36 " "
1 1/4 × 3 × 36 " " "	34 " "
1 1/4 × 4 × 36 " " "	41 " "
1 1/4 × 4 × 38 " " "	45 " "
1 1/4 × 5 × 36 " " "	48 " "
1 1/4 × 5 × 38 " " "	51 " "
1 1/4 × 5 × 40 " " "	54 " "
1 1/2 × 4 × 36 " " "	49 " "
1 1/2 × 4 × 38 " " "	52 " "
1 1/2 × 4 × 40 " " "	55 " "
1 1/2 × 5 × 36 " " "	56 " "
1 1/2 × 5 × 38 " " "	60 " "
1 1/2 × 5 × 40 " " "	64 " "
1 1/2 × 6 × 36 " " "	64 " "
1 1/2 × 6 × 38 " " "	68 " "
1 1/2 × 6 × 40 " " "	73 " "
2 × 4 × 36 " " "	58 " "
2 × 4 × 38 " " "	62 " "
2 × 4 × 40 " " "	65 " "
2 × 5 × 36 " " "	63 " "
2 × 5 × 38 " " "	67 " "
2 × 5 × 40 " " "	72 " "
2 × 6 × 36 " " "	75 " "
2 × 6 × 38 " " "	78 " "
2 × 6 × 40 " " "	85 " "

WEIGHT OF COMMON AXLES.

7/8 inch, weigh about	33 lbs.
1 " " "	41 "
1 1/8 " " "	52 "
1 1/4 " " "	64 "
1 3/8 " " "	83 "
1 1/2 " " "	100 "
1 5/8 " " "	116 "
1 3/4 " " "	146 "
1 7/8 " " "	161 "
2 " " "	188 "
2 1/8 " " "	210 "
2 1/4 " " "	249 "
2 1/2 " " "	310 "
2 3/4 " " "	375 "
3 " " "	475 "

WEIGHT OF DIFFERENT FIGURES OF WROUGHT-IRON AND STEEL.

RULE 1. *For Round Iron.*—Multiply the square of the diameter in inches, by the length in feet, and by 2.63, and the product will be the weight in pounds, avoirdupois, *nearly*.

RULE 2. *For Square Iron.*—Multiply the area of the end of the bar in inches, by the length in feet, and by 3.36; the product will be the weight in pounds, avoirdupois, *nearly*.

RULE 3. *For Square, Angled,* T. *Convex or any figure of Beam Iron.*—Ascertain the area of the end of each figure of the bar in inches; then multiply the area by the length in feet, and that product by 10, and divide by 3; the remainder will be the weight in pounds, avoirdupois, *nearly*.

RULE. 4. *For Square Cast-steel.*—Multiply the area of the end of the bar in inches by the length in feet, and that product by 3.4; the product will be the weight in pounds, avoirdupois, *nearly*.

RULE 5. *For Round Cast-steel.*—Multiply the square of the diameter in inches, by the length in feet, and that product by 2.67; the product will give the weight in pounds, avoirdupois, *nearly*.

PART III.—PAINTING DEPARTMENT.

THE PRINCIPLES OF COLORING IN PAINTING.

"COLORS are distinguished by artists as *pure, broken, reduced, gray* or *dull,*" etc. " The *pure* colors comprehend those which are called *simple,* or primary—red, yellow and blue; and those which are formed from their mixture in pairs (binary compounds) which are termed *secondaries,* viz.: orange, violet, green, and their hues."

The broken colors are formed by the mixture of black with the pure colors, from the highest to the deepest tone.

" A *normal* color is that color in its integrity, unmixed with white, black or any other color.

Tints, shades and *hues* of color are terms used in speaking of colors. Charles Martel, from whom we quote, says:

" These modifications may be comprised under two kinds, one in which a given color is modified by the addition of a small quantity of another color (*hues*); the other where a normal color is modified by the addition of white or black (*tints* and *shades.*)"

" Red, yellow and blue are termed *primaries,* because they are the source from whence all other colors are derived by mixture."

The mixture of primaries in *pairs* are termed *secondaries.*

SECONDARIES.

By mixing two parts of red and two of yellow produces orange. By mixing two parts of yellow and two of blue produces green. By mixing two parts of red and two of blue produces violet.

SECONDARY HUES.

By mixing three parts red and one of blue produces violet-red. By mixing three parts red and one of yellow produces red-orange. By mixing one part red and three of yellow produces orange-yellow. By mixing three parts yellow and one of blue produces yellow-green. By mixing one part yellow and three of blue produces green-blue. By mixing one part red and three of blue produces blue-violet.

By adding to red, 1st, blue, we produce violet and its hues; 2d, yellow, we produce green and its hues; 3d, white, various light tones of red; 4th, black, various dark tones of red; 5th, gray, various broken tones—red-grays or browns; 6th, green, gray

By adding to blue, 1st, red, we produce violet and its hues; 2d, yellow, we produce green and its hues; 3d, white, various light tones of blue; 4th, black, various dark tones of blue; 5th, gray, various broken tones—blue-grays; 6th, orange, gray.

" Normal gray is black mixed with white in various proportions, producing numerous tones of pure gray."

Lamp and gaslights throw out yellowish-colored rays, causing a great many light colors to appear of a different tone than when viewed by sunlight. Certain shades of green and blue are not easily distinguised by lamplight. A blue fabric will appear to be green, or of a greenish cast, caused by the yellow rays falling upon it. Green being formed by the mixture of blue and yellow, whatever contributes yellow to blue, as in the case cited, or by mixture of pigments, the hue will be green.

When colored rays fall on a colored surface which is lighted by diffused daylight, the colored surface is changed; the effect being similar to that produced by adding to it a pigment of the same color as the colored light. When red rays fall upon a black stuff, they make it appear of a purple black; white stuff, they make it appear red; yellow stuff, they make it appear orange; light blue stuff, they make it appear violet.

COMPLEMENTARY COLORS.

"The color required with another color to form white light, is called the complementary of that color: thus, green is the complementary of red, and *vice versa;* blue is the complementary of orange, and *vice versa;* yellow is the complementary of violet, and *vice versa;* because blue and orange, red and green, and yellow and violet, each make up the full complement of rays necessary to form white light."

The foregoing, it must be borne in mind, result from experiments with a prism of flint glass. When a ray of sunlight is passed through a prism of flint glass, and the image received on a sheet of white paper, numerous rays of different color will be noticed. Red, yellow and blue, called primary colors, are made visible. The red rays will be modified by the pure yellow, which, by mixture, produce orange. The yellow rays, being modified or mixed with the blue rays, become greenish, and increase until we arrive at pure green; this hue becomes bluer until we arrive at pure blue.

Place before the eye a bright red object. After looking at it a few seconds, direct the eye upon a sheet of white paper; a faint green image will be seen. By looking at green, red will be called up. Blue will excite the eye to see orange, and orange blue. This is termed successive contrast. In placing colors near each other, it is of the first importance to the painter to bear in mind the above fixed laws. The coach-painter may derive some useful hints from the foregoing. In ornamenting and striping, bear in mind that colors that are complementary purify each other.

The effect of placing white near a colored body is to heighten that color. Black placed near a color tends to lower the tone of the color. Gray increases the brilliancy and purity of the primary colors, and forms harmonies of contrast with red, orange, yellow and light green.

If green is juxtaposed with black, its complementary red added to the black makes it appear rusty. Orange, yellow, blue and violet associate much better. Red and blue, or green and blue, placed together, are less pleasing than red and green, or blue and orange, where no other colors intervene to separate them.

Take for example our "starry banner," which is composed of red, white and blue. If, instead of separating the lower edge of the blue field by a white stripe we should use red, the effect would be coarse and offensive to the eye. Why? Because the complementary of blue is orange, and the complementary of red is green, which produces a confused effect. Therefore, red and blue not being complementary, do not purify each other. By the contiguous white stripe, the colors blue and red are separated, and, as we have before shown, that white heightens a color when placed near it, the effect in this instance is to separate and purify the colors in question. And so throughout the flag, the white stripes alternating with the red stripes clear up and increase the brilliancy of the whole.

CHIAROS-CURO AND FLAT TINTS.

"There are two systems of painting: one in *chiaros-curo* and the other in *flat tints*. The first consists in representing, as accurately as possible, upon the flat surface of canvas, wood, stone, metal, walls, etc., an object in relief in such manner that the image makes an impression on the eye of the spectator similar to that produced by the object itself. Therefore, every part of the image which in the model receives direct light, and which reflects it to the eye of a spectator viewing the object from the same point in which the painter himself viewed it, must be painted with white and bright colors. While the other parts of the image, which do not reflect to the spectator as much light as the first, must appear in colors more or less dimmed with black, or what is the same thing, by shade.

"Painting in *flat tints* is a method of imitating colored objects, much simpler by its simplicity of execution than the preceding, which consists in tracing the outline of the different parts of the model, and in coloring them uniformly with their peculiar colors."

PERSPECTIVE.

In painting we have two kinds of perspective, the *linear* and the *aerial*.

The first is the art of producing upon a plain surface the lineaments and contours of objects, and their various parts in the relations to position and size, in which the eye perceives them.

The second is the art of distributing in a painted imitation, light and shade, as the eye of the painter perceives them in objects placed in different planes, and in each particular object which he wishes to imitate upon a surface.

When we regard attentively two colored objects at the same time, neither of them appears of the color peculiar to it; that is to say, such as it would appear if viewed separately, but of a tint resulting from its peculiar color, and the complementary of the color of the other object.

On the other hand, if the colors of the objects are not of the same tone, the lightest tone will be weakened, and the deepest tone deepened.

In fact, by juxtaposition they will appear different from what they really are.

The first conclusion to be deduced from this is, that the painter will rapidly appreciate in his model the color peculiar to each part, and the modifications of tone and of color which they may receive from contiguous colors. He will then be much better prepared to imitate what he sees than if he was ignorant of this law.

The eye, after observing one color for a certain time, having acquired a tendency to see its complementary, and as this tendency is of some duration, it follows, not only that the eye of the painter, thus modified, cannot see correctly the color which he had for some time looked at, but also another which might strike them while this modification lasts. So that, conformably to what we know of mixed contrasts, the eye will see, not the color which strikes him in the second place, but the result of this color, and of the complementary of that first seen.

Suppose a painter has to imitate a white stuff, with two contiguous borders, one red the other blue; he perceives that each of them is changed by virtue of their reciprocal contrast; thus, the red becomes more and more orange, in proportion as it approaches the blue, as the blue becomes more and more green as it approaches the red. The painter, knowing by the law of contrast the effect of blue upon red, and reciprocally the red upon blue, will always reflect that the green hues of the blue, and the orange hues of the red, result from contrast; consequently, in painting the borders simply red and blue, reduced in some parts by white or by shade, the effect he wishes to imitate, will be reproduced.

The effect of placing gray upon a yellow ground, according to its contrast, the pattern will appear of a lilac or a violet color. The painter wishing to imitate this object can reproduce it faithfully with gray. The painter who is ignorant of the law of contrast of colors, and the reciprocal influence of one color upon another when placed near each other, would attempt to imitate the colors as presented to the eye; blue and red, for instance, if placed near each other would appear different in color to the eye than what they really are. The sight of red would call up its complementary green, which, being added to the blue, would tinge it. If, in imitating the pattern, the painter should add green to his blue, he would have an exaggerated pattern. Or, should he add orange, the complementary of blue, to his red, a like result would follow.

When a color is put on a canvas it not only colors that part of the canvas to which the pencil has been applied, but it also colors the surrounding space with the complementary of that color.

Thus, a red circle appears to be surrounded with a green aureola. A green circle, with circular rays of red. An orange circle is surrounded with blue. A blue circle with orange. A yellow circle is surrounded with violet. A violet circle with a yellow aureola.

To place white beside a color is to heighten its tone. Put black beside a color and it weakens its tone. Gray placed beside a color renders it more brilliant, and the gray is tinted by the complementary of the color.

Hence, if the painter wishes to imitate this gray, which appears tinted with the complementary of the pure color, he need not use a colored gray.

To put a dark color near a different, but lighter color, will heighten the tone of the first, and lower that of the second.

A light blue placed beside a yellow tinges it orange, and, consequently, heightens its tone. While the darker shades of blue would weaken it, not only hiding the orange tint, but to very sensitive eyes appear of a greenish hue.

The Greek painters, whose palette was composed of white, black, red, yellow and blue, executed some very fine pictures.

It has been remarked that it was probably the method of Titian to paint with one color, such as umber and white, and when the chiaros-curo, or light and shading, was complete, he glazed over it the color of each object.

The instructions we have given under the heading of "The Principles of Coloring in Painting," if properly understood, will be found of incalculable value to the painter.

We will conclude by giving a list of the colors in general use, which may be purchased of any respectable color house.

OIL COLORS IN PATENT COLLAPSIBLE TUBES.

Whites.—Flake white, Nottingham white, blanc d'argent, permanent white.

Blues.—Antwerp blue, indigo, Prussian blue, permanent blue.

Yellows.—Brown ochre, chrome yellow, chrome deep, chrome orange, Italian pink, king's yellow, Naples yellow, light; Naples yellow, deep; orpiment, patent yellow, Roman ochre, raw sienna, transparent gold ochre, yellow lake, yellow ochre.

Browns.—Asphaltum, bone brown, burnt umber, brown pink, cappah brown, Cologne earth, mummy, raw umber, Verona brown, Vandyke brown.

Greens.—Emerald green, olive lake, terre verte, chrome green, verdigris.

Reds.—Burnt sienna, burnt Roman ochre, crimson lake, Indian red, Indian lake, light red, scarlet lake, Venetian red, Chinese vermilion.

Blacks.—Blue black, black lead, ivory black, lamp black.

Mediums.—Megilp, sugar of lead.

Extra Colors.—Burnt lake, vermilion, cobalt, French ultramarine, Indian yellow rose madder, pink madder, Rubens madder, brown madder, Mars madder, Mars violet, Mars scarlet, Mars yellow, Mars orange, oxide of chromium, malachite green, mineral gray, lemon yellow, orange vermilion, ultramarine ash, carmine, cadmium yellow, burnt carmine, purple madder.

PRACTICAL APPLICATION OF THE FOREGOING.

Having selected the colors named, and provided a palette-board, palette-knife, and a piece of canvas, tin or board, and a half dozen pencils, proceed to experiment in the mixture of colors as laid down in the above table. A small quantity only will be needed of each. Having placed the three primaries—red, yellow and blue—on the palette, proceed to mix them according to the table, placing them on the canvas or prepared panel in the order named above. A space of one inch square laid off on the canvas will be sufficient, being careful to keep the colors as pure as possible. To do this it will be necessary to keep the pencils clean. When the pencils are to be changed from one color to another rinse them out well in turpentine, or wash them with soap and water, wiping them dry on a piece of soft rag. It may be found more convenient to use only turpentine while at the work, as it frees the hair more readily of color, and leaves the pencil in condition to continue the work without any delay. When you wish to lay the work by, cleanse the pencils thoroughly with soap and water, clean off the palette-board, and put everything snugly away for another time. A short time spent in this manner, as opportunity offers, will repay the trouble and expense, and be of incalculable benefit. Tube colors will be most convenient if you wish to use oil colors. Water colors would answer the purpose, but, as the coach-painter needs no water colors in his work, it is better to practice with oil colors.

White and black represent light and shade. By adding white to any color we produce various light tones of that color; by adding black to any color we produce various dark tones. All light colors bring forward, according to their strength, every part of the pattern being painted, while dark colors are employed to represent the more remote parts; in other words, those parts of the object which reflect the most light would be painted in light tones, and the more remote parts of a deeper or darker hue. Place before you a cylinder of wood, painted, say red, and varnished. A white stripe will be seen at the highest point on the surface, running the length of the cylinder, parallel to its axis. The more remote parts will appear of different shades of red, as less light is reflected. To paint the object named on a plain surface, so as to appear in relief, it would be necessary to mix three or four hues of red, placing the lightest in the center, the deeper tones to the right and left of the first, and the darkest to the extreme right and left. By drawing a fine line of white through the center, and a pale reddish tint at each side for the reflected light, always to be seen on circular objects, you will have produced, on a plain surface, the appearance of a cylinder in relief.

It is by light we are enabled to distinguish colors. The light of the sun, or solar light, is white, yet subject to modifications from different states of the atmosphere. The prevailing colors in nature are what are termed cool colors, and are tempered so as to present to the eye nothing injurious. Green, which abounds in almost every natural landscape, and the ethereal blue above us, are cool in color. In the woods in summer you may notice a great variety of green, but none are dazzling. The close observer of nature will observe in his rambles that grays are abundant. The trunks and branches of many trees, the dead twigs scattered about over the ground and also the rocks that jut out here and there partake of this tint. Rich browns are found on the old decaying trunks of fallen timber, and by light and shade among the foliage of the trees may be seen a great variety of olive greens. At this season of the year the forests put on their gayest attire. We will be compelled to admit, too, that Jack Frost has a good eye for color. See the rich yellows, olives, scarlet crimson and other hues he has spread over the landscape. The lovliest sight is to stand on some rustic bridge, and look forward over a smooth meandering streamlet, where the distant view is inclosed by a heavy growth of timber. The drooping branches kiss the water, and sway gently as if ashamed of their own reflected beauty.

The gray of the morning has its charm, and the golden hues of sunset, dripping at times by the overflow of their own loveliness, are modifications of natural light. The waters reflect the sky and surrounding objects, repeating their colors. Thus we find but little in nature of a glaring cast.

THE ART OF COACH PAINTING.

In the following treatise we do not promise to lay down rules to govern the painter under *all* circumstances, or to set up *our* notions in opposition to other painters, who have, through years of toil in the paint room, discovered modes of working, probably different from our own, equally as good, and, it may be, better.

Now it cannot be expected that every manufacturer of carriages will be able to furnish the painter with rooms well adapted to turning out first-class work. Employers with small capital are compelled to put up with many inconveniences, and they find the painter louder in his demands for *more room* than either of the other branches. We have often heard the remark made, that the "*painters* are the most troublesome set of fellows in the shop; that they are never satisfied," etc.

We will not deny the charge, but merely say in reply, that they are the *subjects* of more fault finding than the other hands, and expected, in many instances, to perform *miracles* almost. This is especially true in small shops through the country, where they require a painter and a green boy to rush out the work, new and old, giving as good a finish on new work as a first-class shop in New York, Philadelphia or Cincinnati, in which the work is divided out among different grades of workmen.

THE PAINT SHOP.

The paint shop should be roomy, well lighted and ventilated. Bodies and gearings should not be painted in the same room if it can possibly be avoided, and the rough work on bodies should be done in a room separate from the coloring room.

In a well-regulated paint shop will be found a good assortment of brushes, suited to the different kinds of work; paint pots in abundance, two paint mills, a marble slab on which to mix colors, and also a stone to be used exclusively for making putty on; water buckets, sponges, chamois, palette knives and putty knives. Light tressels, set on castors, for light bodies; and heavy tressels, with two wheels and pole, for heavy bodies.

Screens, covered with heavy paper, or enameled cloth, to protect varnished work from flying motes, and from the unsightly marks made by flies in summer.

An assortment of colors, wet and dry lead, whiting, lump and ground pumice stone, English rubbing stone, leather shavings for cleaning the paint mill, sand-paper, brooms, and a ley kettle for dirty paint pots.

Shelving sufficient to hold the cups that are in use, and also small drawers to hold loose dry colors to prevent wastage. On the walls should be placed two or three wooden boxes—one for old sand-paper, one for new sand-paper, and one for odd nuts, bolts, screws, etc., which will gather during the course of the year.

Wrenches, hammers, screw-drivers, etc., etc., should have each an appropriate place. Narrow strips of leather should be tacked up along the windows, and at different places through the room, in which to place the dusters, to avoid the necessity of throwing them on the floor, or placing in a position where they will be easily thrown down in the dirt.

Light stools for holding the paint and varnish cups, made broader at the bottom than at the top, so as not to be easily upset; wheel-boards or props, covered cans for varnish brushes, and wire stands to lay the brushes on when varnishing.

Having thus given a general idea of the wants of the *voracious* painter, we will now enter into the details, and place before the reader the every-day workings of the paint shop in the order in which we have stated them in the commencement of this article.

LIGHT AND VENTILATION.

We know of nothing more annoying to a painter than a dark room to work in. In mixing various shades, or compound colors, one should have good light to insure the tone required. But as the paint stone may be placed close by a window, and fixed permanently on a bench (as it should be), the compounding of colors would not be the most serious objection to a dark paint room.

In the busy season, when the painter is often compelled to work on some jobs at a distance from the windows, many imperfections may be overlooked on the surface, and in touching up old work, the wrong shade may be given to bruised places for the want of sufficient light. These trifles may seem hardly worthy of notice, yet every painter is aware that much trouble is occasioned thereby.

In striping a gearing or body, ornamenting, coloring, rubbing varnish and varnishing, one should have good light. If a shop cannot be lighted from the sides, then there should be sky-lights, for *good light* a painter *must* have to insure work which will not look as if it was grained when it is subjected to the severe test of the clear sunlight.

The paint room should have a means of ventilation which would continually carry off the noxious vapors arising from poisonous lead and colors and the fumes of turpentine. As not one in a hundred of the carriage shops are thus ventilated, the foreman in the shop should see to it that the windows are opened daily in the winter season, and a current of fresh air allowed to pass through the shop. Good light and ventilation would add greatly to the health of the painter, and secure to employers the services of men skilled in their trade, who, under the present system, are compelled to seek other employment, through failing health, at that period in life when their services are most valuable.

Bodies and gearings should be painted in separate apartments, as it is impossible to keep the tires from coming in contact with bodies as the carriages are moved about from time to time. Bodies are also liable to be spattered with lead and filling, and are often bruised by apprentices, who, at the dinner hour, enter into a general friendly combat, by throwing at each other the remnants of their dinners, dusters, or anything else that comes handy.

Boys cannot be expected to think much while engaged in their sports, and it happens that serious damage is done by them.

The rough work on bodies should be done in a room kept for that purpose; the process of rubbing out work demands the use of a great deal of water, which, mixed with the filling, flows over the floor, making it wet and disagreeable to those who may be working near by, and not unfrequently finds its way through into the room below, annoying those who are there at work.

In small establishments it will be useless to attempt a complete division of work, yet, if you have but one room, it will be better to divide *that* up, by keeping bodies at one end and gearings at the other, than to have everything in confusion.

BRUSHES.

Paint and varnish brushes are numbered as follows, viz.: from No. 1 up to No. 6, and from 0 (naught) up to six naughts or ciphers. No. 6 is the smallest; the sizes increase till we arrive at six naughts (000000).

For carriage work, from 0 to 0000 are about the sizes required. The smaller sizes may be found convenient occasionally, but should not be used where larger sizes would serve as well. It is a waste of time and labor to mince around over work with small *tools* or brushes, when larger ones could be substituted.

Smaller brushes are manufactured, numbered from one up to 10, which painters call *tools* or *sash tools*. Nos. 8, 9, 10 are the sizes best adapted to carriage work. Just here we will state that painters invariably call the small brushes *tools*. Those used for striping and ornamenting are called *pencils*.

The above-named brushes are made round and oval, and filled with different qualities of bristles. The ordinary paint brushes contain the inferior and coarser grades of bristles; the varnish brushes the selected and finer qualities. Certain grasses are sometimes mixed in with the bristles, making a worthless and expensive brush in the end.

There are also *flat* bristle brushes of various sizes, which may be used to advantage on bodies and gearings.

The larger size for bodies, the lesser for gearings. The binding or banding, on a paint or varnish brush, is an important item.

Brushes are bound with cord, wire, metal stamped in imitation of wire, tin and copper. The *oval* and *round* paint and varnish brushes are bound with cord, wire, imitation of wire, and copper. The *flat* bristle, fitch, badger, Thum's half elastic, bear-hair, and camel-hair, with *tin*.

Wire and copper bindings are the most durable. The cord-bound brush may be rendered more serviceable by giving the binding and the back part, where the handle is inserted, a coat of oil lead. This coating will protect the cord from rotting. Paint brushes should be suspended by the handles in water, covering the bristles and no more.

VARNISH BRUSHES.

Painters differ widely in their estimates as to the best varnish brush for coach-body work. We occasionally find those who prefer the old-fashioned bristle, and claim for it all the good qualities to be found in any other make. We will not quarrel with them, as we have seen first-class work done by the use of the brush named. We think, however, that less skill is required in laying on a level coat of varnish with the flat brushes.

The black sable and badger (both flat) have found favor with a great many painters within a few years, especially on light work. One serious objection to them is, when applying rubbing varnish, the hair is too soft to spread on the varnish without thinning it down more or less with turpentine. Still, they are not to be cast aside as worthless, for they are capable of producing beautifully finished work.

Thum's half-elastic (flat) is now the favorite brush for finishing coats, and may be used for the rubbing coats also. It combines in one brush the strength required to spread the varnish; the point is as soft and fine as the fitch or badger, and, therefore, will not leave marks after it.

After having used the black sable, or fitch-hair varnish brush, the painter will be apt to think the Thum brushes rather clumsy, and object to them on that account. After using them for a month or two, and then attempting to varnish with the flat fitch or badger, he will wonder how he had finished work with so weak a brush.

Having mentioned the different kinds in use, it remains with the workman to select those best adapted to his wants.

PAINT BRUSHES.

For carriage work the medium sizes should be used. For lead and rough coatings on bodies a larger brush is required than for carriage parts. Body brushes should be kept separate from those for carriage parts, and not used, as is too often the case, at one time on a body, at another on the gearing. Gearings wear a brush out hollow in the middle of the point, which unfits them for laying on a level coat on a body.

We think it cheaper in the end to buy oval or round varnish brushes for lead and filling coats, as they wear longer, and do the work better from the start. When they are worn sufficiently to be fit for color they will render good service until the bristles are worn down too short for any purpose.

A new paint brush should be "bridled." This is done by covering it with a piece of leather, extending from the heel to about midway of the bristles, and stitched together, drawing the leather tight enough to keep the bristles straight. By this means a new brush may be put into immediate use on rough coatings. When the end is worn down, cut away the leather, and the brush will preserve its shape.

DUSTERS.

Of these there are also two or three qualities. The cheaper kinds will answer every purpose for the lead and rough coatings. The black are best for colored and varnished surfaces. Neither the varnish nor color duster should be used on lead, nor the lead duster on color or varnish coats. In the varnish room, more particularly, it is absolutely necessary to keep a clean duster. Flat and round dusters are each useful to the finisher; the flat one is best suited to narrow places, and will avoid bruises which may happen in the attempt to dust out with the round brush.

The duster used for finishing should be washed out occasionally with soap and water; and from day to day, as work may require finishing, it should be rinsed out in clean water, and dried before dusting off the work.

MOTTLERS OR BLENDERS.

These are flat camel-hair brushes, bound in tin, which are now used by all coach-painters in coloring bodies, and not unfrequently on carriage parts. Those with short handles we have found the best. Sizes ranging from $\frac{1}{2}$ inch to 3 inches are as large as the coach-painter requires. The 1 inch, 2 inch and $2\frac{1}{2}$ inch are full large for buggy, barouche and rockaway bodies; but for heavy coaches, with close quarters, the 3 inch should be added to the set. The $\frac{1}{2}$ inch blender is useful in touching up on bodies where a narrow place is to be colored, as it lays the color on thin and level, avoiding laps and ridges; also, in blacking off irons, etc., on the carriage parts.

Color, properly applied with a blender, will be free of brush marks, and the surface gained by rubbing out the body improved. They are worthless as varnish brushes.

PENCILS.

The small brushes used for striping, ornamenting, lettering, etc., are called *pencils.* The stripers are made of sable, camel and cow hair. Sable hair is of two colors, viz.: red and black, either of which is superior to any hair now being used. The red sable hair is somewhat the finest, and is probably the best for ornamenting pencils. For large, broad stripers we think the black ones are superior.

For fine lining either kind may be used, but *we* prefer the black.

Camel-hair pencils work very pleasantly for broad lines, and are preferred by some first-class stripers. The hair being weaker than the sable, they should be shorter. Those made from cow hair are worthless.

ORNAMENTING PENCILS.

These may be had in quill, *without* handles, or tin-bound, *with* handles. Those without handles are in goose, duck, crow and swan quills. The sizes are various—suited to the most delicate touching, or the larger ornamental work. The best are those with handles, tin bound. The ornamenter will need both flat and round pencils—round pencils for laying color on heavy, the flat ones for thin and crispy touches.

They should be kept perfectly clean, greased, and laid away in a box, so arranged that the points will not become bent. When the smallest quantity of paint dries in the hair, the pencil will not work well.

LETTERING PENCILS.

Sable and camel hair are both used; they should be from $\frac{1}{2}$ inch to 1 inch or more long. Buy them in quills, making handles to suit them. The shorter ones for filling in, after the outlines are traced.

PAINT POTS.

Pint and quart cups are the most convenient sizes for general use.

The shop should be well supplied—but it is seldom the case—and painters resort to empty oyster cans, fruit cans, mustard boxes, etc. The coarser paints may be kept in empty whitelead kegs, cut down and supplied with rope or wire handles. Cups for fine colors should be kept clean on the edges. When they are set away, cover them up with paper, which will keep out dirt, prevent the paint, in a measure, from skinning over, and last, but not least, be a notice to *meddlers* that the cup and color *are in use,* and are to be *let alone.*

Quart cups are best for varnishing on carriage parts. Pint cups are as large as is necessary for body work. All varnish cups should be kept perfectly clean. To do this, it will be necessary, after using, to rinse them out with turpentine, and with a stiff-bristle tool work up the varnish that collects on the side of the cups, and with a piece of hair, soap and water, cleanse them thoroughly, rinse them in clean water, wipe dry with chamois skin, and hang them up in a clean place. If you have got "stuck," and haven't time to clean the cups, put them in water until the following day.

The finisher should have at least three varnish cups; four will not be too great a number. If the employer will not furnish them, then buy them yourself, and you will not lose anything by it; for it will be folly to attempt to get on a clean coat without a sufficient number of cups. One will be needed for clean varnish for the panels, one for the clean second-hand varnish which is generally used for the top, arch, rockers and inside edges, and one for a wiping cup. The fourth cup is required in case of any accident wherein it may be necessary to cleanse the brushes, or remove the varnish from a panel, or panels, when it is dirty, or where a run or a heavy flow has appeared.

It is a very good plan to have a set of varnish cups made with false bottoms filled with lead, which prevents them from being easily overturned and when, in the process of varnishing, it may be necessary to wipe out the brush while the left hand is in an awkward position relative to the work, or when it may be actually in use, it will be found very convenient. The cup sitting firmly, the painter may wipe out the brush without using the left hand to steady it (the cup).

In the carriage varnish room, cups made in the manner described will be found very handy and economical. In varnishing a carriage part, the cup is not held in the hand so much as on body work, and the brush is wiped out more frequently while the cup sits on the stool.

Besides the cups named, the painter should provide himself with several small vessels for holding striping color, and the panel colors, which may be needed to touch up the panels during the progress of the painting.

PAINT MILLS.

There is no better *friend* to the painter than a good paint mill, and no greater cause of vexation than to be compelled to use an old worn out one. It is a tedious operation, at best, to grind out a cup of paint by hand, and although the work generally falls on an apprentice, or some rough hand, still there is always dissatisfaction manifested when the paint mill is ordered up. There are three or four kinds of mills manufactured, Harris' patent being the most popular.

They vary in size and price, and, therefore, are suited to the wants of large or small factories. Where several painters are employed there should be two paint mills; one kept exclusively for grinding colors, and another for lead, filling, pumice stone, etc. By this means, the colors are not so liable to be soiled by mixture with lead color, or other rough, heavy paints, which clog the mill up so rapidly. For it is useless to attempt to keep a mill perfectly clean in the hurried season, where the different hands and boys grind out drop black, vermilion, green, lead color and filling, on the same day, and all in a hurry.

Scenes like the following are common in a shop where there is but one paint mill:

Jour.—Johnny, what have you got in that mill?

Johnny.—Dark lead color, sir, for my gearing.

Jour.—Well, hurry up and run it through, and clean the mill out nicely; I want to grind some carmine.

Johnny hurries up, not in the grinding out of his dark lead, but in the manner of cleaning the mill. The mill gets a wipe and a promise, and the jour gets in a passion, it may be, and soundly berates the "cub" for leaving the mill so dirty. The apprentice replies very tartly, and for a short time, at least, there are unpleasant feelings on both sides.

The paint mill should be fastened solidly to the bench on which the "stone" rests, and near enough to it to allow of the paint being put in without being compelled to walk half way across the shop.

A tin scoup *should be* found in every paint shop for conveying the color from the stone to the mill. With the scoup *all* of the paint can be taken from the stone, also the turpentine which the painter may use in cleaning it.

The mill should have a close cover, to be used while grinding out colors, etc., as a protection to the health of the painter. The mill, in motion, generates more or less heat, which, added to the odor from the mixed color, is sometimes sickening, and at all times offensive.

When the mill is run by steam, the speed should not be much greater than when turned by hand. For, being *continually* in motion the mill soon gets *hot*, and heats up the color, causing it to become like *liver*, and we think prevents drop black, and probably other colors, from drying well.

The color also hardens in the grooves on the grinding surfaces, which prevents the color from running out; if turpentine is added it may destroy the body of the color.

When the mill becomes hot so that the color will not drop, the belt should be thrown off until it (the mill) has cooled.

To avoid accidents, there should be a wrench made to fit the regulating screw, to obviate the necessity of placing the hand in a dangerous position. We have seen serious accidents from carelessness in this particular. One, in which a young man had his fingers crushed between the cogs, requiring a surgeon's attention. It was a loss of several weeks' time to the young man; the doctor's bill however was very generously paid by his employer. The price of a suitable wrench, in this instance, was not less than $10.

THE PAINT STONE

should be a marble slab, about two feet square, set in a wooden frame, and securely fastened to the bench, at the left of the mill. It should have a cover to protect it from dirt, and from being daubed up by having paint and varnish cups set on it. A well-kept "stone" is an unusual sight among carriage painters; not because there are not those in every shop who are tidy and would like to have everything in order, but men grow weary of cleaning up after the slovenly ones, and, in time, add their share to the dirty appearance.

In a well-regulated shop there will always be found a man whose business it is to attend to the grinding of paint, and the work connected therewith. It is folly to depend on apprentices, and, in a large shop, we do not think it is right to demand of a boy the labor which is severe even on a stout man. There is always work enough about a paint room to keep a hand busy, should he not be employed at the paint mill. He may sweep the rooms, attend to fires (in winter), put on priming coats on wheels and gearing, clean off grease, sand-paper, etc.

Putty should not be mixed on the stone used for colors. There should be a *putty* stone kept for the purpose. The palette knife, mill duster, oil, turpentine, japan, leather shavings, etc., should have a place near the mill and stone.

WATER BUCKETS.

Of these there should be a good supply. Those used in the varnish rooms should be kept free from grease by having a wash place, or buckets kept expressly for washing in. It is a very filthy practice to use oil on the hands, then catch up the soap and wash in a bucket used for washing off bodies, or gearings, preparatory to varnishing. Those who are so thoughtless generally *wipe* their hands and face on the chamois, thus filling up the measure of their filthiness.

The shop that furnishes but, say one water bucket to two or three hands, may get along after a fashion, but will, during the year, lose the price of a cart load of them, by the hands waiting on each other.

SPONGES.

These should be selected with care, and as they are in such constant use by the painter, it is economical to buy those of a good quality. They differ somewhat in color and size—the "cup-sponge" being the smallest and finest, the larger sizes being of "globular, conical and cylindrical forms," and varying in softness and strength. Those of a pale yellowish tint we have found the most serviceable.

The dirty brown sponge, which is of a reddish cast inside, is generally rotten. Buy those which will be of a convenient size without cutting, or that will need to be cut only in two.

The body sponges should not be used on the carriage part, and *vice versa*. It is a filthy practice to wash the face, neck, or person with the sponges designed for the surface work, as they become greasy, and may be the means of ruining a piece of work.

The sponge is a very curious and interesting animal or vegetable, *we* do not pretend to know which, as that question seems to be in dispute. It is found far down in ocean depths, clinging to the rock and shells, and is gathered in the Mediterranean and in the West Indies. Under a powerful microscope, the sponge presents a singular appearance, and it is affirmed that it is actually pinned together by metallic pins, with heads and points similar to those used at the toilet. These microscopic pins are set at every conceivable angle, thus securely holding the parts together.

CHAMOIS.

"The shammy," called the "wash leather," by some painters, is used for drying off the surface of work, after washing it with water.

The "chamois" is an animal of the antelope kind, the hide of which furnishes a very superior quality of leather, and was probably used originally for the purpose above stated, as well as for many others. But the skins now in general use are not of that kind, and might be more appropriately named *sheepies*, as they are what is termed alum-dressed sheep-skins.

They are different in quality; the best being compact, yet soft. The cheaper kinds are spongy and almost worthless. The shammy gives off furze when it is first put into use, which may be kept down in a measure, by wetting it in water, wring out, and hang up until dry. Repeat this once or twice before using it.

The shammy should be kept clean and free from grease. The hands and face should never be wiped with it. When it will not dry up the water readily, spread it out on a clean, smooth piece of board, and with soap, water, and a piece of curled hair, scour it well on each side, and rinse it in clean water until the soap and dirt are removed.

Those used in the finishing rooms are to be kept for that purpose alone. They improve by use, and are generally more highly prized when worn to shreds. When a good piece of skin has two or three small holes worn in the center, it may be preserved for a considerable length of time, by sewing up the holes with sewing silk. Place the edges together, whip them over neatly, then open the shammy, and pound the seam down flat. We have saved a favorite piece of leather for months of usefulness by this method. There is no fear of it scratching the surface.

THE PALETTE KNIFE.

This instrument is generally made of steel, should be highly tempered, thin and flexible. They are also made of ivory, horn and bone.

The carriage painter requires but one kind, as the colors used are mostly of that nature, which are not affected by the action of steel. A steel palette knife, with a blade from nine to twelve inches long, will suit for all the purposes of mixing the colors used for bodies and carriage parts.

Those with blades about five or six inches long are best adapted for handling striping and ornamenting colors. Ivory, horn and bone are used by artists who wish to keep the colors pure that contain any portion of arsenic. Realgar, orpiment and lemon yellow are of this nature—the steel knife causing them to assume a greenish hue.

PUTTY KNIVES.

The putty knife is one of those indispensable tools with the painter which should be well made and rightly tempered. The carriage painter requires quite a different one from the house painter.

They should have enough *spring* in the blade to give slightly under the pressure required to fill holes, or to plaster grain. Those made flexible only at a point near the handle are a *nuisance*.

The blade should spring from the point through its whole length, not a great deal but sufficient to be perceptible. One thus tempered will do the work far better, besides relieving the wrist and hand from a good deal of the strain on them.

For the ordinary puttying on carriage work the knife should be, say one and a quarter inches across the bevel at end of blade. Sizes ranging from half an inch to two and a half inches are useful at times. The half-inch knife will be found convenient in puttying narrow places on bodies.

It is well to have three sizes. One, say half an inch across the beveled end of the blade; one, an inch and a quarter, and another two inches and a-half. The smallest and largest size here given you will probably have to order made. The half-inch blade will be useful in puttying narrow places, the inside edges of rockaway door with drop lights, coaches, etc. The largest size for plastering large panels, which economizes time and labor, as well as making a better job. The medium size will answer all ordinary purposes.

The point of the blade should be kept free from dents or nicks, and when it wears out hollow in the middle, grind it off even. The edge requires to be sharp, but not in the sense that we speak of a pocket-knife; it should be rounding to prevent scratching, or cutting the surface. Keep it free from putty and paint, handle and blade.

TRESTLES.

The trestles required by the painter differ from those best adapted to the other branches. While the wood-worker, blacksmith and trimmer wish their work to stand solid until finished, the painter must be continually moving. He must daily be moving bodies in and out, up to the windows for light, and back from the windows when coated or rubbed. If the trimmer wants a body, light or heavy, he generally notifies the painter, who is expected to carry it, by help of other painters, or roll it, if he has a suitable trestle, up to the trimmer's bench; and thus it goes from day to day, move! move! move! Now, to lessen this wear and tear on the painter, he should be supplied with wheeled trestles without stint. True, the light buggy bodies may be handled on barrels, to which we are very partial. But barrels will not answer the purpose when bodies are to be moved from one room to another.

We prefer the square framed trestle, set on four heavy castors, for bodies, other than six-passenger rockaways and coaches. They should be well braced with iron on two sides at least, and the braces placed so as to admit of the steps having ample space to play in. It not unfrequently occurs that a *new* body has to be finished with the steps attached, and it is found difficult to set the body solidly on account of the steps or back braces of the same coming in contact with the top bars of the trestle giving a rocking motion. But as new work is generally painted with the steps off, the manner of bracing mentioned will be found of greater advantage on old work. In the hurried season it is a waste of time to take off the steps, which has to be done unless the trestles used admit of the body sitting level.

Trestles for heavy work are constructed differently, varying according to the ideas entertained as to which style is best suited to the wants of the painter. Those made long enough to support a coach-body, back and front, save a great deal of trouble in bolstering up the body with plank, boxes, etc. The front is made the highest, or adjustable, that it may be regulated to suit the variation in the height of fronts; it has two wheels, and is operated by a pole.

Another kind, which is framed only to the size of the bottom of the body proper, will answer every purpose, and take up less space, which is no small item in the paint room. It should be made very strong, and have two wheels fifteen inches in diameter, with two inch tread, if made of wood. It requires a pole, or the pole may be dispensed with by bolting the body and trestle together, and the job moved around by the dash, or a stick screwed across the front. The body may remain on the trestle until it is ready to be hung off.

For light bodies we think there is no trestle equal to those made with a *turn-table* **on** top. The top of the trestle need be only eighteen inches square, and the height two feet six inches, when completed. It may be set on castors, or not, as the painter **may** choose. . *With* castors, the body may be rolled back to a suitable place after having been finished, without the aid of another hand.

SCREENS.

The screen is a device for protecting finished work from flying motes or dust, which may be raised thoughtlessly by any one about the shop, the unavoidable accidents resulting from high winds, and the marks and blotches caused by flies. They should be light, yet so braced as to prevent sagging down when placed in position. The frame may be covered with enameled cloth, or heavy brown paper. The latter may be had at any paper warehouse, of various widths, and of any length required.

Having covered the frame, attach wings on each side, tacking the edge of the paper to the frame, using strips of leather or enameled cloth to tack through, to strengthen the paper. The end wings may be separate, and used or not, as circumstances may require. Eyelets, strengthened by leather, and small nails in the frame to attach the wings, complete the description.

The buggy screen should be at least five feet long, and three and a-half or four feet wide. Larger sizes, for covering the back and hind quarters of heavy bodies, need only side wings. The painter should have also small screens to cover the deck panel on standing top, and non-shifting, falling top buggies.

There are a few varnish rooms in the larger cities which may not need anything of the kind, but the majority of them throughout the country are very defective, causing imperfectly finished work.

In using a screen to cover a buggy, first varnish the inside of seat and front; then set a box or empty varnish can on the seat, and lay the screen over, allowing it to rest on the box or can, and on the dash (if on) or balanced by a weight laid on top. The wings being folded, when a side is varnished, let down the wing carefully, and follow on around the body, covering up as you finish. Buggies, with deck panels behind, which are so difficult to bring out clean, should be turned over and the panel varnished from beneath. In this case, varnish the *inside* of seat and front of body first; turn the body over and lay the screen on. Commence on one side, following around, finishing the deck panel last, of the body proper; the outside of seat being the last part varnished. The wings may now be let down gently, and the body be safe from accidents.

Standing-top work, or non-shifting falling-tops, with *deck panels*, should be *finished* by completing the body before attempting to varnish the deck panel. Having fastened the small screen to the prop blocks, raise it up, and varnish the panel under the screen; let it down gently, and set the body in a position to prevent any draft of air passing across the panel. On round cornered work, let the screen down after the back is varnished, before completing the remaining side. We speak of the round cornered deck panel.

When the screens are not in use they should be hung up, and before using them they should be well dusted off, *outside* of the varnish room.

Coaches, rockaways and other styles having back and quarter panels to be protected (especially from flies), should, after being varnished, be covered by setting the screen up behind on boxes or a low bench made for the purpose, and the wings be brought around the quarters, and tacked at some convenient point. The body should be in such position that it will not need to be moved after having been covered.

The deck panel, on the front of heavy work, should be varnished under the screen not finished, and the screen laid on afterward. This panel should also be varnished last of all, except the sides of the boot or front, and the dash if paneled.

By this means the deck panel will not catch any dust that may be falling while the painter is moving the body about in working on the rest of the job.

CANS FOR VARNISH BRUSHES

should be made of tin, with a close-fitting hinged lid. They should be wide and deep enough to hold two sets of brushes, suspended by the handles, the points clearing the bottom at least an inch, and the brushes not allowed to touch each other. Or the can may be made with a flange on the outside, the cover fitting down on to it, and having a handle on top to operate it.

WIRE STAND.

The wire stand is used for laying the varnish brushes on to keep them from getting dirty, and also as the most convenient place for them while varnishing. Take stout wire and bend it into the form of a small bench with four legs. The top wires should run lengthwise, four in number.

The brushes are laid on with the points extending over the outside wire. The wires being few in number, they cannot collect much varnish, and are readily cleaned. It is light, always ready for use, and, in fact, the *best* varnish stand we have ever used.

WHEEL BOARD OR PROP

is used by the carriage-part painter to elevate the wheels. Is made of wood, and of sufficient length to raise the hind wheel about four inches from the floor, and notched down to adapt it to the front axle. It should be broader at the bottom than at the top, to prevent it being easily overturned. The line of weight should pass through the middle of the board; to do this, it is only necessary to have the highest point in the middle, notching down on each side. Made thus, it will be almost impossible to upset it. The bottom of the board to have a V-shaped cut in it.

As wheel boards cost but little, it is well to have a full supply of them. Those used in the finishing room, if planed smooth and coated with lead, will be easily dusted or washed off before using.

It is not advisable to finish gearings with the single wheel boards, although a great many painters do so. The objections are that but two wheels can be raised at a time, and in the handling of the gearing during the process of varnishing there is greater liability to accident, such as the gearing running backward or forward, by an undue pressure on a wheel while varnishing it. It is a better plan to raise the gearing clear of the floor, by means of barrels or light trestles, made expressly for this purpose.

The front trestle should have a bearing for the axle to play on, to admit of the wheels being moved forward or back, when the painter has occasion to work on the inside of the wheels or gearing. When the wheels are all raised clear of the floor there is no trouble shifting wheel boards; the washing, dusting off and varnishing are all done to better advantage, and the wheels may be turned occasionally until the varnish is set, which will correct a tendency to runs or sagging of the varnish from between the spokes at the hub, or at the felloe.

The Sarvin patent wheel is more liable to collect the varnish between the spokes, and allow it to flow out when the wheels are at rest, than those which are made with staggered spokes, or spokes with space between them at the hub. So that it is best to have the wheels all elevated, that each one may be turned half way round, occasionally allowing the flows or runs to settle back again.

Having given a detailed description of the tools and appliances to assist the painter in his laborious work, we now approach a more difficult part of our instructions, viz.: the compounding of colors. But before we use the palette knife, it will be proper to give the names and qualities of the pigments used.

The carriage painter does not need a great variety of paints or colors to complete a vehicle, so that it will be acceptable to those of refined taste. Should he employ only black, without a stripe or ornament, he way produce a piece of work which will please the eye and satisfy the majority of persons. In our opinion, dark colors are far more in accordance with good taste than the gaudy ones which have been in use of late years. But we must be governed by the fashions, very often, without regard to our better judgment.

COLORS.

The colors generally found in a respectable carriage paint shop may be divided into three classes, viz.: those more commonly used for the rough coatings, such as—1st. Keg or wet lead, dry whitelead, whiting, yellow ochre and redlead. 2d. *Ground colors*, or those which the painter uses in combination with other pigments, as chrome yellow, Indian red, raw umber, chrome green, Prussian blue, lamp-black, drop-black, etc. 3d. *Panel colors*, as carmine, lake of various hues, ultramarine blue, verdigris, milori green, etc., etc.

Keg or wet lead is of the first importance in the paint shop, for it is not only used as the foundation, but enters largely into the mixture of various colors used, as drabs, straw and stone colors, etc.

In the mixture of rough stuff or filling, whitelead gives elasticity and life to the ochre, and when properly used, forms the tenacious part of the under coatings. But oil whitelead should not be used where there is not sufficient time allowed for it to dry thoroughly.

After a good foundation has been secured, and smooth coatings of lead are desired, which will sand-paper smoothly and leave a pleasant surface to color over, the dry whitelead should be used.

It may be mixed with or without oil, a small quantity of japan, and thinned with turpentine. In making putty it cannot well be dispensed with.

We prefer it to wet lead in the mixture of drabs, straw and other light colors, for the reason that the color works smoother, and is not so liable to skin over, or hang on the sides and edges of the cup.

Whiting is not used to any great extent among carriage painters, dry whitelead answering every purpose. Whiting and whitelead make good putty, and probably it would be far better if carriage painters would use it more than they do.

Yellow ochre, when used by carriage painters, enters almost exclusively into the mixture of rough coatings. It is a kind of clay, working and drying well, but of late has been superseded by the introduction of mineral paints which furnish a less porous surface.

Redlead is a good dryer, and might be employed to advantage in the mixture of rough coatings, thereby dispensing with the use of so much japan. If japan could be depended on as a safe article to mix paints with, it would be quite different; but as it is, nowadays, the painter must bear all the blame of cracked and blistered work, when it is too often the fault of the japan.

Redlead needs to be ground in oil where it is used for a foundation for vermilion.

GROUND COLORS.

Chrome yellow is seldom used clear, except for line striping. There are different shades of it as well as qualities, the best being the cheapest, having more body. Lemon and orange are all that the carriage painter requires; with these he can mix up any hues needed by the addition of reds.

With white, yellow forms straw colors of different kinds; with red, orange tints; and with blue, green, lighter or darker, according to the proportions used.

In using yellow with blue or black, to form green, the painter should be careful to select a yellow which is free from red. An orange chrome will not produce, in combination with blue or black, a clear green; and, on the other hand, where a green tint is not wanted, no color should be used wherein there is a tendency toward yellow. Any color partaking of a yellow cast, when mixed with other dark colors, except reds and browns, tend to form greenish tints. If beginners will bear this in mind, it will save them a great many failures in the attempt to produce clear tones.

INDIAN RED

is a very strong color, and always of service to the carriage painter, especially in forming the ground work for transparent colors, such as lakes of a reddish or purple cast, carmine and the darker shades of Bismarck. Mixed with lamp-black it forms the most durable under-coatings that can be obtained, where a brown is needed.

With white a flesh color is formed, making a good ground work for American, Chinese, or English vermilion.

It covers and works well for broad line striping on black grounds, but is not rich enough in color of itself, except on cheap work. By adding lake, a richer tone is secured, which will cover well, and save the time and expense of glazing, when it is important to gain time.

It serves well as a shade for vermilion on letters, scroll work or shaded striping.

On straw color and other light grounds, it assumes a darker appearance in contrast, which may be corrected by changing the color by the addition of vermilion.

The best Indian red is of purple cast.

RAW UMBER

is used to a considerable extent, and is indispensable to the painter. With blue and yellow, it forms a very pleasant range of modest greens.

In combination with white and yellow, gives drab tints, or stone color, which may be saddened with black, or enriched with vermilion or lake. In mixing a light striping color, which may be too much of a raw, yellowish cast, a small quantity of raw umber will correct the defect. It is a useful color in shading on white, and may be classed among our most useful colors.

CHROME GREEN,

Used clear, has a coarse and offensive appearance to the majority of persons, and is seldom used by the carriage painter, except in combination with other colors, and even then very sparingly.

Mixed with Prussian blue, drop and lamp-black, the greens obtained are dull, and not to be compared with those from the mixture of Prussian blue with Dutch pink; but as a ground work it answers every purpose. And greens, like lakes and other transparent colors, have to be glazed to obtain richness. With white, various delicate pea-green and other tints may be made. It can be heightened with yellow, forming brighter tones of green, or saddened with black.

It will be found useful in correcting a color which is too red, and mixed in small quantities with white, gives a purer appearance to that color, although it is inferior to the addition of ultramarine blue for the same purpose.

LAMP-BLACK,

although possessing a greater body than perhaps any other black, is not suited to the wants of the carriage painter, for the outside black coatings on good work.

Its wearing qualities are superior, but the color is not intense enough to make a passable piece of work. It may be used as a foundation coating, but this plan is not followed by the generality of painters; being, naturally, of a greasy nature, it is not safe for the *carriage* painter to use much oil in mixing it.

DROP-BLACK

is of a deeper tone (or should be, which is not the case with a great deal of it of late,) and, therefore, is used for the outer coatings. Drop-black, when of a dull brown shade, may be partially corrected by the addition of Prussian blue, in the proportion of one part blue to three of black; but this, in the sun, is apt to show changeable, the Prussian blue giving it that tendency. Where the best English drop-black can be had we would advise that it be used clear. White and black forms grays of different degrees according to the proportions used. Grays may be enriched with yellow or red, added in only sufficient quantity to change the tint. In ornamenting, these colored grays will be found of great value.

CHROME YELLOW AND BLACK

produce green, which is of a somber hue, compared with chrome yellow and blue; still they are used a great deal, and are better suited to the wants of the carriage painter, in that they are not so grassy.

Black saddens all colors to which it is added, and is useful in toning them down where any other color would only increase the difficulty in which the painter often finds himself when attempting to match a color.

Black added to any color of a yellowish cast tends to form green; added to any color of reddish tone it produces brown. It darkens blues, but sullies the colors, and so on with other colors, it really saddens, and, we might say, robs the color to a certain extent of its purity.

Black may be truthfully named as the most important color in the hands of the painter. A carriage would look in shocking bad taste were it painted all white, or yellow, or red, or green, or blue, with no relief by striping; but with black, *plain black*, what style of vehicle may not be turned out which will be at least acceptable, and not attract the remarks of the majority of persons as to its unsightly appearance?

PANEL COLORS.

Carmine is a red far surpassing vermilion in richness of tone, yet similar to the best deep English vermilion in height of color. It is often adulterated with vermilion by color men, which, of course, injures its purity. *Pure* carmine will dissolve in aqua ammonia, leaving no sediment.

The various grades of carmine are numbered from No. 1 on up, No. 40 being the quality used by the carriage painter. It is a slow drier and should be mixed with either varnish or a pale japan. Raw oil and sugar of lead will answer where the work to be painted is in no hurry. Carmine has not sufficient body to cover, and therefore must be laid over a ground work. It partakes of the color of the ground, giving a variety of rich hues. It tones down the brightness of vermilion, and in turn vermilion brings out the full richness of the carmine.

When laid on Indian red it is proportionally darker, over dark brown it produces a deep, rich tone. It should be painted over a ground mixed of durable colors; laid over rose-pink, or any similar cheap and perishable color, will be time and money thrown away, which includes, of course, a good reputation.

Although classed with the panel colors, we would not advise its use for panels, except on very light buggies, and then but sparingly. For producing a rich, bright red stripe it has no equal.

Madder carmine, made from the madder plant, although not in general use for surface painting, is a better drier and retains its freshness and purity longer than the color from cochineal.

Carmine is of great service in ornamenting. Mixed with asphaltum, verdigris, delicate greens, olives, drabs, etc., etc., it imparts a warm tinge without injuring the color. With white it gives a delicate pink. White, Naples yellow, and carmine, a flesh color tint. Washed over green it gives a warm shade. Mixed with vermilion in different proportions, the ornamenter may produce a very rich center to any ornament requiring a red field. It is indispensable in flower painting.

LAKE.

Of this pigment there are several qualities. Those in common use are English purple lake; Munich, Florence lake; chatamuck or carmine lake, and rose lake. There are finer qualities than those named, and other colors, as royal purple, and those of a brown shade, which are seldom required. The light purple may be imitated with a good quality of lake, ultramarine blue, and white, and the browns with umber, yellow and lake. These gay mixtures are not durable, and do not find much favor.

English purple lake will bear some raw oil in mixing it; the rest are better without oil.

All these transparent colors need to be laid on a ground work similar to their own color when "wet up." Two coats of color, and one of color and varnish, should cover on a solid ground work. But if an extra coat is needed, put it on rather than have the panels *grained*.

Munich or chatamuck lake may be used as a substitute for carmine, if you wish to practice a little deception, or do the work according to the price paid. The colors mentioned above should be ground very fine, so that their full strength may be obtained, thus reducing the actual cost. Fine paint will spread over a greater surface than that which is coarse, to say nothing of the trouble and vexation of having a transparent color rub down, showing specks.

Grind your lake heavier than opaque colors, and do not mix up more than the quantity needed for the work in hand. Use a clean cup and clean brushes. If you pick up a paint cup that has had some other color in it, and half clean it, then snatch up a dirty brush and attempt to lay on a pure, solid color, you will be disappointed. Do as little patch work as possible in laying on transparent colors. Go over the panels at once, finishing all while yet wet.

Where the panels are very large use a broad brush and work rapidly, blending in the edges while wet. Lay your color on "up and down."

ULTRAMARINE BLUE,

when pure, is a very durable color. It is prepared from the mineral, *Lapis lazuli* called ultramarine (over the sea), because it resembles the beautiful blue seen in a very clear atmosphere above the waters.

We are not partial to it as a panel color unless it is used clear on a heavy body. Mixed with the red lakes it tones them down without serious injury to the purity of the color.

It requires a ground work, which should be as near the color you intend to lay over it as possible.

For clear ultramarine a dark lead-colored ground will answer, or take Antwerp or Prussian blue and white, and mix a tint that will match the ultramarine when mixed. It will be found a difficult color to handle by the novice.

The only secret in laying it on solidly is to have sufficient varnish or boiled oil in the color to prevent it "flying off," or, in other words, drying too dead.

The brushes and cups should be perfectly clean, to avoid muddying the color. A good quality of this blue is so pure in tone that it possesses a beauty unsurpassed. Two coats, and one of color and varnish, should cover solidly. Ultramarine, like other transparent colors, should be kept air-tight in the cup when not in use, and the ground work be free from roughness or grit. If your color is not clean and free from skins or grit, or if the ground color be coarse, you may expect trouble when you begin rubbing your varnish coats.

Specks will appear which will not be easily touched up and hidden.

Ultramarine, in combination with white, forms sky-blue tints; with white and black, steel colors; with red, purple and violet. In painting clear white, ultramarine should be added to counteract the effect of the oil or varnish. It will give to whitelead a purer appearance.

VERDIGRIS.

Rust of copper, an acetate of copper. It is very transparent, of a bluish color, and requires a ground work. On the panels of a heavy carriage it gives a rich tone of green. It is not used to any great extent as a panel color—greens not being desired by the majority of persons. Milori green, or green lake, is a color resembling chrome green when dry, but is much richer than chrome green when mixed up. It has a strong body, covers well, and with white forms pleasant light greens for sleigh or other fancy work. As a ground color for emerald green or verdigris it will serve a good purpose.

COMPOUND COLORS, OR THOSE COMPOSED OF TWO OR MORE COLORS MIXED TOGETHER.

The carriage painter is not required to produce a great variety of colors, hues or tints, the panel colors being generally of the proper shade without any addition, and when a change is required, for bodies especially, it results from simple mixtures, such as have already been noticed. On carriage parts, of late years, there has been a run of drabs, and umber-colored browns, which have called on the painter for the exercise of an unusual amount of skill in their mixture, and the mixture of striping colors to harmonize. *Some* beautiful effects have been produced, while a *great many* glaring and offensive combinations have annoyed sensitive eyes.

As there cannot be the slightest mixture of any two or more colors, without producing a change of hue, it follows that there is no limit to number of hues and tints which may be produced. But as only a certain class are acceptable to the eye, the painter need use only a good selection.

Normal or pure gray.—White and black in various proportions.

Colored grays.—Red and green, blue and orange, and red, yellow and blue.

Straw color.—White, chrome yellow and raw umber; white, yellow ochre, vermilion and raw umber, or white, yellow ochre and vermilion.

Light buff.—White and yellow ochre.

Deep buff.—White, yellow ochre and red.

Salmon color.—White, yellow and vermilion, or white, yellow, vermilion and lake.

Flesh color.—White, Naples yellow and vermilion; white, raw sienna and light red.

Orange.—Equal parts of red and yellow.

Gold color.—White, yellow, red and raw umber, blue or black.

Pearl color.—White, black and vermilion, or Indian red.

Lead color.—White, blue and small portion of black.

Stone color.—White, yellow and umber; black, umber and yellow.

Canary color.—White and chrome yellow.

French gray.—White and drop black.

Tan color.—Burnt sienna, yellow and raw umber.

Linen color.—Black and white, warmed with umber in the shadows, or heightened with yellow and reds, where in the painting any part should approach the skin.

Pea green.—White and chrome green.

Sea green.—Prussian blue and yellow, Prussian blue and Dutch pink, Prussian blue yellow and Dutch pink.

Citron.—Green and orange, or which is the same, yellow, blue and red—green being formed by yellow and blue, and orange by the mixture of red and yellow.

New York red.—Carmine and vermilion.

Chocolate.—Black and Spanish brown.

Umber-toned drabs.—White, raw umber, lemon yellow, and lake; white, burnt umber, and orange chrome; white, umber, yellow, red and a little black.

Olive.—Umber, yellow and black; raw umber and yellow; green, umber and yellow.

London smoke.—Burnt umber, yellow, white and red; Vandyke brown, yellow, white and lake.

Lilac.—White, carmine and ultramarine blue.

Purple.—Ultramarine blue, whiting and carmine, blue and red.

Violet.—Blue, red and black; or blue and red.

Wine color.—Purple lake and ultramarine blue, lake blue and black.

Bismarck.—Burnt sienna, yellow and lake; Dutch pink, burnt umber and vermilion; Dutch pink, umber, yellow and vermilion or lake.

Dark rich brown.—Vandyke brown, burnt sienna and lake.

Green.—Blue and yellow, mixed in different proportions, give an endless variety of greens; drop black and yellow produce a green well suited to heavy carriages.

Changeable color.—Clear Prussian blue; it must not be color and varnished.

Maroon.—Carmine, yellow and burnt umber, or crimson lake and burnt umber.

The foregoing list will enable the painter to mix up about all the colors required in coach painting. A great many shades may be made of each of those given, and combinations formed which will be acceptable. The fashions do not always allow the painter to use colors that may be to himself most pleasing, and he is often pained by an order to paint and stripe a carriage with glaring and raw colors; but these coarse effects soon grow into disrepute, and we find that on the average the colors used by carriage-makers are in good keeping.

OILS, JAPAN AND VARNISH.

Oil is an unctuous substance, animal or vegetable. There are what is termed fixed and essential oils. The fixed oils most commonly used are linseed (flaxseed), poppy and nut oils. Turpentine is an essential oil. In the use of oil the painter desires to add a wearing quality to the pigments used, knowing full well that if his paints are mixed without a fatty substance there will be no adhesiveness imparted, and but little power to resist the action of the atmosphere in its changes.

Oil is used in a less or greater quantity, according to the kind of work to be painted and the time allowed for completing the work in hand. The house painter may, with safety, coat outside work with color mixed in oil alone, as he aims, or should, to protect the walls and other exposed parts from changes of atmosphere. Using no varnish to finish with, the coatings have the full strength and flexibility of the oil to keep them firm, and are not so liable to crack.

In coach painting the case is quite different. Oil must be used sparingly, else the body of paint laid on will not dry thoroughly; then, when it is inclosed by varnish, there is a brittle surface imparted, which checks the drying of the oily coats for a time resulting in cracks.

Linseed oil is considered the best for general purposes, on account of its strength and flexibility.

Oil is influenced in drying by the colors with which it is combined, some of which hasten while others retard it. For carriage painting a good quality of raw linseed oil should be used, as it works pleasantly, dries dead when not used in excess, and is free from that gumminess too often found in the use of boiled oil. Raw oil, simmered over a gentle fire for two or three hours, adds to its drying qualities, and if certain oxides and salts of lead are added, they impart a power to dry very rapidly.

Striping colors, mixed in boiled oil, work kindly, and dry well over night.

It has been stated that the drying of oils depends on the following conditions: The presence of oxygen, which, by an incipient combustion of the hydrogenous oils, fixes them, whence, whatever contributes oxygen to oil dries it, as pure air, sunshine, etc.; hence, all the perfect oxides of metals dry oils. Imperfect oxides, by extracting oxygen from oil, retard drying; hydrogenous substances are ill dryers in oil.

The best dryers are those which contain oxygen in excess, as litharge (which is an oxyd of lead), sugar of lead, minium (the red oxyd of lead), massicot (also an oxyd of lead), manganese, umbers, sulphate of zinc, or white copperas or verdigris.

Oxygen was named from its property of generating acids; it is the supporter of ordinary combustion.

Hydrogen is a gas constituting one of the elements of water. A moment's thought on the part of those who have not given this matter any attention will fix in the mind the reasonableness of the foregoing statements. We might make this very plain by saying that oxygen is fire, and hydrogen water.

JAPAN.

Japan, which is used by the carriage-maker only as a dryer, is made by boiling linseed oil with one or more of the articles named above.

It is too often the case that our japans are made so dark that they sully the purity of fine transparent colors. The manufacturers of japan are not at fault in this matter, as the demand on them of late years has been to produce dryers which would not fail to fasten the paints in a few hours. Hurry, hurry, is the word; and, as a quick and powerful dryer is necessarily dark, the painter must have recourse to varnish, or boiled oil, in the mixture of transparent and delicate colors. Japan, as we generally find it on the market, cannot add any tenacity to the paints in which it is mixed, and should be tempered with oil where the painter lays on but one coat a day. Where two or three coats a day are applied, it is best to use no oil, but depend on the varnish coats to hold all solid.

An article called japanner's gold size, better known to English than American painters, is made as follows: Powder finely of asphaltum, litharge or redlead, each one ounce; stir them into a pint of linseed oil, and simmer the mixture over a gentle fire, or on a sand bath, till solution has taken place, scum ceases to rise, and the fluid thickens on cooling.

VARNISH.

Any viscid, glossy liquid is a varnish; but as the carriage painter has no use for those kinds, which are made in alcohol, or any other liquid which does not impart a gloss and good wearing quality, we need not mention them here. All that a carriage painter requires is to be furnished with a solid-drying rubbing varnish, and a tenacious, pleasant-working and durable finishing varnish.

Varnishes made by the use of oils vary so much in their natures that it is bewildering to even think of them. That there is a difference in the quality of gums used, which add to or detract from the good quality of varnishes, there can be no doubt, but when we are told that Zanzibar gum alone is used, that the oil was the purest and best, that the whole process of making the varnish was in the keeping of old and experienced hands, and then, after trying the varnish, we find it worthless, we are at a loss to account for it. The wearing of any varnish depends on the quality and quantity of the oil used, and the time allowed it to ripen after being racked off.

Rubbing varnishes are required to dry firmly in from two to five days; consequently less oil enters into their composition. A good wearing rubbing varnish should not be rubbed until the fourth or fifth day after being laid on. When rubbed, it should not sweat out (become glossy) soon after, even in hot weather. Slow-drying rubbing varnish, when allowed to stand a day or so after having been rubbed down, will sweat out in hot weather, and should again be run over with the "rub rag" and fine pumice before another coat is applied.

Rubbing varnish that sweats at all times, soon after being rubbed, is liable to crack, and should not be used.

By the use of hard-drying varnish the painter is enabled to level his work down and prepare for the last coat or finishing varnish. This last coating must be of an opposite nature to that on which it is laid if great brilliancy is sought after, and as its surface must ever be exposed to the action of heat and cold, sunshine and shower, it must possess an elasticity, or oily nature, that will resist these changes for a great length of time.

The English varnish manufacturers possess a secret which seems to be hidden, so far, from others; for, while oils may be used in equal quantities by two varnish makers, and the same quality of gum, the result of the two separate days' work are far different.

Finishing varnishes are considered good or worthless by the finisher according as they work freely, are free from particles of gum, flow out well, and *generally* remain as they are left when the job is completed. This much having been secured, the wearing properties are almost always found to follow.

Varnishes are subject to many changes after being spread on a surface, and occasion the painter a vast amount of trouble and vexation.

We cannot give a satisfactory reason why varnish should be so sensitive; for, if we could, we need not labor for a living—the secret would be worth a fortune. We will, however, give a few hints which may be of service to the painter.

IRREGULARITIES ON SURFACE OF VARNISH.

Varnish is subject to various changes after having been applied to a body or carriage part. It crawls, runs, enamels, pits, blotches, sags down and hangs on the lower edges of the panels, smokes or clouds over, has the appearance as if dust had been dredged on it through a piece of fine mull; or, as happens with carriage parts, gathers up and hangs in heavy ridges or beads along the center of the spokes.

These irregularities will happen at times with the best quality of varnish in the market, and while being used by workmen of undoubted skill, surrounded with everything necessary to insure perfect work, proving, conclusively, that atmospheric influences alone often cause irregularities on the surface, known by the terms mentioned above. But, while this may be true, there is no doubt that the finisher may be to blame in some cases for an imperfect job.

Extreme heat in the summer, and cold in the winter, each exert an influence. Heat, by melting down the varnish, or, when charged with electricity, as when a storm is brewing, causing "pits." In winter, the cold retards varnish from setting equally, and may occasion "enameling" and "blotches."

By "crawling," the painter means that his varnish, immediately after having been brushed on, does not cling to the surface equally, but shrinks away from it in spots. This feature, presenting itself before the panel or space is completed, does not occasion much trouble. The application of water, by means of a sponge, or a damp chamois, will remove the tendency at once. Spots on the panel already coated may be corrected by blowing the breath on them, which will contribute enough dampness to make the varnish cohere when again brushed over. A glossy surface must be haired or rubbed down with ground pumice and water. Varnish crawls more frequently when the surface to which it is applied is cold. Whether the act of brushing produces electricity, or the surface was already excited by some mysterious agency, we are unable to say. The varnish is repelled by some means, and behaves as when poured on a piece of glass that has been slightly rubbed with paper.

"Runs" should not *always* be charged to the finisher, although we admit that, in the majority of instances, he is to blame. We have seen varnish that would sag, or "run," when applied in the usual manner by the most skillful workmen; sometimes on a body after a few hours, and on carriage parts after several weeks had elapsed. The varnish referred to was too soft, not possessing the property of increasing in volume on coming in contact with the air, and supporting itself by adhering in a continuous body.

When a heavy coat of varnish has been laid on, and any portion of it slighted by the brush, the part slighted will be very apt to run, because the varnish has not been blended in with that which immediately surrounds it; it hangs alone, and somewhat heavier, and seeks a level or space to incorporate itself with the other varnish, but, except in a few instances, fails to find it.

A finishing varnish that will accommodate itself to the space required to swallow up a run, without assistance from the painter, is prized very highly. Runs under the moldings, and out of corners, are occasioned by the manner of laying on the varnish.

A surface cut up by moldings into large and small spaces should not have an equally heavy coating, as the smaller spaces will not support as much varnish as the larger ones; and where the moldings join, forming corners, allowance should be made for the tendency of the varnish to collect, and by its own weight follow along, guided by the moldings, and at length sag down. It is better to lay on a heavy mass of varnish on the large spaces, and afterward work it up toward and on the moldings, than to begin on the moldings first.

"Enameling" is an appearance as if a piece of enameled leather had been pressed on to the surface of the varnish. The term is used by carriage painters only; or, at least, it originated among carriage painters.

A coat of varnish, put on another which is not hard, will often shrivel up, because it has no chance to set evenly; the under coat being checked in drying, the outer coat will sink in, and the two coats combine, forming, as it were, one coat, neither of which can dry properly.

Rubbing varnish, added to a finishing varnish, will act in the same manner, unless they be thoroughly incorporated by remaining together for several hours.

Rubbing varnish drawn off at the bottom of the can will be apt to enamel.

PITTING.

"Pits," or pock-marks occur under the following circumstances:

1st. When japan is added to force the drying.

2d. When a quick-drying varnish is added to one of a high grade.

3d. When a storm is gathering in hot weather, before the rain falls, varnish that has been applied will often pit.

4th. Steam rising from a wet floor, in a close room, and having no means of escape.

5th. Difference in the temperature (in winter) between the varnish and body, or the room cooling off at night before the varnish is set.

6th. Laying on the varnish too heavy without properly brushing it through.

7th. When the walls attract dampness in summer.

A varnish that pits will first appear dull; the next change that follows will be collapsing of the surface.

When a job has pitted, take a varnish brush, and, beginning at the top, lay on a heavy coat of turpentine, and as the varnish is softened up, scrape it off with a piece of harness leather sharpened on the edge like a chisel. Scrape the varnish up and put it in a cup. Continue this until all is removed, then with soft rags and clean turpentine wipe it over again; the stickiness of the turpentine remaining on the surface remove with a mild soap. The pumice stone and rag must follow, and when the job is perfectly clean, and again touched up, a second trial will probably produce a perfect finish. It is a waste of time to attempt to rub out the pits.

"Sagging down" off the panel is wholly the fault of the varnish; it is too oily, and will not remain "where it is put." *Remedy*—send it back.

"Fine dirt." When a varnish is laid on clean, and looks full and rich when finished, and the following day appears as if it had been sprinkled with dust, it lacks body, either from being too new, or from the process of manufacturing.

A high grade varnish, that has become very cold while in the tanks, will sometimes act thus, we have been informed, but we cannot speak from our own experience.

A very heavy coat of varnish, on a body or gearing, or a very light one, neither produce the best work; a medium coat should be the rule.

Having overlooked the cause of varnish clouding or smoking over, we will here give attention to it.

Rubbing or leveling varnishes, and the best quality of finishing varnish, each presents this appearance at times. In answer to a question we asked an old experienced varnish maker, some years ago, he replied, "that there were some qualities of gum that would invariably cloud when spread on a surface, and that it was an evidence of inferiority in the gum." We will not deny the statement, but our observations in this direction do not bear out its truthfulness.

Finished work clouds over while standing in the wareroom or repository, because it is kept from sunlight and air, and is more or less exposed to the sulphurous vapors from coal smoke—the best English varnishes forming no exception.

Exposure to sun and frequent washings will restore the gloss in a great measure; but when a carriage has stood for several months, the better plan is to re-varnish it.

PAINTING COACH.

Having previously spoken of the tools and appliances used in the paint room, and of colors, oil, varnishes and japans, we will take up a coach and carry it through from the priming to the finishing coat of varnish. We select a coach, for the simple reason that it will cover all lesser styles of work. The bottom, top and inside painting or slushing is generally done in the wood shop, so that when the body arrives in the paint room, those parts are one coat in advance of the remaining surface.

As the top and inside of the panels require more attention than some manufacturers give to them, it will not be out of place to mention them in passing.

The painter cannot be held responsible for green timber in a top, which causes the covering to rise up in ridges or blisters, nor for carelessness on the part of those who draw on the muslin, in not rubbing it down to prevent air from getting under it; but he should feel sufficient interest in his work to see that the inside of the top shall have a good, heavy coating of slush or oil lead, to preserve the wood from dampness, and on the outside of the top that it is properly primed—not with dirty or skinny paint, but clean, smooth lead. When this is dry, the nail-holes and low places should be puttied with a firm drying putty, which will bear blocking down with sand-paper, leaving the top as level as possible. When this is dusted off clean, apply a heavy coating of smooth lead in oil, with sufficient varnish in it to hold the lead together, so that when the muslin is drawn on it may be rubbed down, and, as it were, bedded into the paint. The roof will then be perfectly air-tight, and in good condition for filling in. If, on the other hand, the muslin is drawn on over a rough, uneven coating, the highest points will be rubbed through before the surrounding surface is brought down level; the water gets under, and puttying will not prevent the cloth from raising up after the coach is exposed to the weather.

The inside of the body should be well coated, which is a great protection to the panels, especially in the process of rubbing out the rough stuff. The priming coat should be mixed of the best pure keg lead and oil, with only a small quantity of dryer, and allowed at least one week in which to dry. This first coating is a very important one. It should be thin; we do not mean that the paint should be very thin, but that the lead should be well worked into the nail-holes and grain of the wood, leaving only a thin film on the outside. A well-worn, springy brush is the best. When this coating is dry, sand-paper it carefully, and apply the second coat of lead, using less oil. The third day thereafter putty the nail-holes half full. The second day after puttying apply the third coat of lead, mixing to dry firmly, using no oil except what the keg lead furnishes. When dry finish puttying the nail-holes, and plaster up the grain that may be very open, leaving no putty on the outside; or, rub into the grain lead mixed up heavy.

THE ART OF COACH PAINTING.

The body has received three coats of lead, and is puttied. It may now stand two, three or five days, as time will permit. When again taken in hand, sand-paper off any putty that may be above the level of the surface; dust off and brush on a level coat of lead, which must also dry hard and firm. Each coat of lead should be laid on as level as possible, and be made to do its share of filling the grain. The roof should not be slighted, but the bottom and inside of doors may have one coat less. The body may now be set aside for three or four days, when it will be ready for the rough stuff.

The rough coatings should dry very firmly, possessing only sufficient elasticity to bind them to the surface. The first coat will bear a trifle more oil than the remaining three or four, and should stand about four days; the other coats can be put on every second day. Four coats and a guide coat should fill up well. The foundation coats fill better by being laid on cross-wise alternately.

If English filling is used, mix one part of keg or oil lead and four parts English filling; mix up stiff in a good brown japan and raw oil. The oil should be in the proportion of one-fourth to three-fourths japan. Varnish may be added or not, as the painter may think best. We have serious doubts as to its adding any to the good qualities of rough stuff. It tends to hold the particles of paint closer together, and probably makes the rough stuff rub tougher. But it cannot be doubted that varnish is more liable to crack when exposed to the weather than oil, and to whatever extent varnish is added to paints, to that extent is it more liable to crack. But the common practice is to add varnish to rough stuff, some painters holding that American and others that English varnish is the best. In this country, where painting is done (by steam) we were about to say, it cannot be expected that it will wear any great length of time.

The time allowed for filling in a coach body, as laid down in the preceding number, may be extended as far as circumstances will permit. We do not think, however, there is any positive gain in the wearing of the foundation coats by allowing weeks to intervene between coatings. When a coat is hard, it is ready for another; and it is far better to have the body filled and set aside, than to divide the time between coatings, and probably be compelled to rub out the body before the last coat is firm. The English method of using a large proportion of oil would require considerable time for each coat to harden; but as our practice is so different, there can be no necessity for it. The first coat of rough stuff should be applied somewhat thinner than the others, and, in some instances, would be improved by being mixed one-half white lead. It should also be made more elastic than the succeeding ones, as it will then take a firmer hold on the "dead" lead coat over which it is placed, contributing a portion of its elasticity to that coat, and also cling more firmly to the hard-drying coats which follow.

The first coat should cover every portion of the lead surface; be well brushed in yet not allowed to lie on heavy along the moldings or in the corners. It should be fine, and laid on smoothly, as it is better in rubbing out to leave on as much of it as possible. The remaining coats should be applied reasonably heavy, laid off level, and kept from lapping over the edges, or rounding the sharp corners, thus destroying the sharp, clean lines given by the body-maker.

Any defects noticed, while filling in, should be puttied; ever keeping in mind that the perfection of finish aimed at is secured only by care at every step taken. The guide coat is a mere stain, put on thin, and intended to assist the rubber in detecting the unevenness of the surface.

The body having been filled in, may be set aside to harden; or, if the smith is ready for it, now is the best time for him to take it, as any dents, burns, or other mishaps may be remedied without being detected. When it has returned from the smith shop examine it closely. If there are any bruises, putty them; any burned places which will show, scrape the paint off to the wood. Prime the bare spots, and putty and fill them as well as time will allow. The rubbing or leveling should be in the care of an experienced man. Moldings should invariably be rubbed out before the space surrounding them (to prevent cutting a "gutter" or groove along the panel) and corners and narrow spaces before the adjacent surface.

Use English rubbing stone for the main part of the leveling, finishing off with pumice stone. It is preferable to begin on top, and follow on down, so that the filling water may not run over and dry on any part that has been finished. Water should not stand for any length of time on the inside of the body; and when the rubbing is completed, wash off clean outside and in, drying off with a "chamois" kept for the purpose.

The body, when dry, is to be carefully sand-papered with 0 or No. 1 sand-paper, the corners cleaned out, and a coat of dark lead put on. When dry, scratch over the lead with fine sand-paper, which will make the lead appear of lighter color; the low places will then be apparent, for they will show dark. Putty up any imperfections with putty, mixed of lead and varnish, and when dry, face down with lump pumice and water; follow with fine sand-paper, when the surface will be in condition to receive the color coats. The coat of dark lead may be omitted, and the color applied immediately to rough stuff; but us a coat of lamp-black or drop-black must be substituted we think advantages are in favor of the lead coat.

The colors are to be ground very fine, kept clean, and spread on with camel-hair motlers (blenders); sizes ranging from one inch up to three inches. If the panels are to be painted different from the other parts, lay on the black first; for if any black falls on the panel color, it occasions some trouble, and will sully the purity of a transparent color. By repeatedly turning the brush over, there is less liability to accidents of this kind. Each coat of color should be haired down. Two coats, and color and varnish, are sufficient for opaque colors, and the same for those that are transparent, if we do not count the ground-work. Each coat of varnish should have not less than three days to dry—four or five days would be better. The first coat of rubbing varnish may be applied thinner than those following (when color and varnish has not been used) to prevent staining the colors. When the first coat has been rubbed over lightly, the moldings, or edges, separating the panels, may be blacked. The black should be glossy by the use of varnish; this is to bear up on the second coat of varnish over the moldings.

The pencils used on moldings should be large enough to cover them at once, and the color run on, avoiding laps, except at or near the corners. When the moldings are dry the body is to be washed off clean, and the second coat of varnish spread on. Avoid the use of turpentine, if possible, but if the varnish is heavy and dark, the amount of turpentine introduced sufficient to make it flow pleasantly, will not injure it. The half elastic and fine bristle brush are better adapted to the working of heavy varnish than the sable or badger. In varnishing a coach, you may begin on the roof, bringing the varnish to within two or three inches of the outer edges. Next, the inside of doors, etc., then the arch. When these are completed, start on the head rail on one side; lay the varnish on heavy, and follow quickly to the quarter. The edge on the roof, which was skipped before, is to be coated and finished with the outside, thus preventing a heavy edge. Continue around the body, finishing the boot last.

The frames, and other loose pieces about a coach, should be brought along with the body, and not *slighted*, as is too often the case. The frames are most conveniently handled by the use of a device made similar to that which secures a swinging mirror—a base and two uprights, stoutly framed together, allowing space for the frame to swing.

It is held in position by two pointed iron pins; one stationary, the other movable, which is made of ¼ or 5-16 round iron. One end bent, forming a crank; a thread cut on the axis, which must work in a washer the same as a bolt and nut. One side of the door or glass frame is pressed against the stationary point, and the other side screwed up until it enters the edge on that side, holding the frame securely.

While varnishing, the painter may examine his work, and set it any angle, either to detect dirt, or have the varnish correct a tendency to flow out of corners, etc. When a frame is finished, it may be set aside, and another inserted. The different rubbing coats are applied in about the same manner, and upon them depends the whole beauty of the finishing coat. If the former are wavy, from not being leveled down, or contain runs and grit, the finisher will despair of producing a perfect finish; in fact, the last coat will only bring out more distinctly all imperfections.

"Well then," we hear you say, "how shall we lay on the rubbing coats, and how level them, and remove waviness and runs?" The question is more easily asked than answered, for it is difficult to write down that which is acquired only by years of practice. We will, however, make the venture, first stating certain points:

1st. The varnish should not be patched on.
2d. On all large connecting surfaces, more varnish should be applied than it is intended to leave on.
3d. When a narrow space connects with a larger panel, the whole must be completed together.
4th. The varnish should be brushed "up and down."
5th. In rubbing or leveling, all brush marks and faults must be corrected.

Having previously stated that the beauty of the finishing coat depends almost entirely on the perfection to which the rubbing or leveling coats are brought up, we will now endeavor to explain the points assumed, first directing our remarks to the use of rubbing varnish.

1st. *The varnish should not be patched on.*

By this we mean, that the varnish, when applied to any part of the surface of a body, should be brushed over a certain part, previously decided on, which will, when completed, form a connecting whole, without laps, and not, as is the practice with many painters, lay on the varnish only the width of the brush at a time, and that so lightly that when a fault appears the varnish is found to be set too much to be worked and blended together. The attempt to rebrush the defective part, be it a run or heavy lap, will be found of no avail. The unsightly joint, if such it may be called, will not unite with the rest of the varnish, and nothing but the strong arm and skill of the rubber will remove it. Just here we would say, that *a first-class varnish rubber is one of the most important hands in a body room.* Finishers fully understand their value, but do not, we think, as a general statement, give them, *openly*, their just dues.

2d. *On all large connecting surfaces more varnish should be applied than it is intended to leave on.*

We will take, for example, the upper back panels of a coach. The back light, or window, and the drip molding over it, divide the surface into what we might term four spaces; two vertical portions, and two which are horizontal. To lay on a level coat over this portion of a body, the varnish should be put on quite heavy. No attempt should be made to level the varnish until a sufficient quantity has been put on over the whole surface to insure against its setting.

Having applied it as stated, run through it quickly with the large brush, and with a smaller one carry a portion of the varnish upon and underneath the drip molding.

This first laying off of the varnish is designed to connect it over the whole surface; and while the skillful varnisher will spread it with a view to having about an equal quantity over every part, he will not care so much for that as to be certain that he has a sufficient quantity applied to insure it against setting before he has properly manipulated it.

His next care will be to remove a portion of the varnish above and beneath the drip molding, and around the back light, to prevent it from sagging, after which he again addresses himself to the panels, this time with a view to the proper leveling of the whole. His quick eye detects those portions where the varnish is heaviest, and by up and down or by cross brushing, as it may require, the varnish is leveled off. Again the tool (small varnish brush) is passed around, or under the moldings, and then the panels are brushed horizontally over every part, and finished by vertical strokes.

During this whole operation the painter must work quickly, and aim to finish the portion in hand before the varnish sets, so as to allow it to "flow out." In the process of leveling off the brush should be occasionally cleaned out, by wiping it on the wiping cup; for, as we stated before, more varnish has been applied than it is designed to leave on.

Varnish laid on in this manner will show no laps; for, having been finished together over the panel, it will flow out together, and appear as one undivided whole.

3d. *When a narrow space connects with a large panel, the whole should be completed together.*

We mention this separately from the second proposition, because it is the practice with some varnishers to finish each part separately, thus giving a broken connection, as where the head rail space joins the back quarter. As the remedy has been laid down in the plan of varnishing the upper back panels, it will need no further explanation here.

4th. *The varnish should be brushed "up and down."*

That is, it is to be brushed vertically. "Up and down" is the language of the shop, and, of course, it comes natural to use it. Now, a panel an inch or two wide, a narrow belt, or space, cannot be so brushed; but experience has proven that larger panels require to be laid off in this manner.

The advantages are, that a greater quantity of varnish can be laid on with less liability to run or sag; the varnish flows better, and the result is, better work is produced.

5th. *In rubbing, or leveling, all brush marks and faults must be corrected.*

When the first coat of varnish is being rubbed, it is neccessary to merely remove the gloss, as in the attempt to remove grit, etc., the color might be disturbed. The second and third coats will bear to be well rubbed. The second coat will hide the defects of the first coat, and the third improve on the second, thus preparing the surface for that season of anxiety—the preparation for, and the applying successfully of, the finishing coat.

Before this interesting period in the history of every carriage has been reached, much hard labor of body and mind must be endured. Runs, bruises and burns may have to be corrected. Runs on the second coat are more easily corrected than when they appear on the third, as they get two coats, which is also true in regard to other defects. If, after a job is varnished, a run is discovered which cannot be corrected with the brush, it (the run) should be rubbed over as soon as the varnish is hard enough to bear it, allowing it to dry until the body is rubbed for the next coat. By this means half the run is destroyed, and the next rubbing will remove it entirely. If this course is not pursued, the run will be softer than the level portions of the surface, and will not come down under the leveling block.

In leveling the coats of varnish, use pieces of cork, wood or lump pumice stone, cut to suit the various spaces, and covered with two thicknesses of cloth. Block down both small and large spaces, using a somewhat coarse, though even grade, of ground pumice, rubbing across the brush marks. Pumice stone should be used freely, and the strength of the arms and shoulders laid out on the work. Brush marks, and other defects, *must* be rubbed out. To do this, it is not necessary to draw the finger over the panel every moment to look at something you know not what. The wet panel will show the ridges far better than the part dried off by the finger. By drawing the rubbing block across the ridges, they will show whether they are rubbed enough or not. If you have laid on a proper second coat, you need not be afraid to lay out your strength on it. Cut it down well, slighting no part. To rub close up to the moldings, and in sharp corners, have a stick of hard wood cut at one end in the form of a knife blade, and place the cloth over it (one thickness) placing the sharp edge up to the molding, which will, by a few strokes, remove the grit that may be there. The coarse pumice will grow finer in the process of rubbing, which will answer to finish with. Or a fine grade may be used, kept for the purpose. No pumice should be allowed to dry anywhere in the corners.

Rub next to the moldings *last*, for there is more danger of rubbing through, as there is less varnish here than out on the panel. Having slighted no part, wash off clean, touch up where any sharp edges have been cut through, and apply the third coat, which should be as carefully handled as the finishing one. When this has stood a few days, give it a light rubbing, and let it stand until ready for the finish. It will then be well dried and cut down nicely; the grit, if any, will not tear out. If the third coat has been laid on properly, it will not require so much rubbing as the second. The pumice should be fine and even, so as to avoid scratching. The sharp edges should not be rubbed off bare, for it is preferable to slight them in the rubbing rather than to be compelled to touch them up. The practice of having to touch up the edges all over a body, previous to laying on a coat of varnish, is a very foolish one, and may be avoided in a great measure by rounding them slightly when the body first comes from the woodshop.

The after coatings will sharpen them sufficiently not to be detected; but where this precaution has not been taken, slight them in rubbing, as the third coat should be well nigh as perfect as the last or finishing. The experienced rubber will look carefully over each panel before putting on any pumice; and where there may be grit, or any slight defect, he will give these his especial attention.

Having leveled the surface, less pumice should be used, and finally scarcely any. The panels should be finished off by passing the rub cloth throughout their length, bringing them to a polish. A body that has passed through the hands of a first-class rubber will, when washed clean and dried off, present a beautiful appearance. The finisher, as he surveys it, will feel a sense of pleasure, and at once decide to spare no pains in the effort to complete the job. In our largest and best regulated shops the finisher is not a man of all work—one day with a pot of lead in his hand, the next coloring; now in the body room, and then in the gearing room. Oh, no, nothing of the kind. He is a finisher in the truest sense, and is not even required to rub varnish. Still further, he does not varnish the roof, arch and inside edges of the body. He has *help* in the finishing room, which varnish all except the panels. These he attends to, and through his skill, attained by daily practice, the finishing coat is put on so nearly perfect that, to painters who have never had like advantages, surprise and mortification will be mingled as they gaze at the beautiful work, and remember their own wavy and dirty jobs.

The method adopted in laying the finishing coat is similar to that in the use of rubbing varnish. The varnish is applied heavily, leveled by repeated brushings, and carefully examined during the operation to detect any foreign particles that may appear.

A picker is used, made of whalebone, sharpened to a point, or any other device that will remove the particles of dirt or gum. Having brushed the varnish throughout a given surface, let it stand a few moments, when the bubbles will evaporate, thus leaving the hard particles remaining, which must be removed. The finishing strokes are given lightly.

As it is beyond the power of words to give a perfect description of the manner of laying off this last coat, we will have to leave the inexperienced to gather whatever may be of value from the foregoing, and by care, good taste, and an unbending purpose, supplement our hints by *practice*.

In conclusion we would remark, that to secure a perfectly clean piece of work, the *room, body, cups and brushes, and the clothing of the varnisher himself, must be scrupulously clean;* for, without these precautions, it will be madness to make the attempt.

THE CARRIAGE PART, OR RUNNING GEAR.

Having previously directed attention to the body only, we will now devote a certain amount of space to the carriage; for in all well-appointed shops the bodies and carriages are painted in separate apartments by a distinct set of hands.

This division of labor is not only productive of more perfect work, but it is more profitable to the employer, and, we may add, gives better satisfaction to the employees.

As, in the body room, it is requisite to portion the different parts of the work among different hands, so also in the carriage room there must be a systematic division of labor. We have known employers who would sacrifice a part of their profits on a foreman getting twenty-five dollars a week, to satisfy their whim that a foreman should not be afraid to sand-paper, lead and putty, grind paint, and do other dirty work which could be done as well by an apprentice or a half-way painter, earning from three to seven or eight dollars a week.

In the carriage room, then, there should be those who attend to the rough work, others who color and varnish, and one or more stripers. (We are speaking of factories where several painters are employed.) To secure good and profitable hands, men should be selected to perform those parts best suited to their natural tastes.

A man who may have no taste for striping may make an excellent finisher, and it is not uncommon to find a good striper who cannot, on a carriage part, lay a passable coat of varnish.

The carriage part, presenting only narrow surfaces, and the greater part of these convex in form, does not require the time or labor that has to be bestowed on a body. A most important part of carriage part painting is to have the work well primed; when the priming is entrusted to careless or inexperienced persons it is seldom that the wood is coated as it should be. Where the spokes join the hubs and felloes, bare spots will be left, giving the water a chance to soak in and scale up the paint. The after coatings of lead, color and varnish will protect the wood as long as they remain firm; but when they crack, the water will search out the slighted spots, scaling the paint off down to the wood.

Oil lead for priming has, of late years, been replaced to some extent by the use of different kinds of what are termed wood fillers, which penetrate and fill up the pores of the wood. The carriage part, after having been primed, should go into the smith shop; and when it returns, if there are any greasy finger marks, remove them, and, after a careful sand-papering, dust off and apply the second coat.

Putty on the second coat of lead, sand-paper when dry, and give a third and sometimes a fourth coat. Whatever plan is adopted, the aim must be to fill the grain perfectly and produce a serviceable and a perfect finish throughout.

The carriage part, being more flexible than the body, and required to bear up the weight as well as perform all the rough part of the work, it cannot be expected of it that the painting should wear as well as that on the body. Almost every portion of the carriage part is brought under heavy strain. The wheels are subjected to severe concussions at times, and the felloes and outer ends of the spokes to the grinding and bruising of gravel and paving stones. The springs, being composed of separate leaves or plates, and required to be ever changing their position according to the state of the road, the coatings of paint and varnish cannot long remain as perfect as when they were finished. The face edges of the leaves being rounded, there is formed longitudinal grooves between them, which retain more paint and varnish than the faces; and when the springs are in motion it is apt to flake off.

The coatings on a carriage part should be possessed of sufficient elasticity to cling firmly to each other, and not be applied so "dead" as to form only an enamel over the priming. The first and second coats should dry glossy, the third and fourth coats "dead," the colors to be laid a coat every day. If both broad and fine lines are to be put on, give the last coat of color sufficient varnish to bear hairing off. On this coat run the broad lines, and when dry give a coat of rubbing varnish. Rub this coat with ground pumice stone and water, and run on the fine lines. Finish on this with a medium heavy coat of American or English Elastic Gear Varnish. All the parts of a coach or carriage should be equally well finished; for nothing detracts so much from the beauty of a piece of painting as to see a well finished body and wheels, while other parts look rough and slighted.

Those parts requiring to be much handled in hanging off should not be finished until they are on and screwed up, when they may have a touch of the rub cloth, and receive a coat which will not be soiled by finger prints.

Thus, by care and forethought, the job may be sent to the wareroom perfect throughout, giving satisfaction to the employer, and he in turn will feel warranted in recommending to a customer a piece of work equally well finished wherever attention may be directed.

GROUND AND STRIPING COLORS AND VARNISHING.

The ground colors for the carriage part vary according to the freaks of fashion, therefore do not always follow the color of the panels on the body. During the "late unpleasantness," when money circulated freely, and the gay trappings of the military created a taste for flashy colors, we had quite a run of "staring" patterns to follow. The cavalry suggested yellow, orange and gold; the artillery, red; the zouaves, in their uniform, copied from the barbarians, gave us crimson, and it may be that we insensibly preferred blood colors, imitated with carmine, because our daily papers teemed with the recital of gory battle-fields. Whatever occasioned our wild and crude taste, no one will deny that it then existed. We are now returning to colors more subdued in tone, and less striping is being put on.

The method generally adopted is to paint the carriage parts one or two tones lighter than the color of the panels of the body, except where the panel color is of a hue that will not admit of it. Certain shades of green, blue and red may be used on panels, but would not, when made a tint or two lighter, be suitable for a carriage part. Dark browns, claret and purple lake would not be open to this objection, because, to the majority of persons, they are colors which are pleasing to the eye in both their deep and medium tones.

When the panels are to be painted green, blue or red, and the painter wishes to carry these colors on to the carriage part, it is the better plan to use them for striping only, and this but sparingly—the ground color to be black.

A carriage part painted black may be made to harmonize with any color used on the body, as the striping colors can be selected so as to produce any desired effect. Brilliant striping can be brought out on dark colors only, while if the ground color be light, recourse must be had to dark colors in striping in order to form a contrast. On heavy work we prefer dark colors; on very light work, light colors, rich in tone, are appropriate. The carriage part should not detract from the appearance of the body; that is, there should be sufficient contrast between the two to bring out the beauties of the body. A plainly finished body will appear to better advantage on a showy gearing, and a richly painted body on one that is not overwrought. Ground colors, of any hue that may be given, will accord passably well with a black stripe; but with the exception of white, straw color and similar delicate tints, drop-black of the best quality should be used, for lamp-black will appear gray or of a greenish cast. Lamp-black is cleared up, and appears dark enough on the light colored grounds above mentioned.

From black the painter may ascend through colored grounds by various mixtures. Browns are formed by the admixture of reds with black, and are dark or light according to the tone of the red and the proportions mixed. Drop or lamp-black, and Indian red, and vermilion, are used for ordinary purposes. Other shades of brown are formed from black and burnt sienna, Vandyke brown and burnt sienna, burnt umber and black, etc., and a richer tone of each is secured by the addition of a small quantity of lake. Lighter shades of brown require yellow to heighten them. These cannot be mixed where black is an ingredient, as yellow and black produce green.

Burnt umber, Vandyke brown, and others of this class, must be used, and when mixed with reds and yellows, give a range of brown tints well adapted to carriage parts. Burnt umber and orange chrome produce a handsome ground work, and when striped with drop-black and gold is very attractive. A very pretty style of striping for a dark brown of the above mixture would be to mix the broad line a shade or two lighter than the ground color, fine lining it with black and gold, or black and red, orange or tan color. Black striping may be omitted, but, in our opinion, to the injury of the ground colors.

Lakes, of the different hues manufactured, form a richer type of colors, and may be used pure, or mixed together when lighter tones are required. English purple lake, by adding to it carmine, Munich lake, and others which are brilliant, preserves the richness so desirable. Vermilion and inferior reds destroy their beauty.

Drop-black, used for striping on these colors, is very pleasing. Rich brown shades, either dark or light, relieved by fine lines that harmonize, are also very handsome.

Greens of a dark shade are used more or less, but this color is not so much admired by the majority of persons as to warrant its use to any extent on carriage parts. Olive-toned greens are considered the best adapted to coach work. A bright green is offensive to the large majority of persons; a dark, rich green, admired by but a few. Milori green, glazed with verdigris, or Emerald green, and striped black, should be the limit adopted, as the *cheap* mixtures are unsaleable.

Tan, straw, corn, Jonquil yellow, and all very bright tints, give a clean, fresh and light appearance, and are well adapted to light work. They are generally striped with black, browns and lake. Neutral tints may be used with good effect also.

In striping carriage parts, the painter should use bright colors very sparingly when he is ordered to put on the full amount of striping. A fine line placed on the face of the spokes and hubs, and distributed over the inside carriage, would bear to be far brighter in color than when each side of the spokes and the face, the hubs, the felloes, on both sides, etc., are to be striped.

Colors, such as clear chrome yellow, chrome green and cheap reds, should not be used on good work.

Mellow down chrome, and other bright yellows, with white, or give them an orange cast with red. Greens should be mixed of a pea green tint rather than those of a yellowish green. A green tint may be toned down with carmine or lake, and still preserve sufficient of a green cast. For striping white, use French zinc white, cremnitz, flake or china white. The first mentioned will be found to grind easily under the palette knife, and produce softer tints than those whites made from lead. The proper arrangement of colors having been noticed in our articles on the "Principles of Coloring," it will not be necessary to recapitulate them.

VARNISHING.

The coatings of varnish contribute largely to the durability as well as beauty of a carriage part. The ground and striping colors are shown in their purity only after they are varnished and have a good surface; and the after test of wearing depends on the quantity and quality of the varnish applied.

Every carriage part should have at least two coats of *clear* varnish. The first coat of clear varnish to be applied over the color and varnish; the second, a finishing coat, possessing a good body and wearing qualities. Ground pumice and water must be used to cut down the varnish, otherwise the finishing coat will be robbed of its beauty.

In laying on the finishing coat, apply it neither very heavy or very light. Put on a medium coat—a thin one will appear gritty and rough; one too heavy will sink in and grow dim. Use the best American or English elastic gear varnish for finishing.

STYLES OF STRIPING, AND THE MIXTURE OF STRIPING COLORS.

The variety of styles in striping carriage work are limited to a few in number, when we discard those which lessen rather than increase the beauty of the work, for a painter may spend two or three days in *hand* scrolling and striping a job, and after all his pains produce only a display of labor, without having added any real beauty.

In calling to mind the different patterns that have been and are being used, we notice, first, the fine line. In times past it was considered in bad taste to stripe a buggy or carriage with a heavy line, and to have *then* attempted to sell a job striped on both sides of spokes, felloes and other parts, would have laid the carriage-maker open to severe criticism. When the demand arose for a greater display, the fine line was enlivened by being dotted at the ends of each stripe, or in the middle, using the same color as that of the stripe, no striping being put on the sides of the spokes. We have, since then, the following styles:

A fine line on each side and front of spokes.

A fine line dotted at ends, or in the middle of stripes.

A fine line dotted and finished at ends with fine scroll.

A fine line finished with large open scroll.

Two fine lines of the same color, running parallel to each other, representing a broad line edged.

Three fine lines (triple stripe), representing a broad line edged and centered with fine lines.

These prepared the way for broad lines on the spokes and other parts, giving us, eventually, first, a single heavy line on the center of spokes, felloes, hubs, etc., and calling out the genius of the painter in studying up variations, among which we notice the following:

A quarter or three-eighths broad line of a modest color.

The same, centered with a contrasting fine line.

The broad line, centered and striped on front edge.

Broad line, centered and striped on both edges.

Broad line, centered, and fine lines placed at a distance from the edge.

As these gradual changes took place in the quantity of striping run on, the desire arose for gaudy colors, and modest and retiring colors were in a great measure abandoned, ushering in straw, orange, red and crimson, and these were followed and accompanied with a perfect shower of gold leaf. The flash style still obtains in some sections of the country, it being a difficult matter to sell work plainly finished.

Broad lines having been introduced on the spokes and every other presenting face front or back, over the gearing, the only change that could be made was in the manner of running the stripes.

1st. They were started at the hub and carried to a convenient distance from the felloe, the end squared by cutting them off with the ground color, and then followed

2d. Centering, edging and distance fine lines, all cut off an inch or so from the felloe.

3d. The same, except that the edging lines were connected across the end of broad stripe, as were the distance lines.

Then the stripes were carried from a point an inch or so distant from the hub to the same distance from felloe.

From hub to felloe, the entire length of spoke, then centered, edged and distanced.

Carmine glazed on vermilion was introduced, and had almost unbounded sway for a time. Half and half, as it was called, was short lived. This style consisted in striping one-half of each spoke, etc., with a color contrasting with the ground color.

The ground colors ranged from white through almost every mixture the fertile brain of the painter could invent, and often the puzzle was to select a suitable color with which to stripe a gearing.

Bodies were introduced with "bowls," "cut-under," or "wheel-houses," and paneled off so that different colors could be brought near each other. Striping both heavy and fine lines became fashionable, and on the bowls and cut-under a vast amount of labor was expended. During the late war there appeared to be a frenzy for display, and the efforts of carriage architects and artists to keep pace with it well nigh exhausted their ability to produce anything new, handsome and striking.

The compounding of the colors used for striping during these years which witnessed such a variety, may not be uninteresting. We will begin with white, and endeavor to mention the greater number used.

White Broad Line, which was mixed of clear china, flake, zinc, or cremnitz white. The last named possesses the purest tone.

Silver Color—White and a minute portion of drop or blue-black.

French Gray—White and drop-black.

Straw Color—White and chrome yellow, No. 2; white, chrome yellow and raw umber; white, chrome yellow and vermilion; white and yellow ochre; white and Oxford ochre.

Corn Color—White, yellow ochre and umber; white, yellow ochre, umber and a little black.

Light Buff—White and lemon chrome; white and yellow ochre.

Deep Buff—White, orange chrome and raw umber; white, lemon chrome and vermilion; white, chrome yellow, No. 2, and light red.

Cinnamon—White, yellow, light red and umber; burnt sienna, yellow and umber.

Canary Color—White and lemon chrome; poonah yellow glazed on white; white and drop ochre.

Cane Color—White, ochre, a little umber or black.

White being the principal ingredient in all of the light colors made use of, it will be readily noticed that a great variety of tints and shades may be produced by the quantity of any color that may be added. Therefore in compounding light colors white should be first placed on the palette or stone, and the colors added in small quantities, until the required tint is produced; and it is likewise necessary to compare the tint with the ground work on which it is to be placed. If the ground color is dark, the tint will appear, in contrast, to part with some of its color, and must be strengthened.

White, mixed with the various pigments termed yellow, gives tints lighter or darker, according to the tone of the pigment employed. White and light chrome yellow, mixed and compared with white and medium, or deep chrome, present considerable difference; and when the darker yellows are made use of, the change is still greater. The same is true of other colors mixed with white. Zinc white produces softer tints than whitelead.

Pink—White and vermilion; white and light red; white and Indian red; white and carmine, etc.

Flesh Color—White, vermilion and Naples yellow; white, Indian red and ochre.

Orange—Vermilion and lemon chrome; carmine and lemon chrome.

Drabs—White and raw umber; raw umber and yellow; white, raw umber and red; white, umber, yellow and black.

Blues—White and ultramarine blue; white and ultramarine blue, glazed with clear ultramarine; ultramarine blue, glazed over straw color; ultramarine blue, glazed over white.

Lilacs and Purples—White, vermilion and ultramarine blue; white, carmine and ultramarine blue; carmine and ultramarine blue.

Pea Green—White and chrome green; white and verdigris; white, blue and yellow.

Grass Greens—Chrome yellow and Prussian blue; chrome yellow, blue and black; chrome yellow, Prussian blue and Dutch pink.

Browns—Burnt sienna and yellow; burnt umber, yellow and vermilion; burnt umber, yellow and lake; burnt umber and orange chrome; burnt umber, yellow and Indian or light red; Vandyke brown, burnt sienna and yellow; drop-black and Indian red, black and vermilion; black, vermilion and carmine. A great variety of shades may be had by varying the proportions of the colors named, and by the use of others not necessary to mention.

Reds—Clear English Trieste vermilion; English vermilion and carmine; English vermilion, glazed with carmine; English vermilion, glazed with Munich lake; carmine glazed on Indian red; carmine, glazed on dark brown; carmine and Indian red.

The foregoing list comprises the majority of the hues and shades made use of; while in the attempt to find something new, other colors were accidentally formed, the mixtures of which would puzzle the painter to mix again, were he called upon to do so.

Green Leaf was very freely used, degenerating at length into bronze and paint imitations of gold leaf.

On all dark and medium grounds the light contrasting colors, either as fine or coarse lines, were used; on light grounds, various shades darker than the ground color. Wherever fine colors, either for ground color or striping, were employed, the finished work was greatly admired; while imitations condemned the work on sight.

In the proper selection of striping colors it would be well to remember the effect produced by the mixture of their complementaries. The eye, when directed at a color, is, after a few seconds, excited to see a faint image of another color. Green will call up red, and red green. Blue will excite the eye to see orange, and orange blue. If, then, you should place green and blue side by side, a confused effect would be produced. It would be similar to mixing green, red, blue and orange on the palette, which will at once convince the painter that a poor color would be the result. Red and blue, or green and blue, are less pleasing than red and green, blue and orange. Red and blue and green and blue should be separated by white. Yellow, orange, blue and violet associate well with black, while green would give to the black a rusty appearance, on account of the eye being excited to see red.

The painter should experiment with different colors, and keep a memorandum of the mixture of those which associate best, and keep samples of every job he finishes. These will also be of value in assisting a customer in selecting, and in speedily settling any objections he may make when his carriage is finished.

In conclusion, we would remark, the purity of striping colors depends greatly on the color of the liquids with which they are mixed. Dark brown japan should be seldom, if ever, used. Varnish—varnish, oil and japan, oil and sugar of lead; boiled oil; raw oil, and redlead simmered over a gentle fire for an hour or two, or any other light-colored mixture. Carmine and lakes requiring a more powerful dryer than any other colors.

Panel colors vary but little from one generation to another, so that the coach painter readily acquires the art of compounding those shades which are produced by the mixture of a few different pigments. Carriage parts and striping colors present a wider field for the exercise of good taste in coloring, but these, too, are subject to the caprices of fashion—a season of display being often followed by that of the plainest sort of work.

During the continuance of the instructions given, we have touched upon the paint shop, as it should be, calling particular attention to light, ventilation, the tools and devices requisite, the proper division of the work, the care of brushes, pencils, etc., economy in the use of stock, the nature of oils, japan and varnish, compound colors, etc., and also carried a coach through from the priming to the finishing coat of varnish. Wherever our descriptions of the working of the paint room may seem to be slighted, it must be remembered that language is too poor to express certain manipulations which are acquired by daily practice, and which cannot be communicated to another except by or through his powers of imitation. It is one thing to sit down and write out an article explanatory of laying the finishing coat of varnish, and quite another matter for he who reads the explanation to possess himself of that indescribable something which is the result of large experience, and which the writer has no more power of communicating with pen and ink than if he scribbled his paper without properly forming a single word. While this is true, valuable hints may be given which will tend to greatly improve those who have had no opportunity to see work finished in good style, and in the anxiety to bring their work as near perfection as possible, they will adopt a mode of handling the brush and varnish peculiar to themselves—different in non-essentials, it may be—yet producing what each one desires, a clean, level coat of varnish.

In conclusion, we remark that it has been far from an easy task to present the "Art of Coach Painting" in the form adopted; and, defective as it may be, they who attempt a similar exercise will probably learn that very simple things, done by the aid of tools, are often the most difficult to explain with pen and ink.

ALPHABETS.

In this connection is given three alphabets, viz.: "The Unique," "The Ornamented" and the "German Text." Of the "Unique" style we may say the letters are novel in design, and when singly placed or arranged into words, are very attractive. The letters are of an ornamental block pattern, being formed as if made of narrow strips of wood, beveled on the ends, and the pieces then placed in a position to form each letter.

Their beauty or oddity, whichever you please to call it, consists in their irregularity, relieved by a light line to the left. The proportions may be ascertained by measurement with the dividers. Square the letters, then take in the dividers the width, and ascertain how many times it is contained in the height. Having secured the proper proportions, you may enlarge them to any required size.

By drawing horizontal lines through the different corners or angles of each letter, the exact position of every part of them will be disclosed, and in enlarging, these lines will greatly assist in securing to every letter its precise form.

UNIQUE ALPHABET.

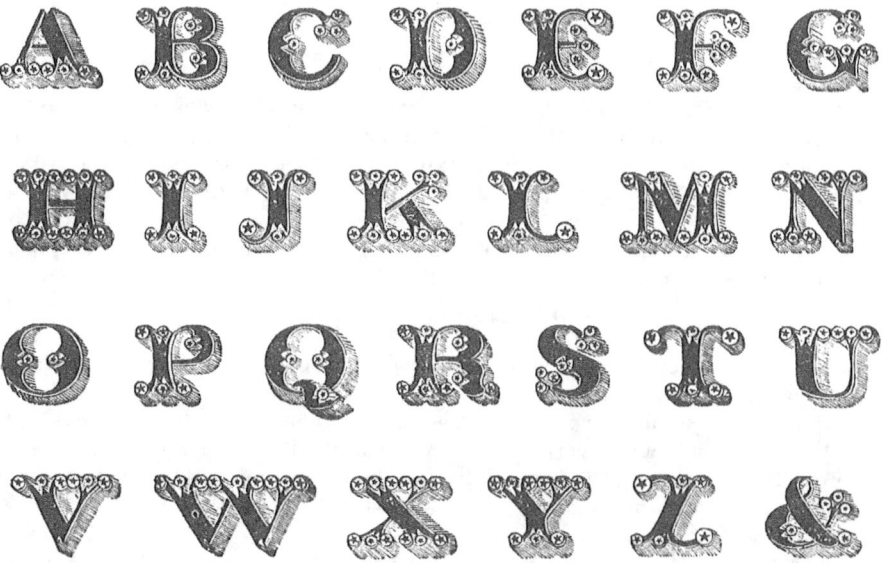

ALPHABET ORNAMENTED.

GERMAN TEXT.

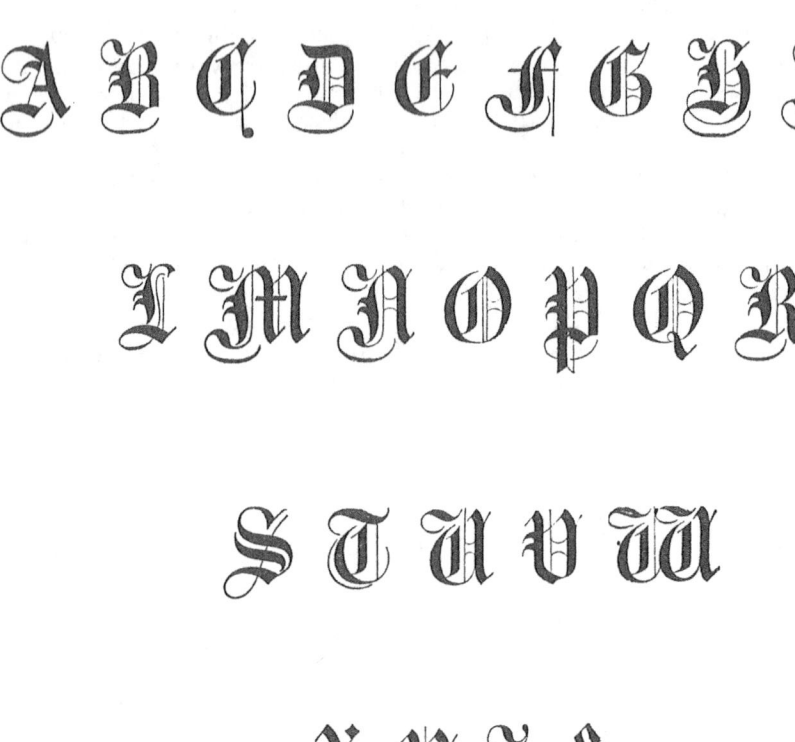

The second alphabet, which is elaborately ornamented, is well adapted to the requirements of the carriage painter. These letters may be enlarged if desired, and by preserving the same style of ornamentation, will produce either an attractive initial or monogram. The third alphabet, called "German Text," is unquestionably the most appropriate, and by the union of curved and straight lines the most beautiful of all alphabets. The letters are susceptible of still greater ornament than they possess naturally, without fear of rendering them illegible.

INITIAL LETTERS.

An initial letter well painted is not, in our estimation, a whit behind its rival, the monogram. The initial, 'tis true, stands alone, and can have no relief except by being accompanied with scroll work; but this is not a serious objection, as it makes up in distinctness what it lacks in variety, when compared with the monogram.

The practice of painting the initial letter on the side panels of buggies and carriages has been kept up for a number of years; in fact the initial is at all times allowed to be a proper style of ornament.

The variety herewith given will suffice to acquaint the painter with the value of single letters as ornaments, when they are properly enriched. The initial may be used also beneath a crest.

ENLARGING LETTERS.

To enlarge letters of any kind, inclose them by straight lines, which shall touch their height and width. This process will give you the squares of the letters. Divide this figure by two diagonal lines, carrying them beyond its corners to any distance you may please, and then mark above the figure mentioned the size to which you would have the letter enlarged; and through the point marked produce a straight line, which shall intersect with diagonal lines, and from these two points of intersection complete the outline of the enlarged figure.

This enlarged figure will be in exact proportion with the smaller one. Continue the horizontal lines to any required extent, and this fixes the height of all the letters. The width of the enlarged figure will give the square of the remaining letters, within which they should be laid off; remembering to make M and W somewhat larger than letters A, B, N, P, etc.

The most expeditious method is to use the compasses. If your letter is one inch high, and two-thirds of an inch wide, and you wish to enlarge it, take in the compasses its height, and step off one, two, or more times its height (or any fractional part of it); the height of one letter gives the height of all on the same line. The enlarged width of the letters is obtained by increasing them in the same ratio as the height perpendicularly was increased.

INITIAL LETTER B—GERMAN TEXT.

A form in which an initial letter may be ornamented. The dart and the center line of the letter divide the whole equally, and in producing this pattern were the first lines laid down. The outline of the letter followed, the two main curves which stand perpendicularly next, and from these the remaining scrolls were added.

In the attempt to paint this on a panel the novice would probably fail to produce a good ornament, for the simple reason that he would be compelled to mark in all the scrolls, and as it is impossible to give them their exact place on the panel, he would confuse the design or lose sight of it. The skillful ornamenter would mark on the main curves only, and afterward fill in according to the spaces that remained.

Lay in letter, dart and cord, gold. The rest with a purple tint, mixed of flake white, ultramarine blue and carmine; the lights with pink, made of carmine and white.

D—GERMAN TEXT,

possesses all the grace and beauty of lines that could be desired in a single letter. Its character as an ornament also must enhance its value with he who may be the possessor of the letter D as an initial.

In painting this letter on a panel it would be better to make the hair lines somewhat heavier than our engraving shows them, and correct also any defects in the curves. Observe the proportions carefully, and endeavor to re-produce them. The letter would appear extremely neat in gold, shaded with asphaltum and lighted with white; but as colors now are all the rage, we advise the use of a color agreeing in tone with that of the striping on the carriage part. If the carriage is to be striped light blue, carry the color up on to the body by painting the letter blue, and so on with other colors.

G—ROMAN ORNATE.

It will be noticed that the form of this letter is that of the Roman letter G, the only difference being in this, that the plain letter has been varied by the addition of leafing. The attempt on the part of the carriage painter to produce designs of this character on any letter of the alphabet will be rendered easy if he but remembers and puts into practice what we have just stated. Should he attempt to produce a full alphabet, the ornamental work on the several letters should be similar, for it would be out of character to have for instance the main stem of the letter divided as here shown in G, and the remaining letters or a portion of them drawn differently. In designing a monogram, full license is given to vary the form of the letters in order that each may be more easily deciphered, but, even with monograms, the English have of late adopted the plan of having the letters as near alike as their difference in distinctive shape will admit. An initial letter depends alone for beauty on the variety of form and colors that may be given it, and in order to secure these we ornament either the center, the upper or the lower half, and use different colors on the parts respectively. On this principle we coat G in its upper, a shade or two lighter than the under part, the colors to be selected so as to contrast well with the ground color on which it is to be painted. We may use light and dark tones of red, yellow and blue, or mixtures composed of one or more of these, with white. In using reds, the dark varieties form a natural shade for the light ones. Vermilion may be shaded with Indian red, Tuscan red, and any others darker than vermilion. Bright yellow, by those naturally darker and so on. Where richness in tone is required add carmine or lake.

TWO STYLES OF LETTER P.

The first one as here placed being smaller and of a plainer design will not require the skill requisite to paint the second. This letter will demand an eye quick to perceive, and a hand skillful to execute, in order to preserve its upright position, it being so completely clothed with curved lines.

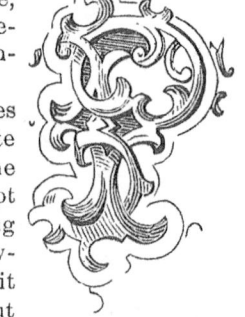

To acquire skill with the pencil, requires practice with the pencil; but the first requisite is to be able to mark out plain letters with the lead pencil or chalk. Having done so, attempt to vary the form of the plain letter by adding curved lines and scrolls. Having the knowledge of the plain letter in its proportions, it can be carried in the mind and applied without visible form or substance to the pattern on the panel, and this power of seeing a correct form and applying it, is that which gives the ornamenter his reputation as a skillful artist. From what we have said, the reader will perceive that in order to acquire a quick perception, and a hand which will obey the will implicitly, labor must be performed both of mind and hand. Wishing will not secure it; sighing after it without practical application will accomplish nothing.

By referring to the letter it will be noticed it is composed of circular and oval figures. The shank of the letter has curves springing out to right and left, both above and below the middle of the shank and the upper curved portion of the letter, instead of being joined to the shank, as in a plain letter, has its curved lines carried over, and twining around the upright portion.

Here the letter R is used as a single letter, accompanied by fine penciling, also beneath a crest. The latter method will be found to produce a very attractive ornament.

LETTER S.

Its natural form is graceful, being composed of curves bearing in opposite directions, and which fall into each other, presenting to the eye a continuous, but varied line. The ornamentation also falls into the usual shape of the letter naturally. The upper and lower ends of the letter terminate in three stems, covered by three-lobed leafing, and the main stem of the letter is preserved in shape by appearing to grow out naturally from its outer and inner edges.

Lay in the letters gold, over which work out the design with transparent colors, suitable to the panel color on which they are to be painted. If colors alone be employed, the panel color may be taken as a part of the coloring of the letter. For instance, if the panel color be dark brown, lake, blue or green, mix up lighter tints of either one, as the case may be; and considering the panel color as the darkest shade, lighten up from it. This method, when successfully carried out, will leave but a thin film of color on the surface, a matter of no slight importance, where the letter is painted on the last coat of rubbing varnish.

Some object to ornamenting on the last rubbing coat of varnish, because the finishing varnish is liable to draw over the ornament. But this objection is founded on the timidity of the finisher, he fearing to give the ornament a slight rubbing, when preparing the body for the finishing coat.

The letter V will doubtless please by the novelty of its ornamentation. The body of the letter retains its natural outline almost wholly. From the upper part of the left stem springs a scroll which curves downward reaching to the middle of the letter, and growing out from the first is a second scroll which serves to ornament the lower part.

Lay in the letter in color in harmony with the principal striping color, deepening the tone of the color on the stem of letter as shown by the shade lines. The leafing should be made out with light, medium and dark tints blended into each other, so as to avoid the scratchy appearance which an opposite method produces. Lake and vermilion, dark green and light green, dark brown and light brown, dark blue and light blue are used according as they may suit the work in hand.

MONOGRAMS.

To those who are skilled in the art these studies may be of no benefit, but to those who wish to learn we trust the instructions will be valuable. In drawing monograms

A B C D E F

you are at liberty to proportion the letters in any combination to suit your own taste. You may lengthen, widen or shorten one or more of them, or join two together, as we often see Æ in the word Ætna. The patterns A B C and D E F are formed on a parallelogram (or long square), measuring five by seven-eighths of an inch; or rather, we should say, that each letter is so formed.

By cutting out a piece of paper, five by seven eights of an inch, and laying it on either of the letters, you will perceive the variation from the right or straight lines. G H I are formed on a square of one inch. Here H is the main letter, because G and I are made subject to it; that is, H was formed on the square mentioned above, and the other letters drawn afterward. The letter H is proportioned so as to give room for the letters G and I to be properly displayed, still keeping the design as compact as possible. In painting a monogram you may use different colors for each letter, avoiding too great a contrast between. White, yellow, orange, etc., are advancing colors; browns, blues, etc., are retiring colors. Warm browns, olives, and blue, purple and greens, of a proper tint, follow yellows and reds into shade. Suppose you wish to paint for practice the pattern G H I. Having secured the outline, distinctly marked on a piece of patent leather, or other material; take burnt umber and vermilion, and mix up three hues,

G H I

the lightest place on G, the next tone on H, and the darkest on I; when sufficiently dry, separate the letters where they interwine with a wash of asphaltum, and light G with pure red, H with the color of G, and I with the color of H. Other colors may be used in the same manner. Practice alone will make the eye sensitive to the gradation of colors, tints and shades necessary to produce a pleasing ornament. The above is a very simple set palette, but will, to the new beginner, prove useful in preparing for more difficult tasks.

The monograms J K, L M and N O P are also drawn from square lines.

J K and L M, being ornamental, will require some delicacy of handling. N O P are

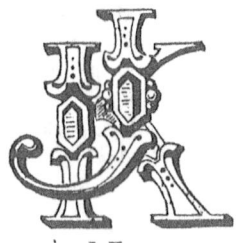

J K L M

plain, but locked together so as to form a pleasing variety.

We gave a simple palette for painting the monogram G H I. Three colors were selected, viz.: vermilion, burnt umber and asphaltum. We will now add flake white, chrome yellow, No. 2, and carmine. By adding white and yellow to burnt umber and vermilion we obtain a brighter tint of each, and by the addition of a small quantity of carmine to asphaltum we produce a rich, warm brown for glazing.

Having secured a distinct outline of the monograms, take flake white, chrome yellow, No. 2, and vermilion, and mix up a bright straw color and place it to the right on the palette board. Next take burnt umber and vermilion and mix up a red brown. Take a portion of the straw color, and form two or three tints by adding to it red-brown in different proportions. Place them on the palette with the lightest to the right, the darker ones ranging toward the left.

N O P

We now have straw color, two or three deeper tints formed by adding red-brown; vermilion, red-brown, burnt umber, carmine and asphaltum. For convenience in description we will number the tints, 1, 2 and 3.

Coat J K with No. 2, separating the letters, where they cross, with tint No. 3, and define the leafing, etc., with red brown on the shadow side. When dry, high light with straw color and deepen the shades, where needed, with asphaltum tinged with carmine, which forms a rich brown glaze.

L M and N O P we will leave to the taste of the student. A pleasing effect is always produced where the tints employed in ornamenting are mixed of a few colors, but differing in strength.

Thus, from a dark brown you approach the high lights through light brown tints, orange, yellow or straw color, and so with other colors. Practice alone will make perfect. By using few colors at first and mixing them among each other many useful lessons may be learned.

Q R S. T U V. The patterns here given will not be found difficult of execution. T U V, to be properly displayed, will only require a distinct outline of each letter, after which size them in, and lay on gold leaf. When the gold is laid, separate the letters where they come in contact, or where any part of one letter laps over that of another, by a wash of asphaltum enriched with carmine. Imagining the light to come from the left, let every part of the pattern on that side be lighted.

Q R S

Put on the shades first; where a leaf is to be brought out, merely give the form of the leaf on the shaded side; where a circle is to be displayed on any part of the letters, sweep around the right hand side, and as much of the bottom as would naturally fall in shadow. Carry out this plan throughout the minutest details.

When this is finished, take a clean, fine-pointed sable pencil and *flake white*, or *flake white* tinted with *chrome yellow*, No. 2; or *flake white* tinted with *light red*, and finish the details in the particular form each part requires.

To light and shade an ornament or monogram in this manner requires practice in pencil drawing. The painter must be prepared to use the sable pencil with as much confidence as the lead pencil, and fully understand the pattern he attempts to paint. When every stroke of the pencil is applied with

T U V

precision the ornament or monogram will have a clean, sharp outline in every part.

Q R S to the new beginner, will be found more difficult to handle. Q and R should be laid on first. Great care should be taken that the spaces be left sufficiently open, so that each letter may be properly shown. S.—The beauty of this letter will depend on the gracefulness of the sweeps. The best pencil for outlining this letter, or in forming curves so small, is a fine line sable, three-eighths or half an inch long. Lay in Q with *gold leaf*, except those parts covered by R and S.

Paint R a rich pink, mixed of *carmine* and *flake white*. Lay the color thin, and light the letter with the same color used heavier.

A dark green will have a good effect in painting S. Outline the letter, and fill in with perpendicular lines as shown. Those parts to be lighted should be touched with lighter green. The form of the outside border or shield should be laid in with gold, and shaded and lighted, as before stated.

W X Y

Z &

W X Y are simple in form. The letter X was first drawn, and by the addition of a stem to its upper half the letter Y is indicated. W was then placed in its position as a plain letter, and on this the ornamental character was worked out.

Lay in the letters with burnt sienna and white, high light with white or straw color, and separate the letters with a wash of asphaltum and carmine. Paint the outside border with light blue of different tones, the deepest shades with purple, mixed of ultramarine blue and carmine, or scarlet lake, high light with white. Z & may be laid in with gold, and shaded with asphaltum, so as to represent the pattern as given.

—A P or P P A.—This pattern forms a neat combination, one that will be entirely free from confusion of lines when painted on the panels of a carriage. The ornamentation is of a character that adds sufficient richness of lines, without being liable to the charge, "over-wrought," or "fussy." Paint P P dark blue, lighted with light blue and white. A.—Gold shaded with verdigris, darkened by adding asphaltum, and lighted by the use of straw color: or reverse the process by making P P gold, and A dark blue, lighted and shaded as before stated.

A P

PS A drawn within a parellelogram (long square), measuring one inch by seven-eighths, the letter A having been marked off by the two inclined lines necessary to form the sides of the Roman letter A. The ornamental character was given by curves made according to taste, yet governed by the foundation lines in preserving the position the letter shows in the cut. S followed, being so placed that a portion of it should form the horizontal bar, made use of to connect the sides of an A, and also to form the upper curved portion of the letter P, which is completed by the stem being placed against the middle part of S.

P S A

Lay in A gold, shade with burnt sienna or asphaltum, and light with either white or straw color. The upper division of S light blue (ultramarine blue and white), lighted with orange (carmine and yellow). The lower portion of S purple (ultramarine blue, carmine and white), also the stem which forms the letter P.

From the light blue of the upper portion of S follow down to purple on the lower part by shades of purple, increasing in depth, and light with orange.

U S A.—Formed to represent unity and strength, and the American colors. U is partly formed of the "Bundle of Rods," the lower portion of the letter being dependent on letter S for its curved shape. S carries the stars representing the States, and A the stripes on the national flag. Color U in imitation of yellow pine or oak; S with light and dark shades of blue; the stars white or gold, and A red and white.

T R A.—A complicated, delicate pattern, well suited for small panels, being drawn according to rules previously given. The penciling must be carefully executed, so as to define each letter perfectly, and the colors be such as will contrast strongly, or otherwise the monogram, when finished, will be illegible.

U S A

S A P, arranged so as to admit of being painted rapidly; that is, each of the letters are quite slender, therefore requiring but single strokes of the pencil. The letter S would be formed by beginning at its upper portion with a light stroke, and, as the body of the letter is approached, gradually pressing down on the pencil, thus producing at once the heavy portion. The extreme ends of the letter require but one stroke each on the parts that form them.

T R A

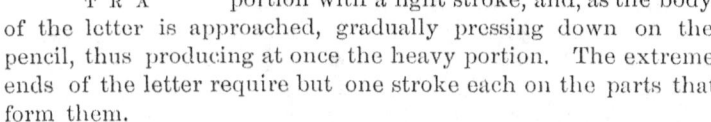

S A P

A and P require no more labor than S—we mean at to the filling in of the letter S; for either of them may be produced by single strokes of the pencil. In coloring, we suggest a pale pink tint for the upper half of S, its lower part a pale olive green. A, light blue, shaded with dark blue. P, vermilion and carmine, the whole lighted with canary color, used sparingly, but with very decided touches.

C. B.—GOTHIC. Somewhat difficult to combine in a monogram of three letters, unless the selection be made from the whole alphabet, with perfect freedom to use any three letters that will combine easily. One gothic, one antique Tuscan, and a plain Roman letter, combine more readily, and therefore produce a more satisfactory design.

C T B—Block letters, drawn in oblique perspective, forming an entirely new style of monogram, and one that is very pretty. T is the main letter; B depending on it for its front portion or shank, and C for its support in the position given it. By squaring each letter, the foundation lines will be apparent. When drawn correctly, C and T would find their vanishing points above and to the right, and B to the left, on a horizontal line, which represents the supposed position or level of the eye above the base line on which the letters rest.

C B

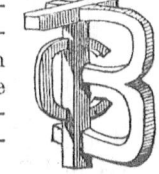

Without attempting to enter into an explanation of the principles of perspective, we will give a very simple method for drawing like patterns:

C T B

Produce two horizontal lines, parallel to each other, say two inches apart, and on the lower line as a base erect a perpendicular line one inch in height, and to the right and left of this, on the upper horizontal line, mark two points, say two and a half or three inches distant. Now draw lines from the top and bottom of the perpendicular line, converging or vanishing at the two points previously marked. These lines will inclose two oblique spaces, within which the object must be drawn. If, to the right and left of the perpendicular line first drawn other lines be produced parallel to it, depth will be obtained, which may be thrown into the form of a cube in perspective, and this can be divided into letters.

N I B furnishes a design of block letters placed in parallel perspective, and it will serve for practice in drawing, if not otherwise valuable. By laying the straight edge to the beveled outlines of the letter B, the vanishing point may be obtained, which will be at the intersection of those lines at a point above and to the right of the letters. Through the vanishing point draw a horizontal line; all straight lines, forming the width of B, must be drawn parallel to the horizontal line; also the lines of N and I, which give the thickness of those letters.

N I B

C R

C. R.—In this we have two letters intertwined in a careless manner, as may be seen in nature in clambering vines. It is not intended that they should be formed with great nicety, but rather as if they had accidentally united in the form given.

Lay in gold or color, giving the curves rather an irregular form.

C A V

C A V—Lay in C with dark blue, lighted with light blue and chrome yellow, No 2. A with Tuscan red, lighted with vermilion and orange, and V with olive green, lighted with a bright tint of olive green and white. Separate the letters with a wash of asphaltum.

If the painter has the opportunity offered of putting the monograms on the third coat of rubbing varnish, he may trace the ornaments on the panels, and work them up nicely; but should the carriage have been already familiar with the wareroom, apply the ornaments to the finishing varnish without rubbing it down.

C I N

C I N.—Drawn within a space ⅞ by 1⅛ inch. Lay in C tan color, shaded with burnt sienna, glazed with asphaltum to form the darkest shades; high light with white, toned with burnt sienna. I in dark and light shades of purple, lighted with pale orange. N lake, lighted with vermilion; or paint the upper half of each letter with the colors named, and the lower half in dark hues of the same color. Where the latter method is adopted, care must be taken to blend the colors on each letter so as to avoid an appearance as if the letters were cut in two. The coloring should be such as will define each letter throughout its extent, otherwise there will be difficulty in deciphering the monogram, which should on carriage work show some skill in the combination, and yet not approach to the grade of a puzzle. Either of the three colors named, mixed in different proportions, will produce sufficient contrast to make a distinct separation of the letters, and the color of the lighter be suitable for the high lights on the darker ones.

E P.—In this pattern a plain letter is placed in connection with another letter, which

PAINTING DEPARTMENT.

E P

is ornamented. By this means the pattern is not confused; the plain letter E gives full force to the ornamented letter P.

F E S.—Lay in F E with gold, shaded with asphaltum and verdigris, mixed so as to form greenish tints, varying in strength to suit the number of shades required; high light with straw color, mixed of flake white and chrome yellow, No. 2; put in the dots with carmine, shaded with carmine, darkened with black. The high lights, clear white. Outline S with a dark, rich brown, filling in the body of the letter with the same color by use of perpendicular lines, as shown in the engraving. Light with tan color. Form the diamond shaped center with clear carmine.

F E S

H R G—will require delicate handling. H and R lay in olive, lighted with yellow, toned down with olive. G, lake, lighted with vermilion. The dark markings on G may be done with lake and black, forming a rich brown.

H · R G

S N G.—Drawn within a space ⅞ by 1 inch. The ornamentation of the letters G and N is peculiarly adapted to producing a pleasing effect, when painted on the panels of a carriage. S is made very plain, and as applied in this design, serves to bind together the letters N and G. Outline N with dark brown, using light lines on the left side of the details of the letter, and heavy lines to form the shaded parts. Put in the diagonal lines with vermilion, and when the colors are dry glaze over with carmine. G may be done in gold leaf. The center glazed with verdigris, and the ends of the letters washed with verdigris, darkened a trifle by the addition of asphaltum. Paint S olive green, the parts receiving light with pale light green. High light G with pale pea green, and insinuate a small portion of red on the letter G.

S N G

A C G.—In this design we have made use of a pair of compasses, the calipers and an iron hand screw—three tools very familiar to the wood worker. The compasses furnish an approximation at least to the letter A; the calipers to the letter C, and the iron hand screw to that of G. As considerable license is granted to the designer, in the selection and formation of the letters composing a monogram, we have availed ourself of this license and brought into service the tools named. We do not claim for the design any beauty, but merely a slight degree of novelty. A carriage wood-worker, happening to possess initials agreeing with the above, in the order we have placed them, or either of the changes that may be made, might adopt this style of monogram for his coach door (?), or if he be fond of outward adornments, may have a breast or coat pin, made in silver or gold, and thus advertise himself, as do some members of secret societies.

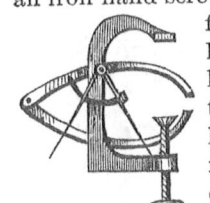

A C G

J E C.—Lay in all gold, then wash over E with verdigris, and C with carmine, except the star center. High light J and the center of C with straw color, and E with a pale green—white and verdigris. This will make a beautiful contrast, and be readily appreciated by admirers of monograms.

S K J.—This pattern was drawn on a rectangle, one and a quarter by one inch. Lay in each letter different in color, using darker tints of

J E C

S K J

each letter where any shading is required.

Be careful to secure a clean and distinct outline of each letter, and in the after-shading endeavor to preserve the outline free from raggedness, or an appearance as if the painter was nervous. Keep the colors subdued in tone, and depend on the half-lights and high lights to bring out the design to the degree of sharpness required.

M E V.—Monogram in scrip, presenting a neat appearance. This style is acceptable to the majority of persons. It is difficult of execution, as the beauty of the monogram consists in the perfection in which the letters are formed. This pattern is by no means perfect, but will serve to give an idea of combining the letters made use of, also hints as to the employment of others. In painting scrip, use a cutting-up pencil, about an inch in length and one-eighth inch thick, one that carries a fine, true point. The color should be very fine and mixed to flow freely. The curves require the use of a free hand, directed by confidence in the mind that they can be made. The pencil handle should be held lightly between the fingers, and allowed to revolve in forming the curves. To secure an easy and free control of the point of the pencil, the handle should be held at a distance of at least two inches from the point where the hair is inserted. Lay in the letters, orange, blue and gold, high lighted with white.

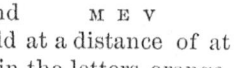

M E V

M S—Rustic Style.—The design serves to add variety to the painter's stock of patterns, and may occasionally be required on carriage work. Rustic letters, when formed into words and sentences, please for a time by their novelty; but the eye soon tires, in the attempt to decipher words painted or printed in a confused style. In designing rustic letters, it is necessary to possess some knowledge of the natural appearance of the limbs of trees, so that the branches, knots, etc., may approach to nature in their form and relative positions. Color the letters gray, shade them with brown and greenish gray tints. High light with flesh color.

M S

N P Ornamented.—This combination will make a pretty monogram. Lay in the letter N with olive green, and P with carmine, slightly changed in hue by the addition of white. Half light N with light olive, and high light it with chrome yellow, No. 2. Half light P with pink (white and carmine), and high light with white. It is quite common in the city to paint monograms on the finishing coat. It is not the best course to pursue, but it cannot always be avoided. Carriages are sent to the wareroom finished ready for sale, and should a purchaser desire his monogram painted on the doors of the carriage he has selected, it would not pay to rub down the panels for the sake of a pair of monograms, neither would it be advisable to add an extra coat of finishing varnish. The ornamenter must possess sufficient skill to be able to paint the monograms without using a pattern traced on the panels. The varnish will require to be wiped over with a clean damp chamois, and the rest-stick, at its upper end, be covered with a damp piece of chamois, in order to guard against bruising or marring the varnish. The colors require varnish in their mixture to prevent them from drying too dead; also, to protect them when the carriage is being washed.

N P

PAINTING DEPARTMENT.

O M L.—In this pattern the letters are comparatively plain, and the spaces large enough to clearly define each letter. O may be colored with gold ochre, shaded with burnt sienna, and lighted with straw color; M with lake lighted with vermilion, and L with milori green, shaded with the same color, slightly darkened with carmine, and lighted with straw color.

O M L

ORNAMENTED MONOGRAM O V I.—The cavalry hat, blue black, lighted with dark lead color.

Wreath, red and white; O, V, I, gold; scrolls, a dark, rich brown, lighted with orange or a light brown tint.

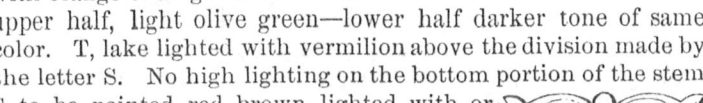

O V I

O T S.—Paint O, upper half, light olive green—lower half darker tone of same color. T, lake lighted with vermilion above the division made by the letter S. No high lighting on the bottom portion of the stem S to be painted red brown, lighted with orange, or lay in all the letters gold, and glaze the colors over the leaf.

Monogram O M T was drawn on an oblong, 1 inch by 1⅛ inches. If the ground color of the panels is claret, or English purple lake, the letters may be painted with the same color lightened up with vermilion and white, forming three distinct tints. On brown, coat the letters with lighter shades of brown, and so with other colors.

O T S

O M T

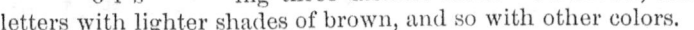

V A T.—This combination forms a pleasing variety, and will afford good practice in the use of the pencil. Lay in the letters as indicated by the shading. The letter V to be darker than A, and T deeper in tone than either V or A. The letters may all be laid in gold and afterward glazed with colors. The delicate vine at the base, colors either a delicate green, tinged with carmine, or purple, red brown, or a neutral tint.

T R A, a very neat design, drawn after copies from celebrated French artists. The letters furnish a very odd yet attractive style. It will be noticed that the stem of letter T covers the center perpendicularly, and that the outer lower portion of A and R are drawn to touch on the same line. The main stems of these letters terminate in twin forms, arranged so as to cross each other at the center of the monogram, and balance each other on either side. In the matter of coloring, we would mention that it is now becoming fashionable to use but one color on the letters, separating them on their edges by a suitable high light. Monograms painted in this manner should be drawn so that the design will not be confused by ornamentation, that is, the main outlines of each letter should be distinctly defined, and the spaces must be so arranged as not to confuse the outlines. The pattern here shown may be colored carmine, and the edges separated by straw color or blue, and the letters be defined by canary color, or a lighter tint of blue than the body of the letters are painted.

V A T

T R A

S M T

SUB

S M T.—In the foregoing cut we give a plain, neat design adapted to the practice of inlaying. The pattern as drawn allows for three different colored woods. Namely—Maples, Satinwood and Mahogany; or Rosewood in light and dark shades. The letters having been carefully cut out, should be laid on the panel in which they are to be inserted and their forms traced. After which the panel must be cut down to a depth agreeing with the thickness of the wood of which the letters are made, then fitted and glued.

S U B.—RUSTIC.—Patterns of this kind may be arranged with willow or grape vine, and tacked down on a piece of panel, forming a good study for the student in drawing. Paint in gold or gray, with dark brown and green markings, in imitation of nature.

ORNAMENTS.

The selection of ornaments herewith presented furnishes as great a variety in size and peculiarity of design as the ornamenter will require as a foundation for producing other patterns. The circle, oval or ellipse, triangle, lozenge and square figures were employed in designing the ornaments, which will be readily noticed, as they are arranged as nearly as possible in the order above stated.

Ornament No. 1. The upper half of the circle may be left out if desired, producing a circular base and a triangular crest. Lay in purple (white, ultramarine blue and carmine); separate the leafing by shading with clear carmine; high light with pink (carmine and white).

No. 1.

No. 2.

No. 2. A garter, with leafing added. Lay in garter gold, the leafing in colors darker than the gold. Keep the colors subdued, and depend on the high lighting to bring out the richness you aim at. Have a pencil for each color so as to preserve them pure; and where any blending is required, use a clean, dry pencil.

No. 3.

No. 3. Color the "Beast" brown, in varying shades, the half lights to be red-brown: the lights, orange: highest lights, yellow. The tongue, red, but kept down dark, except a spot of bright red near the middle. The remainder of the ornament gold.

No. 4.

No. 4. This small ornament, although simple in form, will be found somewhat difficult of execution by the novice.

Such patterns will readily disclose to the painter whether he has mastered the handling of the cutting-up pencil. If in attempting the circular part of the ornament the nerves become unsteady, and create a lack of confidence, the painter should practice until assured that the hand will obey the will.

Lay in the ornament, gold, shaded with asphaltum, tinted with carmine, and high light with a delicate pink, mixed of flake white and light red. The wreaths may be painted blue and white. Mix three hues of blue, placing the darkest at the bottom, or lower part of each band, represented in the engraving as colored. The white bands should not be laid in with pure white, but tinted with black, forming a delicate gray

The high light which would be seen through the center of the wreath, use white, tinted with yellow. The space partly covered with diagonal lines may be left plain, showing only the panel color, or cross barred with a delicate gray, mixed of flake white and black; the tint barely changed by adding light red, carmine or vermilion.

RUSTIC INITIAL LETTER U—COMBINED WITH A GARTER.—Size in the entire pattern, and  after the gold has been laid, glaze over the inner part of the garter with a light blue; the inner and out edges to remain gold. The flying ribbon, pink (carmine and white;) the shading, clear carmine, and carmine saddened with black. Stems of the letter U, reddish brown and pale green; leafing in similar colors.

No. 5, a medallion inclosing a shield, surmounted by a crest; an eagle displayed, and at the base a wreath of laurel. We leave this to the taste of the painter, the engraving showing the strength of the colors to be used on the several parts. The circular band may be omitted if desired.

No. 5.

No. 6. Lay in the "satyr" head, monogram, clasp, and band surrounding the garter, gold 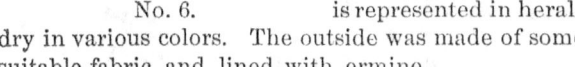 The garter, pink (carmine and white). Shade the gold with verdigris, and light it with pink.

A MANTLE—In which is displayed the garter and a small monogram, I T B.

Monograms are still in fashion, and as the Garter and Mantle have always found favor among Americans, we have combined them, which, if delicately handled, will produce an attractive ornament.

The mantle, or cloak, is represented in heraldry in various colors. The outside was made of some suitable fabric, and lined with ermine.

No. 6.

Ermine is painted in representation of the fur of the animal of that name, with black tufts scattered over its surface. The ornamental painter need not confine himself to this single arrangement for the lining of a mantle, but may vary his colors.

The pattern here shown may be painted as follows: the folds of the mantle, which show the outside fabric, paint with a color of a purple tone, mixed of white, lake and ultramarine blue. Mix up at least three tints, so as to produce on the folds the rounded appearance they would naturally assume, placing the lightest of the three tints at the highest point on the folds, and the other two to the left and right.

The colors should dry slow enough to allow them to be blended where they meet, which gives the appearance of the cloth when viewed in the sunlight. That is, the folds, not reflecting the same quantity of light from every part, would appear darker toward the edges. By blending, the whole is united, and shows but one piece of fabric under its modifications of light and shade. The inside or lining of the mantle, paint a delicate gray, the lights white, the tufts with burnt umber, shaded with asphaltum.

The monogram, garter and cord, on the mantle, gold. Having traced a correct outline of the pattern on the panels, size in those parts to be covered with gold leaf, and when the leaf is laid, proceed with the colors.

When the mantle has been colored and blended in, and becomes dry, high light it with pink, mixed of carmine and white, and touch over the deepest part of the shadows with asphaltum.

The gold cord should pe put on of a uniform width, which is best secured by the use of a fine cutting-up pencil. Remember that the cord is round, and must be shaded accordingly.

Having washed on the shade, produce the effect of the strands of cord with the color you use for lighting the gold parts of the ornament. This is done by diagonal lines placed on the light parts of the cords which were not washed over. Never carry the lights on to the shaded part.

No. 7. In painting this style of ornament, select dark rich colors, which can best be secured by glazing.

The cap crimson, wreath green and gray, lighted with a delicate pink.

The circular part gold, shaded with asphaltum tinted with carmine; also, outside border of shield gold, upper division of shield red, dark and rich in tone, instead of a staring light red, such as clear vermilion gives. Chevron white, not pure white, but a delicate gray, lighted with clear white. The lower division blue, the deep shades of a purple hue.

The leafing at the base paint with a color mixed of burnt umber, yellow and lake; shade with asphaltum, tinted with carmine; high light with orange or vermilion.

No. 7.

The accompanying pattern is after Gustave Doré, the celebrated French artist:

It is a winged ellipse, which produces an odd and still a very pretty design. Lay in all gold, shade the details with verdigris, darkened with asphaltum, and high light with pink (light red and white). The escutcheon may be colored with light brown, carmine and dark brown. The edges of diagonal bar to be dotted minutely with vermilion.

No. 8. This pattern is suitable for a space ⅞ of an inch square.

The middle portion of the semi-circular band is varied by "dancette," or zig-zag work. Lay in all gold; shade with asphaltum, enriched with a small portion of carmine; high light with white. The semi-circular space to be glazed with purple, (carmine and ultramarine blue); dancette work (canary color chrome yellow No. 2 and white) in the center, and graduated toward either corner with a darker tint of yellow; wreath which supports the crescent, blue and white.

No. 8.

No. 9. An odd pattern, but one which will work up nicely. Main outlines drawn by use of the side and end of an oval pattern; scrolling added according to fancy. Lay in the crescent-shaped chaplet, and the outlines of the two diagonal bars gold. The jewels on the chaplet, green and red, alternating, in imitation of the emerald and ruby; diagonal bars red and white on a blue ground. The lower scrolling dark green, touched up with light green; the green to be mixed of an olive cast. Upper scrolling and outline of the remainder of ornament pink (carmine and white).

No. 9.

No. 10. An oval shield, showing the American colors. Lay in the border, eagle and stars, and division lines, gold. Fill in the oval band with a lead color (white, blue and black) of two or three tints; the dark markings blue; the spread eagle and stars lighted with straw color; the field containing the stars, blue. The colors to be put on by cross hatching with a fine-pointed sable pencil. The bars, on lower division, red and white, alternating. Put in the colors by fine lining, instead of filling the spaces solidly. Use carmine and white for red bars, shaded with carmine, which gives a silky appearance.

No. 10.

No. 11. Drawn by using an oval as the foundation or first lines; the after details are merely imaginary, and cannot be laid down by any rule, further than to secure variety of form and graceful curves.

No. 11.

No. 12.

No. 12. An elliptical form, varied by the use of the circle, may be laid in all gold, or part in color. After the pattern is marked on, use a fine line pencil, about an inch long, to form the fine lines, and also to sweep the circles, before putting on the leafing. The fine line pencil should be provided with a handle.

GARTER.

No. 13. The "Garter" has been used in this country as an ornament for carriages number of years, and is still used, more or less. The original design here given produces an attractive ornament. There is a defect in the engraving which should be remedied by the painter when finishing the ornament. We refer to the drooping vines from centers at outside of garter. Instead of perpendicular and short horizontal lines there should be the representation of leaves and small flowers.

Lay in all gold except the *crest*, light and shade as shown. When the shading is dry, wash over the markings on the garter with carmine, verdigris, or any other transparent colors your taste may suggest.

The vine and leafing at the base may be left plain gold, or

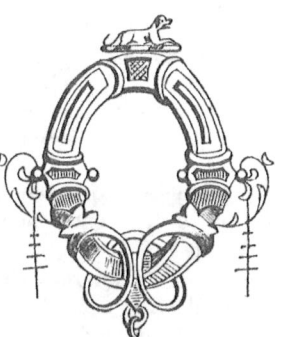

No. 13.

No. 14.

washed over with a delicate pea-green tint. White, gray of two or three tints, and burnt umber will be sufficient to delineate the hound; the wreath, paint with colors to contrast with those nearest to it on the garter.

No. 14. A new and decidedly handsome ornament. Outline the garter with gold. The buckle and slide gold also. Fill in the garter with light and dark tints of blue, and high light with canary color. The floral gorgons, paint in brown shades, and light with orange and clear yellow; a small portion of lake added to these browns will cause them to bear out richly when varnished. Let the medium lights and shades predominate, and the high lights be added by first carefully considering their true positions, and then touching them on with confident strokes of the pencil; which will give sharpness to all the details. The pendant stems with leaves and berries may be colored olive green, and shaded with russet. When the ornament is dry, glazing will improve it.

No.15. The center is filled in with the CAUDCES, a Roman emblem.

The rod, or center staff, on which the wings are represented "displayed," and around which two serpents are twined, signifies *power;* the wings, *fleetness;* the serpents, *wisdom.*

This pattern will please the eye if every part be laid in gold; but there is an opportunity afforded for a display of coloring, which for practice, if for no other purpose, should be seized upon.

In sweeping the curves, pay no attention to the minor details, for on the gracefulness of these "lines of beauty," as Hogarth termed them, the appearance of the ornament must depend.

Having secured the outlines, fill in on them whatever leafing is shown.

No. 16. This pattern is best suited to panels which are wide in proportion to the depth. Will look well on a light wagon, sleigh, or other gaily painted piece of work.

Lay in all gold, or gold and color. Having secured the

No. 15.

outline, distinctly marked on the panel size in the pattern, without reference to details, which should be brought out after the gold is laid. Let the shades be well blended down, and the lights clear and sharp. Escutcheon highly ornamented; *volant* or flying. The pattern is drawn for panels which are somewhat wider than deep, but will answer

No. 16.

for panels of any shape. Lay in all gold except the center of the escutcheon. Having marked the pattern on the panels with great exactness, size in carefully, and when ready apply the gold; burnishing it up well with soft, clean raw cotton. The gold leaf having been laid, proceed to color the center field.

The center may be colored with dark and light tints of Purple, (carmine, ultramarine, blue and white). Blend the tints down with a clean, dry pencil, so that no hard lines may be visible. Wash over those portions of the gold which require shading, with asphaltum, warmed slightly with carmine; high light with pink (light red and white). These touchings should be few and carefully touched.

No. 17.

No. 17. A large and quite handsome pattern after Gustave Doré. It may be termed a winged triangle. The center portion is surmounted by a crown, which is, of course, repulsive to an American. We copied the pattern as we found it, and should any choose to paint the ornament, an American military cap may be substituted for the crown. It will be readily perceived that the body of the figure is in the form of a triangle—its upper portion being enriched by scroll work, and the lower terminating in a single leafing. Painted in colors alone, or colors and gold intermingled, this will furnish a beautiful ornament. The pattern may be divided and employed in ornamental striping.

No. 18. A small pattern adapted to light work, where neatness is sought rather than intrusive display. It is elaborate enough to suit the taste of those who admire very fine pencil work, and the center field admits of an initial letter, or a very delicate monogram. The coloring should be dark and rich, using different tones of the same scale of colors; the high lights to be applied sharply yet very fine, and given their exact place and no other. These small ornaments are

No. 18.

frequently spoiled by the high lights being carried on to those portions which could not possibly receive the strongest light, were the pattern cast in plaster, and then placed in a position corresponding with that which the ornamenter assumes in coloring the ornament. Gold, tan, red-brown and deep rich brown are colors which, when properly handled, produce the richest ornament. White, blue and vermilion are used to a great extent, but they are valuable, more on account of their brightness, and the facility with which a striking effect is produced, than on any richness they possess.

The ornament may be painted with the colors named, mixed among each other, the red and blue forming purples of different degrees of intensity; and by the addition of white, various lighter tints may be obtained. Blue and white will afford varying tints of light blue, and red and white various tints of pink; with purple, light blue and pink, a less glaring and raw effect is produced than if we should employ white, red and blue in their purity. The eye does not require to be affected by a great amount of a color, in its full strength, in order to give the impression that a bright color is present; and in ornamenting, the painter need not load the ornament with bright colors for fear the proper impression will not be made on the eye of the observer. The main portion should be subdued in tone, the full brightness of the colors to be suggested by the highest lights.

SCROLL ORNAMENT.—This pattern is simple in design, and will be easily executed by those who may have acquired a free use of the cutting-up pencil. It is of a triangular form, and drawn by the use of the circle and oval. It may be used as given, or reversed by leaving off the crest. One-half of the scroll part of the ornament, taken lengthwise, may be used for spring bars, head blocks and axle beds; as a whole, it will suit for the flat face of the under part of the spring. It may also be divided, and arranged to suit the corners of the panels.

No. 19.

No. 19. Lay in all gold, separate the parts where required, and produce the effect of interlacing by a judicious use of light and shade. The pencil best suited to this class of ornaments is a fine cutting-up pencil an inch long. Having traced the pattern on the panel, commence by painting the crest, next the main upper left hand division of the scroll part, paying no attention to the leafing or any minor details. It will be noticed that the center line of the heavy leafing is a part of the scroll line which passes from the wreath or ribbon at the top, and is completed at the base, so that, to secure easy curves, this line should be laid in through its whole length, and the leafing or any minor dividing lines be governed by it.

Next lay the other half in the same manner; having secured these main curves, add the details.

Where two fine lines cross each other, the effect may be produced of one passing under the other by simply lighting one of the lines across their intersection, which, by contrast, makes the gold appear darker, thus forming the shade required.

Paint the wreath blue and white, the crest to be merely lighted with the color used for high lighting the other parts.

No. 20. A CHASTE DESIGN.—The Dalmatian, or coach dog, surmounts the whole, adding a lively effect. The initial given in this and kindred forms we think more attractive than a simple monogram.

No. 20.

The coach dog is white, with black spots. Let the color be of a warm gray, the high lights white. Spots dark brown.

The scrolls should be painted different in color from the initial letter.

The taste of the painter should direct as to the colors best suited to the panel color.

BUCK'S HEAD.—Mix up two or three brown tints for the darker portions, the neck in front white, the antlers of a dull grayish cast, touched up with a warm tint, as white and yellow ochre, shaded with black or dark brown. The dark browns may be deepened by glazing with asphaltum, which should be laid very thin.

Glazing cannot be put over wet colors; and as the painter seldom has time, in these days of transferring, to wait very long on his colors, the glazing can be omitted.

Do not mix a clear white for the neck; add a little ochre, giving a yellowish cast, or black, forming a gray, and touch up or high-light with but few strokes of clear white.

This ornament, the "Wyvern," a species of dragon, is formed with the body of the cock, the wings of the bat, and the tail of the serpent. It will require very delicate

handling. It may be laid in gold, and afterward glazed over with brown tints of light and dark tones, leaving those parts clear gold where the high lights are required. The high lights may then be put on with straw color, mixed of flake white and yellow, with a delicate pink mixed of flake white and light red, or with white tinted with burnt sienna.

The shield may be glazed with white, pale green, or any other bright color which will contrast strongly with the colors surrounding it. The leafing, *fleur-de-lis* (pronounced flare-de-le), flower of the lily, should be painted some dark color, so as to properly display it.

It will have a pleasing effect to paint this style of ornament with colors similar to the ground color of the panels; for instance, if the ground color is brown, mix up different tints of lighter brown, painting the ornament all brown (except the shield) of different degrees of strength.

If the panels are purple, use lighter tones of purple, and so with other colors. Be careful to keep the outlines sharp and distinct. Lay on the shades with breadth, blending them together so that there will be no patchy appearance. The light and dark markings to be put on last, with quick, confident strokes of the fine-pointed sable pencil. The base may be laid in gold, and the lower division glazed with asphaltum.

No. 21. A garter inclosing the initial letter V. Garter surrounded by a vine the whole surmounted by a demi or half lion, issuing from a wreath. May be laid in all gold, shaded with asphaltum, tinted with light-red and white, and colors introduced on the wreath and garter; or the lion may be colored true to nature, and gold and colors used on the remaining portion.

No. 21

No. 22.

No. 22 will be found very appropriate. The animal is a species of a dragon, a fabled monster, having the head, neck and wings of a bird, and the body that of a wild beast. He supports a Norman shield, the "fess," or center point, displaying the Maltese cross.

In painting this ornament, secure first a correct outline of the whole, then mix up two or three tints of whatever color you design painting it, having a pencil for each, and a clean pencil for blending the edges, so that there will be no hard lines left on it. Lay on the shaded portions first, and then the half lights, keeping them subdued in tone, so as to allow for the finishing touches showing clear and distinct.

On a claret panel, the whole may be painted in different hues of purple and red; on a dark blue panel, varying shades of blue lighter than the groundwork, and so with other colors. The shaded portions must be distinct, and gradually connected with the light portions by lighter tints.

The dragon or wyvern may be painted an ashen gray, the high lights of a warm tint of a yellowish cast. Outline of shield, gold; the upper division, celestial blue; the lower field, a pale orange; the cross, brown, shaded with asphaltum; the wreath, blue and white; the flying ribbon and leafing in gold.

No. 23. The horse, noble in bearing, beautiful in form, and of priceless value to man, has from the earliest and rudest attempts at pictorial representation, formed a pleasing study for painters and sculptors. He has figured in the quiet and peaceful rural landscape, where he does his master's bidding from the time the soil is turned over until the heavily laden carts are deposited at the barn door with the golden harvest. On the road for pleasure, in the din of battle, or wherever man chooses to place him, he is faithful to the trust reposed in him.

Artists inform us that the horse is the most difficult subject to draw and color correctly among animals. This being the case, the student should not become discouraged if, at the first, second or third attempt to paint the pattern here given, or any other one he may select for his study, should look like anything except a horse.

The wood cut here shown is not perfect, but will answer our purpose. The position of the head, producing the arched line of the neck, will be found as difficult a study as the learner will care to attempt. First, secure a distinct outline, then mix up three or four tints of the color you decide to use, and also place on the palette your color for shadows.

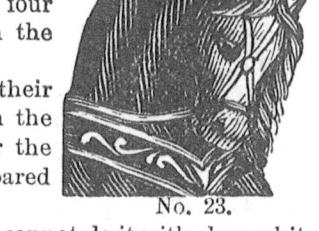

No. 23.

Lay on the shadow tints where your outline indicates their true position, paint them thinly, and proceed to lay on the half lights, remembering that all objects viewed under the effects of light show but a small portion of *high light* compared with that which is graduated off into the shadows.

If you decide to represent a white horse, so called, you cannot do it with clear white. There must be a set of delicate grays mixed, so graduated as to give the true contour of the muscles, etc., finally abandoned for the positive shadows, which are of an amber cast.

Having brought out in gray tones every part of the study, the high lights may be touched on boldly, with a tint not far removed from clear white.

A black horse cannot be correctly shown by the use of black and white. Dark brown and lead colored shades, with the shadows deepened by glazing, and the lights of a bluish tint would have to be employed.

Colors of a deep red, and others of a variety of shades, to be noticed on the glossy coating of the horse, may be imitated successfully by looking closely at the mass of the color in shadow, and preparing the colors accordingly, and not by carrying in the mind the effect produced upon the eye by the highest light and color; for if you begin by mixing the brightest tint, and get your pattern coated with this, you will be at sea in the attempt to produce a natural effect.

No. 24.

No. 24. A ribbon or band, and is suited to very narrow panels, and to belt spaces; and then, neatly handled, forms a very pretty ornament. In practice we find that sizing in the pattern, and laying on gold leaf is the speediest method to adopt, as well as affording the richest coloring, by employing transparent colors over the gold.

In sizing in this pattern, pay no attention to minute details; and where the cords and tassels cross over the face of the band, omit them; being careful, however, to size in any portion that may project beyond the face of band. After having laid the gold proceed with the shading, for which purpose use asphaltum, to which add a small proportion of carmine, with this wash on the shading at the ends of the band, paying no attention to the cords and tassels. Next make out the shape of the leafing shown above and beneath the band. These strokes need to be made with precision, for they bring out only the shaded sides of the leaves. Next put in the shaded sides of the cords and tassels, which are as yet a part of the surface gold. High light with straw color (flake white and chrome yellow No. 2), and when the colors are dry glaze over with carmine, lake or other transparent colors. The gradation of shades having previously been formed with asphaltum, it remains to do no more than wash over those shades with color, when the ornament will appear as if shaded with two or three tones of the color used.

The accompanying original designs are intended for small panels or spaces.

No. 25. Lay in all gold, and over this use colors. The gold will best define the scrolls and right lines; colors to be used on the inclosed spaces. The crowning space to be red; the upper left-hand space white, its

No. 25.

PAINTING DEPARTMENT.   227

opposite blue. The first to be filled in with perpendicular lines, as shown in the engraving; the others by crossed-barred lines of the color suited to the respective spaces.

No. 26. This pattern may also be laid in gold, except the crest, which paint a warm gray, shaded with raw umber, and lighted with white. The space bounded by the upper scrolls to be lined with pink, mixed of carmine and white. The space should not be gilt. The small circles or drops shaded with a light green, the centers touched with a spot of white.

No. 26.

No. 27 A semi-oval form. Having traced it on the panels, mix two or three tints of umber and white, and with these finish the hound. The scroll on which it sits, red and white, alternating. The inner spaces of escutcheon: upper half, three tints of purple, the lightest in center, and graduated toward either side. The edges of contact must be blended away, so as to avoid showing hard lines. The lower portion: three tints of pale green, blended off as before stated. The inner part of borders, three hues, namely: carmine and vermilion, carmine, carmine and black. The dark markings with carmine, saddened with black. The fine scrolling at base with pale green, (verdigris and white), high-lighted with white. Keep the tints and hues pure and clean by employing a pencil for each one.

No. 27.

No. 28. APAUMEE—In heraldry, a hand displayed open and extended, showing the palm. Lay in the ornament with gold and colors.

No. 29. A fancy shield, drawn within a rectangle, measuring 1⅛ inches by ⅞. The crest is about one-third of the length of the shield. The top division of shield is one-fifth of the length. The horizontal bar, at base of upper division, is one-half the width of the upper space, or one-tenth the length of shield; the greatest width is equal to the length from the horizontal bar to the bottom of the shield. The center scroll work is drawn on a heart-shaped outline; the remainder of the markings are portions of circles and ovals. Lay in the outline of shield gold, the eagle in different shades of brown, breast touched up with gray and white; upper part of shield barred with purple lines. Horizontal bar across shield white. Surface surrounding the heart-shaped center, dark brown, or some other dark color, as may best suit the ground color of panels; scroll work at this point light brown, touched up with orange and vermilion. Middle of heart, lake, carmine and vermilion. Surrounding fine scrolling, delicate pea green, touched up with white and verdigris.

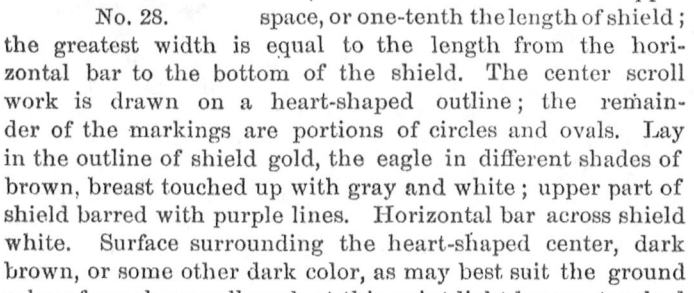

No. 28.   No. 29.

No. 30.

No. 30. The armor, gold; the plumes, crimson; leafing at the base, pea green, touched up with white; shades, verdigris; outline of shield and lower field, gold. The upper space lined as engraved, using light blue (ultramarine and white) in the middle, and shading off into each corner with purple (ultramarine, carmine and white.) Lance-heads, gold; flags, crimson; scrolls across face of shield, brown, lighted with orange

flying ribbon at the base, pea green, shaded with verdigris.

No. 31. An odd pattern. It is drawn within a square, making it suitable for small spaces and panels. Lay in all gold, and introduce colors on the crest, the triangular field at the base and the fine scrolls which are at the sides.

No. 31.   No. 32.

No. 32. This was drawn on a lozenge shaped outline, which may be reproduced by dividing the ornament equally, horizontally and vertically, measuring five-eighths of an inch each way from the center point, and connecting the four points by four straight lines; within this figure draw an oval one inch long and a half inch wide, using the lower half to form the main body of the bottom half of the ornament.

The remaining lines on this portion of the pattern were then added.

The upper division at the center is the half of a circle, half inch in diameter, the side supports being parts of an oval figure.

Having thus dissected the pattern, for the benefit of those who are not accustomed to designing, we proceed to the coloring.

On these small patterns the quickest method is to lay in all except the inclosed body of the pattern, with gold, reserving the inner field for colors; or for one color, blended and worked down softly. And the colors selected should appear rich, not staring. Filled in with clear vermilion, the ornamenter would obtain a staring color, but the same, toned down with lake and carmine, would impart richness. Carmine and white, of two or three tints, will produce a pleasing effect. Various shades of rich brown also suit well.

In using these let the yellow tint be pretty decided where the highest light falls, and follow into shadow with tints of brown. And the darker portions may afterward be glazed with carmine, where great depth and richness is sought.

Glazing must never be attempted over a color that is not dry. It is employed to either brighten or darken a portion of the work, *transparent colors* being used for the purpose. The effect of glazing will be better understood by those who have had no experience in its use, by a few hints we will herewith add. Prepare a panel, and on it lay off several small divisions, and color them differently. Say: white, Naples yellow, chrome yellow, yellow ochre, raw sienna, burnt sienna, vermilion, light and dark brown, light and dark green, and black. When all are dry, mix up a small portion of carmine, and draw a half inch stripe across all of them.

It will then be noticed that the carmine imparts a change of tone to each of the colors, all differing according to the brightness or dullness of the ground colors. The ground colors also changing the original color of the carmine, the white will appear of pink tint. Naples' yellow, a little darker pink, yellow ochre, raw sienna and burnt sienna, yellowish red tints. The vermilion being of a bright red hue, and approaching nearest to the color of the carmine, would be made a trifle darker, and in turn would rob the carmine of a portion of its color; the two combined forming a bright rich tone of carmine, the light and dark browns would detract from the purity and brilliancy of the carmine, still giving very handsome colors. The greens would produce nothing worthy of imitation for general use, but over black a deep rich tone of carmine would be produced. Black, however, being so dark, would require to be covered solidly to show the full effect.

When the carmine is dry, other transparent colors may be used in the same manner; verdigris, purple lake, yellow lake and asphaltum may be employed. But we will return to the ornament, and dismiss it with a few words. Having sized it in, and laid the gold leaf, sponge and wipe the ornament dry. Shade the gold delicately with asphaltum, clear (or tinted with virdigris), and high light with a delicate pink (light red and white). One stroke of the pencil should give the required light or shade on all the details. Mix up three tints of whatever color you decide to paint the inner part, placing the brightest in the center, and the darker tints to the right and left; then with a dry pencil blend the edges of each, so as to efface the lines of contact. When dry, paint the center diamond, in imitation of the emerald (green) and the small circles in imitation of the ruby (deep red). A spot of white or pale green will finish the emerald; and a spot of vermilion on each of the circles will complete the rubies. The barred work on the upper part should be of a gray or purple tint.

SHIELD AND MONOGRAM.—A Norman shield, with the U. S. colors, and the letters S T forming a monogram.

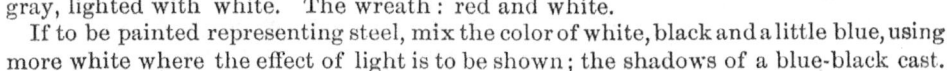

The sides are drawn from a circle about three inches in diameter; the curves at the top by the use of a circle three-quarters of an inch in diameter. The upper division of the shield is nearly one-third its length.

Lay in the outline, stars and scrolls with gold; the diagonal bars, red, white and blue; the monogram to be painted last.

In mixing the colors, red, white and blue, use subdued tints of each; and instead of filling in the spaces solidly, take a fine pointed sable pencil and cross-bar them. S T should be painted so as to show plainly over the bars. Those parts which run on to the red may be painted a lighter red; on a white, a brown, or dark greenish tint; on the blue, lighter blue.

The upper space surrounding the stars will need only a few perpendicular lines from the center each way toward the edges, leaving the ground color of the panel to form the shades. If properly executed, these will give the swell required on that part of the shield.

No. 33. Suitable for large panels on heavy work. Lay in the outer border of shield and connecting scrolls, gold; glaze the border with light blue, leaving the edges gold; when dry, cross-line the border with dark blue. Shade the scrolls with asphaltum, toned with ultramarine blue, and high light with white. The middle portion of shield to be left the color of panel; the form here given it to be represented by outlines only. The wreath supporting crest, red and white alternating. The crest blue, lighted with light blue tints, approaching to white.

BARRED HELMET AND BREAST-PLATE.—Lay in gold shade with asphaltum, high light with straw color. The plume: gray, lighted with white. The wreath: red and white.

No. 33.

If to be painted representing steel, mix the color of white, black and a little blue, using more white where the effect of light is to be shown; the shadows of a blue-black cast.

Brass color, if desired: mix of yellow, light red and black. Light, with straw color, tinted with light red.

No. 34. SKELETON, ADDING OF LEAFING, AND FINISHED SCROLL.—The first drawing shows the foundation lines, or skeleton of the pattern as seen in the third figure, and is, in fact, the frame work, so to speak, over which the design is placed. In designing scrolls, then, the new beginner should early accustom himself to producing lines as a skeleton, over which he is to place such clothing as his taste suggests, and in the study of pat-

terns, of whatever degree of elaboration, he must look at them, not as a whole, but first direct the attention to the center lines, and having obtained their sweeps or curves, take up the details according to their prominence in giving character to the study.

In the simple pattern here set, no difficulty will be experienced in fully understanding it—the skeleton showing distinctly that three circles have been connected by flowing lines, on which are placed leafing agreeing with the parts of the skeleton in which they are placed. We cannot lay down rules to govern the taste in clothing the skeleton, as on the same lines a great variety of patterns may be produced, each one depending on that innate something, called natural gift, or genius, which suggests forms to the mind as if a hidden power was exerted wholly separate from the will. By practice, however, those having no natural taste may acquire a commendable degree of proficiency in designing. Whatever forms are used to enrich the skeleton they should be such as flow easily and naturally with its shape, and the spaces be so arranged as to bring out distinctly each member of the scroll pattern.

Variety of forms and curves must be sought. This is a plain running scroll, that is, it is not varied by intertwining parts or members, and will be sufficient for our present purpose in arresting the attention of those who desire to practice drawing. The second figure gives the leafing without any attempt at shading, showing the outline within which other varieties of form are introduced in completing the pattern. The student must imagine the leafing as standing out in relief, as a piece of wood-carving or a plaster cast would not, under the effects of light, show any positive lines, such as the painter is at liberty to use in bringing out an imitation; we mean lines within the outlines of each leaf. Some of our best scrollers give the form by graduated tints and shades, but this method requires a vast amount of skill in making a sharp and attractive scroll. A medium course seems to be the better plan. On small patterns, suited to carriage work, we generally dismiss them by merely lighting up the scroll with a few touches of a color lighter than the ground color or scroll. The third cut shows the scroll finished, which, if carefully studied and compared with the second one, will disclose the lines and shadows added to finish the pattern.

Scrolls may be drawn to suit any shaped space that may be presented, by taking a pattern of the space with paper, and on it drawing the scroll touching the outlines. If a scroll be drawn within a triangular figure, for instance, and the lines forming the triangle be erased, the form must be triangular; but as the positive outline of the triangle has been removed, its true shape will not be detected by the unpracticed eye. To design a scroll to suit the corner of a panel, or any part of a gearing, it is only necessary to take the shape of the space to be ornamented and proceed as above stated. When completed, rub Indian red on the back of the paper, place it up to the panel, and with a sharp pointed stick or a dull needle mark over the lines, which will leave a copy of the original on the panel, or the pattern may be perforated with a needle or pin, placed in position and pounced on. A skillful ornamenter will put on small scrolls without previously drawing a pattern. He will merely step off with the dividers the length and width of the scroll on each panel, as a guide to the limits of the outlines, and proceed to paint them, the first one painted being the guide for those which follow.

CORNER PATTERNS FOR CARRIAGES AND SLEIGHS.

The annexed engravings furnish the laying out and finish of a plain running scroll, adapted to corner of a panel.

Fig. 1 represents the pattern in outline, the vertical dotted lines showing the proportions the parts or members bear to each other. The dotted lines on the body of the pattern indicate the direction or bearing of the stem or main branch of the scroll pattern, which is hidden to some extent in the finished engraving seen in Fig. 2, and which would not readily be noticed by the uneducated eye. The leafing is added according to taste, yet so arranged as to flow natu-  rally with the main stem, or branch out from it, without abruptly breaking off its connection. In other words, the scroll should be composed as if one should have bent a piece of wire to the shape of the main branch, and afterward placed on the leafing. The leafing would cover portions of the wire, and yet, although partially hidden, its connection would be perfect throughout its extent.

In designing for carriage, wagon and omnibus work, the panels furnish the form the scroll pattern should take, and as the size of the scroll is regulated by the space between the moldings, a moment's thought will be sufficient to direct as to the kind of scroll best adapted to any corner.

On light work and heavy carriages for private use, if scrolling is desired, it should be simple in form, delicate, and of colors that harmonize with the ground painting. It should not be painted in such strong contrast as to call attention to the scrolling alone; but these parts should assist in making up an attractive whole. Omnibus and circus work will bear greater display in the size and gaudiness of the ornaments; but even these must be designed with care, else the effect aimed at will be lost. On the same foundation lines a variety of patterns may be produced, depending, of course, on the skill of the designer.

Figure 2 gives the effect produced on figure 1 by light and shade. All guiding lines are removed, and we have the corner ornament completed.

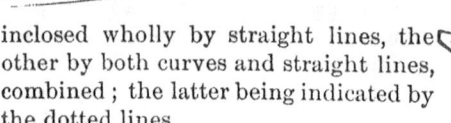

FIG 2.

CORNER SCROLL drawn on the same ground lines as the preceding one, showing that it is not necessary to materially alter the governing lines in order to produce variety.

We here represent two patterns, one inclosed wholly by straight lines, the other by both curves and straight lines, combined; the latter being indicated by the dotted lines.

No. 3.

No. 3. The square pattern (No. 1) may be drawn of any dimensions desired, the proportions in this instance being a square, measuring three-quarters of an inch on one side, divided into fifths, one-fifth taken for each of the two narrow spaces on sides of corner block, and four-fifths for the square or corner block. The size of the pattern in dotted lines is governed by that of the square one, the latter being the proper foundation on which to draw the curved corner piece. In following the dotted lines there will be noticed at A and B two small blocks, which give variety by opposing squares to the curved corner. On the square pattern, as a foundation, it will be readily seen that a great variety may be produced, depending, of course, on the skill of the designer, and having secured an outline similar to the curved pattern, besides furnishing a neat corner piece, simply as a "fine line," it may enriched with leafing, and applied to work demanding greater display. From this apparently simple lesson, taken in all its bearings, may be produced decorations for omnibuses and railway coaches, and (to step out of our usual path) the interior of buildings.

The dotted lines in cut No. 1 formed the skeleton for No. 2—the small squares being omitted, and the scrolls carried over a portion of the narrow spaces, to render the whole more compact. The corner piece should appear light or heavy, according to the size of the "stripe" which connects with it. The stripes connected with the corner pieces should be of the same color, to appear well, although this plan is not always adopted.

Whenever a broad line joins on to a corner ornament, composed of small lines, or very small leafing, it would appear out of character, and the same, were this order reversed. In our practice we aim to have some portion of the corner piece about the size of the stripe which joins it, by this means causing them to appear well balanced. Stripes run on within the space inclosed by the outer lines should be smaller, and of more delicate tints. For if display is sought, the corner piece and stripes should be painted so as to attract the attention, and every other line that may be added assist, but not detract from them. We have drawn a narrow stripe, ornamented on the line of diagonal center, which relieves the corner piece at a point where it is very plain. In the selection of colors, the color of the ground work must first be considered, and a choice made that will contrast well. On dark grounds, a light; and on light grounds, a dark range of colors may be employed.

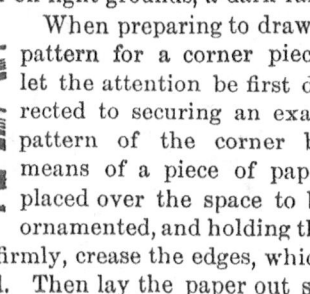

When preparing to draw a pattern for a corner piece, let the attention be first directed to securing an exact pattern of the corner by means of a piece of paper placed over the space to be ornamented, and holding the paper firmly, crease the edges, which give the outline desired. Then lay the paper out smoothly, and with the dividers, carrying a lead pencil, measure the distance you wish the stripes to be from the edges of the panel, and draw the corner piece in conformity with these lines.

No. 1 is an original design suitable for sleighs. The ornamental part is drawn by use of the circle and oval. We have given only the corner piece, and a portion of the stripe to connect therewith. It will readily be noticed that it is only necessary to draw the part we have given and attach it to the upper right hand division to complete the corner piece.

The corner may be laid in gold, and the stripes run on with some dark color; or, the corner piece in color, and the stripes gold.

The center line in broad stripe, as well as the distant lines, should be of a hue which will not detract from the brilliancy of the whole.

No. 2. Paint them according to your own taste. The instructions given for similar patterns may assist the student in some particulars, but we would advise him not to rest satisfied with the few colors named in *our* explanations, but experiment with others, and select from among them those which are most pleasing to the eye.

No. 3. CORNER PIECE AND BROAD STRIPE.—This will make a very neat and attractive corner piece, and broad stripe for sleighs, or other work where showy striping is required. The corner piece should be painted the same color as the broad line, shaded with dark tint of the same color, so that all will harmonize.

The dark color or broad line near the scroll should be painted with

No. 4.

the darkest tint used in shading the scroll, and be put on the stripe when it is dry. It is not continued throughout the length of the stripe, as may be seen by the pattern. The fine line should be of a different color from any of those used on the broad line.

No. 4. Ornamented Broad Line.—Another design for sleighs. This is intended to be placed in the center of the stripes on the dash. On a dark ground, the center and stripes will look best done in gold; on a drab, or other light colored ground, the center may be laid gold, and the striping of a suitable darker color than the ground color, so as to bring out the full brilliancy of the gold.

Fig. 1 shows the surface of a Landau back quarter panel (no moldings are represented). It will be readily seen that the three corner blocks vary in form, each one being governed in shape by the lines bounding it, so that to draw them correctly on paper, or on the panel with the dividers, it is necessary only to decide on the size the small

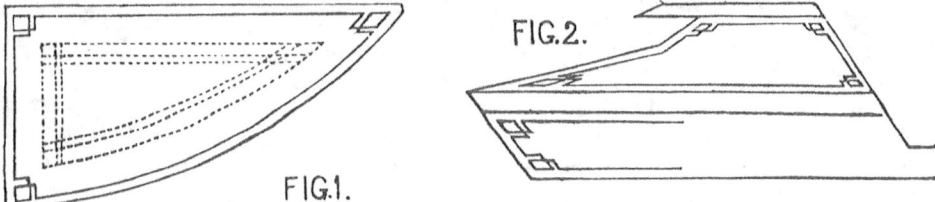

corner blocks should be, and with the dividers step off the proper amount of space on the panel for the distance of stripes from the moldings or edges, and draw the outer lines, and within these lines mark off the size of the blocks and narrow spaces. The dividers need to be set but twice, once for the blocks and once for the narrow spaces surrounding them, as will be proven by examining the smaller figure drawn in the dotted lines. But care must be taken to have the spaces equal, or the size of the blocks will vary, which is shown on the lower lines of the dotted pattern.

Fig. 2 shows the side panels of a very neat style of coal-box body, which we have selected for its variety of angles, and to which also we have applied the very simple rule given above for drawing corner pieces, viz.: the outlines of the panel, when followed, give the precise form for the several corners.

Fig. 3 shows the rule as applied to a panel or space inclosed by straight lines, connecting with an arc or portion of a circle. The three figures alluded to furnish but one pattern of corner piece, except in so far as the variation of the adjacent outlines of the panels affect it.

The pattern represented in Fig. 4 is more ornamental than the preceding ones, yet still retaining the figure produced by

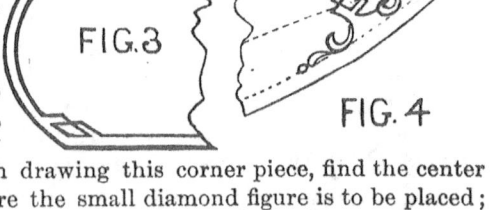

the acute angle of the Landau panel. In drawing this corner piece, find the center of the panel diagonally at the point where the small diamond figure is to be placed; then draw the figure on this center line, and by the use of the circle or the oval add the remaining portion of the design. Having mastered these simple examples, endeavor to originate others, combining the figures formed by straight lines with scrolls of various kinds.

The better plan for the apprentice to pursue in these studies would be to take exact corner patterns of different shaped panels, and trace them on a good quality of drawing paper, and set apart certain evenings for close application in designing. Provide yourself with a drawing board, 18 inches square, a T-square, having a swivel head to produce bevels, a No. 3 Faber pencil, a pair of dividers, and half a dozen fastening tacks (or some paste) to hold the paper firmly in position, and you will be ready, at a trifling cost, to acquire by continued practice a very essential part of carriage painting, and that concerning which but very few foremen, who may be ornamenters, will communicate a single idea.

We would not be understood to recommend the application of the above patterns to the Landau panels, nor even to the buggy body, unless a highly ornamental piece of work is required.

The value of these instructions consists in furnishing the apprentice with the plan of correctly laying down the governing or foundation lines for drawing patterns suitable for carriage work, but more especially for the sleigh, omnibus and fine business wagons. We well remember, when an apprentice, how seemingly mysterious the designing of ornaments and ornamental striping appeared, and how crude our ideas were about correctly copying a complicated pattern. In the attempt to copy, we did not discover that there were certain center lines flowing through the pattern on which the remaining parts were arranged, nor the proportions which one member bore to the other. So, in our ignorance, we groped along, copying details as we proceeded, and soon came to the conclusion that the task was unpleasant and fruitless. But after having received instructions, which resolved the patterns into their first and simplest form, our confused ideas were suddenly changed, and we no longer looked upon a complicated design merely as a whole, but as a whole made up of parts, which were governed by an underlying skeleton or frame work of lines. That which had been obscure, perplexing and fruitless, now assumed a different aspect, and, so far as our trade demanded, we were enabled to produce original designs suited to the work in hand.

This pattern is intended for a light axle bed, and may be put on with the *pen*, the leafing to be filled in with horizontal or curved lines, as may suit the taste of the painter. The use of horizontal lines will answer, provided the painter has the idea of the pattern fully impressed on his mind, and can make these small lines, of the various lengths required, to form the leafing. The scroll and stripes may be put on with gold or bronze. The small center scroll and stripe to be painted a color which will contrast with the ground color of the gearing, but of a subdued tone, to prevent its detracting from the brilliancy of the gold or bronze. Or lay in the scroll and stripes with color, and the center scroll and fine line gold.

ORNAMENTAL PANEL STRIPING

suitable for sleighs. The first pattern given represents dark striping on a light ground work; the second, the reverse. The first requires the use of the striping pencil only, and may be rapidly produced and yet show well for the amount of labor expended. It gives a portion of broad line, and center piece, connected with the distance lines which bar in the broad line. It will certainly be understood that the center piece, when drawn in full, combines with lines corresponding to those on the opposite side, and these lines inclose a broad line, centered with a fine line, as shown in the cut.

In painting, we would first run on the broad lines and follow with the distance lines and center piece, completing the job by centering the broad lines. The broad line should be colored in contrast with the distance lines.

On a white ground, this pattern would look well done in gold and black shaded. The same pattern can be carried out on a curved line by simply marking out the width of the distance lines apart, and within these lines drawing the center piece.

The second cut represents light colors on a dark ground. In this we have also a center ornament and broad line inclosed by comparatively fine lines. The center ornament will take a trifle more time to work out than the preceding one, and the broad line is edged as well as centered. In coloring we may make the broad line and distance lines throughout their extent, the same color, and the center and fine line which edges the broad line the same color. The center stripe on the broad line to be darker than broad line, yet lighter than the ground color of panels. For instance, on panels painted claret: run on the broad and distance lines of light olive, (yellow and raw umber toned with lake), the center piece, and fine lines edging the broad line, flake or cremnitz white, the broad line to be centered with dark olive.

Sample of striping to be used as a belt around a body where there are no moldings the wide stripe and scrolls to be gold leaf, the fine lines in color.

ORNAMENTED BROAD LINE, SCROLL AND FINE LINES FOR SLEIGHS.

We herewith present designs for broad stripe, suitable to the Portland or other styles of sleighs.

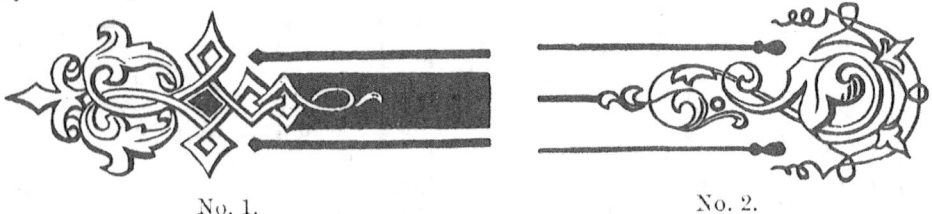

No. 1. No. 2.

The scroll end of No. 1 may be used as a finish at end of stripe, as here shown, or placed as a center ornament to the stripe by running the broad line on each side.

A slight alteration would be required at the outer end of the scroll to make it fit the

No. 3.

stripe. The fine line inclosing the broad stripe being connected with the scrolls should be of the same color. Scroll and stripes, gold. Broad line any dark color suited to the ground color, and the distance fine lines also of a color dark enough to enhance the brilliancy of the gold leaf. On light drab, straw, flesh color, pea green, or any other very light ground color, if gold leaf is employed, it should be back-shaded to relieve it from the brilliancy of the color on which it is placed. The pattern may be drawn to suit any curve desired by taking the sweep of the same as a center line, and on this line re-draw the pattern.

Nos. 2 and 3—Color as fancy may dictate.

No. 4. This scroll is arranged to appear light, the spaces between the parts being large, and the body of the scroll slender. In practice, it will be found that to lay in the whole in gold is the speediest and cheapest method to arrive at a showy effect. Gold leaf furnishes a solid foundation, and requires but little after-work to light and shade

No. 4.

No. 5.

it. In the employment of colors, care should be exercised in their selection, giving preference to those which are dark and rich—the full brightness of the colors to be reserved for those parts which receive the strongest lights.

No. 5. ENRICHED PANEL STRIPE.—This pattern is suited to panel $2\frac{1}{8}$ inches wide, and of indefinite length. In designing such patterns first find the center of the panel longitudinally, and strike a line on the drawing paper representing it. From this line step off with the dividers the half width of panel, on each side, and through the points gained, produce lines parallel with the center line; the space inclosed represents the width of panel. We next decide on the distance we would have the stripes from the moldings. In the cut here given we place them ¼ inch distant, which gives 1⅜ inches as the width from outside to outside of stripes. A vertical line is next drawn, intersecting the lines laid down. From the vertical line, step off on the center line three divisions, 1⅜ inches each. Two of these divisions to be placed (as here shown) to the left, and one to the right of the vertical line. The first division to the left determines the length of the looped end of stripe, and the second the limit of the attached scroll. The division to the right of vertical line secures the length of that member of the scroll. At the points of intersection of the vertical line, with the lines representing the stripes, we have the starting points of the looped end finish. Having secured the proportions the parts bear to each other, design the scroll to suit the taste, and when you have an outline that is satisfactory, go over it with India ink; after which erase all pencil marks, and with a pin or needle perforate the outline; after which operation the pattern is ready for "pouncing" to the panel.

THE COLORING.

On small pieces use rich colors, but so graduated as to allow of but a small proportion of the brightest tones. On large scroll pieces the umber-toned drabs, browns, and olive-greens combine very pleasingly. A good effect may be produced by taking almost any color, and mixing several tints of it, using the natural color as the darkest shade, and the tints according to their strength in the scale ascending toward the highest lights. The latter to be chosen of a color in harmony with the tints previously laid on. As for instance, if the tints are brownish, the high lights should be yellow, or of a color approaching yellow. Practice in the mixture of various tints, and their application to surfaces light and dark will soon suggest the proper degree of strength the tints, tones and colors should be to produce the effect desired.

ORNAMENTS COMPOSED PRINCIPALLY OF SCROLLS.

No. 1. This pattern is suited to narrow belt spaces, or to panels having a greater width than depth. Lay in the crown gold, and also the edges of scrolls where the high light falls. After the gold leaf has been applied, proceed with the colors, selecting those which will agree well with the ground color. In handling these small patterns they may be dismissed with a few carefully given touches, by merely representing the leafing by outlines, mixed of a hue to suit the lighted and shaded por-

No. 1.

tions, and placing each hue or tint in its proper place at once, leaving the ground color to supply the rest. But when depth and richness of colors are sought, repeated coatings and glazing must be resorted to.

The middle portion of the ornament will require the exercise of more skill than the scroll work, but being small, defects will not be readily discovered.

The goat's head and neck may be painted true to nature, representing any one of the varied kinds to be found, or may be somewhat exaggerated in color, if it suits the purpose of the painter in giving a certain effect.

On dark grounds, select the gray and delicate brownish tints. On light grounds, those that are dark.

Having a distinct outline, place the half lights in their proper position, and with a short-haired pencil work the color off toward the shaded parts, the ground color being allowed to form the darkest part. The whole having been coated, with reference only to forming the head and neck, work up the details by colors, lighter or darker, according to demands of the pattern.

The crown may be shaded with asphaltum, tinted with carmine, and lighted with straw color or pink (light red and white). The dots or small circles painted in imitation of ruby and emerald.

TIGER SCROLL.

No. 2. The design herewith given contains simple and easy curves. The largest liberty is allowed in ornaments of this character. The human form, beasts, birds, reptiles and fishes are employed to enrich scroll patterns.

No. 2.

The designer does not follow nature exactly, but may add a curve or scroll here and there, giving the upper portion of the human body, and dispensing with the lower limbs by sweeping them off into graceful scrolls; or, as in the accompanying illustration, make use of the foreparts of a beast, and bring the curve of the back into a scroll line, which, with its members, forms a portion of the scroll pattern.

In the attempt to copy the design, note that its length is twice its width, or nearly so. Square the pattern, find the center, and draw the center lines; then strike the circles and sweeps of the larger portions, filling in on those governing lines the details of leafing, etc.

It may be laid in gold, introducing colors on the scroll work. But as colors are more difficult to handle, we would advise their use. A light colored ground work is the most favorable, as the colors may be selected of gray, purple-gray, drab, brown, light and dark shades, etc., and the scroll patterns be relieved from the ground color by shadow.

Clear vermilion should not be used as a ground work. White, straw color, poonah yellow and umber-toned drabs are the best.

As in a piece of wood carving the individual parts would not reflect one glare of light if the whole were painted pure white, much less would drab, gray, brown, green, or any other color, hue or tint, present this appearance. A brown scroll must be painted in graduated shades and of brown, so with drabs, greens, etc.

Having decided on the colors you wish to employ, mix up three or four hues of each, placing the lightest on those parts to be brought forward, the darker toward the receding portions. Mix the colors to dry slowly that they may be properly blended. Form the leafing with broad lights and shades, using sharp lines sparingly, even on the finishing touches.

No. 3. Ornament suitable for an opera board, sleigh, or vehicle where a large ornament is required. When colors are used, mix up the different tints you decide on, and blend in light and dark shades to form the leafing; afterward high-lighting with a few touches where the pattern needs to be brought out more distinctly.

A scroll will have a more pleasing effect painted as above stated, than if the form of the leafing were worked out by a great many lines put on and crossed, dotted, and retouched.

No. 4. This was drawn for panels of the English Quarter Sleigh. It agrees in outline with the boot, and the scrolls being drawn, open and simple in form, render the pat-

tern easy of execution, and distinctly visible at a reasonable distance. The crest may appear to some to be a difficult task, but a greater part of the seeming difficulty will vanish on securing a correct outline of the whole, and proceed to lay in the lights and shades in broad, distinct masses, leaving the details to be worked up when the full value of the broader masses has been gained. The light on the face and breast in half tone having been laid, proceed to paint in the shaded parts, blending the edges of contact; when dry sketch in the eye, nostrils and other details, and finish each with care.

SCROLLED ORNAMENT.

No. 5.

NO. 5. The lozenge and oval forms were used in producing this design. Previous instruction will suffice as to coloring for the scroll portion. For the medallion with female bust, color the back ground umber-toned drab. The female complexion will require white, Naples, yellow, vermilion and carmine.

The shadows composed of delicate grays; hair, white, yellow, raw sienna, shaded with burnt umber.

TREATMENT OF THE SPOKE FACE.

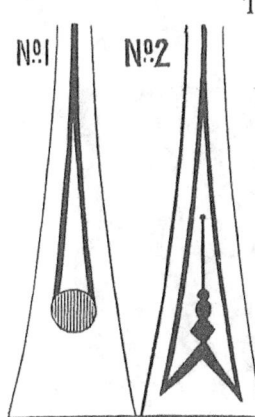

When striping is used to any extent, it is seldom the case that the spoke face is slighted. The space presented is, however, small and rather inconvenient to work at, therefore requires patterns, simple and easy of execution. In the cuts herewith presented are eight styles, either of which will require no unusual outlay of time. No. 1 is painted as shown, consisting only of a circular base, to which are attached the fine lines following up the shape of the spoke. The circle may be set off at a greater distance from the hub than here shown, without injury to the appearance of the face of the wheel. The circle and stripes may be gold, or the circle gold, and the stripes of any suitable color. Although apparently a very simple pattern, it nevertheless looks well when viewed on the street.

No. 2. The triangular space to be laid in in gold. When the leaf has been applied, take either white or straw color, and first draw the center fine line, starting it at the apex of the angle formed at the bottom. This center line will be the guide for the true position of the small diamond and circles which are to be painted over the fine line. Fine lines of a color suited to the ground color should then be run along the outer edges of the gold and up the face of spoke in the usual manner. Those who are familiar with the effect of placing colors on gold will readily perceive that this pattern will show brilliantly. Dark, transparent colors may also be used with good effect, such as carmine, pure, or saddened with blue, brown or black. When gold is employed for the faces of spokes, the greatest care should be exercised in sizing: The distance of the gold pattern from the hub, the length of the pattern, and the extent to which the fine lines are to be carried, should be marked on each spoke with chalk, or, which is far better, a piece of cremnitz white, as the latter will hold a firm sharp chisel edge, or a point, and wear longer than chalk. Having obtained the points named, raise the wheel pretty high, and take your position in front of the center of the hub; a cushion or low stool should be used to sit upon, so as to obtain an easy position, thus relieving the muscles from all undue strain. Have the color, turpentine and pencils conveniently placed and commence by bringing one of the spokes squarely before you; outline the part to be gilt, bringing its upper end to a point, or a size not wider than one of the fine lines will be with which you intend to inclose the gold. Care must be taken to have these points in the center of that part of the spoke where they terminate. Having outlined the pattern, fill in the center with a suitable ornamenting pencil. The sizing having been laid on true, will give a true facing of gold, and much trouble will be saved by avoiding after corrections.

No. 3 will make a neat finish, and will not consume much time in executing it; the center in gold, the striping in color, would be attractive.

It works out very neatly and may be quickly finished. Lay the center gold. The fine lines in color. The lozenge and crescent should be sized in solid, that is to say, form a circle, its lower portion joining on to the upper part of the lozenge, fill them in solid with size, and after the gold has been laid separate them with black or burnt umber. The lower figure is produced as shown by cutting in with dark color, thus showing a small lozenge inclosed by the outlines of a larger one.

No. 3.

Patterns, 4, 5, 6 and 7 will require no more than a passing notice. No. 4 can be produced mainly with the striping pencil; No. 5, fill in the center solid, and cut in the figures shown by the use of dark colors, as recommended for No. 2.

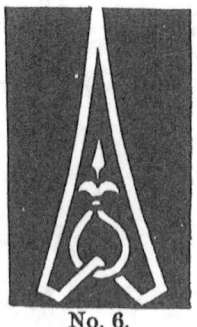

No. 4.   No. 5.   No. 6.

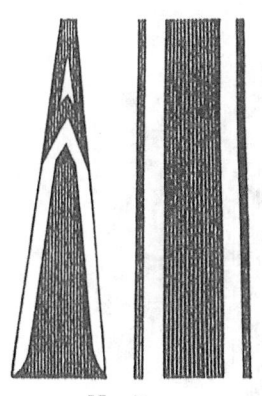

No. 7.

No. 8.

No. 6.—On this the center ornament can be done with an ornamenting pencil; needs no cutting in with dark colors.

No. 7, with its accompanying spoke pattern, has no special value further than this: it represents a prevailing style of striping in this year of our Lord one thousand eight hundred and seventy-two.

No. 8.—Shows three views of a spoke face, in gold leaf, giving the manner of touching up the pattern, in order to produce the effect or finish shown on the third. To new beginners this simple lesson will be well worthy their attention. The first one to the left shows a spoke face in the form the sizing is put on, and also the gold leaf. The second is partially touched up—the third completed; at least it will serve to direct how it is finished, for it is impossible to give a perfect representation in black and white. To begin, raise the wheel, and sit immediately in front of it, so as to bring the spoke vertically before you, the felloe end of spoke beneath. Outline with a good cutting-up pencil, bringing the upper part of pattern to a clean, fine point at the center of spoke. When the outline is finished fill in with a short pencil. When dry enough to gild lay out the leaf and cut it to the shape of the pattern; by cutting the gold with the ends of the pattern, reversed alternately, there will be no waste.

The gold having been laid, proceed to bring out the form of lower middle leaf; also the upper one, of which the original pattern shows no traces. For this purpose use burnt sienna or asphaltum. These are the shading lines, which are to be followed by the lights, as shown in the third cut here given. Use white for the lights. When the fronts are completed, run a fine line on each side, continuing it up the front of spokes a short distance.

ENRICHMENT OF A HUB.

In the annexed cut we give a style for adorning hub by striping and what the painters are pleased to term "carving."

Those who may attempt to imitate it should make up their minds in advance to be patient with the carving, so that the fourth hub may present the same degree of exactness as the first. We do not advise the employment of so much work except on a job where great display is demanded, and even then the colors should not be too obtrusive. There are sections where a Fair job is estimated by the quantity as well as the quality of the work put on in each of the branches, and the painter has to do the lion's share toward making the job attractive to the eyes of the eager crowd; in

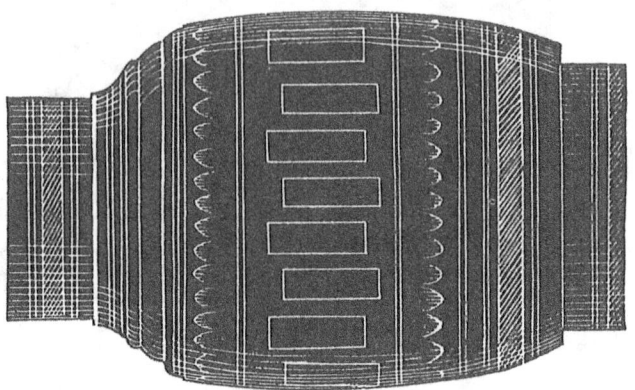

cases of this kind the pattern here set will be of service. As to the striping, the painter must be the judge of that, for the shape and proportions of the hub should regulate him in the number and size of the stripes, for the striping should never destroy the natural appearance of a well-formed hub.

We have laid out the hub for three broad lines, relieved by distance fine lines. The "carving" also relieved by fine lines. Supposing the carriage part painted claret, run on the broad lines with black, and the distance fine lines, gold. The "carving" carmine lightened up a trifle with vermilion; the distance fine lines the same color; and the fine line shown on the concave at front of hub, black. The broad lines should by all means be put on the varnish color, and receive a good coat of clear varnish. When this is dry and has been rubbed down, proceed with the fine lining and "carving." The latter can best be accomplished by the use of the "cutting-up" or "carving" pencil, which is a pencil about an inch long and an eighth of an inch in diameter at the quill, or ferrule. It may be made of either sable or camel hair, and should spring to a fine point when wetted. It is used by bearing down the point and with a quick jerking motion raising the pencil so that its fine point touches the surface. The carving on the hub will be found difficult to execute, because it stands in a reversed position to that which the painter usually paints it; a little practice will, however, overcome the seeming awkwardness. The pencil may be used by putting the point on lightly, and gradually bearing down toward the heavy portion. But this practice does not secure as perfect work as the other method, and, therefore, should not be followed up by the painter who aims at perfection. Great care should be taken to have each stroke of the pencil do its work without after corrections; there should be no wiping out if a perfectly pure groundwork is desired. Let the extra time needed to correct mistakes be employed in avoiding them, and the varnish will then flow out over a clear ground, and tell no tales of slovenliness. It will not be necessary to carry the carving on other portions of the carriage, as this is purely a hub adornment.

TO LAY OUT IMITATION CANE WORK.

Real cane work having been in use, more or less, for a number of years, to give a light as well as genteel appearance to the panels of bodies, the painted imitation soon followed, as it gave all the effect of lightness and ornament without the expense attendant on having real cane inserted. This, too, was a tedious operation, viz.: the laying out and striping of a body; and when "transferring" had attained such perfection in parting with gold and rich colors, as to compel the use of ornaments thus pre pared, we find that it was not long before we were recommended to try "transfer" cane work. And still later we have had another imitation which is far more deceptive, it having open interstices, and being a very exact imitation of real cane work.

As the hand work, put on in oil or varnish colors, is preferred by a great many painters, we wish in this article to speak more particularly as to the manner of laying out the work preparatory to striping. Every painter who has been called upon to do this kind of work is well aware that it is the most important part of the whole operation to have the squares laid out perfectly true; for where there is any irregularity, the diagonal lines will fail to cut the corners of all the squares alike, producing irregularities mortifying to the workman.

We will first direct attention to the use of the DIVIDERS or COMPASSES, and then lay them aside, and show that they are not required at all in spacing off a surface for the purpose under consideration. There are two kinds of surfaces—the flat and the swelled. The first being generally in part or wholly inclosed by straight lines, and the latter by sweeps, varying according to the style of the body.

To lay off cane work on a flat panel in the old way, the painter had recourse to the compasses in stepping off the divisions which formed the squares. The diagonal lines were then added by using a straight edge, so placed that an equal portion of each square was cut across near the corners, forming the octagonal figure desired. The obections to the use of the compasses are, they mar the surface, and also consume more time than is profitable—the compasses' point requires to be pressed into the panel, thus forming minute holes, which mar the surface to some extent. On swelled panels, the molding lines varying throughout the border of the panel, it was found very perplexing to obtain the proper starting points. To obviate the difficulties mentioned, we concluded to try something different, and at length adopted the rigid and flexible rulers, the first to be used on flat panels, and the second on those that are swelled. A few experiments proved that the compasses were not needed, and we abandoned their use.

SIZE OF RULERS.

The width of the rulers determines the size of the squares, and gives the appearance of very open or close cane work, according to the width decided on. Three-eighths of an inch is a medium size, and will answer for all except very narrow spaces, when the width may be reduced to ¼ of an inch, or down to 3-16. They may be made of fine-grained tough ash, this wood being less liable than hickory to spring out of shape; and being tougher than walnut, the edge is not so easily worn out of true by the action of the stile or marker.

These rulers require to be perfectly true throughout their extent, so that when placed against a surface, and lines marked, one on each side of the ruler, the space formed shall present no variation. The *marker* used should be cut wedge-shape, so as to follow closely on the edge of the ruler. On a flat surface use the

RIGID OR STIFF RULER,

placing it against a horizontal line or molding, and lay off the horizontal spaces. Vertical lines may now be marked on by starting from a vertical line obtained by the use of a square. After the first space has been formed, the ruler must be placed an eighth of an inch or so distant, in order to produce the double lines in which the squares are inclosed. Now mark on both sides of ruler before raising it up, and repeat the operation until the outer edge is reached. The squares having been all laid out, it only remains to mark on the diagonal lines, which are easily obtained, the ruler giving the precise amount to be cut off of each square to form the octagonal figure presented by real cane work.

The only difficulty to overcome in laying out cane work on panels having a varying surface, and swept or curved molding lines, is to obtain a line from which to lay off the squares. In this case, we find a vertical line by means of a plumb. Having leveled the body, the plumb line is held above the panel, at some convenient point, and where the string crosses the panel a true vertical is indicated. One side of a square being placed against the string, the other side gives a horizontal line, which may be marked on the panel, and from this lay off as on the flat panel, using the FLEXIBLE RULER. The lines will appear to be inclined on a surface inclosed by swept lines; but this being an optical illusion, proceed with the laying off, when, if it be carefully done, there will be no trouble in getting a perfect piece of work.

Having completed the laying out, the striping may be run on with a fine line pencil, or, which is better, a striping pen. The striping pencil is not so well adapted to the production of perfect stripes as the pen, but will answer the purpose. If the pencil is used, endeavor to have the stripes as near of a size as possible, and all joints well made. When the pen is employed, the stripes are more uniform. On small panels, with a good pen and ruler, there is no necessity for laying out the work, for having secured a line to start from, the ruler may be placed to it, and with the pen, charged with color, the stripes be run on with confidence, as the ruler will bring the work out square.

On large surfaces it is best to allow one set of stripes to dry before attempting to put in others, but when short of time, the ruler may be supported at a distance from the panel by studs, made of cork or wood, and tacked or screwed to the ruler.

Cane work may be laid out the reverse of the plan just given; that is, by first laying off the beveled or diagonal lines, and afterward producing the squares within the diamond figures. The points of these figures should be perfectly vertical and horizontal, otherwise the cane work will be irregular.

The accompanying cut will serve to show how cane work is laid out, and give an approximate idea as to the size of the openings by using a ruler three-eighths of an inch wide.

ENLARGING ORNAMENTS.

An ornament may be increased in size and the proportions preserved with the dividers, proportional compasses, or by the use of squares. In enlarging begin by squaring the pattern.

Next draw diagonal lines from the opposite corners, their intersection giving the center point. Divide the pattern from the center point longitudinally and vertically. As these lines were those first laid down when the pattern was originally drawn, you are put in possession of the governing points in enlarging to any size you desire. If the pattern be one inch square, and you require it drawn to ornament a space measuring twelve inches square, or one foot, it will only be necessary to produce a square of that size; divide it as laid down above, and with the dividers increase each part of the details of the pattern accordingly.

A readier method than using the dividers or compasses is to lay off on the outline an equal number of parts, and throw the pattern into squares; the enlarged figure to have a like number of squares. Then number the lines on each, and sketch in the enlarged pattern, so that the lines and spaces cover or pass through corresponding parts of the small pattern.

A design inclosed by an outline on each of its sides is laid off in twelve or more parts. To enlarge it to one foot lay off the latter into twelve or more parts. To preserve a fine engraving or ornament from pencil marks, use a frame made of pasteboard or tin, the size of the engraving; perforate it all around, at the points of each division, and insert a silk thread, which, when passed through all the holes, will give the required squares. This frame, laid down on the picture or ornament, gives the guiding lines, and does not deface the pattern.

In this kind of work a man must possess a quick eye and a ready hand. Notice the bearing of center lines, and the large masses, and fill in from these, instead of attempting to put in the details as you progress. Scroll patterns will be found more difficult to enlarge than those made up of right lines, or those combined with curves. Scroll work having such a varying outline, and a large portion of it being drawn according to no particular rule, the attempt to produce an exact copy would be a laborious task. But as a slight change of the parts will not be noticed by the majority of observers, the designer may alter, add to or take from certain portions, provided the general appearance remains.

We use the dividers for small ornaments, which are not to be increased a great deal. Taking the length of the ornament if we wish it as large again, and step it off accordingly; the same manner proceed with the width. Then take the bearing of the center lines, sketch them, and by the eye fill in the details.

COLORS.

White Lead.—White carbonate of lead.
Cremnitz White.—White carbonate of lead.
Flake White.—English white lead, in the form of flakes or scales.
Zinc White.—Oxyd of lead.
White Chalk.—Native carbonate of lime.
Naples Yellow.—Compound oxyd of lead and antimony.
Chrome Yellow.—Chromate of lead.
Patent Yellow.—Submuriate, or chloruret of lead.
Massicote.—Protoxide of lead.
Yellow Ochre.—Native pigment.
Sienna.—An iron ore.
Yellow Orpiment or Yellow Arsenic.—Sulphuretted oxyds of arsenic.
King's Yellow.—Yellow orpiment, or Chinese yellow.
Cadmium Yellow.—Sulphuret of cadmium—a metal.
Gall Stone.—An animal calculus, formed in the gall bladder, principally of oxen.
Dutch Pink, and English and Italian Pink.—Yellow colors prepared by impregnating whiting with vegetable yellow tinctures, in the manner of rose pink.
Red Lead.—An ancient pigment, deutoxyd of lead.
Indian Red.—Brought from Bengal, a very rich iron ore; hematite or peroxyd of iron.
Light Red.—An ochre of a russet orange hue. The common light red is brown ochre; burnt yellow ochres afford the best.
Venetian Red, or Scarlet Ochre.—True Venetian red is a native ochre. Colors are prepared artificially from sulphate of iron.
Dragon's Blood.—A resinous substance, brought principally from the East Indies.
Rose Pink.—A coarse kind of lake, produced by dying chalk or whiting with a decoction of Brazil wood.

Naples Yellow is liable to change in composition with the ochres. Prussian and Antwerp Blues, and all pigments of which iron is an ingredient or principle, should be mixed with an ivory knife, steel having a tendency to change their color.

Vermilion.—Obtained from cinnabar, or mercury.

Carmine.—Coloring matter in the cochineal insect, precipitated by the use of acids.

Lake.—Cochineal lake is what may be termed an inferior carmine, it being made from the fluid and deposit remaining after the more brilliant color (**carmine**) has been prepared.

English Purple Lake.—Lake of a slightly purple cast.

Royal Purple Lake.—A very rich purple, of a light color.

Maroon Lake.—A reddish-brown lake.

Magenta Lake.—Lake of a dark purple shade.

Crimson Lake is of a deep red color.

Chatamuck Lake.—A cheap bright red lake, in imitation of **carmine**.

Rose Lake.—Also a cheap lake, of a dark red tone.

Tuscan Red.—A color lately introduced; it is similar to drop lake, and possesses great covering power.

NOTE.—Carmine, being of a brilliant red tone, the painter will readily discover that, by the addition of other colors, it may be modified so as to produce hues in imitation of the different colored lakes.

By adding to carmine vermilion, we destroy its value as carmine, and produce lighter hues of red, richer than vermilion, yet far inferior to the pure carmine.

By adding to carmine ultramarine blue, various tones of purple are produced.

By adding to carmine Prussian blue and black, we have violet.

By adding to carmine black, plum color.

By adding to carmine burnt umber, rich tones of brown or maroon.

By adding to carmine ultramarine blue and white, lilac.

Chrome Green.—Compounds of chromate of lead with Prussian and other blues.

Emerald Green.—A copper green, terrene or earthy base.

Terre Verte.—Ocher of a bluish green.

Verdigris.—Rust of copper; an acetate of copper.

Saxon Green.—A carbonate of copper.

Ultramarine Blue.—Lapis lazuli—azure stone.

Prussian Blue.—Combination of prussic acid with iron and alumina.

Raw Umber.—Native oxyd of iron and manganese, first obtained from Ancient Ombrio (now Spoleto), in Italy.

Burnt Umber.—Raw umber subjected to a certain degree of heat.

Raw Sienna.—An iron ore.

Burnt Sienna.—Raw sienna, burned.

Vandyke Brown.—First prepared by the celebrated master, Vandyke. is, we believe. a bituminous substance.

Asphaltum.—A black bituminous substance, found on the lake Asphaltis.

Lamp-black.—A fine soot collected from the smoke of burning resinous substances.

Drop-black.—The bones of animals, burned.

Ivory-black.—Ivory shavings, burned.

COLORS AND THE EYE.

To the eye alone is the power given to distinguish between the various colors of objects in nature. Within the darkened chamber of this most beautiful and wonderful piece of mechanism the rays of light from objects are collected, and the impressions of color transmitted to the brain. Sightless eyes cannot appreciate the beauties so lavishly spread out in nature, because to no other one of the five senses was this crowning glory given. The blind may enjoy the fragrance of the flowers, sit entranced by strains of delightful music, with the epicure divide the pleasure of a rich repast, and by the sense of touch form very correct ideas of the contour of objects, and through this means may acquire some knowledge of beauty in proportions; but by what method can be communicated to them a just conception of colors? At this point all efforts fail; darkness sits immovable, and pleasing sensations produced by colors are to them as though they had no existence.

The painter's art appeals to the eye and depends on the sense of seeing only as its means of affording pleasure or profit. What would it profit a Bierstadt, after having transferred to canvas the sublime grandeur of the Yo Semite Valley, to place it before a company of blind persons at mid-day? Through the sense of hearing, and by none other, could they be interested by the artist. The relative heights of the peaks, the depth of the valley, the beauty of the waterfalls, might be communicated; but the representation of the scene on the canvas before them would have no greater beauty than that presented to the perfect eye when peering at night in a dark cavern. By the sense of touch the blind would be unable even to trace out the proportions of the several parts of the picture; the size of the canvas and the frame, when carefully handled, would be the only knowledge they could gain.

But the perfect human eye does not in all cases possess the power of discovering beauties where they in reality exist. There are persons, not a few in number, who are incapable of beholding any beauty, even in a gorgeous sunset, and, like the horse or the ox, they merely use the colored rays to light them toward their destination. Others possess a love for colors that are glaring; nothing is pleasing to them except red, blue, green, etc.; a color modified by mixture with other colors, and neutral tints, they pronounce flat, tame. Savage tribes, and the people inhabiting extreme southern countries, are naturally fond of gay colors and highly ornamental apparel; why, we do not profess to know; but such appears to be the case.

The eye requires to be educated in order to be sensitive to harmonies in color, and until appeals have been made to the judgment, aided by contrasts shown between that which is coarse and glaring and the truly beautiful, there can be no great advancement made by the masses.

GOLD PAINT FOR STRIPING.

WINSOR & NEWTON, 38 RATHBONE PLACE, LONDON, ENGLAND, manufacture an article they term liquid gold. It is designed for fine line gold striping, its use removing the necessity of laying on "size." It may be purchased of any respectable dealer in coach painters' supplies.

CLARET COLOR.

ENGLISH PURPLE LAKE, mixed up and applied clear, produces a color which may be termed claret. It may be lightened by the addition of carmine or Munich lake, and still represent claret wine in high light, or saddened into black and represent it in deep shadow. To observe the shades pour the wine into a goblet and view it in different positions with reference to the strength of the rays of light allowed to fall in on it.

OLIVE GREEN.

BRANDON YELLOW AND DROP BLACK produce a very nice shade of olive green, or take umber, yellow and black, the yellow and umber to be first mixed up and the black added in a greater or less quantity, according to the shade desired.

MILORI GREEN.

Milori green or green lake is a color resembling in its dry state the familiar color known as chrome green, but when mixed, milori green by far surpasses chrome green in richness. It possesses a good body, covers and dries well. When mixed with white in varying proportions forms cheerful pea greens. It serves well as a ground color for emerald green and verdigris.

Three shades may be purchased,

OLIVE GREEN,

green and purple; burnt umber, yellow and black; raw umber and yellow, green, umber and yellow. The last named gives various tones, according to the proportions used, and is the cheapest.

Strictly speaking, olive is a green, compounded of blue and a small quantity of yellow and red; but carriage painters do not confine themselves strictly to these niceties. Either of those given above will produce a good color, and as tastes differ so widely, we must leave the decision to the painter.

PURE-TONED STRAW COLOR.

Having prepared a suitable ground work, take either china or flake white (dry), and Oxford ochre. Or, use flake white, lemon chrome, and a very small quantity of vermilion. Delicate colors are more or less influenced by the color of the ground work on which they are laid; so that, to secure a clear, pure tone of straw, and similar colors, the under coatings should be white or nearly so, yet having no tendency toward lead color. A flesh colored ground work would not be objectionable, still not as good as clear white or a pale yellow.

Having purchased the best quality of the colors above named, to prevent sullying them by mixture use no *dark* brown japan as a dryer; oil and sugar of lead, or varnish answers the purpose better. The paint mill and the cup into which the paint is ground, as well as the brushes, should be *perfectly* clean. Having taken these precautions, the color should be applied somewhat heavier than those having more opacity. Two coats and color and varnish should cover solidly.

Use the palest varnishes; the dark, hard-drying rubbing varnish will give a streaked appearance. The finishing coat must not (on a carriage part) be laid on as heavy as for darker colors; for, if there are any laps or runs, they will appear darker than the surrounding colors, thus giving a muddied appearance.

WHITE AND STRAW COLOR FINE LINES.

In mixing white for striping, use French zinc, cremnitz, or china white, as they possess more body than white lead. If the white line is to be run on a dark groundwork add a small quantity of lamp-black to the white, making it a silver color. Lamp-black gives the white more body, and is not detected by the eye. The stripes will also appear of the same shade throughout. A white line when run on a bright color would not bear the lampblack, as the contrast would not be great enough to hide the silver-colored tint. Chrome green or ultramarine blue would answer better.

Straw color, for striping, is certainly a delicate color to handle. As the principal ingredient is white, it is of the first importance to use the best. (See above.) The yellow should be the best lemon chrome. A small portion of vermilion or Indian red will give it more body and not be detected. To prevent the appearance of light and dark portions, the color must be used heavy enough to cover well. The pencil must be filled with the same quantity of color each time, and when the stripe begins to appear thin, run over it again with the pencil partly filled. By this means a thin layer of color is added to that which is already on; the two coats being equal to one put on with a full pencil.

A broad line of straw color requires two coats to produce a perfect finish. The above colors to be mixed in varnish, or oil and sugar of lead.

TO MIX GOLD BRONZE.

Gold bronze may be mixed in boiled oil, raw oil and japan, raw oil and sugar of lead or varnish. Raw oil and japan, with a small quantity of orange chrome, ground very fine, and mixed with the bronze, will help to give it body, and stripe and cover very nicely.

The bronze should be used from a small pan or box lid, and be kept stirred, as it settles rapidly. Use the mixture as thick as possible, so as to secure the particles of bronze in sufficient quantity to have each stripe well covered. Either the coarse or fine line pencil may be used, the only difference being in the quantity required to complete the work.

CREAM COLOR.

Cream color is compounded of white, yellow and red. In mixing all light colors, such as cream, buff, straw, corn, etc., take of white the largest proportion, and tint it with other colors. Cream color may be imitated with white, a small portion of yellow ocher, and a minute portion of red. As varnish imparts a yellow tone to all light color, they should be mixed a tint or two lighter than is desired, so that when finished the color will appear of the proper depth.

TO COLOR A BODY.

Buy a set of good flat camel-hair blenders (Mottler's), and when you have properly dusted your panels, apply the color as thin as it will bear according to its body. Don't mince around over an ordinary sized panel, or even on a large one, but apply the color quickly, spreading it over the panel, and finishing or laying it off while wet. Use as large brushes as possible on the panels. Don't commence on a panel with an inch or two-inch brush when a three inch could be used as well. We have seen painters make three laps on a small panel, when a quick application of the color, with the proper sized brush, would have laid on a more level coat, in half the time.

TRANSPARENT COLORS.

The ground color, over which any transparent color is to be laid, should be of a tone similar to the color of the lake to be used when "wet up," provided you wish to preserve the full richness of the lake, and secure its color exactly.

Drop, Munich and English purple lake, and others of similar hue, may be painted on *dark lead color*, Indian red, *browns* of various shades, and *black*.

As these transparent colors partake more or less of the color of the ground work, it is essential to have the under coatings mixed of colors which are durable

We know of no ground color superior to brown, mixed of *Indian red*, and *lamp-black*. In mixing the ground color it is preferable to have it a shade or two darker, rather than lighter than the panel color.

Lakes, painted on black, are very deep and rich in tone. The ground color should be ground out perfectly fine and painted solidly. If the ground is streaked it will be useless to attempt to make a good, solid piece of painting.

One coat of color, and two coats of color and varnish is the best manner of painting lakes.

BISMARC BROWN.

Bismarc brown may be compounded of various pigments; the richest shades are those made from burnt umber, Dutch pink and lake. Purchase the best English Dutch pink (a kind of yellow lake,) and having laid a ground work of brown on the panels or carriage part to be painted, take out on the stone a certain quantity of Dutch pink, pulverize it, and wet it up; then add burnt umber in such quantity as you desire to have the shade of brown; the addition of a small portion of lake imparts richness. Dutch pink, being very transparent, care must be taken in adding any dark color to it. The beauty of this shade of brown, or Bismarc, will not be fully apparent until the varnish has been applied, varnish adding to its depth and beauty more than to most other colors.

Burnt sienna, yellow and lake; burnt sienna, burnt umber, yellow and lake, in varying proportions, will also give shades termed Bismarc, and as either color predominates, so will the shade of Bismarc be toned.

GOLD STRIPING.

Take the finest gold bronze powder and mix it up in hard-drying varnish, or raw oil and japan, or raw oil and japan and a small quantity of orange chrome, either of which mixtures will answer. Put the mixture in a clean porcelain cup or box lid, and use it therefrom, keeping it stirred up so that the pencil will be charged with each insertion with like quantities of the gold bronze. The particles of bronze settle very rapidly, and unless the mixture is kept agitated the striping will not present a uniform brilliancy.

An imitation of gold striping may be produced by using white, chrome yellow, and light red, toned down with raw umber or black. Take white and add a small quantity chrome yellow. The tint obtained will be too yellow, and by adding a small quantity of raw umber the tone is lowered without serious change of the tint, and a small portion of light red will give the redness of tone required. Black in minute doses will correct a tendency to too great brilliancy. There are various methods of obtaining an imitation of gold, but the colors given will make as perfect an imitation as will be necessary. The painter should imitate gold leaf in shadow if he would avoid a cheap appearance, especially on a carriage part where there is a great amount of striping run on.

FRENCH GRAY.

French gray, so far as we have any knowledge of it, is nothing more than lead color. To make a good color, which would have more purity than the ordinary mixture of black and white lead, use flake or cremnitz white mixed with drop black; a very small quantity of black only is needed, merely enough to turn the white a silvery gray. This mixture will make a cold-looking gray. If you wish it richer, use drop ochre, or yellow and red, merely tinging the gray, but not allowing the yellow or red cast to predominate. On ordinary work, we used to mix only drop black and white lead (lead color), which, after being striped up nicely, and finished, passed for French gray. Varnish darkens it somewhat, so that the painter must make calculation for the shade he desires, by taking into account the difference of a tone or two deeper after being varnished.

DEEP SEA GREEN.

Mix the ground work for deep sea green of drop black and chrome green, or chrome yellow, about one part of green or yellow to three parts black. The color mix of best English Dutch pink, to which is added Prussian blue, in about the same proportion as given for the ground work, viz.: one part blue to three of Dutch pink. The tone may be raised if desired by adding lemon chrome.

LONDON SMOKE.

London smoke is mixed of burnt umber, yellow, white, red, and where in the mixing either of the brighter colors should be too prominent, black will be found useful to tone them down. Vandyke brown, burnt sienna, and yellow may also be used to make an imitation of London smoke. Let the brown shade predominate, but it must be of a yellowish cast.

GROUND FOR ULTRAMARINE BLUE.

Ultramarine blue is painted on a ground work of blue, mixed as near as possible to the shade of blue desired.

If you use the ultramarine blue clear, make a dark ground work; if you lighten it up any with white, the ground work must be lighter. Mix the ground work with Prussian or Antwerp blue and white, either lighter or darker, according to the shade you wish the color. In painting ultramarine blue, having secured your surface in the usual manner, grind out the ultramarine blue in varnish or boiled oil, and apply with a soft, flat, bristle brush. The camel-hair blender may be used on narrow panels, but on large surfaces the bristle brush will be found to work pleasantly, especially where the blue is ground in varnish. Two coats and a coat of color and varnish will be sufficient to cover solidly.

FLORENTINE LAKE.

Florentine lake and vermilion mixed and laid as a ground work for carmine does not produce a foundation suitable for anything but the commonest kind of work, and is hardly fit for even that. It is liable to very strange actions, one of which is to turn gray under the action of water and pumice stone. A case of this kind was recently brought to our notice. The carriage part in question had been color varnished, and in rubbing it down preparatory to laying on the carmine the foundation turned gray. The painter remedied the defect by going over the job with turpentine. After this coating had been applied, and the carmine was laid on, the grayish spots were not visible.

COLOR ITEMS.

CARMINE when used on a carriage part may be dismissed with one coat; if the ground work be solid, and the color laid on by one who is an expert. One coat, however, will not cover perfectly.

A VERY brilliant yellow, suitable for carriage part under a one-man body, may be secured by painting Poonah yellow over a ground of white. Stripe the carriage part with black.

BROWNS FOLLOW YELLOW INTO SHADE.—The carriage part painter may learn from this: if his ground color is dark brown, he may mix harmonious striping colors by using lighter brown shades, and at length arrive at pure yellow for fine lining.

A CARRIAGE part to be painted carmine may be striped with vermilion on the ground work, and when dry, the carmine glazed over all.

LAMP-BLACK is pure carbon, in its uncrystalized or amorphuous (without form) state.

PIGMENT, a paint; color for painting.

TRANSPARENT colors require sufficient varnish, or boiled oil, in their mixture, to prevent them from drying "dead."

IN STRIPING with quick drying color, use shorter pencils than when you have time to use oil color. On carriage parts have short pencils of the same size as your regular stripers, in order that you may be enabled to work more rapidly around the head block and any other places where short curves are required.

NEVER use quick drying color from the palette in a quantity sufficient to stripe a body or gearing. Keep the color in a small pan or box lid, so that it will have but a small amount of its surface exposed to the air.

DROP BLACK is made by charring the bones of animals. The best drop black is made from ivory shavings.

COACH-PAINTERS' GREEN, both light and deep in tone, may be purchased wet, in cans.

A RICH shade of crimson lake may be obtained by glazing clear lake over a ground made of four parts Tuscan red and one part drop black.

RAW UMBER, blue and chrome yellow, produce various tones of tea green.

DUTCH PINK or yellow lake is a most useful color in the formation of delicate drabs. With burnt umber it gives rich shades of brown, and in combination with Prussian blue and lemon chrome, produces decidedly pleasing tones of green.

ROYAL RED is an artificial ochre, composed mostly of iron ore. It serves as a ground work for vermilion, and produces good browns in combination with blue and black.

ULTRAMARINE BLUE added in small quantity will give to white lead a clear tone. In painting white grounds tone the paint with ultramarine blue, and use varnish color, except on the last coat, which should be a pale finishing varnish.

CARMINE and the lakes may be darkened by adding ultramarine or Prussian blue. The former, we think, is the most desirable.

STRAW COLORS, and all others largely composed of white lead, cover better and work more pleasantly by the addition of a small quantity of rubbing varnish.

GOLD COLORED PAINT.—Yellow, or Oxford ochre, burnt sienna, raw umber and white, may be mixed so near the color of gold in half shadow, that a gold pattern upon it will not show in some lights.

NEW COLORS.—Red, brown, and "taca," or tackaranda brown, are among the latest now being used in Philadelphia. They answer a very good purpose, but do not possess any great degree of richness. The red brown has somewhat the appearance of rose lake, and "taca" brown resembles burnt umber.

ORANGE MINERAL is another name for red lead, and is produced by the slow calcination of white lead in iron trays.

TUSCAN RED is a color resembling Indian red, though rather deeper in tone. It affords excellent grounds for carmine and purple lakes.

SILVER WHITE, Dutch pink, and a small portion of lemon chrome yellow, produce a very rich tone of olive drab or amber color.

SUPERIOR BRONZE POWDER, prepared especially for striping carriage parts, may be purchased in twenty different shades.

RAW UMBER is most useful for toning down a color that is too yellow; and, in composition with white, chrome yellow and lake, forms, beautiful drabs.

PURPLE is a color which will be rich or dull, according to the quality of the pigments used. Purple mixed of Prussian blue, vermilion and white, would cover well without any ground work. But *a rich tint of purple* cannot be obtained without using fine lake, ultramarine blue and white lead or whiting. Use about one part blue to five parts of lake, adding white lead or whiting sufficient to change the tint to the shade desired; paint it on a ground work of lead color of a purple cast.

Glazing, when applied to colors, by the carriage painter, signifies the act of laying transparent colors over a previously prepared ground work. Carmine, the various colored lakes, and verdigris have not sufficient covering power or body to produce a solid piece of painting, therefore they require a foundation suited to their respective colors. All painting of this kind is called "glazing." In the Eastern cities painters speak of glazing with putty; by this they mean the act of puttying a surface all over, so as to speedily fill up the grain and pores of the wood. In the Western States carriage painters generally term it "plastering."

Japan Dryer.—Take of linseed oil, 5 gallons; red lead and litharge, each $3\frac{1}{2}$ lbs; raw umber, $1\frac{3}{4}$ lbs.; sugar of lead, $\frac{3}{4}$ lb. Pulverize and mix together the ingredients named, and add them to the oil, and boil or simmer over a steady fire for two or three hours. Remove the kettle, and when the oil has become cooled down to a certain degree, add 5 gallons of turpentine. Having stirred the mixture thoroughly, allow it to remain quiet for ten or twelve hours; then pour it off carefully and can it, when it will be ready for use.

MISCELLANEOUS ITEMS.

Purple, the Emblem of Royalty.—Julius Cæsar was the first person who forbade the use of purple to other mortals, which was indeed but just, for he had sacrificed a million of lives to earn the right to do so. Nero, still more jealous of his blood-colored robe, seeing one day in the theater a noble lady dressed in showy purple, ordered her to be dragged upon the stage and stripped of her dress, and then confiscated her property so that she might not purchase another.—*Ludwig Pfau.*

Colors from Wolfram.—The mineral colors from Wolfram are obtained by decomposing soluble tungstate by means of salts of the metals yielding insoluble phosphates. The tungstate of nickel produces a light green, tungstate of chromium a dark gray, tungstate of cobalt a violet or indigo blue, and tungstate of barium a bright white color. Tungstic acid alone gives a fine light greenish yellow. All these colors may be employed for water or oil color paints. The last is a really desirable and probably quite unchangeable color.

A Body divided by moldings into small panels will not admit of heavy varnishing. The better plan is to lay the varnish lightly and add an extra coat. If four coats in all are generally given on plain paneled work, let the number be five on a body with small panels. By this means runs are avoided and the full amount of varnish applied, the difference in time amounting to nothing worth naming, when compared with the certainty of results.

White Lead.—Until within the present century, the manufacture of lead was confined principally to Holland, and what is known as the Dutch process is now the mode of operation in its production in most of the lead works in Europe and America.

Lump pumice stone may be kept clean and sharp by rubbing it occasionally on a piece of English rubbing stone.

The trade of coach painting is daily growing in importance, and the apprentices of to-day should strive to become well acquainted with the nature of the pigments they must use from year to year.

Test for Japan.—Pour out a few drops of japan on the stone or a piece of glass, and add two or three drops of raw linseed oil. Stir the two together, and if the oil readily combines with the japan, the dryer is of a quality safe to be used on carriage work. If the japan repels the oil, and the end of the stick becomes gummy, the japan is worthless. We have the above from a varnish and japan maker of large experience.

THE importance of having dry, warm feet, during cold weather, should not be underestimated. The painter is required to use water in rubbing, almost daily; and he can not well avoid standing on a floor continually wet. Protect the feet with large, coarse leather shoes, having hard wood or cork soles, if you would avoid severe colds.

THE analysis of linseed oil shows it to contain both albumen and resin, which in their normal state are elastic, but subjected to boiling it becomes brittle by age.

PAINTERS should seldom wash their hands in turpentine, as the practice, if persisted in, will lead to the most serious results, even to the loss of power in the wrist joints.

SLEIGH painting in New England presents a fine field for design. The ground colors used are carmine, lake, brown, light and dark green, cinnamon and a range of light colors composed of varying proportions of white, chrome yellow, umber and lake. On bodies, the broad gold line, edged, is very popular. Fine lines are arranged according to taste. A very large proportion of the ornaments used are "transfers."

A CERTAIN painter in New Haven gives his carriage parts but one sand-papering. He leads, putties, and puts on a coat of Ohio paint, and when all are dry he sandpapers with No. 2 sand-paper. His carriage parts were well filled up and of course presented a smooth finish.

AN old body that is badly cracked and ordered to be re-painted without burning off should not be coated with oil lead, nor puttied with a slow-drying putty. Put on quick-drying coats and quick-drying putty.

WHEN a new apprentice is taken into the paint shop, his first introduction will be to sandpaper, and a gearing in lead color. Now to make of him at once a good hand at sandpapering, set him at work; first at the hub between the spokes; then let him proceed with the hub front and back; next the inside of fellows where the spokes are inserted. Having finished the parts named the remainder of the wheel will be completed with but little fear of its being slighted. On the inside carriage set him at work on the most difficult parts; training him to leave the plain surfaces to the last

COLD WEATHER PENCIL GREASE.—Take beef tallow, and having rendered it, add a small quantity of sweet oil, stirring the mixture well together while warm.

THE "shammy" (chamois), when greasy from constant use, may be cleaned very speedily by laying it out on a piece of board with a smooth face, and rubbing it with white curled hair and soap, first on one side and then on the other. Rinse through two waters, wring out dry, and the leather will again grasp the wet panel pleasantly.

PROBABLY at no time in the history of carriage painting in the United States has there been so many methods employed in painting, and so great a variety of articles put on the market for the painters' use, as in the last three years.

BRING your elbows together and hold them up before a looking glass at least once a day, and then—don't spare the soap.

HEAVY FELT renders excellent service when employed in rubbing or leveling down varnish. Mr. Landers at Newport, R. I., brought it to our notice.

DON'T delay washing the paint-shop windows early in the Fall. Nor the glazing of any broken or defective lights of glass. If you use stoves for heating, see that the grates are in good condition, and the pipe sound. A steady heat and safeguards against fire are items well worthy of attention.

ARTIFICIAL lump pumice stone is being used in Philadelphia, which cuts rapidly, furnishes its own grit, and does not clog; neither is it liable to chip off and produce scratches by getting under the stone.

If you chose to throw an empty lead keg into a hot fire, you may witness the effect of heat on the white lead adhering to the staves. Pure metal will drip out at the grate.

Painter's Colic.—For the benefit of those who may be affected by that much dreaded disease, we give the following recipe. During ten years' experience I have not found a single case it did not cure:

Iodide of potassium	½ oz.
Water	4 "

Mix and take one teaspoonful in a little water, three times a day, immediately after meals, until all pain is gone. If it cause a burning sensation in the stomach, diminish the dose. The principle on which it acts is, the iodide neutralizes the poison of the paint, and carries it out of the system.

There are at present four different makes of filling on the market, which are in competition with "English filling."

Zanzibar Gum is the most valuable for varnish making. Benguela gum stands next in value, and an inferior gum is known as kowrie.

An English coach painter lately wrote to a brother residing in this country, asking of him that he would ascertain what method American painters adopt in order to produce the brilliant finish which he had noticed on American coaches sent over to England. The reply was: "The Americans build up a firm foundation, free from tackiness, and the finishing coat thereby retains all its brilliancy. *At home you use every coating too elastic, from the priming up.*"

The painter need not be surprised at possessing a trembling hand, when striping or ornamenting, if he be addicted to the use of alcoholic drinks.

A stripe cannot of itself run straight, but if the hand that guides the quill keeps it (the quill) directly on a line with the center of the hair, the stripe will behave properly.

The carriage painter who habitually makes use of the chamois in place of a towel may be set down as a slouch.

The English method of painting requires twenty coatings to complete a coach body. In America our best work does not receive above sixteen coats, and a great deal of work coming from good shops does not receive over ten coats.

A fine line of clear white over a dark ground is not easily put on so as to avoid the appearance of laps and varying tints. By adding a small portion of lamp-black the difficulty is overcome.

We are acquainted with a lady in Kentucky who, although blind from early childhood, had the rare faculty of distinguishing colors by the sense of feeling. When she called at a store and demanded goods of a certain color, she could not be deceived by placing before her any other color. Her eyes were open, but devoid of color.

Use pencils of a good length for drawing long stripes, and shorter ones for those parts which require short curved lines.

Rough Stuff.—Take one part each of red led and umber to three parts white lead and ten of ochre. Mix stiff in oil or rubbing varnish; thin with turpentine.

PAINT CRACKING.

What is the cause of paint cracking? You may ask a dozen painters that question and get about as many different answers. A says it has got too much oil in it; B says there is too much japan in it; C says there is not enough oil in it; D says that it dried too quick, and so we might fill pages, if we were to undertake to give all the different ideas that are advanced by different painters. We have had more than an ordinary chance to find out the different views of different painters, as we not only have had a number of years of experience in the shop, but have traveled among the coach shops both east and west, amounting to several hundred in number, and we must say that it is astonishing to see and hear the heartrending ideas, theories and scientifics that are advanced about this one thing. We hardly know where to commence or where to leave off on this question. As we stated above, to undertake to give every one's idea it would fill many pages and make our article so long that we would tire the reader, so we will confine ourself to our own ideas.

Our idea about paint cracking is, there is more paint cracks from the use of oil and hasty work than anything else. Some, in fact a good many painters, persist in mixing their paint very elastic all through, thinking that they will have a tough, elastic coat like rubber, that will give to the swelling and shrinking of the wood, without cracking and scarcely break apart, if the panel of a buggy were split in two. Well, we will admit that in that way they can get a very elastic coat, and if it would remain so and never dry, it would be the thing, but the paint will dry some time. Any material will contract in drying. Their elastic body of paint will continue to dry and contract until its elasticity gets to its utmost limits, then it will give away and spread apart in great gaping cracks. To paint up a job with elastic coats of paint, it should go through a very long process, and the different coats be put on very thin and not put on a coat of paint until the previous coat is perfectly dry. Putting on a number of heavy coats of any kind of paint or rough stuff as fast as they are fairly set, will cause cracking of the worst kind either before or after varnishing. Paint is too often supposed to be dry when it is really not half dry. Six months of good drying weather would be a short time to get a coach body ready for the varnish on the elastic or tacky principle, and this is not all there is to contend with, in the oil process; the oil will sweat through the varnish and cause it to lose its brilliancy or luster—cause it to have a dull, greasy looking gloss.

The quick process, or flat coating, can be hurried so that it will crack, and crack badly. Our few limited ideas about obviating cracking are all summed up in a very few words, viz.: let every part of the wood be thoroughly primed with a good fresh priming; prime inside and out, so the weather cannot act on the wood; let the priming get perfectly dry; then mix every coat of lead and rough stuff with japans and varnishes that will dry firm and hard; put enough in to bind the paint well, and no more; have every coat dead color; do not put on the coats too heavy; let every coat get bone dry before putting on another; put on enough of coats to fill the grain of the wood, and make a perfect surface, and no more; then you will have on a body of paint that is firmly bound together, and thoroughly dry, and when paint is thoroughly dry it can shrink no more (it only shrinks when drying); and if it don't shrink it cannot crack, and in this kind of a body of paint there is no moisture or oil to sweat out and destroy the luster of the varnish. Painting of this description will not crack until the joints of the wood begin to give away and let the water and damp atmosphere in and swell the wood along the edges of the joints, and cause the paint to crack from the swelling and shrinking of the wood. Varnish may crack on top of the best painting.

If the paint is not well protected by varnish it will perish in time, sooner or later, owing to how well it is protected. The ravages of time will destroy anything that is temporal.

A NATURAL PALETTE.

The palette used by painters is for the purpose of holding their colors while engaged in portrait, landscape or ornamental painting. It is generally made of wood, to which successive coats of oil have been applied until the grain of the wood is filled. White porcelain palettes are also employed, which are peculiarly well adapted to showing the strength of the colors. The carriage painter, who makes any pretentions to ornamenting, is not unfamiliar with the palette, although too many neglect to provide themselves with one, employing, instead, a piece of glass or patent leather. Now, nature has provided a palette of translucent horn, which every painter may find placed conveniently nigh. True, it is not large enough to hold many colors; in fact, not more than one color can be conveniently placed upon it at the same time, but in ornamenting, this little natural palette will be found of great service, by obviating the necessity of holding the wooden palette and mahl, or rest-stick, both at the same time. The carriage ornamenter seldom holds the palette balanced on the thumb, lest he may inadvertently strike it (the palette) against the panel, and bruise its surface; he must, therefore, reach down to take up each penciful of color, and thus wastes time. The natural palette is not open to this objection. Another use to which this palette may be put, is that of sketching. Hogarth, England's great caricaturist, availed himself of its ever-presence, and on several occasions secured a fleeting expression of countenance which would otherwise have been lost. The natural palette of translucent horny substance, ever-ready and at times very convenient, is nothing more nor less than the *thumb nail.*

TO PAINT A LIGHT CARRIAGE IN THREE WEEKS.

To finish a light carriage in three weeks, the body should be ready to rub on the eighth day after the priming is put on. First day, priming; third day, second coat of lead; fourth, puttying up; fifth and sixth, three coats of rough stuff; rub out on the ninth. Use quick colors, putting on the first coat of varnish on the tenth. Use clear varnish, laying it on heavy. The second coat on the thirteenth, and the finishing coat on the seventeenth.

The carriage-part will, of course, be completed during the painting of the body, and therefore has not been mentioned.

We have allowed for but two coats of lead, and three coats of varnish, which will insure a passable job in the hands of a first-class painter.

PULVERIZED PUMICE STONE.

Pulverized pumice, in a perfectly pure state, that is, unadulterated with quartz, barytes, or whatever else may be added to give *weight*, cannot always be purchased, unless one buys of a firm that has it ground under their own eye, and are honest enough to give you *pumice stone*, and nothing else. The process of grinding and bolting pumice is a tedious one, requiring an amount of care and attention little dreamed of by those who have not inquired into the matter closely. The pumice, after it is broken up and ground between burr stones, requires to be passed through bolting cloths of a degree of fineness suitable to the grade of pumice required. That which is suitable for leveling down varnish is called No. 1, and that for the final rubbing on a body, medium and polishing. This last is passed through silk bolting cloth of exceeding fineness, in order to free it from the slightest particle of coarse grit; and when unadulterated is an article the painter will highly prize.

The coarser grade should be free from any grit larger than the mass, else this will tear ugly scratches in the varnish, and the utmost care is necessary in preparing this grade to see that the bolting cloths are perfect, and that by no carelessness any foreign substance should accidently get into it.

Pumice may be ground pretty coarse, but if it be uniform in the size of its particles, it will not tear the varnish if in the hands of a careful rubber. Coarse pumice stone, when first applied to the panel, should not be swept around in a circle, but rather worked across, and up and down, not bearing on very hard. It will soon become finer, when the strength of the arms may be safely put upon it.

A large per cent. of the pulverized pumice on the market is rendered next to worthless by the addition of quartz, barytes and other substances which are mixed with it to add to the weight, the pure pumice being very light, and taking too much in bulk to satisfy the wants of grasping dealers. The pure is worth from eight to twelve cents, while the adulterated may be had for two or three cents less on the pound. To avoid scratches, buy the pure pumice, and having taken out a sufficient quantity to last for a time, in a box prepared to hold it, see to it that it is kept free from dust or the flint from sand-paper.

While on the subject of pumice, we have thought it not out of place to add some facts connected with it.

PUMICE, OR PUMICE STONE,

is a substance frequently ejected from volcanoes, the name being given originally from its fancied resemblance to *foam*. In the lump it certainly has very much that appearance. It is the lava which is belched up from the deep recesses of the earth, and often in such enormous quantities as to cause the destruction of the property and lives of thousands who live within its reach. Lava is classed into the light and dark varieties, according to the predominance of the two minerals, feldspar and augite, the former producing the light colored, the latter the dark varieties; the colors being gray, white, redish brown and black.

Feldspar consists of silica, alumina and potash, and is one of the essential constituents of granite, gneiss, mica-slate and porphyry, and enters into the constitution of nearly all volcanic rocks. Augite consists chiefly of silica, magnesia and lime, with oxyd of iron, and sometimes oxyd of manganese. Lava, after being ejected, is influenced by the conditions under which it may be placed. When cooled under pressure it becomes compact, and, in some cases, changed into solid rock. But the opposite is the result when it is allowed to cool in the open air. It then presents the form and structure of the article of commerce which is familiar to every painter. It is porous, and apparently made up of parallel fibres, owing to the parallelism and minuteness of the crowded cells, and supposed to be produced by the disengagement of gas, in which the lava is in a plastic state. It is specifically lighter than water, and large masses are found sometimes floating in the midst of the ocean. That which we import comes from the neighborhood of Mount Vesuvius.

LESS JAPAN AND MORE OIL.

During the summer less japan is required than at any other time during the year, and where the use of oil is of real value in any coat, the quantity may be increased with safety. What are termed quick-drying colors, both for surfaces and striping, do not need to be mixed to work so short. Air and sunlight come in, and contribute oxygen in increased quantities, and if the painter studies his own pleasure, he will allow the weather to assist him. Where the habit has been acquired of dosing colors largely with japan, it is often the case that the state of the weather is not taken into consideration—the same quantity of drier being added in July as in January. A body is to be coated with quick-drying color, which usually means color ground in japan alone. The thermometer stands at 95° in the shade, and now the painter prepares his cup of color (say black) by mixing it with japan and turpentine, and a trifle of varnish. The heated atmosphere searches in the cup for something to work on, and the turpentine says, here am I, a light volatile fluid, ever ready to fly off. And when the color is being spread it *does* fly off to a great extent, and then the excess of driers causes the paint to work very unpleasantly. A small quantity of oil added would have corrected the matter, by allowing the color to work smoothly, and spread on evenly. It is not best to lay on color in a draft of air; close the windows and open them again immediately the body is completed. Quick-drying striping colors are still more annoying when mixed in japan and turpentine. The color will not run from the pencil smoothly, and if the work in hand requires to be varnished the same day, the workman may in his hurry become greatly irritated by the continual dragging of the pencil. We have been amused at times, in hot weather, to see good painters fret themselves over a hurried piece of striping, because they had been so thoughtless as to mix their striping color in japan alone; whereas, had they taken into account the rapidity with which the heated air acts as a drier, sufficient oil would have been added to render the operation pleasant.

We would suggest that quick-drying striping color, for fine lining, be ground stiff in oil, and then turpentine and japan be mixed, and kept in a small vessel, so that in the operation of striping each pencilful of color may be supplied with the quantity of drier requisite to insure its drying. By this method the mass of color remains soft until the work is completed. Fine lining, run on in the morning, will be ready to varnish over in the afternoon, if the work be properly aired. The oil will assist in binding the color, rendering it less liable to work up under the brush. Tube colors are the best and cheapest for fine lining, and contain sufficient oil for the purpose named, and in keeping the driers separate from the color, we are freed from the annoyance of having it continually hardening on the palette.

ENGLISH BLACK JAPAN.

This is an article which was introduced into carriage shops in the United States twenty-three years ago, to our certain knowledge, but how much earlier we are not prepared, to state. When first imported it was employed almost exclusively for the iron work or bodies. The advantage possessed over "color and varnish" was, it contained no grit or grains of color, was intensely black when laid over a dark ground color, flowed out full and rich and dried rapidly. Subsequently it was used to intensify the color of all black work. It replaced the coat of "black varnish," and entered into the first and second coats of clear varnish in a quantity sufficient to tone them with black, the purpose being to secure a black which, when finished, should not, in a clear light, appear of a greenish tinge. Black japan, when applied instead of black varnish color, repuires to be handled very carefully, for if it is cut through the spots will look green, and cannot be retouched so as to make a finish equal to the original color. Black japan requires varnishing with " durable varnish," as it possesses poor wearing qualities.

CONCERNING SLUSH.

It is the practice with some painters to take the "slush" which has accumulated and pour it in a kettle, adding raw linseed oil sufficient to soften up the whole mass, and then boil it until the skins, etc., have discharged their fatty matter. Afterward strain the product, and use it for any purpose for which it is fit. It will dry well, and may be mixed with rough paints, to be used on such parts of the work as do not demand anything better—as the insides of bodies, the roofs of standing top work, bottoms, etc. We do not claim that any profit is derived from boiling down "slush." The oil, firewood and time, computed against the value of the product, would not make a very good showing.

The paint shop need not be annoyed by having a great amount of "slush" on hand, if the foreman, or some other man, will see to it that in the daily use of colors and varnish each hand is economical. Black or any other color of which considerable is used should be ground (in japan only) in a large vessel, and from this removed to the ordinary paint cup, in about the quantity required for coating the work in hand. Turpentine, oil and varnish should never be thrown into the "slush" keg. Cans should be provided for each of them, and when either are gritty or soiled, pour them into the cans, allowing them to settle, and then use them in mixing rough paints.

As in the every day workings of the paint room there will be of necessity a certain amount of "slush" made, provide two half barrels, and in one scrape only the dried skins, which cling to the cups or pots, and in the other the more liquid refuse. This last may be used up about as rapidly as it gathers; the other is next to worthless.

The "slush" keg, if it could speak, would disclose many mishaps on the part of hands in mixing colors; for, unsightly as it is, it has ever served to swallow up and hide from view these failures, and therefore can claim to be somewhat of a friend to the painter, and he in return is not unmindful of its real worth.

GRAIN RAISING ON NEW WORK.—It is not unfrequently the case that new work which stands for two or three months in a dark wareroom develops the unsightly appearance of the grain of each kind of wood of which the body is composed. The ash may be plainly distinguished from poplar, and their junction will be clearly defined. This is no evidence of cheap and hurried painting, as it occurs on work that has been six months in course of completion. Sunlight, fresh air, and repeated washings will correct the defect.

RESOLVENTS FOR REMOVING PAINT.

Resolvents are not to any great extent resorted to by carriage painters in the removal of old paint. Doubtless the majority of painters have experimented with chemical mixtures claiming to be a great improvement over the old plan of softening the paint by heat, but we think our experience will agree with others in this, that while a chemical mixture may perform its work well and speedily in so far as dissolving the paint is concerned, it also goes further and attacks the wood, leaving it in a furzy and gummy condition, on which it is difficult to get paint to dry firmly. House painters use *chemical burners* to but a limited extent, although their work is not required to be so perfect as that of the coach-maker. The charcoal furnace appears to be the favorite means with them of laying bare the wood, and when properly made it certainly performs its part well. The furnace is simple in construction, being nothing more than a small sheet-iron box, with handle. One person can do the work, but generally two persons are employed, one to hold the furnace, and another who is skillful with the chisel or knife, to remove the paint and regulate the distance to which the furnace shall be held from the paint; an important item; for when the body of old paint becomes very hot it vitrifies, and when again it becomes cold it clings with greater tenacity than before it was heated. The furnace would be the most economical device in a shop not supplied with gas. But shops lighted with gas can have nothing more economical, speedy and free from deleterious effects than that of a gas jet. All that is required is a few yards of rubber tubing arranged to suit one of the gas pendants. The tube can be held in the left hand and knife or chisel in the right. The jet or flame is then made to strike the surface, and when the paint is warm through, the knife is used quickly in removing it. The gas jet does not give out a great amount of heat, and therefore does not injure the joints of the body, or draw out the oil from the wood to an extent that delays the after painting. In removing paint, the heat should never be so intense as to quickly raise it up into blisters; it is only necessary to warm it through.

PAINTING BODIES.

First—Let the priming be "patent filling" or lead color, as best suits the whim of the painter, but if the latter, little or no japan should be used, unless the job is in a great hurry.

Second—Let the second coat, whether of red or rough stuff, be of sufficient elasticity to adhere and be durable, but not to that extent as to prevent its drying in at least two or three days; and so with subsequent coats.

Third—Putty should be used very sparingly and as little left on the surface as possible.

Fourth—Every coat of paint or varnish, whether elastic or otherwise, should be thoroughly dry before applying the next coat.

It appears to us that this point is of more importance than all others. It matters not how good or what kind of material is used; if one or more of the under coats is not in a proper condition to receive its successor the job will fail to stand; it will crack in some cases; in others it will "sweat out" and become dull, or show the grain of the wood. We have seen bodies painted in four weeks that did better service than some others that took several months, simply because the paint was prepared to dry, and did dry before the next coat was applied.

We don't wish to encourage the plan of hurrying the work through, but it is a well-known fact that many jobs are spoiled by overexertion to make an extra durable piece of work. The reason is obvious. The paint is put on that never dries thoroughly, consequently the failure.

BASKET WORK.

Of course, in a department devoted to painting, we refer to the *imitation* only of basket work. Should any propose to try their skill at imitating the basket work as used on pony phaetons, a hint may not be out of place. The first point to determine will be whether the copy shall be basket work in straw color, or some shade of brown. However, whatever color is decided on, the ground work should be of a tint agreeing with the shaded sides of the basket work. Straw color in shadow may be mixed of white, yellow ocher and raw umber. On this groundwork lay out with chalk the form of the basket work, and bring out the features of the willow as interwoven, by using brown, or burnt umber for the darkest markings, and the natural color of straw on the parts receiving light. Proceed with the other colors on the same plan. A piece of basket work should be used as a model, and its peculiarities carefully studed.

SCREENS.

Screens, as a protection to finished bodies—especially during the warm weather, when the air is filled with flying insects—are not appreciated as fully as they should be. The varnish room may be perfect in every particular as to excluding dust, and be well ventilated, and yet a single house fly may mar the beauty of an exposed panel, and cause vexation, tedious delay and extra expense.

We are aware that objections are urged against them; some say they concentrate the heat about the body, and render the varnish more liable to wrinkle, or to set with a dull appearance; others, that they collect dust, and are difficult to keep clean enough to place above or around a body.

To these objections we reply, that it is not necessary to inclose a body by placing the screens close up to the sides. They may be placed twelve inches or more distant, which will give a free circulation around the body, and at the same prevent flies from alighting on the panels—for it seems to be the nature of these little tormentors to dislike a dark, confined place. In a poorly arranged carriage shop, where the varnish room partakes of the general poverty in conveniences, screens are an absolute necessity if passable work is demanded of the painter, for he must battle with dust from within, and dust from without. In a shop of this kind there should be provided single screens, large and small, and others having wings, capable of being folded up and let down, when required. Different styles of bodies require different kinds of screens. Standing-top work, such as Jenny Linds, which have a deck panel exposed, require the screen frame to be made to fit in between the back part of seat and the hind pillars, and when covered should be supplied with end wings. Buggy bodies, with deck panels, may be turned upside down, and varnished in that position, and to protect the rest of the body and the seat, a screen with four wings should be placed on the bottom of the body, and when a side is varnished, let down the wing on that side, continuing around the body until it is finished. Rockaways can be protected by having a screen made of a height sufficient to cover the back panel, and with wings long enough to reach forward and cover the doors, to be there fastened by tacking them to wooden pins, previously inserted in the door-handle holes (provided the doors are not detached). Deck panels, on heavy work, may be protected by large screens having two side wings. In using the screen for deck panels, the better plan is to place it over the panel to be varnished, and raise it sufficiently to give room to work beneath it; varnish the panel, and then let the screen carefully down. In making screens use hard, stiff wood, bracing the large frames diagonally, making them light, yet strong. Cover with a good quality of heavy brown paper, and along the edges, where the wings are subject to the wear of raising and lowering, tack a strip of enamel leather, three-eighths wide. Brown roll paper may be purchased at the paper dealers, of suitable width and any length desired. We know from actual observation that in many small shops throughout the country there is trouble with the finishing coats, which the introduction of good screens would speedily dissipate. The vexation, delay and expense attendant on the limping promenade of one fly across the panel would be greater than the cost of a full set of screens, therefore every varnish room that is not finished in the best manner should be supplied with them.

TO PAINT CARMINE.

Carmine being very transparent, it is of the first importance to obtain a solid groundwork on which to paint it. After the groundwork is secured it will be found somewhat difficult to make the carmine cover well, without showing streaks, which would look in the sunlight as if it had been grained. When the surface has been gained take *dry* white lead and Indian red, and mix up a pink or flesh-colored tint. Grind it out very fine, and apply with a flat camel hair mottler (blender). Two coats of this will be necessary, except on very small panels. Hair off each coat to remove any minute particles of grit that may be found on the surface, or, which is better, take 0 sand-paper, and after having rubbed the sanded faces of the paper together, use *without doubling it*, going over the surface lightly. If the color is dry it will not cut through; this will remove grit which hair could glide over.

Next, give the body a coat of English vermilion, if you desire a *brilliant* shade of carmine, and Indian red, if you wish a more subdued color. Take of the best carmine, No. 40, and mix in oil with sugar of lead for dryer; grind out fine, and add a small quantity of varnish to prevent the color drying off dead, or mix in varnish; some painters prefer the latter. Apply the carmine with a blender, if you have one of the *best* quality; if not, use a bear-hair or soft bristle brush. The color *must* work freely be brushed on rapidly, and finished while wet.

Patching the carmine on the width of the brush at a time is the main cause of so much streaked work. Put on the color freely, spread it and finish up and down. If the panels are too large to finish at once, make as few laps as possible.

One coat, and a strong coat of color and varnish should cover solidly. When the color and varnish coat is dry, take off the gloss with a sponge and water and fine pumice, rubbing lightly to avoid disturbing the color.

Put the first coat of rubbing varnish on comparatively thin, rubbing it lightly when dry. The next coat will bear a good rubbing, without fear of disturbing the color, or showing any red specks.

Carmine may be painted on dark browns, or even black, but the richness it possesses is shown to best advantage on a groundwork which is light.

QUICK DRYING ROUGH STUFF.

Take equal parts of whiting, dry white lead and ocher, mix them up with a good drying japan, and run the mass through the mill, using sufficient turpentine to facilitate the grinding. When ground out, add rubbing varnish and dry ground pumice stone in the proportion of about a teacupful of each to a half gallon of rough stuff, and thin down with turpentine. When ready to apply to a body, it must be brushed on quickly, with a short, stiff brush. If you desire it, three coats a day may be put on, and the body be rubbed the second day after; or, put on one coat a day, and rub the body the day following the application of the last coat. Where two or three coats are put on the same day, brush the first coat on thin, the second heavier, and so on. Keep the rough stuff covered with water when not in use.

PERMANENT WOOD FILLING.

Directions for its Use—Bodies of carriages dispensing with lead, but using rough stuff: When the body is received from the woodshop, rub with lump pumice (should the panels not be smooth), dust off, and apply *every thin and even coat* of "permanent wood filling," dark or light shade, according to the work. Allow this to dry two or three days, then apply a coat of rough stuff, made elastic with oil, varnish and japan gold size used in equal parts. When dry, putty up where needed, and when the putty dries add one or two coats of ordinary hard rough stuff, and rub down, when dry, in the usual way; over this color and varnish, as in the old method of lead painting.

Another is to dispense with the use of both lead and rough stuff, by putting on two coats of the "permanent wood filling," then a coat of elastic color; putty on this, and when dry finish as under the old method.

GEARINGS.

Sand-paper smooth, dust off, and apply *a very thin and even coat* of the "filling" to the wood parts; when dry, sand-paper smooth, dust off, and apply *a very thin coat* of "filling" over all, wood-work and iron; when dry, putty up, and when putty dries sand-paper lightly, to remove dust and runs; then dust off, apply an elastic coat of color, ground in oil and varnish (no dead colors should be used on gears), and finish in the usual way. One coat of the "filling" on the springs and iron work is sufficient, and it should be as thin as possible.

For repair work, the parts renewed may be finished as quickly as by the lead method.

PAINTING IRONS.

If there is any part of a carriage which the painter should pride himself on, it is that the irons show a finish equal to the other parts. In our largest and best managed shops men are employed who pay special attention to the irons, and when a body is hung off, and the irons all screwed up to their places, the finisher cleans every nut that requires it, leads, blacks, and color and varnishes, and finally coats them with wearing body or gear varnish. Those parts of the body loops, or other irons which require handling, should not be finished until they have been securely fastened in their respective places.

Irons on a body should be primed, leaded, puttied, and have one coat of filling or rough stuff; then sand-papered, colored and varnished the same as the body. If it is the custom to remove the body loops when the job comes from the smith shop, finish them except the last coat of varnish; rub them down and put them on the body before it is ready for the finishing coat; touch up the nuts and bare places, and finish when you do the body. English or American black japan may be used instead of color and varnish, as they produce a more intense black, but they will not answer for finishing on good work—black japan possessing poor wearing qualities when not protected by wearing varnish.

On cheap work the irons may be speedily finished, by giving them a coat of lamp-black color and varnish as soon as they come from the smith shop, and finish with a heavy coat of *old* drop black, color and varnish.

COAGULATION OF MIXED PAINTS.

A pigment unmixed cannot "thicken up." That point gained, we must look in another direction for the cause. A pigment mixed in water will not become "fat," for the simple reason that water possesses nothing of a fatty nature, and the action of the atmosphere has no further effect than to draw off the water in the form of vapor, leaving the paint a dry, cracked mass. A pigment mixed in turpentine alone would possess but slight adhesiveness, on account of the volatility of the turpentine, and if the pigment so mixed be allowed to remain exposed to a heated atmosphere, the turpentine would soon evaporate, leaving the paint destitute of moisture, dry and flaky.

A pigment mixed in oil alone, and allowed to remain exposed to the air undergoes changes quite different from that mentioned in the preceding examples. The oil, by absorption of oxygen from the atmosphere, undergoes chemical changes, which tend to alter it from a liquid to a solid. In shop parlance, it "grows fat," and should you let it remain without the addition of anything, it would in time become a resinous mass.

The Remedy.—Avoid grinding colors in oil during the "heated term," using only the dryer and the least possible quantity of turpentine to aid the passage of the color from the mill. When ground out cover the color with water to render it air-tight; and cups containing lake or carmine add a cover of enameled cloth, tied tightly around the rim of the cup. The color thus protected is put in a cool place. When a coat is to be applied requiring the addition of oil, remove sufficient color for the work in hand, and place the remainder under its former protection from the air. The paint-mill generates considerable heat while in motion, and often ruins paint containing oil, before it has all passed through.

Paint-mills, actuated by steam-power, run so steadily that after a few revolutions the paint will exude smoking hot, causing drop black and other paints of a fatty nature to become "livered" and worthless, and, in fact, working unfavorably on every other kind of mixture. Paint-mills run by steam-power should be inclosed in a water tight case, and, besides, be timed down to a slow, regular motion. By this means the heat generated would be kept in subjection, and good results follow.

TO FASTEN THE HAIR IN STRIPING PENCILS.

Pencils, by constant use, are like everything else, liable to wear out and become worthless. A well-made pencil, however, should not give off the hairs after having been used but two or three times. At first there will be a few loose hairs, which is nothing unusual. Pencils are often spoiled in greasing them, preparatory to laying them away, by using hard grease, which must be pulled on the pencil between the thumb and finger. This stretches the hairs, also loosens them, and when put into service they pull out. A drop or two of glue, or shellac varnish, put into the quill, will cement the hairs together, and prevent them from drawing out.

When a pencil curls, draw it gently over a piece of warm iron, which has been greased.

RED LEAD AND UMBER VS. LAMP-BLACK FOR THE LEAD COATS.

Red lead and umber are excellent driers, and answer a very good purpose when mixed with white lead, but are inferior to lamp-black, because they add nothing to the tenacity of the lead coatings. Lamp-black is of a greasy nature, and when pure almost indestructible. It also has the property of smoothness in the working of white lead, which is not true of red lead and umber, when substituted for it and employed as the carriage painter is required to do.

The coats of lead on a body or gearing need not be made lead color, as some are in the habit of doing, but by merely changing the lead to a pearl or silver-gray, it will dry firmly and wear equal to work filled with any other mixture. In using read lead or umber, use less japan and more oil—one-fourth japan and three-fourths oil, and one part each of red lead and umber to three parts of lead, thinning down with turpentine to suit the work you may be doing.

FACING A BODY.

To *face* a body use lump pumice stone and water, going over the surface lightly, so as to level down the putty and lead, and also to force the lead into the pores of the filling. *The facing is done after you have put on a coat of lead*, and puttied up any dents, bruises or other defects noticed on the surface after the filling has been rubbed. The facing is only done to prepare the body for a more perfect surface, and while it is undoubtedly the best way to fill up the pores of the rough stuff, many shops do not practice it; some because it takes a little more time, and others because they have never seen a body so prepared. We would advise the use of English rubbing stone instead of a file, to keep the pumice stone sharp and free of lead; by rubbing the pumice stone on a level piece of English stone the pumice will take up some of the grit, and work more pleasantly. In speaking of a body in "filling" or "rough stuff," we say *rub out, rub down*, or *cut down the body;* after the filling or rough stuff is cut down, *sand-papered, leaded,* and puttied where needed, we use the word *face*, because the rubbing is only done on the face of the surface gained by the former rubbing of the body.

PAINT CRACKS.

Paint may crack because too much oil was used; and again may crack on account of the oily principle being dried out.

The cracking in the first instance mentioned would result from the under coatings not being thoroughly dry when the carriage was completed. Afterward, as the drying proceeded, space was required for the evaporation going on, which manifested itself on the outer surface.

When paint or varnish has been long exposed to the action of the weather, the oily principle becomes dried out, the resinous parts only remaining, which contract, producing cracks.

Japan, in which gum shellac forms a part, is very liable to crack, because the gum swells in dissolving, and contracts a great deal in drying. This is true, to a certain extent, with other gums.

Paint cracks are similar to the mud in a creek bottom. The summer freshet rises up over the soil, saturating it; the water falls, leaving the pasty mass exposed to the sun's rays, which draw off the water by evaporation. The surface would at first be crusted over, and as the heat continued, the under-wetted mass, by evaporation, would break through the surface crust, and find vent in every direction, generally breaking through in straight lines, which incline toward the circular form.

On patent leather the circular form of cracks is very perfect.

STRIPING PENCILS.

The red and black sable are preferred by the majority of painters, especially for fine line pencils. The sable hair being more springy, it bears the pencil up better toward the point.

For broad lines, on carriage parts, a good quality of camel-hair pencils will work very pleasantly, and is preferred by some good stripers.

PUTTY.

For new work, take keg lead and work it into as much dry white lead as will form it into a stiff dough, adding a small quantity of good brown japan as a dryer. The dry white lead should be well pulverized and pounded into or with the wet lead until it clings together, forming a tough mass. This putty should be allowed from two to four days to dry.

Quick-drying putty, for general purposes, is mixed with japan and dry white lead.

For bruised places on panels, after the surface has been obtained, mix the putty of dry white lead and varnish, and a small proportion of japan. It will dry in twenty-four hours hard enough to bear leveling down with lump pumice stone. Deep bruises should be puttied two or three times, allowing a day for each to dry.

For smooth putty, which does not require to be sand-papered, take one-third whiting or dry white lead, and two-thirds lamp-black, mixed in oil. This to be used to correct the defects in the iron work not fitting closely, as, for instance, under the arch, or cut-under, or any other curved surface. It clings more tenaciously than putty mixed in the usual manner; and while it would be far better to have no putty in such places, the painter must nevertheless putty up to keep the pumice stone and water out, if for no other purpose.

Bedding putty for quarter lights or glass frames. Take whiting or dry white lead, pulverize them finely, so that there shall be no grit remaining. Have the stone clean. Mix up half whiting, or dry white lead, and lamp-black in japan, forming a pasty mass. There is no putty used which should be so free from grit, as it often happens that a costly glass is broken by the presence of a grain of flint.

Black putty is the ordinary putty used for nail holes, etc., to which is added lamp-black (dry) sufficient to darken it. It assists the painter in completing hurried and old work, as it is more easily covered by the color coats. It is very convenient to use about new work where moldings and irons have to be put on, as it may be touched over with quick black, and the moldings and irons colored and varnished, making a passable finish.

Quick putty.—To fill small bruised places which may have been made on the surface while preparing to finish a piece of work, burnt umber and whiting mixed in japan. Whiting will not stain or wash up as readily as dry white lead.

QUICK-DRYING PUTTY.

Dry white lead, mixed in rubbing varnish, dries firmly over night, so that it may be leveled down with lump pumice stone. It must be kept under water, and taken out in small quantities to prevent wastage; or use dry white lead, red lead and whiting mixed with japan. For puttying bruised places on a body which is about completed, whiting and lamp-black, mixed in japan, will dry quickly, and not be so apt to stain when touching up on it.

In puttying as well as painting, where a job is in a great hurry, recourse must be had to hard, firm-drying putty; but in using it, it will be found the better plan to putty the holes twice; for instance, if the job is in a hurry, and only two coats of lead are to be applied, putty the holes half full on the priming coat, and finish puttying them on the next coat of lead. By this means the putty dries through, and will not be so liable to shrink.

SPONTANEOUS COMBUSTION.

Any substance that will take fire and burn is a combustible, and combustion is a burning or the act of burning. Flame must be applied to produce ordinary combustion, while spontaneous combustion is produced without the application of flame, the word spontaneous signifying voluntary—acting by its own impulse. Damp goods lying piled up will sometimes catch fire, caused by fermentation from heat and damp. Carbonic acid gas is formed, which is attended with combustion. Greasy rags, especially those containing oil and lamp-black very readily ferment and throw off inflammable gases. Lamp-black mixed with linseed oil is also very liable to spontaneous combustion.

The palette or putty knife when oily should not be thrust into dry lamp-black, as the oil may cause spontaneous combustion. The carriage painter will see from the foregoing that no small amount of responsibility rests upon him if he be acquainted with the above facts, and still allows greasy rags to accumulate in the paint room, and lie for any length of time stowed away in a corner.

PAINTING BUGGIES.

"One who can paint a Buggy well, should be able to get up a good job on heavy work."

In the above quotation it is implied that he who can bring up light work to a high degree of finish has the necessary qualifications to immediately change his hand, without having had any previous experience, and produce a good finish on a coach. Our experience in the shop proves the contrary to be true. It may appear unreasonable to those unacquainted with the facts that such should be the case, but nevertheless that does not alter the matter in the least. The process of filling and carrying forward a buggy or light rockaway is the same as that on a coach, in so far as the nature of the coatings is concerned, and their application with paint and varnish brushes, but the great difference between the two classes of work is in the greater size of the coach panels, as well as the various positions they present. In order to become skillful as a heavy-body painter, the mind must be continuously occupied with heavy work until a certain degree of skill and patience are acquired. Those who have spent years working on light work, find that when a heavy job is taken in hand they dread not only the size and shape of the panels, but they feel this lack of patience. On light work one may have in hand several bodies at a time. To clean off a body after it has been rubbed out, does not consume a great deal of time, and thus the painter becomes habituated to short tasks. A few hours on one body, then a few hours on another, by this means he acquires a habit which has the appearance of moving things along rapidly. Now admitting that the workman is quite skillful, suppose we hand into his charge a coach. There it stands. Its dimensions are such as to require space sufficient for three or four light bodies, and there are frames and extra pieces wholly dissimilar to anything on light work; and then it is a ponderous affair. To handle it at all requires an entirely different kind of trestle, and there are no conveniences for handling doors and windows.

Having made the necessary arrangements for moving the body into any desired position, the light-body painter will shrink from the task of cleaning the body ready for the first coat of color. The color coats may not be a source of much anxiety, but when the varnish coats are to be applied, a feeling of distrust in one's ability to lay them on properly will so operate on the nerves as to make the hands tremble violently. We have heard first-class painters on light work assert that they detested the sight of a coach in the shop, and the feeling was but natural, for they felt certain that they had neither the skill nor patience to complete the body in the manner it should be done. A most signal failure in painting a coach body came under our notice a few years ago, wherein a light workman with no experience on heavy work had taken up the idea that he could dash off a coach body with as much certainty in result as he could that of light work. A coach body was given in his charge, and he went at it with a will, but from the laying on the color until the third coat of rubbing varnish each step rendered the body in a worse condition. The body was at last placed in other hands, and after a large amount of extra time and trouble had been expended, the finish was rendered passable. The fault did not lie in a lack of intelligence and ability as a painter on the lighter class of work, but having had no actual experience in painting heavy work, he had not mastered the secrets connected with it, which secrets are acquired by practice, and cannot be communicated by language to the hand of the novice.

Light and heavy work, then, it will be seen, are not to be classed as one and the same kind, which could be further proven if necessary, by bringing forward the wood worker, smith and trimmer, each one of whom would testify that large experience is required in order to fully cope with the difficulties presented in the construction and finish of heavy work.

STRIPING PENS.

This instrument is capable of producing lines of extreme fineness and perfection. It should be of the best quality. The large size German silver lining pen is the proper kind to purchase. The points of the new pen will probably be too sharp, as it is made to be used on paper, and not on a varnished surface. To put it in good working condition when the points are as stated, it is necessary to round them off by rubbing them on an oil stone. This operation requires the greatest care. In order to get the points perfectly true a magnifying glass (every painter should own one) will be found of service in detecting the slightest variation. A straight-edge is required to guide the pen when straight lines are needed, and curved patterns for sweep lines. Either oil or distemper colors can be employed, oil colors, however, being the only kind which the carriage painter should use.

The compasses pen should be used for panels bounded by curved lines. Considerable practice and patience are necessary to put the operator on good terms with these sensitive but valuable instruments. But having obtained the mastery, the advantages gained over the hair pencil will be at once apparent in such work as the imitation of finely woven cane, plaid work, and any other style requiring the repetition of lines which should be of the same width.

The striping pen cannot wholly replace striping pencils made of hair, but when skillfully handled, on certain kinds of work, the hair pencil can never approach it in perfection of lines and the rapidity with which they may be produced.

PRACTICE WITH THE PEN.

The striping color should be ground out in a small vessel, as it will have to be used somewhat thinner than for ordinary striping. The color may be mixed with or without oil, but it works more pleasantly with a small quantity in it.

Fill the pen with a camel-hair pencil, wiping off any color that may flow outside with a piece of cloth, which will not give off lint or furz.

When the pen has been set, and the striping color reduced to the proper consistency, so that it runs freely, making the sized stripe desired, do not alter the set screw. If the pen refuses to allow the color to flow out, pass a slip of writing paper, from the inside, through the point of the jaws, which will remove any minute particles of varnish which may have been taken up in its track across the surface. Wipe the point on a piece of silk, soft old chamois or the hand, to avoid furz or lint.

The pen should be thoroughly cleansed when the job of striping is completed, and laid away in a secure place, where there will be no possibility of anything striking the points. After a few weeks' practice in lining, fine scrool work may be attempted, such as is used in some sections of the country in corners of panels, or on the spring bar and axle beds of gearing, to connect with fine stripes.

Larger scrolls can be formed by outlining with the pen, and filling in the body of the leafing with curved lines of the shape each leaf may require, forming the shades by leaving the ground color clear. With the color properly mixed, and the pen in good order, a hair line may be run on a panel from three to four feet in length before lifting the pen.

GOLD LEAF.

The metal gold has been known from the remotest times.

It is the *sol* or sun of the alchemists, who represented it by the circle, the emblem of perfection. When pure it is nearly as soft as lead, and is the most malleable and ductile of all the metals, but inferior to many in its tenacity. It is not affected by air or water at any temperature. Perfectly pure gold is denominated gold of 24 karats (a karat is a weight of four grains), or fine gold. Gold containing two parts of alloy in twenty-four, is said to be 22 karats fine. Perfectly pure gold is too soft for use as coins, vessels, ornaments, etc., and is therefore alloyed with copper and silver.

As to the origin of the idea of foliating or beating gold, we have been unable to search it out. As the metal has been known from the remotest times, doubtless its malleability was early discovered; this would lead naturally to folliating or beating the ingot, it being a more economical method of using the precious metal, and lessening the cost when employed for ornamental purposes. The Bible mentions beaten gold as early as 1491 B. C. The Lord directed Moses concerning an offering from the people, in order to erect a tabernacle. The offering he was to take was gold, silver, brass, fine linen, dyed goat skins, etc. Among the directions to Moses as to the materials to be used, and the form of the sacred furniture for the tabernacle, we find the following: "And thou shalt make two cherubims of gold: of beaten work shalt thou make them, in the two ends of the mercy-seat." The golden candlestick also, "with his shaft and his branches, his bowls, his knops and his flowers," were to be one *beaten* work of gold. (Ex. xxv. 18, 31, 36.)

It is not clear what was meant by beaten work, for other portions of the sacred furniture were covered with *plates* of gold, which must have been drawn out under the hammer, or rolled by some form of pressure applied. Of one fact we may rest assured, that at that early day the malleability and ductility of metals was well known. The people had in their possession silver chargers, golden bowls and spoons, which were probably of Egyptian manufacture, and their offerings were taken from among these, cast into ingots, and then beaten or rolled, according as the command directed.

The Egyptians were experts in foliating gold, they being able, it is said, to surpass us at this day in reducing the leaf to an extreme degree. Mummies have been exhumed that had been buried three thousand years, the finger nails of which were coated with gold leaf of a very fine quality and exceeding thinness, the gold leaf being a mark, probably, of the wealth of its possessor in his life-time.

GOLD BEATING

is the process by which gold is extended to thin leaves used for gilding. Attempts have been made to apply machinery to gold beating, but though very ingenious, their application is very limited. Most of the gold leaf is still beaten by hand, as follows: The gold is first cast into oblong ingots about three-fourths of an inch wide, and weighing two ounces. The ingot is flattened out into a ribbon of about 1-800ths of an inch in thickness, by passing it between polished steel rollers. This is annealed or softened by heat, and then cut into pieces of one inch square; 150 of these are placed between leaves of vellum—each piece of gold in the center of a square vellum leaf, another placed above, and so on, till the pile of 150 is formed. This pile is inclosed in a double parchment case, and beaten with a 16-pound hammer. The elasticity of the packet considerably lightens the labor of beating, by causing the hammer to rebound with each blow. The beating is continued until the inch pieces are spread out to four-inch squares; they are then taken out and cut into four pieces, and squares thus produced are placed between *gold beaters' skin* instead of vellum, made into piles and inclosed in a parchment case, and beaten as before, but with a lighter hammer. Another quartering and beating produces 2,400 leaves, having an area of about 190 times that of the ribbon, or a thickness of about 1-200,000ths of an inch. An ounce of gold is thus extended to a surface of about one hundred square feet. A still greater degree of thinness may be attained, but not profitably. After the last beating, the leaves are taken up with wood pinchers, placed on a cushion, blown out flat and their ragged edges cut away, by which they are reduced to squares of $3\frac{1}{4}$ inches; 25 of these are placed between the leaves of a paper book, previously rubbed with red chalk to prevent adhesion of the gold, and are sold in this form.

LAYING GOLD LEAF.

In handling gold leaf, much depends on the kind of work in hand. Scroll work, and all large patterns and plain surfaces require only that the book be opened and a whole leaf put on at once; this is repeated, allowing each leaf to overlap the other about an eighth of an inch, and when sufficient has been applied to cover the work, rub it down with raw cotton or rabbit's paw, being careful not to get any size on them. Small gold ornaments, less than the size of a leaf of gold, we gild direct from the book, any pieces of gold worth saving is allowed to drop back into the book when we remove the loose gold. To do this, without wrinkling or crushing the remnants, the leaf must be handled gently, and be broken away from the size around the edges, placing the book directly under the ornaments and catch it as it falls. For striping and lettering, the leaf should be cut a trifle wider than the sizing, and applied with a "tip" or from the leaves of the book, cut of a proper width. To handle gold leaf, it is well to provide a gold cushion, gold knife and a tip. The cushion is designed to hold the leaf while being cut, and during the operation of applying the leaf with the tip. The knife is a thin blade of steel, set in a handle, and made for the purpose and for no other. The tip is made by inserting camel's hair between two pieces of pasteboard, and should be two and a half or three inches wide. The gold is removed from the book to the cushion by turning back a leaf and exposing the gold, then invert the book and press the gold on the cushion. If it should wrinkle, a puff of the breath directly over it will straighten it. The knife should be made to do its work by one cut. The tip may be drawn across the hair or beard, and excited by friction, or be slightly greased.

In using the tip, take up as much gold as it will carry, first placing on the size the piece at the outer end of the tip, and following down toward the heel. By practice, leaf may be laid on very rapidly with the tip, especially on fine lining, but there are other methods employed which work fully as well if not better. The most approved of these is to take an empty gold book, and draw a leaf of it across the hair or beard, (or apply a minute portion of viscid matter in any other form) and press this greased leaf upon the gold. It will adhere throughout its surface and may be cut to any size required, and applied more rapidly than by using the tip. Every leaf of gold requires to be thus treated, when they may be laid together and several cut at once.

Striping and ornamenting in gold requires practice in the mixture and use of sizing, as the gold leaf is rendered brilliant, or dull and lifeless, according as the sizing is clean, thin and evenly laid, or gritty, thick and laid on with overlapping edges. *Size* made of either varnish, or fat oil, should be run on very thin, and the leaf be laid when the size retains but a slight tack. Varnish size must be gilded while the tackiness is more decided than that of fat oil, for varnish, after it begins to set, soon forms an outer pellicle, or thin skin, which incloses the under body of the varnish, and prevents the proper adhesion of any kind of leaf, or even bronze. Fat oil dries slowly, and retains a tackiness sufficient to take leaf, even when it requires pressure to cause the leaf to adhere. When the gold has been laid, it may be rubbed down with clean raw cotton, until it shows a bright, even surface.

Gold powder or bronze may be laid on with a camel-hair pencil, and to do it only requires that the bronze be mixed in a vessel containing as little acid as possible.

In the every-day routine of the shop we use raw oil and japan, japan gold size, boiled oil and varnish, either of which will answer the purpose. We prefer, however, raw oil and a small portion of japan, in which we rub up a little orange chrome and add the bronze in the proportion of two-thirds bronze to one-third color, or use the bronze clear, mixed in an oil drier or varnish. The Bessemer gold paint is now very popular, and comes prepared for immediate use.

Bronze is not suitable for ornamenting, but striping and fine scrolls will admit of its use.

Having prepared your work for striping, obtain a small stoneware or porcelain vessel, mix the bronze as directed, and when using the striping fluid stir it occasionally in order to keep the particles of bronze afloat, for the pencil must take a like quantity of the mixture charged with the same amount of bronze (as nearly as possible) at every dipping, to produce stripes of equal brilliancy. Do not be disappointed if, when the work is finished, you find it inferior to gold leaf. Bronze used in this way answers a very good purpose, saving the time and trouble of sizing in; but to assert that it produces as good work as gold leaf would be to mislead those who have not seen it. Bessemer's gold paint is put up in packages, containing two ounce bottles; one containing fine gold bronze, the other a liquid in which to mix it. Price, $1.25. Fine gold bronze is worth $1 an ounce.

SIZING FOR GOLD LEAF.

Use fat oil, with sugar of lead for a dryer, if you desire a good wearing job. You may add a small quantity of chrome yellow to the fat oil to assist in tracing the lines, or the oil may be used clear, which, we think, is the better plan.

For all ordinary purposes on carriage work a size mixed of varnish, to which a few drops of fat oil is to be added, will be found good enough. Fat oil prevents the varnish from drying out on the edges and other parts, and will save the painter a great deal of trouble.

Sizing should always be put on as thin as possible to prevent the gold from being "*drowned*," that is, sinking into the size. The size should have tack sufficient only to to take the gold.

The leaf may be laid with a "tip," by cutting the books to the sizes required, or by turning down the leaf of the gold book to the size you wish; then pass the finger nail across the gold leaf and apply it to the sizing; rub down the leaf with cotton, or rabbit's paw.

There is no economy in attempting to cut the leaf to the exact size of stripes or other work; let it be a little full, and you will obtain clean edges and a perfect piece of work.

TO PREVENT LEAF FROM ADHERING.

The white of eggs will effectually prevent gold leaf from adhering to the surface surrounding sizing, if laid on heavy enough to form an even thin film. If it is a varnished surface, we rub the varnish down as level as it will bear, and having ejected the white of an egg into a clean cup, add a small quantity of water to it, sufficient only to cut the albumen, without destroying its body. Then with a clean piece of sponge mix them, by dipping in the sponge, filling and squeezing it out, until the albumen is thinned out to a ropy consistence; this, when applied, leaves a thin film on the surface after the water has evaporated.

When an excess of water is used, the adhesive part of the egg is destroyed. In hot weather, when varnished surfaces "sweat out" soon after being leveled down, the egg size should be used somewhat heavier than under more favorable circumstances.

Egg size should be laid on some distance outside of the scroll pattern, letters or ornaments, so as to avoid the possibility of the leaf adhering should it be necessary to let it overlap the pattern.

ANOTHER METHOD.

Take ball liquorice and dissolve it in water, and with a flat camel-hair brush size over the portion you would gild. Make the solution weak. The fluid may be prepared and kept in a bottle ready for use. Made thick will preserve ornaments or work to be repainted.

Whiting and water, mixed and applied with a sponge, and when dry brushed over with a duster, to remove the superfluous particles, may be made use of to prevent gold from adhering, but we would not recommend its use, because it is apt to cloud the work, by fastening itself to the varnish; and further, it requires a great care in removing it, especially on a carriage part.

TO PAINT LIGHT BUGGIES.

FROM PRIMING TO THE FINISH.

The Body—This may be primed with either oil lead or permanent wood filling. The latter requires no mixing, and when properly applied furnishes a very elastic and durable *priming* coat.

When lead is used mix it in oil, and a small quantity of japan, adding only sufficient turpentine to cut the oil slightly. Coat the body inside as well as outside, for there should be no portion of the body exposed to the action of the water used when rubbing down the rough stuff.

The priming coat should be allowed from three to six days to dry.

When thoroughly hardened, sand-paper the surfaces lightly, and lay on the second coat of lead, which should contain less oil than the priming coat, and be spread on somewhat stiffer. Set the body aside for three days, after which give it the third coat of lead, mixed to dry firmly.

PUTTYING.

The third coat of lead having been allowed two or three days to harden, the puttying should follow. Mix the putty of keg lead and dry lead, adding red lead and japan as dryers, or japan alone. Having placed a certain quantity of wet lead on the "putty stone," incorporate the dry lead by use of the palette knife and putty hammer, adding dry lead gradually and pounding the mass until it is rendered tough enough not to adhere to the hands. Putty for nail holes should be worked up very firmly by the process just named, and should not contain an excess of dryer, as this would cause it to shrink. Deep holes, if puttied twice, will insure a thorough and equal drying of the putty. But this is not necessary when the lead coats have been given their full value in assisting in the work of filling the holes.

The open-grained ash portions of the surface should be "plastered" or "knifed in" with soft putty, allowing no putty to remain on the surface. The putty should be allowed from two to three days in which to harden, after which the body should be carefully examined, making use of old sand-paper and the putty knife in removing any putty remaining on the surface.

ROUGH STUFF.

"Coralline," "Tully's Filler," "Reno's French Umber," "English Filling" and yellow ochre mixed with white lead, are employed to fill the grain and give a smooth and level surface. The first two named require no preparation except thinning with turpentine to the proper working consistency. "Reno's French Umber" is furnished in a dry pulverized state, and requires the addition of no other pigment in order to fit it for the work of filling. "English Filling," and the ordinary mixing of yellow ochre and white lead are so well known as to require no special mention.

NUMBER OF COATS.

Give the body at least three coats; four is the usual number for first-class work. The fourth coat, however, need not be carried over any surface but the main outside panels. A "guide coat" is generally put on after the body is filled up, its office being that of furnishing a guide to the "rubber"—as it plainly indicates by its color any low places which the pumice stone has not yet reached. On a dark colored rough stuff yellow ochre may be employed for the guide coat, while on a rough stuff of a light shade Indian red or brown may be used.

The guide coat requires to be nothing more than a stain, therefore is to be mixed with varnish and turpentine, or japan and turpentine, and brushed on quite thin.

The body should now be allowed to stand *at least* a week, when the rubbing may be attempted.

Pumice Stone and *English Rubbing Stone* should be used. The English stone cuts faster than pumice, and is not so liable to become clogged up. It is used principally for the first rubbing, the pumice stone following after, which gives a more perfect surface. Select pumice stone that is light and porous; it will cut better, and not leave scratches. Keep it clean by rubbing it on a flat file, or a piece of English rubbing stone.

Commence on the moldings, rubbing the panels afterward. Where the panels are rubbed out first, the process of rubbing the moldings is apt to cut a groove or channel along the panel, which will show when the body is varnished, if not before.

Do not allow the water to stand inside for any length of time. Sponge it up repeatedly, which will save the joints and corners from injury. When the rubbing is completed, sponge the body off clean, wipe it dry, and let it stand until the following day, when it will be ready for fine sand-papering and a coat of dark lead. When the lead is dry, hair or sand-paper it lightly, which will show any defects in the surface, the dents, or holes appearing darker than the surface disturbed by sand-papering. Now putty up, mixing the putty of dry white lead and varnish, keeping it in water to prevent drying out. When the putty is dry, face the body down with pumice stone, which will force the lead into the pores of the filling and level down the putty, giving a perfect surface.

Next, go over the surface with No. 1 sand-paper; clean out the corners with a dull pocket knife, and proceed to coloring. Mix all colors, except the lakes, with one-third oil and two-thirds japan (the lakes may be ground in varnish or boiled in oil), thinning down with turpentine. Put on one coat a day; hair off each coat to remove grit, etc.

Three coats of clear color, or two coats and color and varnish, will color if properly applied. Three coats of rubbing varnish, and one of finishing, should make a first-class job.

Let each coat of rubbing varnish stand from three to five days before rubbing the body. A *good wearing* varnish will not bear rubbing any sooner. To level down the surface, we use a piece of cork, or pumice stone, beveled on the edges to admit of its being brought up close under the moldings, and to prevent rubbing them. Place over this two thicknesses of cloth, and with fine, even-ground pumice and water, cut down the varnish until dirt and brush marks are removed. Rub the first coat of varnish very lightly; the second and third will bear a great deal more if the varnish has been properly laid on. Keep turpentine out of the rubbing varnish, unless it is very heavy and hard to work.

Finish with English Varnish. Harland & Sons give the best satisfaction, and is now preferred to Noble & Hoare's, because it is not so liable to curdle or draw up. We cannot advise as to the best make of rubbing. There is such a diversity of opinion among painters about varnish that one would have to pronounce them all best for they all have their favorites.

Brushes.—For laying on color, use the flat camel-hair blender (mottler), sizes ranging from half-inch up to three inches, and for finishing there is no brush equal to "Thum's" half elastic.

THE GEARING OR CARRIAGE PART.

To Paint a Gearing so that it will not Crack or Peel off in a Reasonable Length of Time.
—1st. See to it that the wheels and other wood work are well seasoned, and that at the time you prime there is no water or dampness on the surface, especially of the wheels. Mix the priming coat of keg lead, thinned down with raw linseed oil, and add a small quantity of japan sufficient to dry it in about four or five days. Where you have plenty of time, use no japan. Brush the lead well into the grain of the wood, leaving only a thin coat on the surface. (Thin painting wears best.) Be very careful to work the lead well into the wood between the spokes at the hub; also where each spoke joins the felloe. Carelessness at these two points is the just cause of a great deal of complaint; the water having a better chance to lodge and soak in at these places, we generally see the peeling commence at the points named.

Mix the second coat of lead with less oil than the priming, but still sufficient to bind the paint to the metal parts of the gearing (or carriage) which now receive their first coating. Don't paint over rust or grease. See that the iron work is free from everything that would hinder the proper drying of the lead. Lay on the second coat of lead somewhat heavier than the first, leaving a good body of it on the wood parts, driving it rather thinner on the iron work. Let the job stand forty-eight hours. Now putty up with a firm drying putty; let this stand for two days. Sand-paper putty down with No. 2 sand-paper (a coarser sand-paper will tear out the putty), leaving as much lead on the wood as possible. Dust off, and apply a coat of dead lead, ground out fine. Lay this coat on smoothly, and of a body sufficient to cover the wood perfectly; when dry, sand-paper with No. 1 sand-paper, being careful not to cut through. You may now apply the color.

For all ordinary colors, mix them with one-third oil and two-thirds japan, thin down with turpentine as the progress of painting may require, laying on but one coat a day. Two coats of color, and one of color and varnish, with only enough varnish to admit of hairing the gearing off, is the best mode, we think. After two days, hair or moss off the varnish, and run on (if to be striped) the broad line. Next day lay on a medium heavy coat of rubbing varnish. The gearing should now stand four or five days; then rub down with ground pumice and water, and fine line according to the taste. The following day lay on a heavy coat of English varnish.

METHOD OF VARNISHING.

In order to be able to do a good job of varnishing, one must have had no small amount of practice in laying on the rubbing coats, and of rubbing them down and preparing for the last or finishing coat. He must be sensitive to the most minute particles of pumice stone or dust that may be lodged anywhere on the body. The careful finisher never allows pumice stone to dry in the corners or along on the bottom edges at any point, but with the water tool searches it out, knowing that one little crevice overlooked may ruin the entire coat of finishing varnish.

The room, body, cups and brushes, must be perfectly clean, and used for *finishing only*. Two sets of brushes and not less than three clean cups should be used. The panel brushes to be used on the panels and nowhere else. The second set of brushes are for the rockers, inside of seats, doors, and other places.

In one tin cup draw off varnish for the panels; in the second, that which you require for the other surfaces—the third to be used as a wiping cup. Never wipe the brush you are using on the cup from which you are taking varnish. Provide yourself with a wire stand to lay the brushes on, and a light stool to hold cups and brushes. The sponge, shammy and dusters should *never* be used on any other work. If the varnish room is not well arranged and fitted up, have screens made for covering the work.

See to it that the varnish room is kept quiet and free from the visits of any of the hands, or even the employer himself. Should any one take offense at the rules you have laid down, let them see that you are not to be moved from your purpose; defend the varnish room from intrusion until the work is safe from harm.

Pour out sufficient varnish or leave the cork out of the can so as to give the varnish air at least half hour before using. Begin by going over those surfaces separate from the panels. Lay on the varnish in sufficient quantity to admit of its being spread, working down and leveled without dragging on the brush.

Having spread the varnish over a surface, allow it to rest a moment or so and evaporate, then put the brush through it back and forth and "up and down," and let it rest again; in the meantime wipe out around the moldings, then brush the panel horizontally from top to bottom with lighter strokes than before, run the tool along under the moldings, go over this with the large brush, and from the bottom of the panel stroke it upward, blending the varnish off toward the upper part near the molding or upper edges. Do not brush under the moldings as a finishing stroke.

Finish every coat of varnish "up and down," except on very narrow panels or spaces.

Leave on a sufficient body of varnish to flow out full, which will drown the fine specks; the large grit, if any, must be picked out while working the varnish. Use a pointed whalebone, quill, or a small stiff pencil, for picking.

Cleanse the cups with turpentine, then take soap and water, which, with a piece of hair, scour them perfectly clean, rinse them, dry out with the shammy and hang up in a clean place.

VARNISHING CARRIAGE PART.

The carriage part should be raised upon trestles, made expressly for the purpose, so that all the wheels clear the floor at least three inches. Wash off perfectly clean, dust, and then place your stool, with cups and brushes, near one of the front wheels, and you are ready to begin the work. Commence on the spokes, laying on the varnish heavily, and carrying it over as many spokes as the varnish will admit. Finishing varnish should be put over all the spokes, or all except one, left as a guide when finished off. The general practice is to begin at the front part of the wheel, coating first one side of the spokes all the way round, and then the other side, and lastly the back part; the hub may, or may not, be coated at the same time, as suits the convenience of the varnisher. If the varnish has been applied in about equal quantity over each spoke and brushed with a view to leveling it, it will be necessary only to clean out between the spokes, and then draw the brush with a single stroke throughout their length, following up the varnish in the same order in which it was laid on. With the tool straighten up the face of spokes; then level off the hub, finishing it with the point of the brush while the wheel is made to revolve rapidly.

The felloe may be coated between the spokes, finishing each space as you progress, or coat it entire, and finish and clean up the varnish afterward. A soft, flat brush is well adapted to this part of the work, as it lays on the varnish nicely, and does not require attention afterward. The faces of the felloe may also be coated with the same brush, being careful, however, to lay on the varnish by keeping the wheel turned so that the varnish will not drop on the spokes. When a wheel is finished, turn it over occasionally until the varnish is set to prevent it from running. Having completed the wheels, varnish the hind part of the inside carriage, next the front, and lastly the perch and stays; coat the spring bar, springs and axle bed, and then come back and clean out the corners, cross brush and lay off with light strokes with the point of the brush, and so on with the other parts. The shafts should be coated on the bottom as well as top, and the varnish laid off so as to prevent a heavy edge underneath.

A platform carriage is more troublesome to handle, but where suitable devices are at hand to support the parts, the work may be speedily and well done. The wheels should be varnished last, and if arranged so that they can be removed, the inside work can be done quicker and better.

Painters adopt various modes of applying and laying off the varnish, and where equally perfect work is attained, in a reasonable length of time, it is useless to question the method of brushing.

The rule is to lay on a sufficient quantity to flow out and present a solid body, connected throughout every part, showing no laps, which is obtained by working quickly, and brushing the varnish as little as possible.

A very thin coat of varnish will not make a good finish, because it has no depth, and an extra heavy coat will not remain brilliant, because it cannot set or harden through equally. A medium between the two is what is required.

NOBLE & HOARE'S PRIVATE MARK.

On the label attached to each can of Noble & Hoare's varnish is a private mark which gives the date the varnish is made. Thus: 17-3-8 signifies the day of the month, the number of the month, and the year, which would read, the seventeenth day, third month, 1868; 12-2-2, twelfth day, second month, 1872, or March 17, 1868 and February 12, 1872. By observing the marks the age of the varnish may be determined.

THE RUBBING OR LEVELING OF VARNISH.

Next in importance to the laying varnish properly is that of rubbing or leveling it with pumice stone, preparatory to applying a subsequent coat; for it is not unfrequently the case that a body which has been well varnished throughout the rubbing coats, does not produce a fine finish, while on the other hand a body indifferently varnished, when taken in hand by a competent workman, will be so corrected as to give perfect satisfaction.

A body in color, no matter how level its surface, still lacks that beauty of finish which varnish gives, and until the skillful varnisher, and equally skillful rubber, have performed their parts, no great beauty may be expected. A coach body, for instance, is ready for the first coat of varnish. It is the purpose of the painter to apply three coats of rubbing varnish, add then a coat of English, or its American imitation. On the three rubbing coats the main labor must be expended, which shall produce a perfect finish. The first coat being generally put on comparatively light in body, does not require a great deal of rubbing; and just here a careless rubber begins his work of detracting from the beauty and extreme perfection of the finish.

We care not how free from runs and faults this first coat may be; it is in the power of the rubber to damage it. But

HOW DOES HE PROCEE

in order to do it? Our answer is, he takes the body in hand as a piece of work to be performed without having any particular interest in its final appearance. He does not survey the body before applying pumice stone, in order to note whether the different parts of the surface will each bear a like amount of rubbing. He is careless as to the quality of the pumice stone he is about using. In short, he feels that he is merely preparing to do a very ordinary kind of work, that of "getting the body ready for another coat of varnish" The consequence is, the edges are laid bare, and the panels rubbed through to the color at different points, and thus the first coat of varnish is deprived of a portion of its value in contributing to a perfect finish, and in such hands the second and third coats will fare no better.

Wherein, then, does the painter fail in properly leveling the rubbing varnish? We reply, he does so either by ignorance of the proper way to treat the three coats mentioned, or through positive carelessness or laziness, if he understands just how each coat should be rubbed, and still fails to produce a foundation, on which the finishing coat will be borne out in the full beauty which it is capable of producing.

But as work is spoiled oftener through ignorance on the part of the workman, as to manner of leveling down the varnish, we will endeavor to give, in what follows, the proper method of treating the first, second and third coats.

THE FIRST COAT.

This should be laid on comparatively light; and just here we remark that color should be fine, and free from grit, for if the first coat of varnish is laid over coarse, or dirty color, it is robbed of a portion of its value in making up a good surface. The color will always absorb more or less varnish, and unless it is clean, the roughness it contains will project sufficiently to prevent the proper rubbing of the varnish, through fear, on the part of the painter, of producing a speckled appearance. The same is true respecting varnish color when it is applied as a first coat.

The first coat having been laid and allowed to stand until hardened, the body should be brought to the light and examined, to ascertain in what condition are the several spaces and panels, as regards runs, grit in the varnish, and places that may possibly have been missed. The pumice stone should be fine, and entirely free from coarse particles; the sponge, shammy and water free from grease. If, on examination of the body, a run is discovered, which lies above the surface, cut off the top of it carefully, and in rubbing avoid putting pumice stone on it until it is dry. On this coat it is necessary to do no more than remove the gloss, for an attempt to proceed further would endanger the solidity of the colors. When rubbed and washed perfectly clean, the moldings are to be blacked—if not previously done. The color for moldings should dry glossy, so as to bear out well under the second coat of varnish.

ALL SHARP EDGES

should be slighted. The practice of cutting them through and touching upon every coat is a foolish and needless one. If they are rubbed through and then touched up with color they are not in as good condition as when slighted, and the varnish allowed to remain on them untouched. But, if the habit has been contracted of laying them bare at every rubbing, let them remain bare, until the final touching up for the finishing coat; but be careful when you do touch them to have the color very fine, and mixed to dry firmly and with a slight gloss. The second coat of varnish should be laid quite heavy, and especial care taken to have it *clean*. Some painters act as if they thought it of no importance to secure clean varnishing, except on the finishing coat. With such the color is often laid bare in the effort to remove the grit, or at least the surface marred by the scars left on it. The pumice stone employed should be of medium grade, yet of even grain or grit, for, if too fine, time is wasted, and the labor of rubbing needlessly increased. In careful hands a medium grade of pumice will not scratch the surface, and as it becomes finer as the rubbing proceeds it will remove the heavier marks it leaves at first on the varnish.

The most approved plan of rubbing on this coat is to apply the pumice stone freely at first, giving the panel a coarse rubbing all over, then come back and go over it again, using the pumice on the panel, which has in the meantime become finer, and finally polishing the panel by rubbing it throughout its length with the rub cloth and a small quantity of fine pumice. Around near the edges of the panel the rubbing should be at first omitted, for if the same amount of labor is here expended, there will be danger of cutting through, and further it would not be proper to allow the pumice to dry along edges defined by moldings.

Now it is important to know just how far to carry the rubbing on this coat. Painters may be seen who rub a little time and then draw the finger across the panel to look at something they know not what. We speak thus plainly from the fact that no one can know how the rubbing is proceeding toward leveling the varnish on the dry spot left on the panel, where the finger has passed over it. It is only on the panel while wet that brush marks and roughness can be seen.

We will suppose a large panel is being rubbed, the varnish has been laid "up and down." Now the brush marks, if there be any, will follow the same direction. To level the surface the pumice should be applied freely, and the principal rubbing be given cross wise. The pumice and water being thus carried in an opposite direction to the ridges in the varnish, will render them plainly visible. If the varnish then shows defects, either horizontally or perpendicularly, it is only necessary to stroke the rub cloth transversely, in order to note the progress of the rubbing.

When the ridges can no longer be seen, the panel has had sufficient labor expended on it, and may then be polished up with a small quantity of fine pumice. To facilitate the labor, as well as to give the best results, the rub cloth should be backed by a piece of cork, or lump pumice, having a flat face, and the edge beveled. Two thicknesses of cloth are necessary over the block. All body surfaces, small and large, should be blocked down, and the rub cloth, backed by the hand alone, used only in the finishing up of the panels.

Every part of the body having been attended to according to the necessities of each part, and the whole washed clean and touched up where required, the painter should prepare for the third and last coat of hard drying varnish, with all the watchfulness and care he would exercise for the finishing coat; for on the perfection of the third coat rests, in a great measure, the beauty of the finish.

The third coat is to be laid in sufficient body to permit it to flow out full. It must not be teased with the brushes until there is no life remaining in it. Runs and grit must be carefully guarded against. And now supposing that it has been well done, and is ready for rubbing, we will take it in hand. First of all we inspect it, more particularly this time to look after the grit the varnish may contain. The rubbing then proceeds with fine pumice, and is carried but little further than that necessary to cut out the grit. On this coat the finger may be used to good advantage, in watching the the progress of the rubbing, for the principle object is to remove grit. Thus a good body of varnish remains on which to apply the finishing coat. The body being now ready for the finish, the purpose of the above article is fulfilled.

TO REMOVE VARNISH CRACKS.

Aqua ammonia, applied with a sponge, will soften the varnish, when it may be removed with a putty knife or steel scraper made and kept for the purpose. The sponge should be fastened to a stick to prevent bringing the fingers in contact with the ammonia.

This method of removing the varnish will also be found convenient and economical when it is determined to remove *paint* cracks, by *rubbing* them out with lump pumice stone, instead of *burning* off the old paint. The painter is well aware that the most tedious part of the operation is in getting through the varnish. But by the use of *ammonia* the varnish is speedily removed and the rubbing is begun on the coatings of paint, which speedily give way under the action of the pumice stone.

CAUSES OF VARNISH PITTING.

As no shop, from the largest and best appointed down to the most humble, is entirely free from the annoyance of having varnish pit, we are forced to the conclusion that it is caused in very many cases by changes in the atmosphere. But as varnish will pit under such a variety of changes in the atmosphere, and as the atmosphere seems to be capable of such a variety of mixtures, being charged with a greater or less quantity of electricity (which determines the chemical qualities of the atmosphere), we can do no more than give the result of our own observations.

In the first place we would advise to buy the best quality of rubbing and finishing varnishes. Endeavor to obtain rubbing varnish which is not too new; new varnish is more likely to pit and draw up than that which has had six months or a year's age. Avoid using the rubbing varnish too close down to the bottom of the barrel or can, as the dryers and sediment, being there in excess, will, in all probability, cause the varnish to appear dull, shriveled and pitted in a few hours after it has been laid on.

2d. The mixture of two different kinds of varnish will cause pitting, or, at least, shriveling of the surface, when the varnishes are of different natures.

3d. Varnishing over color, or varnish which is not dry.

4th. In summer, when the varnish room is closed tightly, the thermometer indicating 85° to 90°, if the floor is wet down, the steam arising therefrom may cause pitting.

5th. Varnishing in the early spring, fall or winter, in a cold room, where the work to be varnished is chilled, or should the room, varnish and body be of the same temperature, allowing the room to cool down before the varnish has set.

6th. On a hot summer's day, when a storm is gathering, English varnish is apt to pit if laid on before the rain descends.

7th. A varnish room, which attracts dampness, shown by the walls and window panes sweating, will occasion pitting.

8th. High grades of varnish, employed for finishing coats, when laid one over the other, will almost certainly pit or enamel.

9th. Using English varnish, or its imitation, direct from the can, without previously airing it.

10th. Using varnish that is not ripened, or new varnish.

The above may appear to be a formidable array of reasons why varnish misbehaves, and lead the painter to conclude that where there is so much uncertainty, and no positive cure given, that it will be as well to heed none of them, and trust to "luck." Many more might be added had we the notes of painters who keep memoranda of the varnish room, but those given will cover the large majority of cases.

PITTING CHECKED.

Mr. W. B., a leading Philadelphia painter, checked English varnish from completing its work of "pitting," by opening the ventilator, and allowing the heated air to escape. He had two coach bodies in the varnish room, one of which was finished with American wearing body, and the other with English. Shortly after the bodies were completed, in surveying the one coated with English, he noticed that it showed signs of "pitting." He opened the ventilator, then seated himself behind the body intently watching the result. In a few moments the cool air was felt, and under its influence the varnish again flowed out, the pits disappearing, of course. The ventilator required to be closed during the night, and an inspection of the body the following morning showed that the pitting had proceeded again, under the effects of the heated and foul air. The body coated with American wearing body remained perfect.

The value of good ventilation is plainly set forth in the above test. The only difficulty to be overcome is to provide for night ventilation, without risk of damage to the work from storms. The room in which the above named bodies were finished is ventilated through an opening in the center of the ceiling, which serves also as a skylight. A cupola, with swinging shutters, operated by cords, surmounts the roof above the opening in ceiling. This is very effective but far from being perfect. A perfect ventilator will come in good time, for it has been but two or three years since attention has been given at all to upward ventilation in carriage shops, and already we find a goodly number provided with either pipes, flues, or cupolas. Having once been induced to provide the means which are the simplest, a more perfect system will be accepted whenever it has been planned and proven.

CARE OF VARNISH BRUSHES.

All varnish brushes should be kept in air-tight vessels. Those used for finishing to be especially provided for. For the latter the can should be of dimensions adequate to hold two sets of brushes without danger of their touching each other. The ends of bristles to clear the bottom of can at least three-fourths of an inch. The brushes may be suspended on wire in the usual manner, or on wire studs soldered to the inside of can at distances apart suitable to the size of brushes. A piece of heavy wire should be soldered inside at top of can, on which to wipe the brushes whenever they are to be removed. Use English wearing body varnish to preserve the brushes from hardening; being careful to keep the liquid above the bristles. Should the varnish show a tendency to skin over, raw oil should be added. Clean the can out occasionally, for sediment is rapidly deposited.

VARNISH BRUSHES.

There is nothing, probably, that painters have differed more widely on than the best kind of a brush with which to apply the finishing coat of varnish. Our "veteran" painters, to whom we are wont to look up to with no small degree of reference, claim that no brush can be made to equal "the good old round bristle brush." Well, every one to their own notion. We think that improvements have been made in varnish brushes, as well as in many other tools used by mechanics. The black sable and badger hair brushes are far superior to the old-fashioned bristle brushes in many respects, and in some sections of the country are used altogether for the finishing coats. The favorite brushes now among painters who have tested them is the half elastic, made by Mr. Thum, of Philadelphia. The elastic brush is much softer, but we prefer the half elastic, because it has a good spring, and takes better hold of the varnish, when it works toughly, as is sometimes the case. The flat bristle brush, by the same maker, designed for American or rubbing varnish, works very pleasantly for finishing coats and is far superior to any other bristle brush we have ever tried.

"FITCH HAIR IS INVARIABLY USED FOR BODY VARNISIHNG."

Ah! since when did fitch hair varnish brushes come into universal favor among finishers?

We are aware that in some sections, where light work forms the major part of the styles manufactured, the fitch brushes are popular, but to assert that they are invariably used, is to make a very unguarded statement. We would not underrate these soft *hair* brushes, for we know that with them *good* work may be done. But, then, there are such things as *good* work, *better* work, and *best* work, and if we take into the account the manner in which the under coats of varnish are applied, in the use of the fitch hair brushes, we will be compelled to say that the *better* and *best* work cannot be produced except with first-class bristle brushes.

The reasons are, 1st. That with the bristle brush the rubbing coats can be applied without diluting them with turpentine. We will not declare that a small quantity of turpentine is injurious to a varnish which is exceedingly heavy bodied, but the practice of cutting with turpentine the body of all rubbing varnishes, in order to limber down to agree with the limber qualities of the fitch brush, is very wide of being the best method. For, a varnish much reduced with turpentine, loses not only in fullness and brilliancy, but in elasticity, which, we claim, is its wearing quality.

2d. The bristle brush enables the varnisher to work more rapidly, and when coating large surfaces, to apply the varnish over the entire panel and finish it off without forming joints or laps.

3d. It is not so liable to throw off drops of varnish as the fitch, and when varnishing the under panels of coaches, even the friends of fitch hair brushes must resort to the bristle brush, unless they think it fully as pleasant to have the varnish dropping down on their hands as to be free from this annoyance. And further, the bristle brush—Thum's half-elastic particularly—is more easily kept clean.

We lately called on one of our foreman painters in Philadelphia, and it happened he was prepared to apply the second coat of rubbing varnish to an old Clarence coach, the quarter panels of which were solid and quite large. We instinctively glanced at the brushes he brought out, which were a 2½ inch Thum's half-elastic flat, and an inch black sable tool. A hand who was assisting him by attending to inside edges of door pillars, boot, arch, etc., had the same kind of a set. The varnish was a heavy-bodied American rubbing. As the foreman placed himself before the solid quarter-panel, our curiosity was excited to witness his method of applying the varnish. We felt pretty sure that he was a bold, confident varnisher, from the kind of brush he held in his hand, but never having had an opportunity of seeing him apply a coat, we, of course, could not determine what his exact method would be. He used the brush in his left hand; with the tool he dashed on a heavy border of varnish around near the edges of the panel, and with the large brush worked the varnish out to the edges, then spread on a heavy coating, going over the entire panel. He stroked the panel its full length, then cross brushed it, working out into the cup the varnish the brush took up, and finished by going over the panel twice with up and down strokes. He worked in perfect confidence, and was but a few moments in completing the panel. We remarked to him: We see you do no not patch your varnish on timidly. He replied: "No, sir, the timid varnisher can never produce first-class work." Confidence is required in order to put on a sufficent quantity of varnish, and manipulate so as to avoid runs and other defects. Now, we hold that the kind of brushes made use of have somewhat to do with the matter of inspiring confidence or causing timidity. A well-filled, elastic bristle brush, when put into service, takes hold of and spreads with ease a heavy varnish, when applied to large surfaces, while a comparatively weak *hair* brush works directly the opposite. One awakens confidence, while the other leads the painter into a timid and patch-work style.

When reasonably heavy coats of the best American varnish have been applied to a body *pure*, and each coat allowed from four to six days to harden before being rubbed down, and then the finish given with a good wearing body varnish, the whole will form a serviceable protection to the underlying paint, besides possessing a fullness and brilliancy unattainable by using the under coats of varnish thinned much with turpentine. The fitch hair brushes lead to the use of turpentine in excess. The half-elastic bristle brushes remove the necessity for its use, and we leave the reader to choose between them.

HAIRING OFF.

The terms "hairing off" and "mossing off" refer to the practice among painters of removing the gloss of varnish, and also the smoothing of color coats; moss is coarse and otherwise not suitable.

White cow-tail hair is the best, it being long, and when rolled up keeps together in a bunch, and does not shed dirt, or break off into little bits, covering the work.

In painting, we "hair off" each coat of color, except the last one, and on carriage parts and cheap body work, hair off varnish color.

Clear varnish may be haired off, but good work cannot be had except by rubbing down with ground pumice.

WHERE SHOULD VARNISH BE KEPT?

Varnishes are influenced by heat and cold, heat causing varnish that is kept air-tight to become thinner, and the cold to thicken it up.

There is no advantage gained by keeping varnish in a cellar, unless it be that when bought by the barrel and not drawn off into large tin tanks, the dampness in the cellar keeps the barrel from shrinking and causing wastage. Our best arranged shops have tin or zinc tanks in which to keep all varnishes—except English, or its American imitation—the tanks being placed convenient to the workmen, in one of the paint rooms.

BLISTERING.

New work, when run out in hot weather, will blister easily if the last coat of varnish was not hard when first exposed to the direct rays of the sun, or to the reflection from a heated pavement where the carriage may have stood for some length of time.

Work painted during the winter, which has ample time in which to dry, both in the color and the varnish coats, should not blister so readily; still, it cannot be proof against the extreme degrees of heat to which our climate is at times subject.

ACTION OF WATER ON VARNISH.
IS THE MILKY APPEARANCE AN EVIDENCE OF POOR VARNISH?

The cause of varnish turning white when washed, is, we think, the contact of the water with the oil in the varnish. A temporary chemical change takes place, which is, however, removed by the heat of the sun. The practical painter is well aware that if he mixes water with varnish (which generally happens accidentally), the water will assume a whitish appearance on the work he may be doing wherever it is spread over the surface. As soon as the water has evaporated, the varnish will be free from streaks and show its natural color.

Again, take a small quantity of raw linseed oil, and rub it with the finger on a piece of black leather, or any other material; then dip the finger in water, and rub the oil and water together, and you will perceive that the mixture (if mixture it may be called) will assume a whitish appearance.

The action of water on varnishes, especially the finishing varnishes, causing, when washed, or exposed to wet weather, a white or bluish-white appearance on the surface, is no evidence of an inferior varnish. The English wearing-body varnishes, and the best American imitations of the same, present the appearance, to a greater degree than the American rubbing varnishes, because of their greater elasticity.

BLACK, TURNING GREEN.

A black body appears green after a time, because the black surface is covered with varnish coats, each one of which contributes a portion of its yellow color to the black, and as the black absorbs a portion of the varnish, its color is gradually changed to a greenish hue. Black and yellow produce green, whether in the mixture of those pigments on the palette, or by any other means whereby the two colors are so brought together, as for one to be added to the other. We view the black groundwork through a superimposed yellow media, and the black color is lost. If we cut down the varnish to the color, we will find it is still black, proving that the varnish—especially the English—has made the change from black to a somber green. The remedy would be to stain each coat of rubbing varnish with black, and finish with but one clear coat of varnish.

LENGTH OF TIME TO ALLOW NOBLE & HOAR'S VARNISH TO HARDEN.

Noble & Hoar's varnish should not be handled (that is to hang off a body) in less than 48 hours after having been applied. In favorable weather the carriage should not be put into service in less than two weeks. The wearing varnish, we may say, never dries. The surface hardens to a certain extent, but so great is its elasticity, that pumice stone will adhere to it during the process of leveling it down, even after one or two years of hard out-door service.

VARNISH ITEMS.

VARNISH CUPS should be kept as clean as when they were new. Wash them off with turpentine (using a short bristle tool) as soon as you have finished work for the day. Then give them a good scouring inside and outside with soap, water, and a piece of curled hair—don't miss the handles above and beneath, nor the bottoms of the cups; rinse them in clean water, wipe them dry with a chamois and place them bottom up on a wire rack, in a closely covered box or a cupboard.

THE body painter who depends much on the varnish coats to produce a level surface, will never hold the position of a first-class painter. The surface of the body should be brought up as near perfection as possible, before the color is applied, and the color be laid level and entirely free from grit. The varnish coats will then improve the surface to a surprising degree, for they will be required to do no more than add to the beauty of the surface, while over a rough body their value would in part be lost.

AT Goddard's shop, in Boston, each coat of rubbing varnish is allowed from one week to ten days in which to harden, before it is rubbed. The rubbing coats are mixed, half American and half English varnish. The varnish cuts down toughly and slowly, and is by no means pleasant to work on. No shop in America allows so much time for finishing a job as the one named.

OUR best varnishers finish off with "up and down" strokes, on all panels where it is possible to do so. By this means a heavier body of varnish can be left on the panels, with but slight liability to runs and waves. The opposite method requires that the varnish be worked down very flat, leaving but a very small quantity on the panels.

IF a coat of varnish is not rubbed down level, and freed from all grit and scratches, it may not be expected of the next coat that it will be perfect.

VERY cheap varnish is dosed with a material in great favor with Ole Bull and all other violinists, which they use to prevent the bow from slipping.

VARNISH should be blocked down, for a "rub cloth" cannot be held level under the pressure of the fingers.

A VERY cheap varnish must, of necessity, be dear at any price, for it is certainly more profitable to give the customer a varnish that will wear well, and save one's self the trouble and expense of repainting thus doing two jobs for the price that should be received for but one.

THE CAPTIVE FLY.

While recently on a friendly visit to one of our coach painters, the annoyances to which the finisher is subject, especially during warm weather, were freely discussed. Ventilation, dust, and the annoyance from flying insects, came in for a share in the conversation. Our friend related a circumstance which occurred in New York some years ago, which will serve to show how much damage may be done by a single fly on a body unprotected by a screen, and the vexation and delay occasioned before the damaged surface could be restored. In looking back over annoyances past, that which excited anger at the time often becomes food for merriment in the recital; and so it was in this case: A fine coach built to order approached completion, and the purchaser was anxious to hitch to it at the earliest possible moment in which it would be safe to do so. The day arrived on which the finishing coat was to be laid on the body, and as a guard against possible accidents, one of the employers searched the varnish room carefully, and, as he supposed, killed or removed all the flies in the room. The body was varnished, and the finisher before retiring from the shop took a last look to see that all was right. The varnish was clean, full, and in every way satisfactory. But morning came, and with it also came the employer—before mentioned—quite early to examine the coach body. There it was, perfect with the exception of one-quarter panel, on the middle of which a fly had lit, and been captured by the sticky varnish. In its struggles for freedom it had blowed the varnish around its body, and fly and fly blows were a fixed defect.

What expletives were indulged in, in that upper room, as the employer stood gazing at the lifeless insect, does not appear; but on our friend the painter arriving at the shop, he was greeted by the aforesaid employer with a most doleful story of how matters were with the body up stairs, ending his remarks by bitter curses on the fly that had escaped him on the previous day. Together, boss and jour ascended to the finishing room, and there began the consultation as to what should be done, resulting finally in the decision that the job would have to stand two or three weeks before it would be in a condition to admit of re-varnishing the panel. The body was hung off, the coach lowered down into the wareroom, and day after day cold water was applied to the defective panel until it became hard enough to take a light rubbing. After four or five careful rubbings the scars disappeared. The panel was again varnished, and the job at length turned out.

How true it is that the weak things have been chosen to trouble the strong. One fly—in the case above cited—caused additional expenses and most vexatious delay, and made three persons at least very ill-natured—the owner, the builder and the finisher. And now we ask which one of the sister branches in the coach shop is liable to be tormented by so small and insignificant a creature as a single house-fly? And is it at all surprising that the painter should be a very sensitive creature—a man who steps softly—is offended at dust and insects, no matter where he may be, and dislike the noise and clatter of hammers—that he is, in short, a high strung, sensitive creature?

VENTILATION.

The renewal of fresh air to the paint and varnish rooms has engaged the attention of a few carriage-makers throughout the country, but we question whether a shop in the land has made provision for thorough ventilation. As a general statement our paint and varnish rooms are situated in the top story, which on high buildings has the lowest ceiling. The upper story is chosen because it is supposed to present few liabilities to dirt from the street, as well as to have the painters above rather than beneath either of the shops, to avoid dirt and dust from this source, also. The paint shop then, according to present usage, must occupy the top story, which being directly under the roof, and having the lowest ceiling of any room in the building, is doomed to be the hottest place in the factory. And in these days of flat shed roofs, covered with the barbarous tar and sand, which under a July sun become almost red hot, the paint and varnish rooms have become the dread of the workmen, as warm weather approaches. A few shops are partially ventilated, and in these the finisher is not compelled to breathe over and over again *all* the poisonous gases that escape in the room; and so far so good, but as no provision has been made for supplying fresh air near the floor, the thorough ventilation of the room is not secured. A room cannot be cleansed of foul air and supplied with that which is best suited to the lungs and to the varnish without having the means above for the heated and rarefied air to escape, and its place (or the vacuum formed) filled by pure outward air flowing in near the floor. Heated air rises toward the ceiling, and if it there finds a means of escape, will pass out into the open air, but if retained in the room, there can be no inflow of fresh air, and the room soon becomes filled with poisonous gases, very deleterious to health. We have been made sick at the stomach in a few moments, when working in a close varnish room having a low ceiling, by merely mounting a stool to varnish the top of a coach. The stratum of air near the ceiling was so vitiated that it was impossible to breathe it without immediate bad effects.

The objection raised to thorough ventilation is that the current of air passing through the room would be fatal to the finishing coat, by introducing dust into the room. We think the objection wholly unfounded, for it would not be necessary to keep up a continuous draft while laying on the varnish, but pure air might be admitted three or four times during the day. During July and August the finisher shows plainly in his haggard countenance the enervating effects of a poorly ventilated varnish room, and the finishing varnish itself shrivels under the streaming atmosphere before it had time to set.

Of nothing are we more positive than this, namely : that English and fine American finishing varnishes require during hot weather to be laid in a well-ventilated room, experiments proving that high grade varnish is very sensitive and dislikes heated foul air. Two-thirds of the trouble with English varnish during the hot weather arises from want of proper ventilation, and so summer comes and goes, year after year, dreaded by both employer and painter, and the true cause of the vexations and expense of re-varnishing is overlooked.

Well says an employer who had spent considerable money in making his varnish room air-tight : "My painters were always charging the imperfections in the varnish to the lack of an air-tight room and now, having gone to a great deal of expense to secure one, I am advised to break through the solid walls and put in ventilators." Even so, for the varnish room must be kept as cool as possible in the summer, and at all times free from poisonous gases.

Varnish in the can becomes much thinner in hot weather, when put in the coolest place to be found about the shop, but when the body and varnish are subjected to a heat of from 85 to 100 degrees, in an air-tight room, is it at all strange that the varnish should fail to flow out and appear full and rich ? Let the varnish and paint rooms then be remodeled so as to admit of a continual change of air, and we shall soon fail to hear the coach painters' trade charged with being the most unhealthy of the four branches, and through its annoyances with being the least profitable.

WORTHY A PASSING THOUGHT.

The atmosphere of the paint room is unwholesome at all times—the pallid countenances of the majority of painters giving positive proof of the fact, and in summer fully as much so as in winter, although in the former the windows are open a portion of the time. The winter months improve the general health, and enable the painter to take more nourishing food, and this assists in warding off the evil effects of the paint room.

During hot weather the appetite fails, and, as the evenings are the pleasantest part of the day, the temptation is strong to keep late hours. Insufficient sleep and want of appetite weaken the system, reduce the flesh, and render the individual more susceptible to attacks of colic, or derangements having that tendency. Now, it is worthy a passing thought from every carriage painter in the land to look after his health during the "heated term," and inquire into the best plan for preserving it.

Without assuming any guardianship over men who are of age, we would, nevertheless, throw out a few suggestions, which, if heeded, will result in no harm or special inconvenience to them; and 1st, we would caution the painter against the practice, which is common, of bringing his dinner into the paint room and allowing it there to remain during five hours at least, and, at the noon hour, sitting down and eating it in the same room. If he *must* carry his cold bite, because he cannot afford to go out and obtain warm, untainted dinners, then he should provide a proper place to hang the dinner bucket or basket outside of the work-room; that was *our* practice for several years. We carried dinner from late in the fall until spring had fairly opened, and then, during the warm weather, took the noon meal at a restaurant. The change of clothing, the walk, and the warm dinner, assisted in giving a variety to the day's toil, and, on returning to the shop again, we felt in good spirits, and better prepared to close down the windows, if necessary, and enter the suffocating atmosphere of the varnish room. We were led to the practice of airing the dinner basket by the repeated assertions of our better-half that bread left in the basket retained a strong odor of the paint room.

The sense of smell would not be so keen with the painter as to make such nice distinctions; but persons whose olfactories are not blunted very readily pronounce sentence against anything contaminated by paints, oil and varnish. Any portion of the dinner not eaten should be thrown away, for, in no case, should children be allowed to get possession of it. How common it is for little children to want to peep into papa's basket when he returns and sets it down at home. At such times they prize a dry crust more highly than tempting sweetmeats; but it is far better to disappoint than to allow them to partake of the tainted bread. 2d.

NEGLECT OF THE PERSON,

or uncleanness of the skin. During the mild seasons there is no excuse for any one not possessing a clean body. A basin of water, a sponge, soap and towel, will answer the purpose, if it be impossible to secure better convenience; but at whatever cost no painter should fail to take at least three baths during the week. We are bitterly opposed to the practice of some who seem to think it a waste of soap to wash above their wrists, or below the pit of the neck, and, without exaggeration, we have known painters to carry colors on their arms, near the elbows, for weeks together; such filthiness must hasten the undermining of the constitution by inviting disease. Over the surface of the body are innumerable pores, which carry off matter no longer required by the system, and when this operation is checked disease ensues. Cleanliness of person, then, is one of the most essential matters for the painter to attend to; for, with a good supply of wholesome food and a clean skin, he is well fortified against disease. 3d.

LOSS OF REST.

As we stated before, there is a strong temptation to sit up late and enjoy the night air. The practice is injurious, and, besides, occasions ill feelings and loss of time in the morning. The practice of setting up late is more common among single men, for they seek out jovial company and take no notice of passing time. Twelve, one and two o'clock are not unusual hours for retiring. The consequence is, they are in deep sleep just at that time in the morning when they should be up and stirring about. A hurried breakfast, or none at all, a cold bite hastily put up for dinner, and a feverish excitement to get to the shop before the call to work, or within fifteen minutes thereafter: all this flurry comes from loss of rest, from the lack of a system in the affairs of each day and night. The painter, then, needs wholesome food; he must be cleanly, and he must not be deprived of at least six hours of rest *during the night*. Nature demands this, and if disobeyed for any great length of time will surely punish the guilty one. 4th.

THE EXCESSIVE USE OF TOBACCO.

In the three foregoing items we have spoken of the abuse of certain natural laws under which every individual is held. The use of stimulants we look upon as an artificial means of weakening the system, and, therefore, stimulants may be wholly abandoned without a particle of injury to the persons in the habit of using them. Tobacco chewers and smokers should curb their appetites, and be satisfied with a small amount of the delightful but noxious weed. When the painter indulges in tobacco to excess, heartburn, nervousness and lack of ambition and strength to go through with the day's work ensue. A person thus suffering is a very miserable creature, for he not only suffers bodily pain, but he also suffers from the knowledge of the fact that he has been the foolish author of his uneasiness; and while the use of tobacco is more or less injurious (we know it by experience) at all times, we know it to be much more so during the summer; therefore, we would advise the painter to use it very moderately.

We have directed attention in the above to the four leading enemies of the painter's health, namely:—Unwholesome food, uncleanness of person, loss of rest and the use of stimulants, tobacco especially; and while individuals following the other branches may not, with impunity, defy the laws of health, they are not so susceptible to injury as the painter, because they breathe purer air during the ten hours in the shop. Painters, you who are suffering from a feeling of lassitude, suppose you examine into your mode of life, and, if any remark we have made covers your case, try the experiment of reforming in that particular. To counteract, then, the effects of the paint room in hot weather, do not destroy the appetite by drinking, smoking or chewing. Take a good night's rest, and keep the person cleanly.

LITTLE THINGS.

In no branch of the carriage business is it more important that attention should be given to what may be termed *little things*, than in the paint shop; and we may be not wide of the mark to go further and say that it is attention to little things that gives superiority of finish to the painting in one establishment, and the neglect of small matters (so called), which stamps the work of another establishment as inferior. Were it possible to find two paint shops, built exactly alike, having like conveniences for doing work, and these two shops were then opened and placed in charge of two foremen, one of whom was scrupulously nice in little things, and the other careless, we would find, at the end of six months or a year, should we visit them, that the careful foreman had added greatly to his stock of little conveniences for doing work nicely and with ease to himself and his help, and that cleanliness and order were stamped on everything about him. His work would be well and carefully finished in minutest details.

Should we now pay a visit to the shop under the control of the man who considers that great nicety is womanish, and be always planning to render things more convenient as an intolerable bore, we would not be surprised to find a work-room in disorder, the hands fretful, the stock being wasted, and the finished work mussy.

But you may inquire what am I to understand by little things? We answer: On entering the shop in the morning, greeting the hands with a smile and a kind word, which puts them on good terms with you and themselves for the day. Should an accident happen, even though it be chargeable to the lowest apprentice, do not get into a passion and abuse the boy, because he is a boy, while had the same occurred with a hand but a grade below you, you would have said nothing. Do not put an apprentice to an inconvenience in his work to save yourself a few steps, for by your example you teach him to do the same. A foreman under whom we worked some years ago had complete control of his journeymen and apprentices, and all through politeness and kindness. As an instance in point he borrowed a wheel-board for a few moments from a new apprentice, and on returning it set up the wheel as it was when he had asked the accommodation. This was a trifling matter in itself, and yet it made such a lasting impression on the boy that he was ready at any time to accommodate his foreman or his fellow shop-mates.

These are among the little things in the govornment of the shop, and we will now notice others connected with its mechanical workings, wherein the foreman should direct as to what he considers essential, and demand of each one under his direction that he do his share toward preserving and keeping in its proper place whatever he may provide. We may begin at the paint stone. Here the careless workman might consider it but a trifling matter that he leave but a *little* paint adhering to the surface at the corners; and another with the varnish cup in his hand think it but over nicety for his attention to be called to the fact that on setting his varnish cup on the stone a ring of varnish from the bottom of the cup was left there. The *little* bit of paint left on the stone from time to time soon amounts to a good deal, until there is barely room in the center to mix up a small cup of paint; and the practice of setting cups on the stone also adds to its filthiness. Correction is only required at the beginning; the small amount of paint not being left, the greater is never seen; and thus we might direct attention to every portion of the work room. But in the varnish room we may see more plainly the effects of carelessness in minor things. The neglect to prepare the room at the proper time, to wash off the body perfectly clean, or to have the cups and brushes in proper condition are of trifling importance to the thoughtless finisher, yet on attention to these small matters depends the beauty of the finishing coat.

The apprentice who has the reputation of being careful and extremely nice will, when free, be selected to take a good position at good wages. And a journeyman bearing a like character will seldom be found out of employment. Take care, then, of what the heedless term trifling matters, and the weightier matters will take care of themselves.

MENHADEN OIL.

"The Menhaden is a North American fish, striking the coast near Cape Hatteras and sporting in all the coast waters as far north as Maine, where it leaves for the deep waters of the ocean. The fish usually appear in May and leave in October. The menhaden oil is extensively used, being an adulterant of both olive and linseed oils. The fishery and production of oil is prosecuted along the coast and require a capital of more than a million of dollars."

Well, that will do pretty well. The olive oil we may use on our dish of crabs, or, at the proper time, on salid, is, after all, fishy. Again, we learn from the above choice scrap that "Menhaden oil forms a component of *much* of the linseed oil sold." *Pure* linseed oil is known to be the fattest of all the oils manufactured, and, from its powers of resistance to atmospheric influences, the best adapted to the painter's use when used either on house work or by the coach painter. But if *much* of the linseed oil sold is adulterated with fishy fat, and much of it adulterated with cotton-seed oil, another adulterant, pray tell us what quantity of pure linseed oil is furnished the coach painter, in the use of which he is expected to produce work of great durability. What a sorry spectacle the painter presents in these days of vile compounds in the way of oils, japans, varnishes, colors, etc. Let us see of what his materials for coating work are composed. We will begin with wet or keg lead. The employer pays a good round price for that which is nicely labeled " pure lead." The expert painter removes the head of the keg, and inserting the palette knife, raises up a quantity of the so-called *pure lead*, and instead of its dropping off "ropy" from the palette knife, as in years gone by, the lead is flaky and short, like first-class pie crust.

Barytes, a heavy earthy substance, is present. The oil in which this pure lead is mixed is *pure* linseed oil, provided the "menhaden fish oil," or "cotton-seed oil," form no part of it. The turpentine used in thinning the paint may have a strong smell of benzine. The japan drier cannot be made without using oil, and who knows but that the menhaden fish fat and benzine may not form component parts of this important article, to say nothing of that *crack*-generating ingredient—gum shellac. Varnishes are dependent on pure, refined linseed oil for their good wearing qualities, and why may not varnish contain fish or cotton-seed oil? And then, as to colors: Vermilion is adulterated with read lead, and often we find that that which is purchased for the best English vermilion will, when mixed, fall to the bottom of the paint cup, as heavy almost as if it were red lead ground in japan. Carmine called No. 40, which should be the best, often shows the presence of vermilion. Ivory drop-black, which, when pure, is intense in color, and covers and dries well, will often be furnished only in name, it being replaced by a color which looks gray, when compared with a good quantity of lamp-black, and as for covering well or drying properly, it is entirely innocent. Ultramine blue and other colors also come in for their share of adulteration; ground pumice stone, when pure and bolted to a proper degree of fineness is costly, and being very light, requires so much to weigh an honest pound that we find there are those who do not scruple to dose it with barytes to add weight, leading the varnish rubber to exclaim at times, "what *can* be the cause of this pumice cutting so slowly? it slips over the surface like soap."

Gold bronze is innocent of gold; silver leaf is only Dutch metal—and so on to the end of the chapter: there is difficulty of obtaining any article that is pure; and in the face of these facts the poor painter is charged very often with being a botch, because his "painting does not stand." Perplexed beyond measure, and unable, it may be, to give a satisfactory reason why all this trouble should follow him, he may call for a settlement, and be off to some distant point in the hope of retrieving his reputation for turning out good work—if he does not take a bolder step and quit the trade forever. The body maker soon learns to distinguish at a glance between the qualities of the timber he uses; and whether it be well seasoned or not. The glue, canvas, screws, brads, etc., are not of so mysterious a nature as to cause him any trouble. The smith readily detects any poor qualities in the stock he uses, the trimmer has the same advantage, and the employer soon acquires an aptness in judgment concerning these three branches, which compels him to agree with the statement of the workmen. But how different in regard to the painter's supplies. The lead is marked "pure lead," and he takes it for granted that it must be pure; the oil he cannot analyze, and as for varnishes, they are a liquid mystery.

The wood worker may say to the employer, this panel stuff is not thoroughly seasoned, and therefore unfit for working, when the answer will probably be: "You must use it. I can furnish no better at present, and this work must be pushed ahead." No responsibility, then, attaches to the body maker, should trouble ensue. The smith and the trimmer have a like advantage. But the painter cannot say of the varnish that it will not wear, before he has applied it, nor can he always state positively that the colors and the vehicle in which they are mixed will not produce first-class work.

The paints and varnish are applied, and at length, when the work is completed (and it may be spoiled), the painter has no previous words of warning to fall back on, but must stand speechless; or should he put in a plea in his own defense, will fail to convince his employer, for the simple reason that he is ignorant of the real quality of the stock furnished him. What we have said in defense of the painter we trust will be carefully weighed by those who are hasty in charging their painters with being wholly at fault, whenever they fail to produce the most perfectly finished and most durable work.

TRIMMING DEPARTMENT--PART IV.

METHOD IN THE TRIMMING SHOP.

Without a system of some kind, it is impossible to conduct any business in a correct or economical manner. A *method* must be adopted; some rule or set of rules laid down by which you should be governed in your business relations, and which rules could be understood by those doing business with or for you, whether as a merchant, a manufacturer, a foreman, a superintendent, or as a plain journeyman. A system will be of advantage, and will eventually tend to simplify and render easy that which has otherwise been difficult and troublesome. And in our trade, how necessary it is that workmen should adopt a given routine in their daily round of duty. Whether as piece workmen or as day workmen, this will be found advantageous.

Now take two men equally matched as to speed, on a job. One of them does everything by guess or chance. One time he will commence at this end of the job (to trim); at another time he will commence at the other end; his tools are scattered about his bench in happy confusion; he has only a very indistinct idea of where anything he may need can be found; and when he gets in a hurry (ne is seldom out of one), it is worrying to see and hear him. He rushes from his job to the bench to look for some tool he needs, but in the litter he cannot get his eye on it. He frantically turns everything over, and cannot find it. Perhaps it dropped on the floor or under the bench. Down he goes on his knees to the floor. Can't find it. He peers under the bench—*it ain't there*. He tries among the litter again—it is not there. He sings out, in a fretful manner, "did any one take my claw-tool?" No one took it, and back he rushes to his bench again, and takes another GOOD look. This time he (perhaps) finds it. He grabs it up, makes a dive at the nail he wants to draw; but he has lost so much time hunting up the claw-tool, that he does not take time to do it right, and ten chances to one but he digs the top of his finger off, and slaps down the offending tool. And thus he goes on all the way through; always in a hurry, always anxious, always on the hunt for something mislaid or forgotten, alternately losing and finding.

Now, the man along side of him has a much easier way of getting along. He has no trouble about trying to remember where he left his tools, or where this or that article is. He has no rushing to look for this or that; no diving under the bench, and raking up the scraps for lost things; no pettish queries as to who has this or that tool; or where is this or that tool? No such a thing. *He* has no need for any trouble, simply because he knows where every article can be found. He can always lay his hand on anything he wants to use. And how does he do it? Very easily, because he has a "place for everything," and takes time to have "everything in its place." The time lost by the careless *no-method man* in hunting up his stray tools is greater by far than is lost by the methodical workman in keeping *his* conveniently handy, not to speak of the worry and annoyance saved by having things where he can find them. And then, again, in the *method* or *routine*, or want of either, evinced by each. Now, the careless man is always forgetting to paste out this or that part of his job until just when he happens to want it; then he rushes it through, dries it at the stove in a hurry, and while he is doing something else in an equally big fluster, he is reminded of the piece he left drying behind the stove only when he smells it burning. And thus he goes, always in a hurry, always in trouble ; and he really does not get on as fast as our placid man of method, who never allows himself to be put out, and who has a system or order which he follows on each job. He begins with a certain part, and follows it up all the way through the job. Remember, hurry is not speed.

METHOD APPLIED TO A NO-TOP ROAD WAGON.

Begin by fitting the enameled cloth for the foundation of the fall; then fit the cushion, bottom to the seat, and get the correct flare of the seat with the cushion facing patterns; now cut and fit the facings, and paste them out; advance them as far as you can, because you will need them before anything else; next paste out the fall, and if it is to be sewed into the cushion front, advance it as far as you can ; now cut and fit the carpets ready for binding, also the pocket to go under the seat, and now, while the pasted-out-stuff is drying, you can be working to advantage in covering the dash, and when it is done, if the pasted stuff is not yet dry, you can work along at binding the carpets ; then get the tips put on to the shafts, and trim *them*, so that the painters will not be delayed by waiting your convenience to do them ; now stitch the cushion front and the fall, and make up cushion complete, and to prevent the possibility of mistaking one cushion for another, or putting a wrong cushion in a job, mark the number of the job on a piece of paper, and paste to the bottom of cushion ; now make up the shaft and body (check) straps; put them on and put in the carpet, whip socket, seat pocket, and finish up, if possible, so as to get that job entirely off of your mind, and thus leave you mentally free to commence another job without being compelled to carry a part of the last one on your mind; by doing this way you will have more time and be in better condition to think of improvements in the piece of work you are about to commence.

And here we must call attention to the habit some trimmers have of stuffing a cushion as hard as they possibly can. We have seen men cramming (not stuffing) the moss and hair into an unfortunate cushion until the corners were bursting out, and still they crammed away for "dear life," like they could never get enough into it, or ever get it to shape until it was as hard as a brick. Now there is no need of all this hard stuffing ; if a cushion is properly made and stuffed fairly, it will keep its shape ; a stuffing stick is hardly to be tolerated when stuffing a cushion, unless to lift the hair into the corners between the false top and top proper—use your hands, and you will judge better where the stuffing is needed.

METHOD APPLIED TO A TOP BUGGY.

In this class of "shifting rail" work it is rather difficult to lay down a fixed rule of progression, because different shops have each a rule of their own, as to how the trimmer shall manage to work at the job. Some shops always give the body to the trimmer, and let him keep it until it is trimmed ; then other shops will not let the body go to the trimming room at all, and we have to put our tops on the rail alone; with this plan there are some advantages for the shop, because the painter can have the body painted, while the rail and top are getting tailored ; but there is an evident disadvantage to our branch in this arrangement; there is in fact a want of *certainty* in trimming a top on a rail fastened to *anything* but the seat and body to which it belongs; for no matter how careful or particular you may be in setting your rail to a trestle, it may be wrong; while if you have the body, you know just what you are doing, and this fact has made itself so apparent, that all or nearly all the large shops in New York arrange for the bodies to be sent to the trimming shop, and this is as it should be. To trim without the body have a trestle made perfectly level, and a board 1 inch thick and 14 inches wide, and as long as the trestle is wide, screwed fast on top of the trestle; then have two sticks 3 inches wide, an inch thick, 4 feet long; in each end of these sticks there is a long slit cut for the bolts of rail to drop into, and when the nut is screwed up, the rail is fastened to the sticks, and kept to its right width; then screw the two sticks to the trestle cover.

The trestle, with slotted sticks, marked A A, and rail fastened to the sticks, and long, enough for almost any width of rail.

Now, with the rail fast to its place, and the buckram for the back ready dry, either set the top, *or get it set*, and while the bows are being dressed up, make up the back,

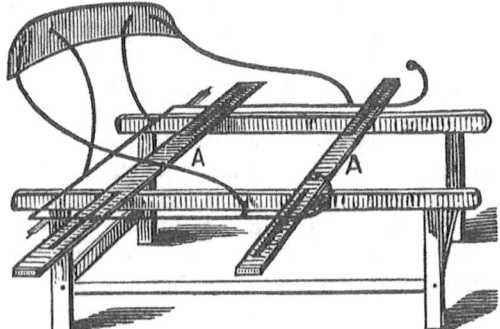

put it in, and then rough out the leather for the top and paste it out; then fit your cushion and fall and carpets; also the seat pocket; paste out the cushion facings and fall, and get them ahead as far as possible; by this time the bows will be dressed up and ready for covering; cover them; nail on your strips to the bows for the head lining to be sewed to; set up your top. To do this *center* all the bows, then make a mark on one bow for the seam line, and then mark all the others the same distance from the center. Use leather straps to draw the front bow to its place; a long one, without a buckle, nailed to bow just below the seam line, and short ones, with buckles, nailed fast to the front of trestle frame. Now put in your head lining, and fit the back and front valence; cover and paste them out. Always put the head linings in before pasting out back valence or inside lace, for the reason that the piece which comes off the width of the cloth in the length of the top is (in roll-up lops) usually sufficiently large to cover the inside lace and valence.

Next put on all your strainers, and if your top leather is dry enough fit it, and while doing so, before you take it off to cut, punch a hole through the side quarters into the bows at the point where the center of the joint-prop should be. Now take off the leather and cut it, then lay it aside and put on the props, and give the length of the joints, and have them made and filed up by the time you are ready to put on your top. Now cover the dash, because the painters will need it to put on the body before the finishing varnish is put on. In fitting the side quarters for the top, if you stretch it lengthwise, before drawing on, only along the middle, about where it will lay over the bend of the bows, it will fit easier and lay smoother along the seam, and if you strike a chalk line along the bottom edge of the side quarter, before taking it off, you will have a better sweep, and the top will fit more naturally without any tendency to draw *upward between the bows.* When your dash is covered, proceed to draw on the muslin for the stuffing of the top, but first nail on a narrow strip of black muslin along where the bottom edge of the white muslin will come, and when the stuffing is all on and the muslin drawn over it, turn up this strip of black, and paste it over the edge of the white muslin. This is for the purpose of hiding the white stuff when the top is let down.

Now bind the back valence and inside lace and front valence; nail them on, but let the back valence lap one fourth of an inch lower than the bottom edge of the inside lace. Now fit up the back stays and back curtain by tacking them on to their places on the top. We are aware that some trimmers are in the habit of fitting up these on the bench; that is, marking them off for the loops and curtain straps, but we do not approve of the plan, because a top is not like a fixed piece of frame work, and it WILL give or spring more or less; we have noticed jobs fitted up in this way which came out anything but right. The best way, as a rule, is to fit *everything where it belongs*. Stitch on your loops, rivet in the knobs in the back stays, and line them as well as the curtain. It is best to tear off a piece the full width of the cloth, just the length of the back curtain; the over width from the curtain will line the stays, and leave a strip for a cushion border. While these are drying, stitch and close your top, and cut off the basting stitches, then hammer down the seam on a smooth, hard board (hammer the wrong side); then rub very hard with a slicker, and, if convenient, get some one to hold one end of the seam while you rub and stretch it up.

Now draw on the leather, commencing in front, then pull backward, then again at the front, and tack down to its place. Now put on your joints, and stitch the edges of back stays, and then nail them on with four or six oz. tacks, at the inner up edge, where the tacks will be covered by the curtain. (Always be careful that the back stays, when up, measure the same across from the bottom inside corner of one to the upper inside corner of the other; if they do not the whole top will look lop-sided); baste on the stays, and with one needle stitch the inside quarters to them; finish the back curtain, baste it on, let it hang loosely; fasten the inside lace to the head lining, either with paste or by sewing. Now fit the side curtains, and then give the joints to the painter; proceed to stitch (or bind off) the top; when done, unscrew the rail from the trestle, and turn the top upside down, and rivet the knobs in the rail. Now paste out the side curtains, and while they are drying make up and finish the cushion and fall and carpets; trim the shafts, and make up the side curtains.

And now finish up all small things as prop blocks, whip socket, apron and straps; but before the painters put on the finishing varnish, nail in the seat pocket. Always FINISH up each job completely, so that no part of it will remain undone, to be an annoyance to you when the job is sold and about to be sent away. Do this way, and many a growl with the boss will be saved.

METHOD APPLIED TO A CLOSE-TOP GIG OR PHAETON.

If it is to be a shifting-top, proceed as for a roll-top. When the body is trimmed, set up the bows; and when the top is in its place, proceed to put in the head lining. Fit on the cloth smoothly *over* the bows, then mark the lower edge of the rail with chalk, on the cloth, from the edge of the back stay line around to a point two inches in front of the prop-block; also from a point about five inches back from the slat-iron prop to the line of the front bow, and four inches above the slat-iron prop. Paste thin enameled leather, four inches wide, on the back part of the head lining, the bottom edge of the leather to come down to the chalk mark already made; but in front it will be necessary for the leather to be wide enough to extend to the mark, four inches above the slat-iron prop; also mark the point of the slat-iron prop. If you intend to slip the head lining on the prop, you should stich a hole in it for that purpose, but it is best to use the *half-bow cap finish*, as it appears to obviate all danger of the head lining tearing out at that point; and we also use the piece of wood screwed to the side of the rail, and projecting out flush with the edge of the seat, for fastening the head linings and leather to. When the lining is in, nail on three strips of straining web along the bend of the bows, or strips of enameled duck will do as well, and two strips from the back bow to the rail, one of them along where the outer edge of the back-stay will be. This strip should have a piece of harness leather stitched to the bottom end of it so as to slip over the prop-iron, the other strip nailed to the bow three inches to the front of the first one, and to the wood on the rail three inches in front of the prop-iron. Now, fit on the leather, rubbing it along where it will lay over the bend of the bow. Nail it back and front over the bend first, then draw it downward, and slightly back to the prop-iron of rail. Cut a hole, and slip it on this; then go to the front bow and draw smooth, and last go to the bottom again and draw slightly downward. Punch a small hole through the leather, into the bow, just where the top props are to be, and when you take off the leather to stitch it, paste a thin piece of enameled leather cut round with edges shaved thin, about two inches in diameter, over the holes. This will save the leather from getting worn at the prop. If the back stays are ready, nail them up and fasten them temporarily to the side quarters with eight-ounce tacks, taking care to allow the quarter to be loose or full; in other words, the side quarter should have one-fourth of an inch fullness from a straight line where it is seamed to the back stay.

Now, take off and stitch the quarter to the stay, then put on the props, and give the length of the joints to the blacksmith, and have them ready to put on when you draw on your top. It is a good plan to have the joints on the top when the back curtain is hung to its place, because you will then be able to hang it evenly. Put on your muslin and stuffing, and proceed as for a roll-up-top.

LANDAU BACK, QUARTER AND FALL.

The back is made with one full row of squares, two rows of buttons, at the bottom, besides the finishing squares (see sketch No. 1); then the swell of the back is carried up to within four inches of the upper edge, 1⅛ inches fullness in the swell, then the top of the back is finished with a very large roll, say 5 inches fullness, thus showing the back with only one row of buttons on the upper sweep.

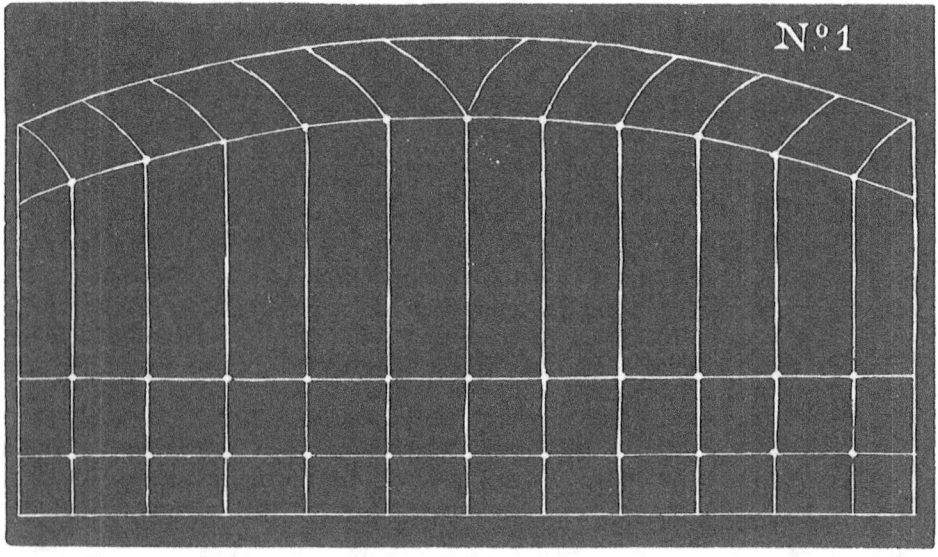

BACK.

The arm pieces are made in a very peculiar way, and the *modus operandi* is very difficult of explanation. In place of the usual arm piece block, a piece of plank 2¾ wide by half an inch thick is fitted in with the usual sweep to it; now fit four pieces of single fly buckram to form, as it were, a *funnel*, the shape of the arm piece desired; then sew seaming lace to the two edges of the funnel, which will show inside of the body; to the lower lace edge sew a piece of cloth in smooth, so that it will cover the bottom of the funnel or cylinder. To the same lace edge blind-sew in a piece of cloth, for the purpose of forming a wrinkled roll on this inner face of the funnel, wrinkled one inch fullness to every three inches of length, and as full the other way as desired, for it ought to be full enough to come out flush with the bottom side quarter. Next blind-sew the other edge of this roll to the other or top seaming lace and stuff lightly with hair, thus forming one roll on the inner face, and having the lower face covered with the smooth cloth.

Now blind-sew another piece of cloth to the top lace, as in the other case, to form *another* wrinkled roll on top of the funnel or cylinder, but the outer edge of this roll is to be finished by nailing to the outside of the piece of plank first spoken of. All this sewing is, of course, to be done on the bench, one side of the funnel to be left open for this purpose. Next nail in the bottom-side quarter, made up in squares, and then nail the side of the funnel which is fitted against the arm board to the board, and over the quarter, thus finishing the lower part; then nail the fourth or top side to the top edge of arm board; next stuff, from the front, this funnel pretty solid, and finish the top roll, which has been left open thus far, into the outside of the arm board, thus completing the arm piece, which

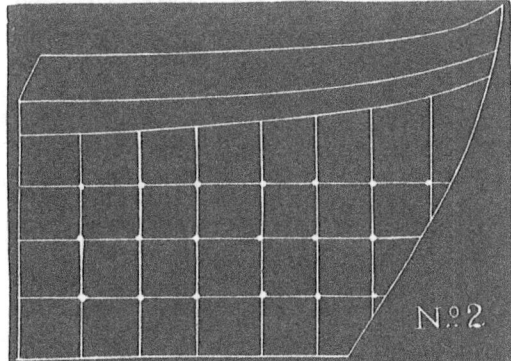

ARM PIECE AND QUARTER.

shows two wrinkled rolls, divided by two rows of seaming lace, and which, we think, looks first-rate.

THE DOOR FALL

is made up on three-ply of buckram, pasted together, but one ply is cut off about an inch from the top to allow the fall to hinge. The fall is made about twelve inches deep; the lower edge is a round or a six-square corner; presume the broad lace bent and corners sewed, and all ready to put on the buckram, which is cut to the shape intended for the fall; lay the lace on the buckram, and mark where its inside edge comes all around the buckram; then mark 1¼ inches from this mark; next paste a piece of carpet into the buckram to come within half an inch of this mark, which is 1⅜ inches from the edge of the lace. Cover this carpet with a piece of cloth, pasting on the buckram; take a piece of seaming lace long enough to reach around the fall, and sew a piece of cloth to it, for the purpose of forming

DOOR FALL.

a wrinkled roll around the three sides of the fall, inside the broad lace and outside the covered carpet, between both, with one inch fullness to each three inches of length; sew this seaming lace and roll to the buckram at the mark, one and a quarter inches from broad lace; gather the other edge with a running string, and sew down and stuff lightly; finishing in such a manner that the broad lace shall cover this sewed edge; next paste on the broad lace, and cover the wrong side with silk or muslin; when dry, stitch both edges, thus finishing a very beautiful door fall.

THE SWINGING HOLDERS

are made up on a single-ply of buckram, to which may be pasted a piece of muslin; cut it out three inches wide, or the width of the broad lace, and sew a piece of cloth to one side of it, to form a wrinkled roll; one inch fullness to every three of length, except where it is bent; at this point more fullness is required; next sew seaming lace to both edges, stretching in the cushion hooks for this purpose; when done, sew a piece of cloth smoothly to one seaming lace, then turn it over, and blind-sew to the other stuff the smooth side with a strip of cotton batting, and the wrinkled side with hair very lightly; sew the frog on eight inches from the top, thus finishing the holder, and saving broad lace; and as the wrinkled rolls can be easily pieced, all scraps should be worked in.

We would also say that the arm piece cannot very well be used on any but open quarters, where the workman can finish the upper edge from the outside of the job.

CLARENCE DOORS.

No. 1.—In the cut herewith given is shown a neat style of trimming for door. It is made as follows: paste out three plies of buckram, and lay off for block or biscuit pattern, leaving space enough all around for a broad lace border, and at the top leave double the space. The top space is formed into a plain cloth roll of the same goods as the job is trimmed with.

In this instance the trimming is brown cloth; the broad lace is silk and worsted of a shade much lighter than the cloth. The diamond shaped and connecting figures are worsted, and are raised.

CARD POCKET.

The card pocket is made of tin and covered with Turkey Morocco, the color of the trimming.

No. 1.   No. 2.

No. 2.—This style, as will be perceived, is somewhat different from the other. The surface of the door is trimmed plain, the fall alone being stuffed. The fall is stuffed in diamond form, and inclosed with a lace border. A drop pocket is shown with a broad lace border; this pocket runs up under the fall to the top, and is there nailed.

TRIMMING DEPARTMENT. 301

INSIDE VIEW OF ROUND-FRONT CLARENCE.

SHOWING ONE-HALF OF BACK, ONE SIDE, ALSO ONE-HALF OF FRONT.

There are some new points in the style, especially in the finish of the arm piece with a roll continued around the top of the back and in finishing the top of back, with

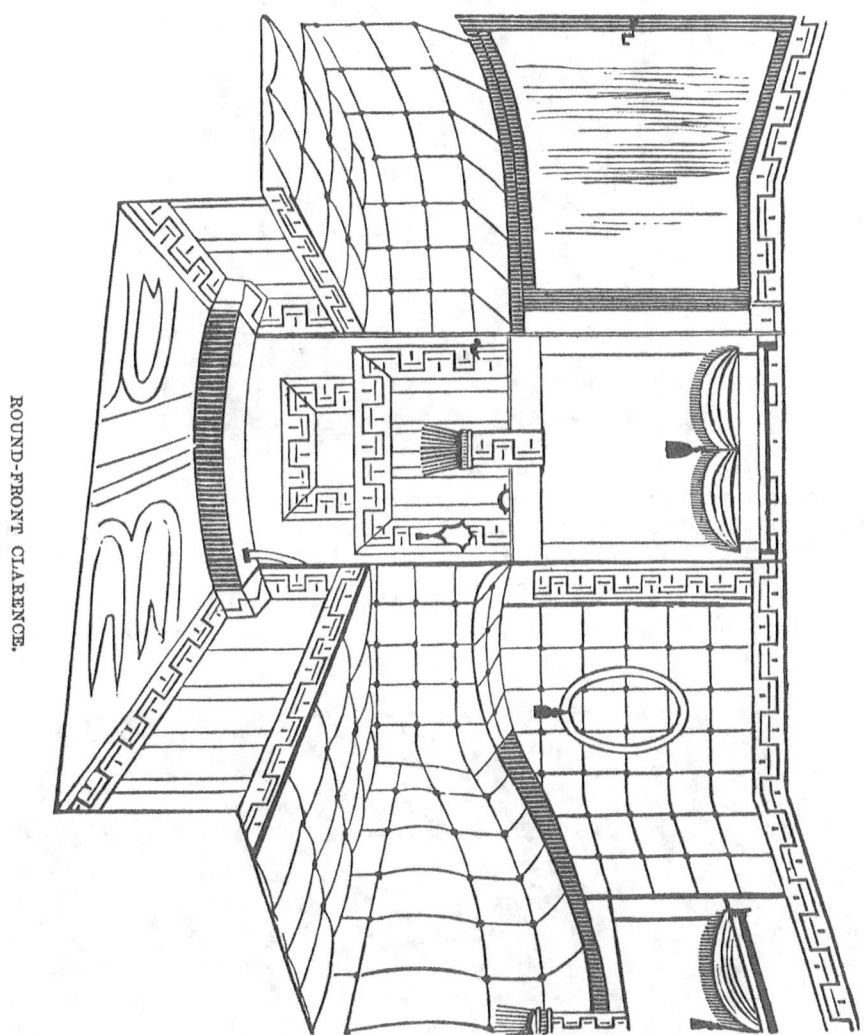

ROUND-FRONT CLARENCE.

the folds flared outward from the center. The doors are sometimes trimmed a little different; for example, the door falls are rounded or six-sided at the bottom, as also the dropping door pocket to match. Covered buttons are most in favor; tufts are seldom seen now. Cushion top made up on a frame, square top. The pattern of lace most in favor is something like that represented in the drawing, and called the Wall of Troy pattern.

FALL FOR COACH OR PHAETON DOORS.

For coach doors, where lace is used, the checker work in the center is made by cutting the cloth or reps into strips, 2½ inches wide; iron down to 1¼ inches, then cut strips of cotton batting, 1⅞ inches wide, lay into the cloth or reps, and baste edges together; then lap them in as in the pattern, or in squares.

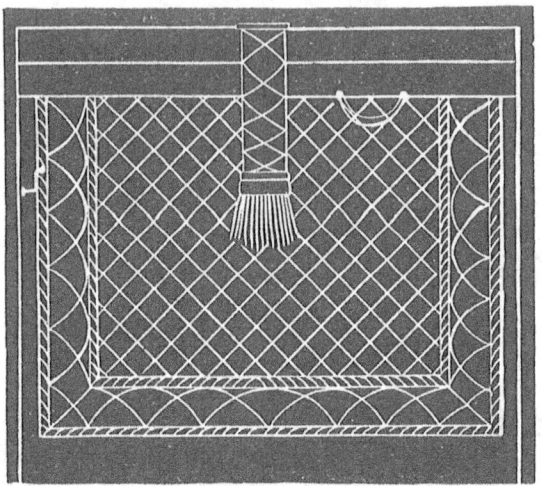

FALL FOR COACH DOORS.

For a phaeton, etc., where the twisted lace is used, make a broad lace of 2 inches in width of collar leather, with a three-edged raiser laid on the middle, and stitch a twisted seaming lace to both edges; then lay it on the plaited work.

TRIMMING FOR LANDAU SLEIGH.

Fig. 1 shows the style of the back, which, being simple in construction, will not need a detailed description. Fig. 2, showing the cushion, front and fall, which are very tastefully ornamented.

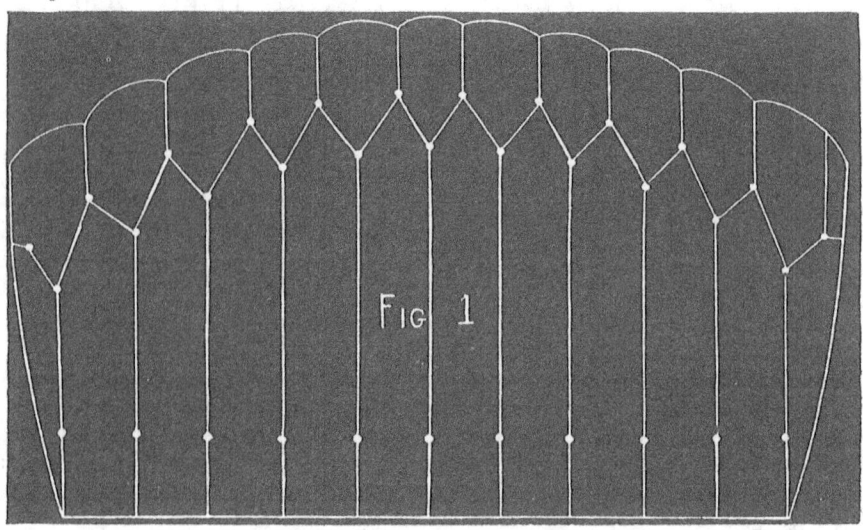

BACK FOR LANDAU SLEIGH.

The back, as you see, is made with rolls and points; foundation for back is made of burckram five thickness pasted together. The reason for using so many thicknesses of buckram is, that the back is not fastened, being made just the size, and forced into its place. It has no false belly, the rolls and belly are stuffed separately, each having one-half inch fullness. The fall is made stiff, and paneled off as shown in cushion squab-top.

FALL FOR LANDAU SLEIGH.

THE SARATOGA.
BACK CUSHION AND FALL.

The back with smooth center, wrinkled roll all around, and plain smooth roll on top and the cushion with a smooth center, and wrinkled roll all around is entirely new, and we think it is the very best style of trimming yet introduced for buggy or phaeton where the drop back is used. It looks light, neat, *is economical*, and comparatively easy to make, taking very little if any more time than a plain back done in squares. The cushion will occupy more time in making than the ordinary plain cushion, but do not think it takes any longer to make it than it does to make a squab-top cushion.

TO MAKE THE BACK.

Use four-ply of buckram, as usual for drop backs. Cut to within three inches of the bottom of the seat, and on the sides out to the edge of the seat; then rounded in to fit against the back of seat in the corners. Mark the upper roll (B) two and three-fourth inches in the center, and to a point, as shown at both ends. *This is to be a smooth roll.* Next, mark two and a quarter inches all around the back for the wrinkled roll ($x\ x\ x$), then make belly for the center panel (D) of moss, *and pretty hard*, covered with muslin or enameled cloth, sewed down one-quarter of an inch inside the inner line of the wrinkled roll. To cut the cloth, lay your cloth (doubled) on the bench, and mark three and a quarter inches from the bottom edge of it; then cut out the piece for the center (D) above this mark. Now, mark three and a quarter inches from the outside edge of the cloth for the wrinkled roll, and cut it that width all in one piece around the space left where the center was cut out. Next, cut the piece for the small roll (B), with two and a quarter inches fullness at the center, so that when made it will project well out beyond the wrinkled roll. Now, draw your cloth belly (D) very tight lengthwise, even bending the buckram at both eds. This is done so that the back may not wrinkle in the center when finished. Next, sew patent leather welt on (with rattan in place of seaming cord) for the inside edge of wrinkled roll; then wrinkle your roll to this welt; then sew down the upper edge of wrinkled roll stuff, and sew patent leather welt on, leaving both ends of welt loose, and long enough to go around the back and meet in the middle of the bottom edge. Next, paste and cover the wrong

THE SARATOGA BACK.

side of the back with enameled duck, letting it project three quarters of an inch beyond the buckram. Now sew the loose ends of your welt to this projecting edge, and then blind-sew the three lower sides of the wrinkled roll to this welt, which makes a very good finish.

Note.—The smooth roll (B) for the top *must* be sewed down to the upper welt of the wrinkled roll before covering the wrong side of the back with duck.

Nail in and finish the top edge with seaming, and rubbed-down welt, the seaming to be on inside edge.

THE CUSHION FRONT

is made of split leather, $1\frac{3}{4}$ inches wide, and pasted on muslin; then two very narrow raisers pasted on $\frac{5}{8}$ apart, *scant;* the ends cut off 1 inch from the ends of facing, then cover with cloth, and paste on a rubbed-down welt neatly between the raisers.

THE CUSHION TOP

has a plain, smooth center, like the back, and a wrinkled roll 2¾ inches all around, which may be sewed in with or without a patent leather welt.

THE FALL

is a plain piece of cloth pasted on enameled duck, a rubbed-down welt ⅞ wide pasted on 2¼ inches from the edge, and another on the outside edge, but this is only half-rubbed down, the outer edge being left loose, so as to turn it over and bind the fall with it. A small, light pocket or card case is sometimes made in the center of the fall, and an extra piece is stitched on the wrong side for the purpose of slipping in a fall stick.

TO MAKE SHELL WORK OR HERRING BONE.

For a lazy back, say three inches wide, mark a perpendicular line on the center of the wood, then two other perpendicular lines, three inches at each side of this center one; next center along the whole length of the back; then step it off with the compasses for the center nails, one inch apart, beginning at the perpendicular line, which is three inches from the center.

Now proceed to form the semblance of a diamond (to be six inches long in the center), with hair, covered with muslin; then cover it with the material to be used; then cut off your cloth, about double the length of the back, from the end of the diamond; center the cloth, and step it off with the compasses, 1⅞ inches apart. This gives you ⅞ of an inch fullness between the nails; the cloth needs to be as wide as are the sides of the center diamond; drive all the tuft nails through the cloth into the wood (do not drive them home), commencing at the point of the diamond, and turning in the edge of the cloth at the first nail.

Now stuff all the little rolls lightly, allowing the hair to stick out loosely to the edge of the wood; next begin at the center, and pull the cloth to the top of the center diamond; the rest of the work is easy, as all the other rolls will lap almost to the place where they belong; finish the edge with seaming and pasting welt (seaming inside).

DROP BACK.

This plan for a drop back makes a fine job; the center is in biscuits, but there is a

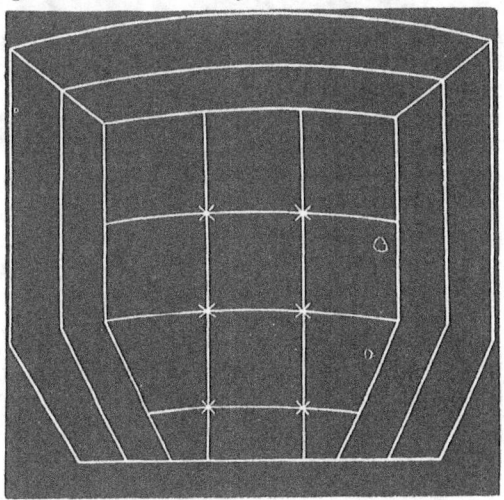

false belly under them; the belly is stuffed before drawing in the buttons, and it has a double roll around it. The false belly makes it have a bold appearance, and the double roll gives it a finish.

Turn a welt on the end edge of the back also, and then sew the roll and cover into the welt, as it makes a better finish. A biscuit back is nothing new, but a back with two rolls and a raised center is not in general use. The rolls are one inch wide, with one and a half inches fullness; the false belly does not extend beyond the first row of buttons on top and bottom of back, and to inside welt on end.

PARK PHAETON SEAT, WITH SQUARE CORNER WINGS.
TRIMMED WITH BLUE CLOTH AND BLUE MOROCCO COVERED BUTTONS.

The back and cushion top is made in squares with blue cloth, also the fall, but it has a morocco welt pasted on, $2\frac{1}{4}$ inches wide from the edge (as shown in drawing); the outer edge bound with a turn-over welt of patent leather, as also the scolloped piece on top of the fall. The cushion facing is of morocco, with rubbed-down welt.

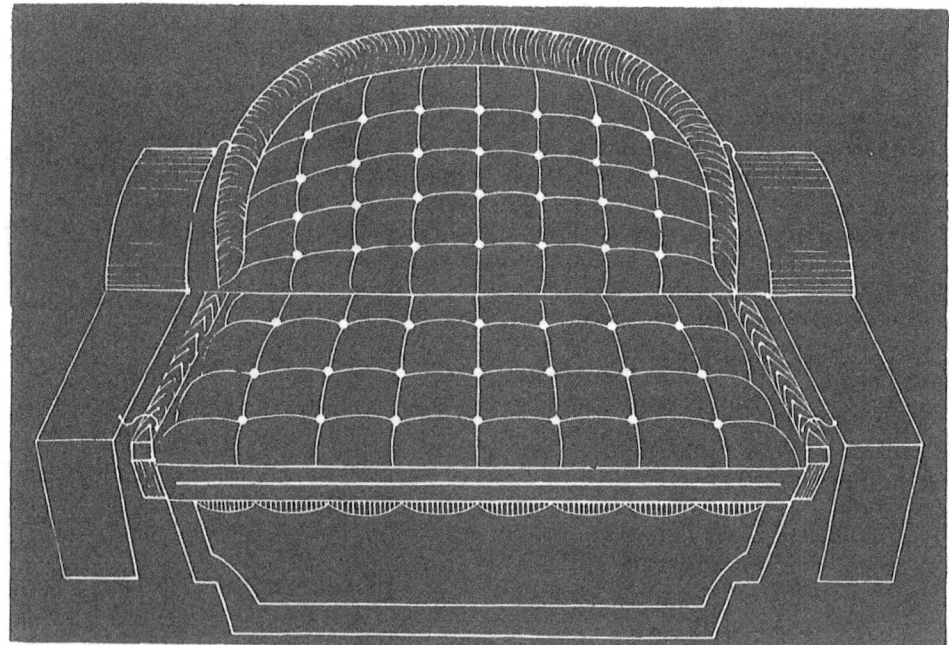

pasted on the center (as shown in drawing); patent leather seaming welt all the way through the job. A wrinkled roll of morocco around the top and both sides of the back, and may be carried to the front pillar on the top of the low arm piece (or duster), or else finish this low arm piece with the herring-bone style, as shown. Carpet to be blue, with a crimson border, about 4 inches in width, all around it. Bind the edge very wide with patent leather. Patent leather rocker covers, with two rows of stitching all around. Make the front facing for the cushion $2\frac{1}{2}$ inches high when finished. The dash to be covered double, and without a flap; but bind the edge of dash all around, as also the edges of the wings with patent leather. The cushion straps to be $\frac{7}{8}$ of an inch wide, and covered with morocco. Buckle them in front in place of knobbing them on as some do; also, remember to blind-nail the outer edge of the rocker covers, leaving binding edge loose for that purpose, and stitch leather to the inside edge.

You will then have no nails to show at all. In this class of work it is necessary to finish very neatly, therefore be very careful to have no *black head nails*, or any other kind seen; if you do, you will mar the general effect of your work. The stitching should be done with silk, for thread will very soon turn gray, and look faded and shabby. The stitching on the dash should be carefully done, and not less than 12 or even 14 to the inch.

CUSHION AND FALL.

The accompanying design for cushion and fall will be found in some respects new, at least the cording, which is made in the following manner. Take light rail leather and cut out strips two inches wide, wet it down and paste in your cordings, which is made of harness leather, either waving or zig-zag (as in design), but not over five-

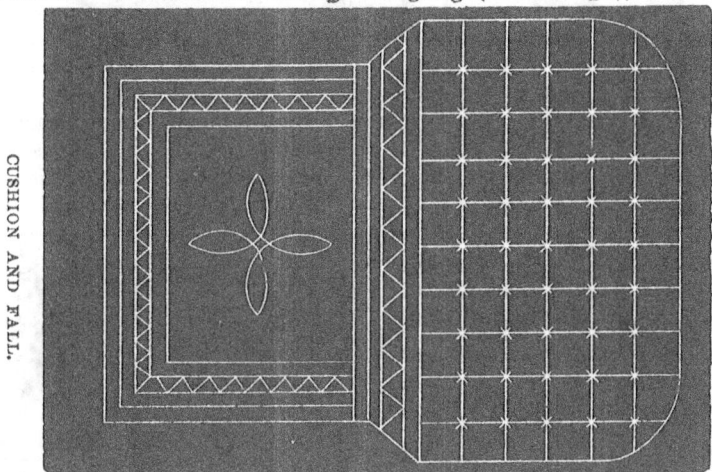

CUSHION AND FALL.

eighths of an inch wide from point to point for the fall (the cushion facing can be made wider). After cutting the shape you want, take off the edges with an edging tool, double the patent leather over the cord and paste it down, creasing it around the cord. Place two rows of one-eighth cord on your fall, one, an inch from outside, the other, two inches above that; now paste the corded welt between the two as seen in the design. Your facing will only have the corded welt in center.

The fall is bound with turn-over welt.

The design in center of fall can be made any shape desired.

DROP BACK, CUSHION AND FALL.

The back is made up on four or five ply of buckram pasted, the first row of buttons to be two and a half inches from the bottom, and at least three inches apart. The second row of buttons to be six inches from the top of the back; there are but two rows in the back proper, and one-quarter of an inch fullness from the top to the bottom row of buttons with one and one-eighth inches fullness between the tufts (to form the rolls); the top of the back proper is finished three and a half inches from the top edge, and a patent leather welt sewed on it to form the lower edge of the tufted roll, as shown.

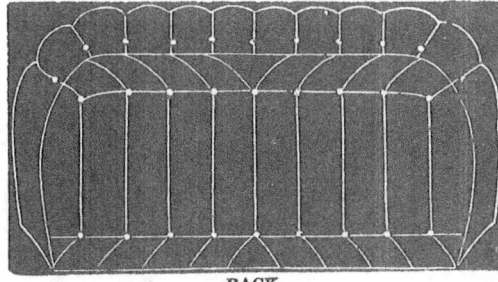

BACK.

The buttons for the roll are only 1¼ inches from the line of the patent leather welt, which finishes the top of back proper, and there is no fullness on the material from the welt to the button holes, but between the roll buttons 1⅛ inches fullness, and from the roll buttons to upper edge make the roll very full, say 2½ inches fullness. Cut a small scollop out of the lower edge of roll at the plait-lines, so as to draw it tight from buttons to welt: from the last button on each side of the roll it is continued down to the bottom of back as a smooth plain roll, the welt finish to be continued all around and down to the bottom points. Bind the outer edge with a turn-over welt of patent leather, and finish the *top on the wood*, with seaming welt on the outside and a rubbed down leather welt inside. Take care to turn the pasting welt inward from the seaming welt.

CUSHION.

The cushion is very simple and is made up very much as a driver's cushion (dickey seat), is made. Allow only half an inch fullness from the front to the back row of buttons (all the way), and the same from each of the middle rows to the outside row of buttons, but from the outer row of buttons, all around the sides, and back of cushion allow 1¾ inches fullness; and from the front row to front edge of cushion one inch fullness, and for the roll between the center row of buttons 1½ inches. Remember there is no welt sewed on to form this outer roll; the fullness given here will form the roll without trouble. The top should be made on a frame, and a stitch at each button hole mark to hold it to shape until the top is sewed in, and the cushion turned and stuffed.

THE FALL

is made up on enameled duck, the three square pieces are of carpet pasted on, and the lines on both sides of them are three-edged raisers cut very narrow, and pasted on, then all covered with the cloth, pasted. The border is of split leather stiffened and bound with a turned-over welt of leather on both sides. The lines are: first, an inch strap pasted on the split leather, and then a three-edged raiser, cut to the full width of what the raiser machine will cut, then pasted on and covered with cloth. When

CUSHION AND FALL.

dry, to be stiched all around the raisers and carpet. The fall may be sewed to the cushion or not, as the case may be: for instance, if the seat panels are lined, you must nail on your fall, and then a strip, one inch wide and bound with leather on both edges, nailed on top of the fall. In this case, use two cushion straps; but if the seat panels are NOT lined, then sew in the fall to the cushion, and with straps sewed in.

The cushion front is made with three raisers on it, the center one to show with patent leather on top, this raiser to be cut to the full size of what the three-edged raiser machine will cut; the others, one on each side, to be cut very small and show only as covered with the cloth. The small raisers can be cut in the raiser machine by simply putting a little wedge under the back of the knife to lift it up, thus depressing the point on the roller; by this means you can cut the raisers to any size, *no matter how small.*

HORSE-SHOE BACK.

This name is derived from its near resemblance to a horse shoe; this pattern, when finished neatly, has a very good effect. The manner of making we will describe: Take buckram, and paste it four thicknesses, and when dry fit it in the body, and then lay out the rolls; put muslin on the center, and stuff as high as the rolls are intended to be. Next, paste out cloth and leather pieces ⅛ of an inch wide, turn the edges in and stich them with white silk, and lay the pieces on the muslin; paste one strip of cloth then one strip of leather alternately across each other until all the swell is covered; then put the rolls around. If you wish to put a figure of any kind on the center of the upper part of back, you may do so without fear of its detracting from the appearance.

The patent leather as seen interlaced in the back and the welting which surrounds the rolls, contrasted with the color of the cloth, produces a rich effect.

BACK, WITH IRON ARM PIECES,

is a very neat, plain and stylish-looking affair, suitable for either roll-up or close-top buggy work.

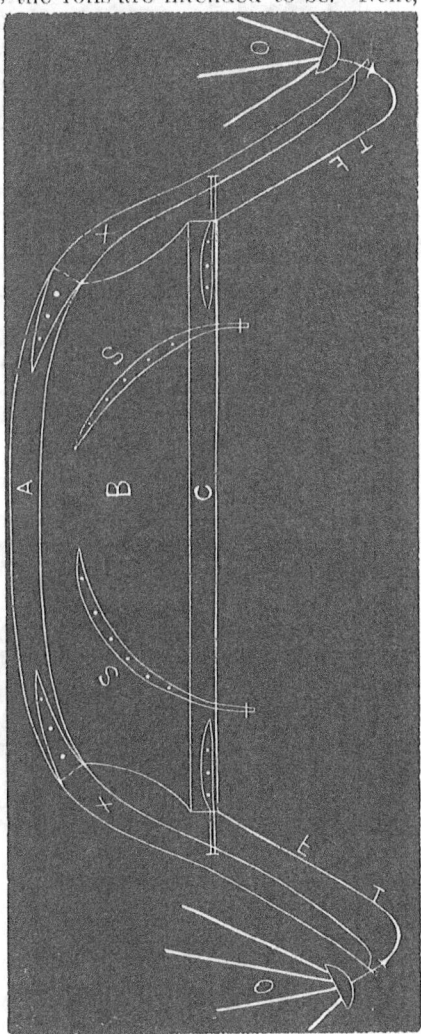

Description.—The part marked A B C is wood, and extends from corner to corner of the seat. F F is like the ordinary rail, but screwed to the wood, as shown. X X are the band-iron arm pieces, continued around with the sweep of the seat, and eyed on to the rail at the "slat-iron prop." The back (joint) prop is forged to the rail, F F. The irons, S S, are screwed to the wood and continued down the ends as bolts to fasten to the seat. O O are the slat irons.

HOW TO TRIM THE BACK, ETC.

The center, B, is trimmed smooth (plain,) without wrinkles or squares, *a la Grecian-bend back*. The space marks A and C are smooth rolls, the lines at each side of A and C to be patent leather welts.

DROP BACK.

This back is made up in two separate parts, and finished with roll around each part, either smooth or wrinkled. The roll is made up of five pieces. After the squares are

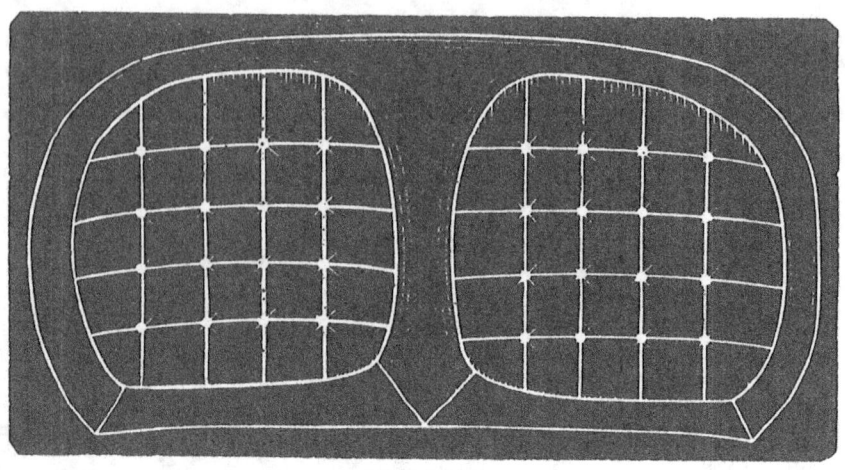

DROP BACK.

finished sew on a patent leather welt, then sew the roll to the welt, and finish the outer edge with a welt also. Sew the roll to the center part first, then all around, as in other roll-backs.

THE "GRECIAN BEND" DROP BACK,

as shown, is suitable for light open-top work, or for close tops, when the shifting rail is used; with this difference or advantage, that if there is an arm rail, the upper roll should be continued around to form the arm rest, without breaking the sweep of the welt. We ought here to remark that although those wrinkled or ruffled rolls were the style in all kinds of goods, yet they look best in leather.

HOW TO MAKE THE BACK.

Four-ply of buckram, pasted together, as usual; mark off the size of the plain smooth center, paste; lay on the hair, then cover with muslin, and sew it down to the shape of the center, taking care to sew inside of the pencil mark; next take the leather, and pull (stretch) it all you safely can, then sew it on, taking care to keep it very tight, from end to end, for fear of it wrinkling; pare it off so as to expose the pencil mark; next, sew down your welt, the edge to the pencil mark; then the ruffled roll; run a strong thread through it, and gather it up to the desired lengths; sew it down, stuffing lightly. The top edge may be finished with pasting and seaming, or blind-sewed to a pasting welt.

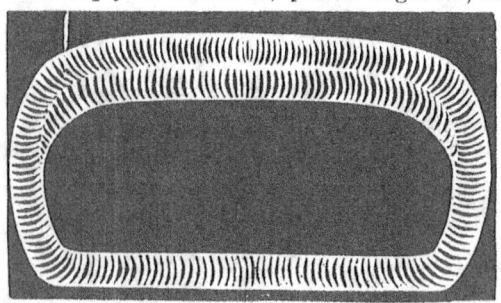

"GRECIAN BEND" DROP BACK

For all parts where there are no very short turns or bends, trimmers will find that *rattan* is much better than the ordinary cord used for lace or welt.

TWO STYLES LIGHT BUGGY SEAT, BACK AND FALL.

On the first the fall is made with five plaits on a scollop welt, the welts all corded. The seat is tufted square, with corded facing. The back is also in squares, with a smooth roll all around; welts all of collar leather.

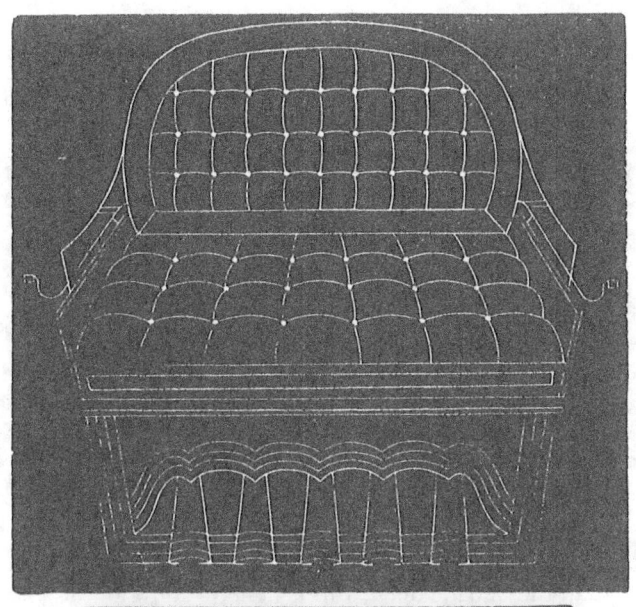

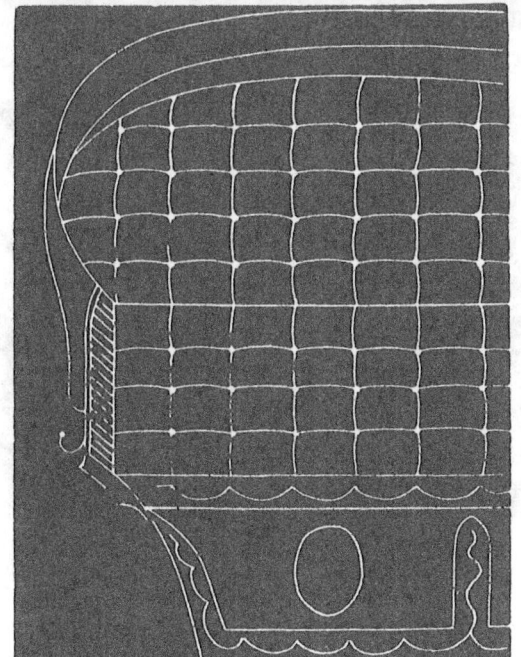

BUGGY SEAT AND FALL.

The second cut represents a portion of a buggy seat and fall. The upper roll, enamel leather; width, one and three quarter inches. The lower roll of bow leather; with one and a quarter inches. The back and seat of enameled; biscuit work in three-inch squares. Bow leather, fall and cushion front, with "new moon" raisers and oval center pieces.

LAZY BACKS.

No. 1 is a very neat style of trimming. Three smooth rolls with leather welt between each roll, and the edge finished with welt also, but the seaming must show on *inside* edge; in laying off all *three* roll backs, let the center roll be marked five-eighths

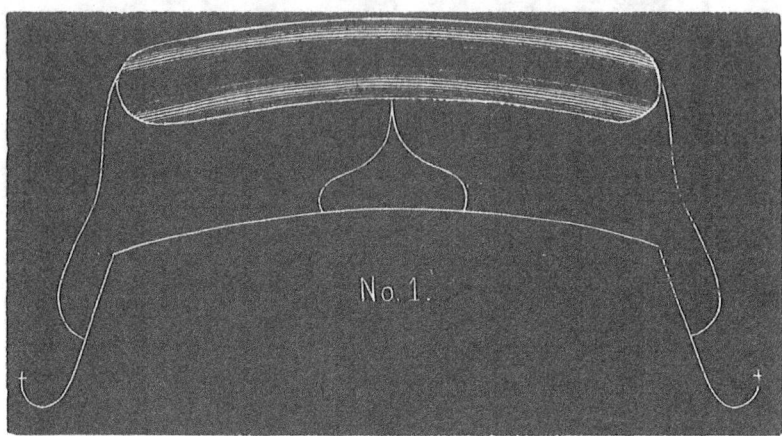

of an inch wider than the outside ones; this is done to allow for the width of the welt, which is always nailed on top of the center roll, and takes up about that space; then, when finished, the rolls will be all of a size, whereas, if you lay them off equal on the wood, the outer rolls will be the heaviest.

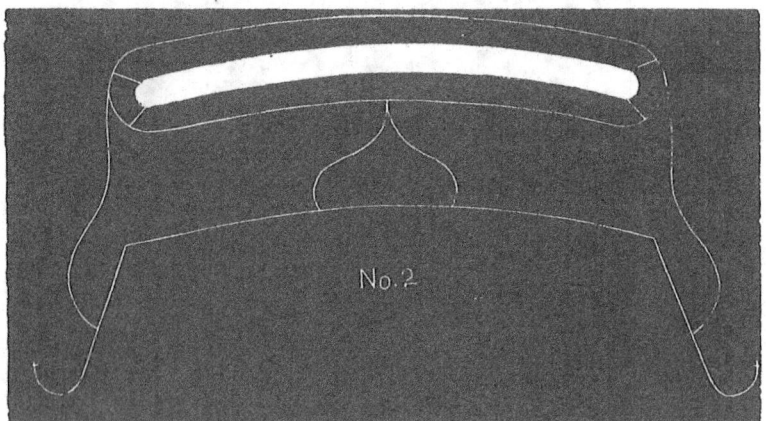

No. 2 shows a three roll back also, with the difference that in this case the outer roll is continued all round the back; this is a very tasty and popular style. You may join the cloth on the corners as marked, either sewed or simply lapped; remember to pull both cloth and welt tight when making up; further, sew your cloth to the inner welt before you nail down, and be sure to line the cloth with a piece of muslin, pasting only on the edges—these three roll backs may be changed by making the outer roll in wrinkles and the center smooth, or *vice versa*, or the center one in small squares and the outside ones smooth or in wrinkles.

No. 3 is best adapted to dog carts and two-seat jobs. This would look well with the center roll plain, and the outside ones wrinkled, or *vice versa*, and will do where the arm rail is made for an arm piece (or rest); in that case the lower roll to be continued

around on the arm piece. Use rattan for your welts in all these jobs, and show as few black (or finishing) nails of any kind as possible. When leather or morocco is used, we would recommend that the center roll be made plain (or smooth), and the outer ones wrinkled with fully one-third fullness for the wrinkles, and *stuff soft* where wrinkles are used. Either of these styles should make a nice job, but the trimmer must see that no irons are fitted to the front face of the wood work of the back, or, at least, that they be put on in such a way that they shall not interfere with or obstruct the sweep of the leather welt; it is best to have the irons let into the back part of the wood work.

END FINISH OF DROP BACKS FOR SHIFTING-TOP WORK.

The improvement claimed is that in the finish of the ends of the back no stitches show outside, and all appearance of clumsiness is obviated. Sew your cord on the end of the back, either before or after the back is made up, leaving the end roll open, of course, so that you can blind-sew the back cover. Then proceed to sew in your cloth, and your back is neatly finished.

NEW BOX LOOP.

Cut out a pattern of wood on which to form your loops; make it, say 2½ inches long and ⅝ of an inch wide, ¼ of an inch thick, tapered down to ⅛ of an inch at one end. On this pattern you can make the loops either 2 inches, or 1¾ inches long, as you may desire, which will leave half an inch of projecting wood by which you can remove the pattern. Take a piece of trimming leather, say 6 inches long and 3 inches wide, double it, and paste the flesh sides together, and form it over the pattern. When dry they will keep their shape; water will not affect them. Being pliable they give with the stay, and are considered by those who have used them far preferable to the tin loop.

DRAG CUSHION AND FALL, WITH PLAITED WELTING.

In the accompanying cut, Fig. 1, is shown a style of cushion and fall used in a light two-seat Drag.

The material made use of is Turkey morocco, of a dark maroon color; the welting and binding of black patent rail leather. The style of the welting is new, and we think very neat; we will describe the method of making it. We have termed it

PLAITED WELTING

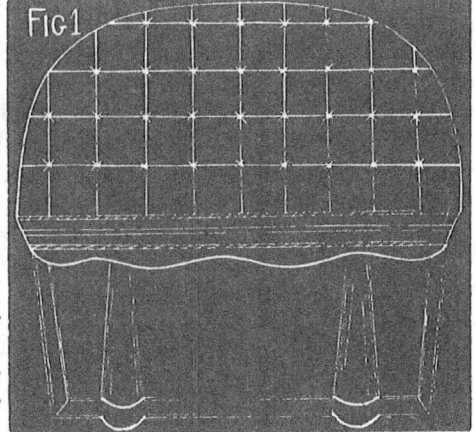

because it is made of either plaited seaming cord or harness leather, over which the patent leather is drawn and creased. The cord may be made of two strands of a tightly-twisted seaming cord forming a braid of two plaits, or it may be made of harness leather. In making the cord of leather we set the gauge at an eighth, and cut off strips enough for the work in hand, after which, with the edge tool, we remove the square edges, then sharpen one end of each of the pieces and draw them through a No. 3 punch. A punch in good condition performs this little operation very nicely. The leather having been rounded by means of the punch, make a two plait; this will give a cord one-quarter of an inch in diameter. The welting leather is then wetted, and drawn tightly over the cord and basted. We now take light tufting twine and set the leather into the crevices between the cords and allow it to remain until it is dry. On removing the twine the leather will remain permanently creased, and have the appearance of a plaited cord.

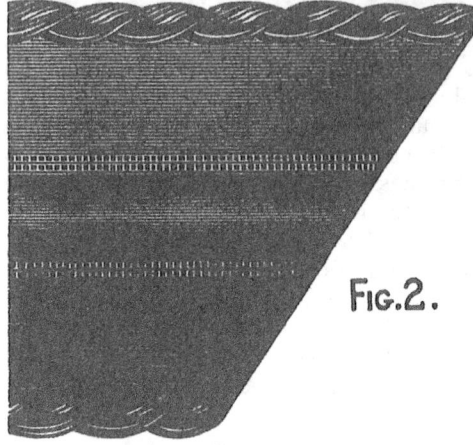

THE CUSHION,

as the drawing plainly indicates, is of the biscuit pattern. The facing or front of cushion is inlaid with a half oval patent leather welt, three-eighths of an inch in width, on each side of which the trimming leather is stitched with either black or orange silk. (See Fig. 2.)

The falls have box plaits, and are also inlaid with patent leather, and bound with turn-over welt on the outside. They have also a scolloped band at the top.

The seats are finished on the edges with "plaited welting." In the above it will be seen there is no effort at display. The trimming of a job in this way consumes considerable time, but when it is completed it is attractive to the eye, and is only a step removed from what might be called a plain job.

IMPROVEMENT FOR TRIMMING TURN-OVER SEATS.

The style of making the cushion like a squab without a border, or of making it with a border, and then nailing it to the seat, always involves a great deal of unnecessary work. In case any of the iron work should be broken, then you have to rip up the whole seat trimming and replace it again; sometimes half a day's work for the trimmer.

Fig. 1.—Bottom of Cushion Frame with the Iron Lip Plate.

The improvement consists in making a wooden frame (or box without top or bottom), to fit the seat, and cut out where the irons of the lazy back or side rail may be; the frame to be 1¾ or 2 inches high, or as high as you may desire the border of the cushion to be. Now have two irons made about 4 inches long, the width of the edge of the frame, with a lip on one edge, projecting out ¾ of an inch, a countersunk pole in each lip to take a ¾ or 1 inch screw.

Fig. 1 gives a sectional view of the bottom of cushion frame. The iron lip is indicated at A; and in Fig. 2 we have a sectional view of the cushion when trimmed; the lip plate A being shown as it appears when ready to be fastened to the seat. Then make up your cushion top in whatever style you see fit. Nail it plainly down on to the edge of your frame; then turn the frame over and stuff in moss, and then draw a piece of canvas over the bottom. Now tuft through as in the ordinary cushion. Next lace or welt (seaming) all around on top of the cushion *in card sideward*. Then blind-nail the facings on to the welt or lace; turn them out, and simply nail them on the bottom edge of the frame; then either stitch up the corners or laps, and paste them. Place your cushion in its place and put a screw in each of the projecting lips; this holds it to the seat, and, in case of broken irons, can be removed in a few moments without ripping, and be replaced in the same way. It is hardly necessary for us to say that the facings should be made up to correspond with the facings of the front cushions, both as to stitching, raisers or other ornamentation.

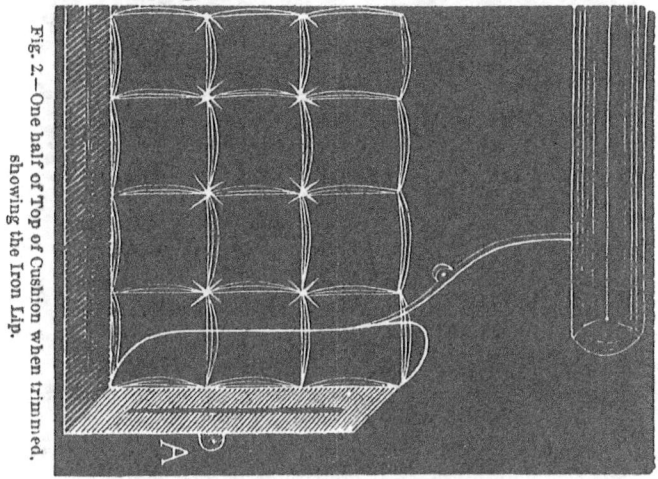

Fig. 2.—One half of Top of Cushion when trimmed, showing the Iron Lip.

BACK, CUSHION AND FALL.

The back is made in rolls and diamonds, as represented in drawing, the top and bottom plates pulled, or flared, outward from the center, the ends and bottom bound with a turned-over welt of patent leather. The back is made up on thick pasteboard, or on five-ply of buckram pasted together, which is cut so as to just meet the top edge of the cushion when completed. The only wood used is the narrow piece usual on light buggies from which the back hangs when finished. The cushion is made up as per

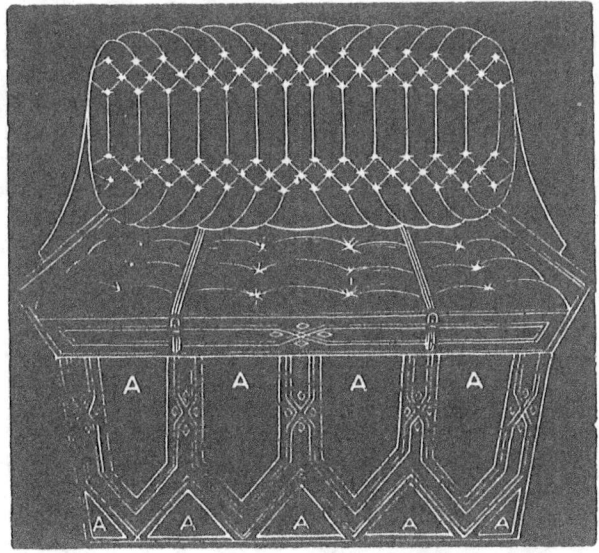

directions for making cushion and fall, with this exception: no patent leather shows on the front, the cloth being merely rubbed in on the raisers as shown; the small diamonds in the center are made of harness leather, and pasted on the same as the other raisers. The fall is made up as per directions for making cushion and fall; the small

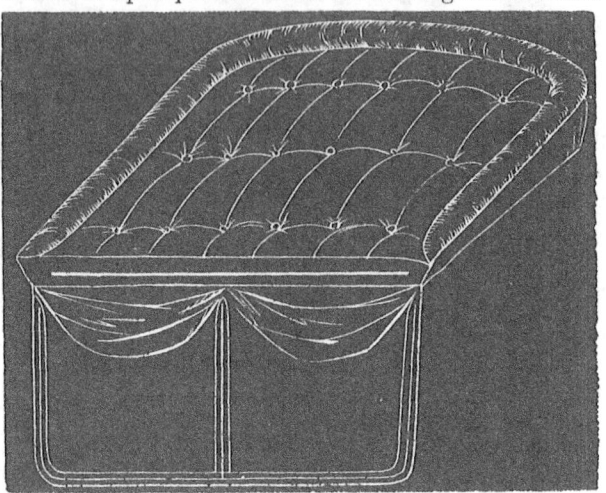

CUSHION AND FALL SEWED TOGETHER.

double lines are raisers made half round or three edged; the spaces marked A are underlaid with carpet. The whole fall shows no patent leather except the binding around the outer edge.

HOW TO MAKE THE FALL.

Take a piece of heavy enameled duck and fit it to the body about to be trimmed. Lay it then on a smooth board; a tack in each corner. Then take a strip of harness or split leather seven-eighths of an inch wide, one-eighth in thickness, and paste it on the duck, one-quarter of an inch from the edge around the two ends and bottom, also on the middle of the fall (see pattern), the top edges finishing to a point, one inch from the top. Next take a strip of harness leather, three-eighths of an inch wide and of a good thickness; split it, or reduce it to a three-edged shape and paste it on the top of the piece (or raiser, as it is called,) already on. If you wish it fancy, you may lay on three-edged strips in any pattern pendant from the top of the fall, but they must be narrower and lighter than the other one. When dry, paste all over with good soft paste, and lay on your cloth, taking care to rub well in on the raisers. When this is dry, stitch along the edge of each raiser and bind the outer edge of a turn-over welt of patent leather, except the top edge. The appearance of this fall can be greatly improved by laying strips of thin welt (patent) leather, three-eighths of an inch wide, edges turned in, to be stitched, on all along the inner edge of the wide raisers, on top of the cloth of course. This makes a very stylish and much used fall for light work; it is cheap, easily made, and looks fine.

HOW TO MAKE THE CUSHION.

Take a piece of heavy duck, enameled, and lay it on the seat; mark all around with a pencil, then cut it one-quarter of an inch larger all the way round. Next take a piece of average split leather, one and three-quarter inches in width, fit it to the *pencil* mark in front of your cushion bottom fitting to the flare of your seat; also, the back of the cushion is made usually about two and a half inches high. Then tack a strip of muslin on a board, paste the cushion front on that, then paste a three-edged raiser on it; when dry, cut the ends of the raisers to a nice point (see drawing), and paste a piece of thin collar (patent) leather on it, large enough to project half an inch round. When dry, take a piece of cloth, one inch wider than the collar leather, and as long as it. Next, measure your three-edged raiser, and cut, or tear, the cloth along the center, one inch shorter than the raiser. Sew this cut neatly along the edge of the raiser, wrong side of the cloth up, and in such manner that when finished and turned over the edge of the cloth will show turned in all around the raiser. Next take a piece of thin twine or cord, and lay it under the cloth around the raiser; paste and turn the cloth down over the twine on the collar leather; rub it well up against the twine, thus forming a cloth facing, with patent leather center and raised welt all around, making one of the most durable and neat cushion fronts ever got up, and one very much in vogue here. When dry, stitch all around the raised welt, outer edge.

The back and side facings are generally made of split leather, or stiffened buckram, two and a half inches high at the back and reduced along the sides to fit the front, and covered with cloth and lined with muslin, leaving cloth and muslin enough over the edge to sew it. The welts use are collar (patent) leather, cut one inch wide. Now stitch up your corners, sew your facing to the bottom until you come to the front corner; here take your fall and sew it in firmly three or four inches, then also sew or stitch in a short strap, three inches long and one and a quarter inches wide, of harness leather, in such a manner that when the cushion is complete it will be on the front of the bottom and under the fall. It is for the purpose of fastening the cushion to the seat in place of cushion straps. For a plain top cushion, cut the top two and a half inches longer and two inches wider than the facings; the top lined with good muslin, the corners well cut off or rounded so as to leave very little wrinkle in the corners. All cushions to be made up with a false top. Stuff the bottom with moss or cut sponge; the upper compartment or top with hair very lightly. The cushion represented has a wrinkled roll, two and one-eighth inches wide round the back and sides; none on the front. This top is made on a frame; very little fullness to the top proper, but to the roll, as follows: one-half as long again as the surface to be covered by the roll, and in the width two and a quarter inches fullness. Three rows of buttons, six in each row, the first row three and a half inches from the end and three inches from the front of the cushion. Be careful to lay off your buttons correctly.

DRIVING CUSHION FOR DOG CARTS, AND FOUR-PASSENGER JOBS.

The peculiarity of this cushion is, that it has a smooth roll (such as used to be on

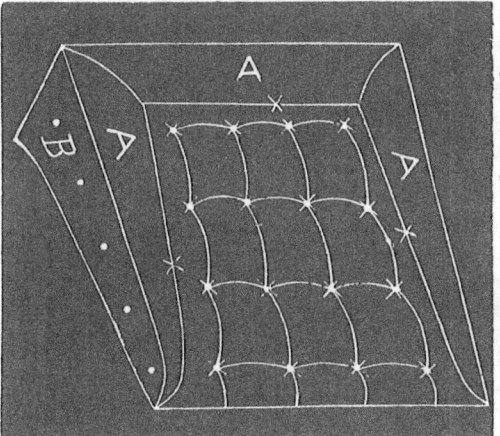

stick-seat rails) around the sides and back of the cushion; of course we mean around the top of it; the roll to be about three inches wide at the back, and gradually reduced in front to one and a half inches, two and a quarter inches fullness in the roll at the back, and reduced in proportion as it reaches the front. Make the back facing four and a quarter inches high, and reduce the side facing to one and a quarter inches in front. The roll is sewed in with a patent leather welt to the false top, and the back corners are fitted and stitched together with a light welt.

TO MAKE THE CUSHION.

Cut the bottom to the size of the driving box or cushion it is to be placed on top of, then mark off the button (tuft) holes in this manner, two inches from the front edge and two inches from the back edge, provided the sides do not flare more than two inches; if they do, you must graduate the distance four inches from the inside edge, two and a half inches from the outer, provided the cushion does not flare out more than one and a half inches; if it does, you must graduate the distance to be marked. Now fit the facings, and sew in the bottom, then fit a piece of buckram or enameled duck to the top edge of the facings; be exact; then mark the size of the roll on this, as directed, and also mark the button-tuft holes to correspond exactly with those in the top and bottom of cushion, and punch a hole large enough for the stem of the button to sink into when tufted. Next there should be four rows of buttons in the top and bottom. Now cut your material for the top proper, and allow, if of cloth, only a full one-eighth of fullness between each button in the squares; if of leather, a very little more fullness. Lay hair lightly on the duck top, and sew the material to it along the mark for the edge of the roll; then take a stitch through the material at the hole in the duck. Next sew the roll (which is already sewed to a patent leather welt), to its place on the duck or buckram top. Now sew in your top as in a common cushion, only leaving the outer edge of the roll loose; turn, and stuff *solid*, and tuft (button) the top to the bottom, and with a twine run a button into each of the holes in the facing, the twine to pass as a basting stitch up through the duck top; then pull tight, thus fastening the sides firmly to the top. Next stuff, and blind-sew the roll to the outer edge of cushion. Two straps about two and a half inches long should be firmly stitched to the cushion bottom for the cushion strap to pass through.

The lines marked X represent the inner welt of roll. A the roll, and B the facing.

THE STAR-TOP CUSHION AND FALL.

WITH WRINKLED ROLL ALL AROUND, FALL SEWED TO THE CUSHION, AND PATENT LEATHER WELT ALL AROUND THE CUSHION STAR.

The top of the cushion is to be made on a frame. The patter for the full size of the top is first marked on the frame covered; next mark the inside line for the wrinkled roll; next mark the star, being careful to have the double points at the back and front of the cushion (as in the drawing). Now to make up the star we always make it in two pieces, allowing one and a quarter inches fullness across the narrow part, and five-eighths of an inch fullness on the lower part. Join the two pieces together; line with muslin; iron down the line marks; put one button in the center; baste to the frame and stuff. Now sew patent leather welt along the outer line of the star, the raw edge of welt outward. Next take the pattern of star, lay it on your cloth, and cut the cloth three-quarters of an inch larger all around the star, thus giving fullness for the top outside.

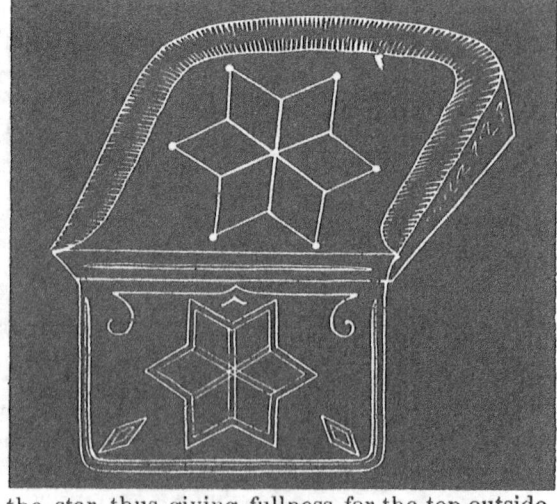

Now line with muslin, and sew down on top of patent leather welt already sewed to the star. Lay hair lightly over space to be covered, and sew this cloth to inside line of the wrinkled roll. Next sew patent leather welt around the three sides of cushion top, designated as the inside line of the wrinkled row. Now sew your roll to this welt, stuff and baste to outside edge.

The lines on the fall are all made of harness leather, raisers pasted on, the cloth pasted over and neatly stitched to the lines, the outside edge bound with turn-over welt of patent leather, or the edge of star may be bound with leather, thus increasing its attractiveness.

TRIMMING FOR LIGHT ROAD WAGONS.

The trimming of this style is cloth, the cushion top made in squares; the front facing (cushion) is two and a quarter inches high when finished; the center of the front along its whole length is made with cloth strips (three-eighths of an inch wide), into crossed diamonds; the edges of these strips stitched with white silk; this is stitched on patent leather, the width of front facing; the edges of the facing stitched with white silk through the facing, thus showing a smooth patent leather edging, to a narrow strip of cross diamond work seven-eighths of an inch wide.

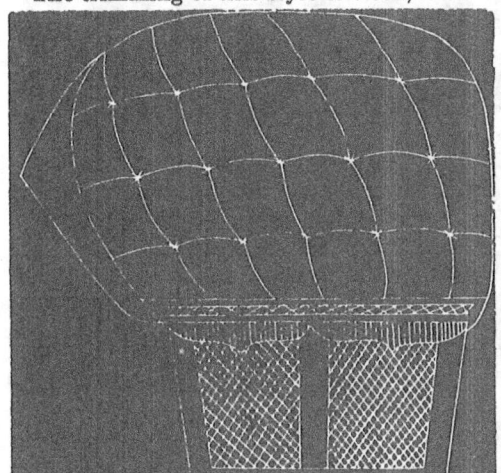

The fall is made to match; it has a border of patent leather one and three-quarter inches wide, around the three sides, and in center the spaces filled up with narrow cross diamond of cloth; their edges also stiched with white silk. The fall is also nailed to the seat, the top of it finished with a narrow scolloped valence of cloth bound with turn-over welt of patent leather stitched with white silk.

This style is plainer. It is done with purple (or wine color) cloth; the cushion top wrinkled in very evenly, and well done; the cushion front one and five-eighths inches high, a plain strip of rubbed-down patent leather welt, three-eighths of an inch wide, stitched on to the center of the front. The fall is simply a plain piece of enameled duck, cut to the size, covered with cloth (pasted on); then a strip of the three-eighth inch rubbed-down, welt pasted on one and three-quarter inches from the edge; then two other strips of the same

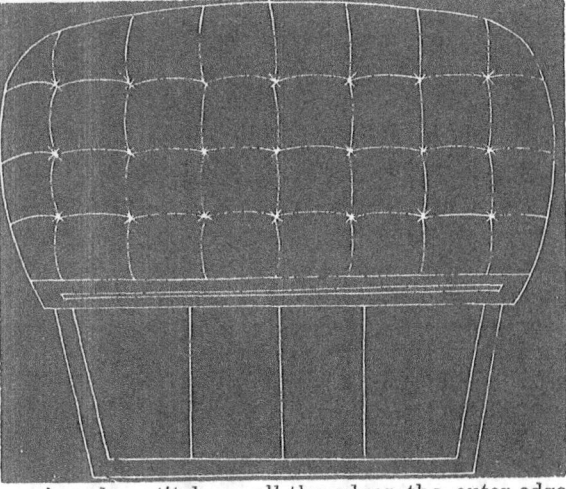

dividing the fall into three panels; when dry, stitch on all the edges, the outer edge bound with turn-over welt. The fall is sewed to the cushion, and a small strap sewed in with the bottom, in the center of the front, to hold the cushion to the seat.

RUSTIC TRIMMING.

FALL.

Cut the buckram to fit the seat, allowing for the rug and oil carpet at the bottom, and roll stick at the top. Paste your corduroy on smooth, allowing enough to turn over at the ends and bottom for a row of stitching. Get the smallest rattan that you can obtain, and flatten it on one side; then cut it off in lengths to suit the fall, leaving a small space between the ends of each piece, as shown at No. 3; scive the ends down to a sharp edge, and cover it neatly with mole-skin canvas, and let the seam come on the flattened surface. You cannot conveniently use bow leather to cover this rattan, as it is too thick and apt to tear out. In sewing the canvas on the rattan, use a small, round needle, and rub the seam down smooth; fasten this rattan on the fall, about 1½ inches from the edge by the means of narrow, covered tin loops, with short raisers on them creased down elegantly. In placing the rattan on, you must mark a line with your French chalk as your guide, and place it directly over it, *seam down*. You must fasten these tin loops by punching holes through the corduroy and buckram, and pushing them 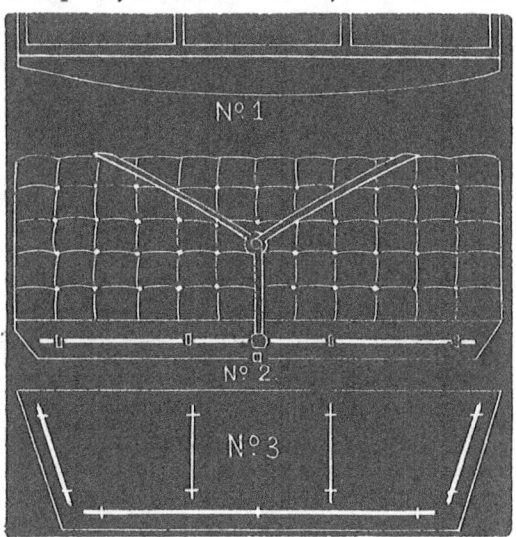 through and bending them down on the under side, as in harness making. Place something flat on the loop, close up to the rattan, giving a slight tap with the hammer, which will secure them properly; then paste a piece of muslin over them; afterward paste your canvas on the bottom; bring the corduroy over the edge and paste it down. When dry, stitch a row at the *bottom* with the machine, and *whip* it on at the *ends*. This is obvious, as the cord or ridges in the corduroy run down; therefore you can stitch straight across the bottom, while you cannot on the ends. In fastening the rattan on the fall, it will require eleven loops, provided you have two center pieces; if but one, then nine will answer.

CUSHION.

The front facing is to be made on the same principle, the rattan to be fastened in the middle of the facing, with four of these covered tin loops, as shown at No. 2; the outside ones to be near the end of the rattan; facing to be 1½ inches wide after the cushion is made up. In making, *sew the bottom in all around first*, then sew in the *front edge of the top;* turn the cushion, and blind-sew in the ends and back, leaving a place open through which to stuff the top and bottom; by this means you keep from bending the rattan; use a bow leather welt, and puff the top in small squares. Cushion strap must be made in the shape of the letter V. A ring sewed in where it forks off about the center of the cushion, and a buckle in front (see No. 2); the forked ends to pass behind the cushion, and nail to the seat. If the job is silver-plated, use a silver ring for the center of the strap, and the billet end to pass through a five-eighth *perforated* silver harness buckle in front. The strap to be made of thin harness leather, covered with corduroy, and bound neatly with thin bow leather.

ROLL STICK.

Before nailing on the stick (which must not be more than three-fourths of an inch wide), take a strip of bow leather; wet; double it and slick down; nail it to the bottom edge of the seat, along the edge of the buckram in the fall; then tack on a strip of corduroy, and nail the stick on over the edge of it; draw your corduroy over the stick, and nail it to the seat. By this means you show a leather welt at the bottom of the roll stick that adds greatly to the appearance of the fall.

DASH.

No flap: cover on both sides; inside with dash leather, outside with bow leather; stitch close to the bars, and draw your bottom thread a little the tightest; by this means you will have the thick and thin leather to meet about equal distances from the surfaces. Don't bind; trim the edges close; put on your dye; afterward ink; polish until it shines. Cut your hand-holds in the shape of the letter D; the valance to be raised with leather, as shown at No. 1.

RUG.—Bind around neatly with bow leather.

APRON.

Make in the ordinary manner with bonnet, with this improvement: Just over the whip socket paste a piece of leather to the apron; take your compasses and mark off a round hole for a whip to pass through the apron into the whip socket, stitch around and cut it out, and when you go to sew your bonnet to the apron, sew in a small patent leather flap to fall over this hole when the whip is taken out. If you do not understand how these loops are put in, any harness maker will show how it is done on harness. A buggy trimmed in this style makes a tasty appearance.

FALL PATTERN.

To make a first-class job, it is necessary to commence with a good foundation, and one that will not make the fall too stiff, but leave it soft and pliable. A fall should be made so that it will fold or bend up and not crease or break in turning up. Duck oil cloth with linen pasted on it makes a nice and pliable job. When it is pasted out and dry, fit it in the body. We give herewith a design of the fall, not claiming it as new, for it is not, but to show a neat style of cording. It has a border around it and two panels. The cording is cut with a knife made for that purpose. The first row of straight cord is one-half inch from outer edge, the second straight cord two inches above; that will leave sufficient room for the serpentine cord to go between, and also for the foot of the stitching machine to pass around.

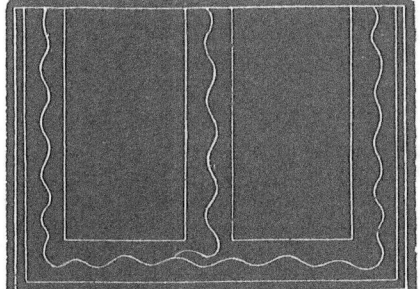

We also give a drawing of the knife that cuts this cording; any good smith can make one; a great deal of time will be saved by its use, and no stock wasted, for the pieces we would call waste pieces make a very nice cording. After the cording is pasted on and

dry, with the edging tool take off the sharp edges to prevent them from cutting the cloth.

In pasting the cloth in the fall, spread the paste on the cloth; it will crease down so much better than spreading it on the fall. After it is dry, stitch all the rows but the outer one; trim off the edges and line the back with muslin oil cloth, and bind the outer edge with turn-over welt, and stitch it. Also before tacking in the fall in the seat, get a stick made $1\frac{1}{2}$ inches wide, and one-fourth inch thick, beveled on one side; slip it between the lining and the fall, and drive several $2\frac{1}{2}$ oz. tacks through the lining of the fall into the stick, which will keep it firm, then the fall is ready to be tacked on lining of seat. Some shops paint the seat, but we don't think they look as neat and comfortable as a trimmed seat. To line the seat properly, cut the back piece with the nap running down; the side pieces, nap running to the front; fit them in and join them in the corners nicely with a welt. Don't line your cloth with muslin, but line the seat separately with muslin, padding the seat lightly with hair or cotton. Our reason for not lining the cloth but putting it on separate is, if you blind-sew in the cloth it will make a neater finish. In trimming shifting-top seats where the irons are set in level with the seat, you can finish them in different ways from blind-sewing them in. Paste out a flat welt; crease it; and when dry, stitch one edge down and loosen the other edge from the paste and blind-tack it on. Be careful and draw it tight, then turn it over and paste it down; drive a few gimp tacks in it, then take some leather varnish on a cushion awl, and drop sufficient on the tacks to cover the heads, and they will not be noticed. For a change it looks well and makes a neat finish.

KNIFE FOR CUTTING CORD.

In the annexed illustration is shown at A a section of knife or punch, natural size, viewed standing erect. This view gives the exact size in width and depth of shoulders, and the size of shank or handle just above the head or cutter. The shank to be tapered up, and made of any convenient length. At B is shown the precise shape of the cutting edge, and represents the impression the edge would leave when pressed gently on a piece of leather. It will be seen at a glance that in the use of the knife it is only necessary to move it forward after the first cut, placing the curved end in position to continue the sweep. In cutting serpentine cord it saves much time and labor, and therefore will be a valuable addition to the trimmer's "kit" of tools.

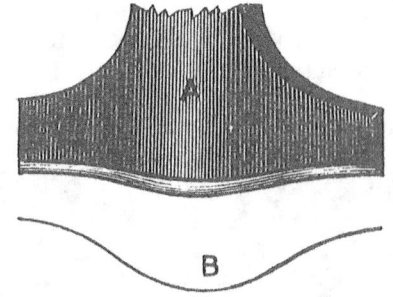

CUSHION HOOK.

The drawing is for full size, and will suit almost any cushion, high or low. They can be made either of steel or iron. To fasten the hook insert the hook into the long slot A, and a strap into the other slot, and fasten to bench, and you will be well paid for your time of making them.

TO PREVENT THE FRONT AND BACK BOWS FROM SPRINGING.

It has been the rule in most shops to put a piece of wood on the edge or under side of the front and back bow to keep it from springing, yet, with all this, the bows will settle when the back stays shrink from use; besides, they make rather a clumsy appearance. The plan I pursue is very simple—namely: before setting your top, take the bows you want for your back and front, and get the blacksmith to fit you a piece of band iron ¾ wide (or as wide as the bow will allow) on the inside edge of the bows, allowing it to run down as far as will allow without interfering with the leather, when covering, and after fitted. Let the blacksmith drill screw holes about 4 inches apart, so that when the finisher sets the top he can screw the irons on. If trimmers will try this they will find that they can get their back stays and back curtain to set better. If the bow is out of square, by fitting the iron to one end it will be very easy to have both sides alike.

FALL FOR DICKEY SEAT OR CLOSE-TOP BUGGY.

The fall can be made up in the usual manner, on enameled duck. The lines around the Gothic are of patent leather, cut three-fourths of an inch wide, and rubbed down edge to edge, thus leaving it three-eighths wide, pasted with thick paste, then rubbed on the wrong side with sandpaper. The other (wide) lines, harness leather strips, seven-eighths of an inch wide, and of a fair thickness, pasted on in the style represented in sketch, and on top of them paste three-eighths three-edged raisers of harness leather. When the harness leather raisers are dried on the duck, paste your cloth (or leather) on top, and crease well into the edges of raisers; then take the patent leather, already prepared, and paste it to the cloth, but close to the edges of the wide raisers, as in sketch. When dry, stich both edges of patent leather and both edges of three-cornered raiser, and bind the outside edge of the fall with a turn-over welt of patent leather.

The top edge of fall, when nailed to the seat, is finished with a piece of strip leather covered with the cloth or leather, same as fall, and bound all round with turn-over welt. This makes a very rich and durable fall.

THE POCKET FALL WITH ZIG-ZAG RAISERS.

This, for doctor's carriages and for the inside back fall of phaetons, looks well, and its convenience must be apparent.

Directions.—The border is of split leather two inches wide, on which is laid the zig-zag raisers, as shown; they are mitered together at the corners and pasted on. Then paste on the cloth or leather; when dry, stitch along the edges of the raisers; next stitch a turn-over welt on both edges, but do not stitch the finishing row of welt yet. The body of the fall is made of enameled cloth, the material pasted on. The pockets are made in this way: the front 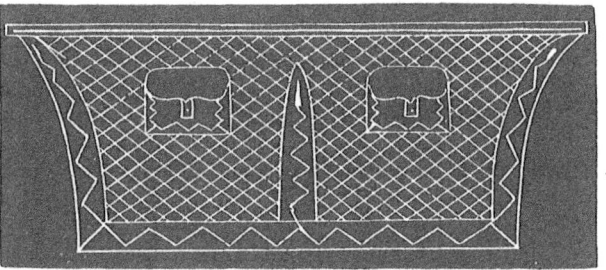 and flaps of split leather covered with the trimming material, and bound with collar leather; the flap fastened to the front with strap and button; the sides of the pocket to be of enameled leather, so they will be pliable enough to fold up when the pocket is closed; size of pocket, eight by seven inches. Now paste your border on body of fall, and stitch the finishing rows of welt.

SPRING CUSHION.

To supply a solid spring cushion, use no wood, nor iron frame, with which to fasten the springs, but yet the springs are required to mutually support each other. To do this, the springs are made square on top and bottom; that is to say, the springs are the same as those now in use, except that the ends of the top and bottom are made square. The square ends are bound together with iron or copper wire, until the whole width and length is filled up, which the seat requires. This will make a mass of springs, which is then sewed to rough linen. The cushion can be made of cloth, corduroy, or leather, in either square or fluted patterns, as may suit the taste of the trimmer, and the spring frame then be inserted, the front and ends properly filled up, and sewed up at the back.

These springs are peculiarly adapted to carriages, where the seat frame is angular.

SQUAB-TOP CUSHION.

The most certain and perfect way is to make the cushion top on a frame covered with canvas or coarse muslin.

1st. Fit and sew your facings to the cushion bottom, then cut a paper pattern for the size of the top; lay this pattern off in squares of about three and a half inches, beginning at the center; of course, you will see that the end rows of buttons will not be too far out to tuft into the bottom. Notice the same for the back row. The front row of buttons should be about three inches from the front; five-eighths of an inch fullness in the squares will do very well.

If the goods you use should be cloth, do not fail to line it, pasting at each tuft hole, then iron (press) down the lines of squares; allow one and three quarters of an inch fullness from the front row of tufts to the front, two and a fourth inches from the back row to back, and two and a fourth inches from the end rows to the outside.

To make up the squab, mark through the paper pattern all the tuft holes, and also the shape; then paste slightly and lay on a thin coat of hair; then lay on the cloth and take a stich with needle and thread, and each hole of the middle rows; then form up this row of squares. Next, set down the back and front rows of tufts and form these squares; then sew down and finish the edges. Now, sew the front of the squab to the facing; then turn your cushion and blind-sew the balance of the squab, leaving the usual place to stuff; stuff the bottom, then lay the paper pattern on the bottom of the cushion, and mark the tuft holes. Now, tuft as usual.

Another way is to lay off a buckram shape of the top in squares; mark out the cloth and baste it all around, plaiting in the edges; then sew into the facing as with the ordidary cushion top, and also stuff and tuft, as in the case of a common cushion.

CUSHION TOP WITH THE FOLDS.

First make a paper, or buckram pattern, the size and shape required for the cushion top; then mark off the squares on it; next, stretch muslin or canvas on a frame, a trifle larger than the pattern; now lay on the pattern and mark its size on the covered frame; also punch all the holes through; also carry the lines of the holes to the edge of the pattern as a guide to plait down the cloth by; then lay off the cloth, allowing one inch fullness between the squares (if the material is silk, less fullness will be required, as silk goods, such as reps, cotelaine, etc., assume the required shape much easier than cloth), and one and a half inches from the front row of tufts to the outer front edge of top; then on the line of the holes cut inward a scant quarter of an inch, thus giving the cloth between the line of holes a slight sweep; now line the cloth with muslin, or, if the cloth be thin, or if the material be silk, lay a sheet of cotton batting on the goods, basting all around the edge, and pasting at each hole; now lay paste very lightly all over the pattern on the frame; then put a light coat of hair on the pasted frame; next take a strong stitch at each of the *middle* row of tuft-holes, then at the next row, but do not pull them tight, or fasten them yet; now stuff to shape, *very lightly*, with the fingers. Proceed in this way until the squares are all shaped, then tie down, and finish the edge. When done, turn over the frame and paste the back or wrong side pretty thickly. When dry, cut from the frame and sew in as in the ordinary wrinkled-top cushion. Be careful to stuff the top very light, and for several reasons, one of which is, that when the cushion is turned and the under part firmly stuffed, the squab top is forced upward, and you cannot press it down again as you might in the ordinary cushion, by sitting on it or trampling it; then if you have stuffed the squab hard, when you come to tuft down you will have the cushion so hard that the sitter will soon be compelled to call in the aid of a chiropidist to cure his corns.

CUSHION FOR NO-TOP BUGGY.

First, the bottom is marked out as is usual to the size of the seat, and a quarter of an inch allowed all around for the seam, then the front facing is cut out and fitted, but should not be more than one and a half inches high, and made in this way.

HOW TO MAKE THE FRONT FACING.

A piece of split leather is fitted to the pencil marks at the bottom, and flared to fit the seat. A piece of muslin is then stretched on a board, and the leather pasted on to it, patent side to the muslin; next, three very fine raisers, three-sixteenths of an inch wide, are pasted on, or else one three-edged raiser. When dry the raisers are covered with a thin welt leather, which must be carefully creased in the raisers; these raisers can be put on correctly by having a piece of hard wood made just the thickness you desire the raisers should be apart. Now take a piece of sand-paper and scratch the welt leather between the raisers so the paste will stick. Next take a piece of the cloth you are trimming with, wide enough to cover the whole facing; past it lightly and then with a sharp knife cut a slit along it for each raiser, only this cut must be one inch shorter than the raiser, to allow for the stretch of the cloth; pull and tack down both ends; force the cloth in between the raisers; then lay the piece of stick before spoken of on the cloth as it now is between the raisers (which show as a leather welt), and strike the stick with the hammer, so as to set the cloth down to its place. When dry, stitch, and you have a front facing with three very small leather welts showing between rows of cloth. It looks well.

The back and side facing should be in one, or without a welt in the corners, and must be fitted to the full size of the seat, the top of the facing to be even with the top edge of the seat; line the japanned side with muslin pasted on. When dry, mark with pencil two inches from the upper edge all around: then take a strip or strips of cloth, three and a quarter inches wide, wrinkle it to the pencil mark (one inch fullness to three inches of length), for the purpose of forming a wrinkled roll on the top of and projecting clear above the cushion top proper. Next sew a welt around the edge of it to the pencil marks; now join your front facing to this one.

The top should be made in squares on a frame, and must fit *only* from the welt *below* the wrinkled roll to the front, if made. Now sew it to the front facing, then *turn over* the cushion and sew in the top to the welt, which is already sewed on below the roll Next sew the top edge of the wrinkled roll, and stuff lightly as you progress; when done, have a piece of cloth ready the size and sweep of the back and side facings, and large enough to cover them. Sew this cloth and also a welt to the top of roll (now the top edge of cushion), and when done turn it over so as to cover the back and side facings; you may either pull it tight or paste it down. Now sew the welt for the cushion bottom to the facing, and blind-sew in both sides of the bottom, leaving the back of it open to stuff from. Now stuff the cushion from the bottom, and tuft to suit the top.

The fall is sewed to the cushion, and two small straps are sewed in with it to fasten the cushion to the seat, while the cushion straps are *mock ones*, sewed in under the roll, and the buckle sewed in with the fall, so as to show against the front of the cushion; these cushion straps are made of light split leather, covered with cloth, and bound with collar leather on the edges. This is a very handsome, stylish, but difficult cushion to make at first; but if we have succeeded in making the *modus operandi* clear enough for practical use, you will have a cushion with a wrinkled roll standing from the cushion proper, and filling the seat so completely as to make it appear as if lined, while the mock cushion straps, and the fact that the cushion does not have to be turned inside out to stuff it, makes the most perfect and beautiful finish we have ever seen.

PATTERNS FOR STICK-SEAT ROLLS.

How to cut the miter of a roll for front and back corner of stick seat, and also a rule by which to cut the leather for a roll on a round (front) corner stick seat.

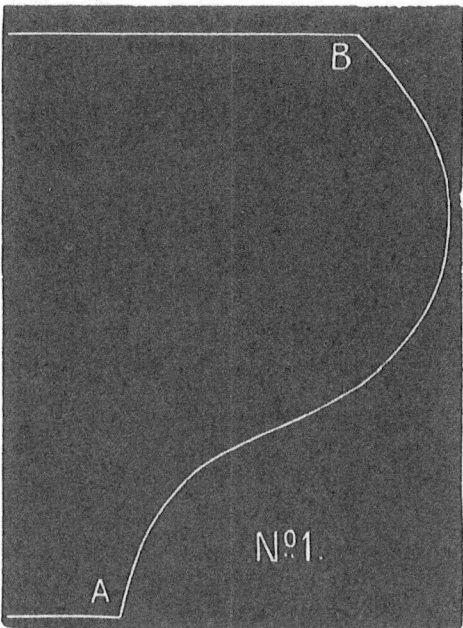

No. 1 is for the miter of a front corner. Presume you wish to put on a three-inch roll. Lay the pattern to the rail, and if the point B touches the top corner, and the point A the lower joint of the rail, *go ahead;* if not, cut a new pattern by this, but altered to reach the upper and lower points of joint of the rail.

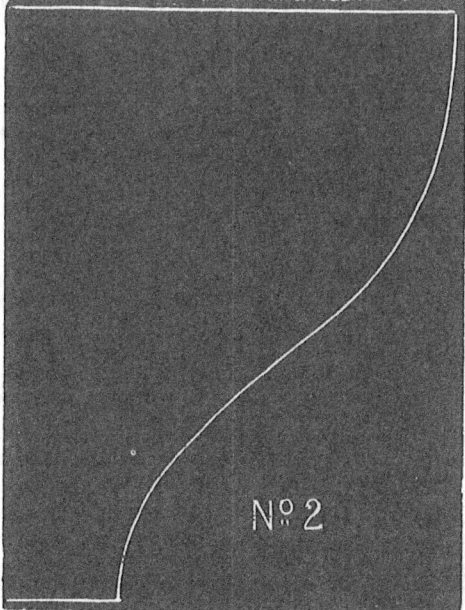

No. 2 is for the back corner, a three-inch roll also; the same directions apply as in No. 1. When properly fitted, baste a narrow welt to one side of each miter, then lay the other side to it, and stitch with two noodles.

No. 3 is a sketch of a round front corner. Say you wish a three-inch roll. Cut a paper pattern of the sweep or bend of the rail, then cut the leather one and three-fourth

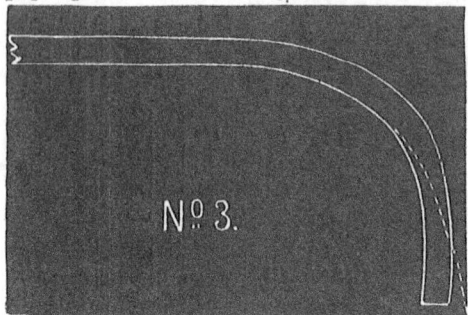

inches, *straighter, or with less bend* than the rail; the object gained by this is, that if you tried to put on a roll with leather, cut the exact sweep of rail, you would have too much fullness, or too many wrinkles on top of the bend. By cutting in straighter you draw all that surplus leather down, and you can thus get a smooth roll on. You must remember one thing, that the putting on of this kind of roll is a little difficult for a "novice."

COVERING DASHES.

We seldom see a dash with flap on it now-a-days, all being covered double; and in shops where several trimmers are employed, they seldom have to cover them, and it is a job that nearly all trimmers like to shirk if they can. But some of us have been less fortunate than others, and for their benefit will give some hints about covering them that may be useful. To make a nice and neat job, collar leather is the best, or, as some call it, grain dash. If you have a side that cuts to advantage, do not cut the dash cover in two pieces, but fold it over at the bottom without cutting, and you have a good finish at the bottom without having any blacking or finishing to do; and dashes seldom have top rails any more. So to prevent the top from soon wearing off, take a strip of harness leather; shave down one edge, and when you draw on the cover, draw the inside piece very tight to prevent it from getting full and wrinkling in stitching; insert the harness leather between the pieces with the shaved edge down over the bar, so it will not make it clumsy, and in finishing off the dash, the harness leather finishes or dresses up very nicely. With that kind of dash we usually make hood aprons of heavy gum cloth. Seam them up on the ends with a welt. The hood should drop down from six to ten inches, with two straps which knob under the bottom. Some make them to slip under the dash valance and knob under the front. It is the easiest and quickest way, but not the best, for it lets the apron down on the knees and makes them damp, and is also more unhandy to fasten on if you are in a hurry to keep from getting wet.

TO PREVENT THE TOP OF DASHES FROM BEING WORN THROUGH BY THE REINS.

In dashes without a rail we hear great complaint of the leather getting soon worn off. This can be prevented by inserting a piece of good, hard harness leather between the back and front covers of dashes, to be stitched in with upper, or outer, row of stitching. The harness leather must, to be right, be cut this way: take a strip three-quarters wide and split it in the splitting machine to a wedge shape, the whole length required. This can be done by placing a small wooden wedge in the machine, the full thickness of the leather to be split. Let the leather lap on the wedge, or sharp edge of the wood, then pull through, and it will split as described.

How to put it into the Dash.—Baste it in with the top row, the sharp edge down. When you pare off the dash, do so rounding it, leaving the thick edge of the harness leather exposed to the friction of the reins. If you adopt this plan, there will be no complaint of the dash getting worn through on the top edge.

THE PHŒNIX SHAFT STRAPS.

Prepare a piece of board 44 inches long, by nailing two strips of split leather along it, leaving a channel between them; next cut your strips of harness leather, $\frac{7}{8}$ wide, as usual, and then split off the face of it in the splitting machine; then take this face, which must be very thin, and lay it on the channel prepared; next stretch tightly a small three-edged raiser, or in common jobs, a piece of seaming cord on it; lay on good paste, and then put the other split of your strap on top; rub down hard with slicker, and when nearly dry, take off and stitch at both sides of the raiser; then cut the strap to scant $\frac{3}{4}$ of an inch wide; edge tool and crease, and you have what we consider the very neatest and most durable strap now made. These straps are only used for trace bearers and whippletrees; use rounded key straps with them.

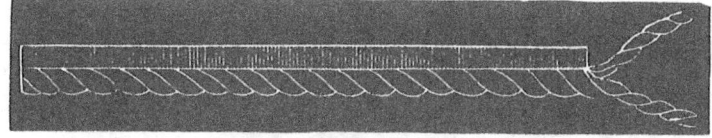

TWISTED LEATHER SEAMING CORD.

Take two pieces of hard seaming cord, and twist together by lapping one over the other; then cut strips of colar leather, $1\frac{1}{4}$ inches wide; wet and paste over the twisted cord, then sew it as in the ordinary seaming welt; next take machine thread, and with a round needle sew a winding stitch over the corded part of the lace, by inserting the needle through the lace on the line of the sewing first mentioned (as in common seaming welt), and bring it around the corded part and insert again, always on the same side. You will have to go over this winding stitch twice, so as to get the thread into all the creases of the cord, which will show through the leather. When sewed, stretch tight on a board, and let it dry. This style of lace is very neat for all kinds of light top and no-top work.

ROUND CROSS STRAPS.

This design for cross straps is very neat and not very difficult to make. You first take two pieces of heavy harness leather of the required length, $1\frac{3}{4}$ inches wide, draw to a thickness, make a groove in the edges for the stitches, lay in two thicknesses of

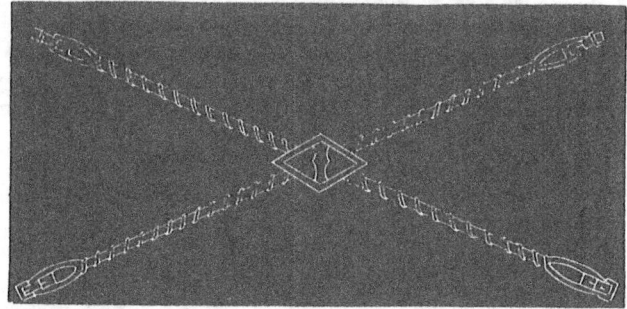

leather and stitch together, leaving the center and ends open for buckles. After rounding the edges, wind a $\frac{3}{8}$ strap of even bridle leather round, like the design, after which paste and stitch a nice piece of wet collar leather around the strap. Then wind heavy tufting twine close around by the cord and let it dry.

CHEAP BACK CROSS STRAPS, FOR A HACK CARRIAGE.

Take two pieces of very thick cane rattan, or rounded pieces of hickory, the required length; slightly flatten both the ends and the centers where they cross. Now, fasten the buckles and straps to the ends of these pieces with clout nails; clinch them through all. Then cover with patent leather, either stitched on or herring-bone sewed; the ends to finish neatly up against the loops of the buckled end straps. For the center, use a piece of ¾ wide rubbed down welt; cross it twice over the center to form a knot. These will be found simple, cheap and strong, and will not slack up like the leather cross straps.

BACK VALANCE AND CORDED ROLL-UP STRAPS FOR LIGHT OR HEAVY WORK.

This makes a very stylish and attractive finish. The back valence is made light, being now reduced to a width of about two and a quarter inches, for roll

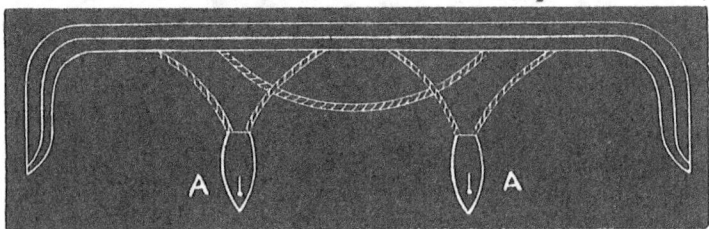

up or close top buggies, a three-edged raiser laid on and stitched about three-quarters of an inch from the lower edge, and the lower edge bound with a turn-over welt of collar leather. The corded roll-up straps are made by twisting two pieces of hard seaming cord together in this manner; stick an awl in the bench, take the pieces of seaming cord, fasten to the awl, then lap them one over the other, hand over hand, as in plaiting, and continue until enough is twisted for your purpose. Then take a piece of light collar leather, wetted and pasted, just wide enough to meet around the twisted cord; sew the edges together over the cord, with a herring bone stitch; now oil it, next take a piece of sewing machine thread (say 3 cord, No. 35), and wind it around the leather, sinking the thread well into the crease formed by the twist of the cord, which will show through the leather when it is all wound; stretch tight on a board and let dry. Then for the knob-hole strap, marked A. They are made of patent leather and collar leather pasted together, cut to this shape and stitched around the edge, and to the covered twisted cord.

Although this sketch is given for light work, this style is used a great deal on heavy top work, such as barouches, victorias, bretts, etc., but with this difference, that for heavy work the cords are put on double, and four straps in place of two, which makes an exceedingly rich and stylish finish.

HEAD LINING IN CLOSE TOP.

Cut the cloth the same as the leather, fit on tightly over the bows in three pieces; if the top has five bows, of course list three of them. Take your French chalk and mark on the edge of the bows, where you sew the listing, and sew it on at the bench; get a small cord, cover it with cloth, and use it for a welt in jamming together your head lining. In tacking in head lining, tack it fast to the middle first, then your other bows afterward, and you will have no trouble in obtaining a nice job.

TO PREVENT HEAD LINING FROM WRINKLING.

To prevent a head lining from wrinkling, tack a strap at the middle of the front bow, draw it tight over the top, and tack it at the middle of the back bow. This is done after the top is set, and before the head lining is put in. It draws the bows together a little in the middle and prevents them from springing or giving, when the muslin and top is drawing on the bows, which would cause the bows to spring a little toward each other and thus loosen the head lining, unless it is drawn very tightly. Of course the strap is taken off again, which do just previous to stitching the top. We also think it expedites the work a little to spring the back bow downward previous to tacking on the back stays.

FITTING QUARTERS.

In fitting on quarters, do not wet or dampen them, yet draw them on tight from front to back, so as to get the fullness out as much as possible, but not tight crosswise; cut them wide enough to turn up to the prop, which should be two and a quarter inches from the bottom edge of the valance; do not sew knob patches in the valance, but drive your knob just below the edge of valance, so they will catch the knob and keep it down; do not place any hair on the bows, for when the top shrinks in a little, while the bow will show, and when you bind the top off on the back part, sand-paper the comb off, and put some paste on the top edge before you tack the binding; then tack the binding on tight, and let it dry before binding it, and you will have a very firm comb.

CLOTH, BROAD LACE,
WITH PATENT LEATHER WELTS AND PATENT LEATHER CHECKERS.

The broad lace is made up in the usual way, except that the cloth is cut through with a half-inch chisel every half-inch of its length, and a strip of patent leather rub-

bed-down welt, a half-inch wide, is drawn through each alternate cut, thus showing a half-inch of patent leather, then half an inch of cloth and so on; it is then stitched with silk, two rows, one along the edge of the leather welt, and the other just outside of the welt, and then a row along the edge of the lace.

INSIDE FRONT VIEW OF ROLL-UP TOP.
STYLE OF LAZY BACK, SHOWING BACK VALANCE AND FINISH FOR THE BOTTOM OF HEAD LINING.

How to make the Trimming for the Lazy Back.—Wrinkled roll all round, and two smooth rolls; wrinkled roll to be one and a quarter inches wide when finished, and the smooth rolls to be five-eighths of an inch when finished. First mark the distance on the wood of the back with compasses and pencil. To make the smooth rolls, take two pieces of cloth the full length of space to be covered, and one inch wider; line with muslin; nail down the upper edge of lower roll to the middle line; then nail the lower edge of the top roll to that, without any welt between; then nail a narrow strip of oil cloth on top of both, to keep the edges well down; next, stuff both rolls lightly with hair; nail down the upper and lower edges in such a manner that you can see the dividing pencil marks when finished; now take a patent leather welt, sewed up, and nail to lower edge of lower roll; cut out the cord at the end of the point; next nail a welt to the top edge of the upper roll; cut the cord out at the junction, and let it lap over the lower welt. To make the wrinkled roll, cut a straight strip of cloth one-half as long again as the upper line of the back, and one and three quarter inches wider; wrinkle, or gather, by running a strong thread through the lower edge; the same for the under wrinkled roll, except to run the thread through the upper edge of the cloth; next, nail with four-once tacks to the welt already laid down; stuff up the lower roll first, then the top; make the top one meet the ends of the lower one in such a manner that, when finished, they shall appear to be in one piece all around; finish the inside edge of the wood with seaming welt all around the outer edge with leather, pasting welt. This makes a very light, stylish, and finished back, showing four rolls, divided by patent leather welts.

SCOLLOPED BACK VALANCE,

AND 'STRAIGHT-EDGED LACE FOR BOTTOM EDGE OF HEAD LINING.

How to make the Valance.—Take a piece of common thick split leather as long as the bows are wide, and wide enough to reach from the top edge of the back bow to the point where the lower edge of the head lining will be; fit it carefully in this position, marking the outer edge of the back bow on the leather; next, cut it five inches from the bow mark; cover the reverse side of leather with cloth; when dry, scollop (see pattern); then bind each scollop with a turn-over welt, cutting the welt short off at at the junction of each scollop; next, take a piece of patent leather, rubbed down to three-eighths of an inch wide, and loop up between the scollops (see pattern); clinch, with two-ounce tacks, the top of loops to the valance; pare the ends, so that no lumps will show through the back curtain. To make the lace for finishing the edge of head lining, take a piece of heavy enameled duck about four inches wide, and as long as the space to be covered; cover with cloth; when dry, cut into two strips (one for each side), each two inches wide; bind the top edge with a turn over welt, and pare off; then bind the lower edges the same way, but leave the binding on the back wide enough to reach the top edge of the lace; do not pare off; next, supposing your head

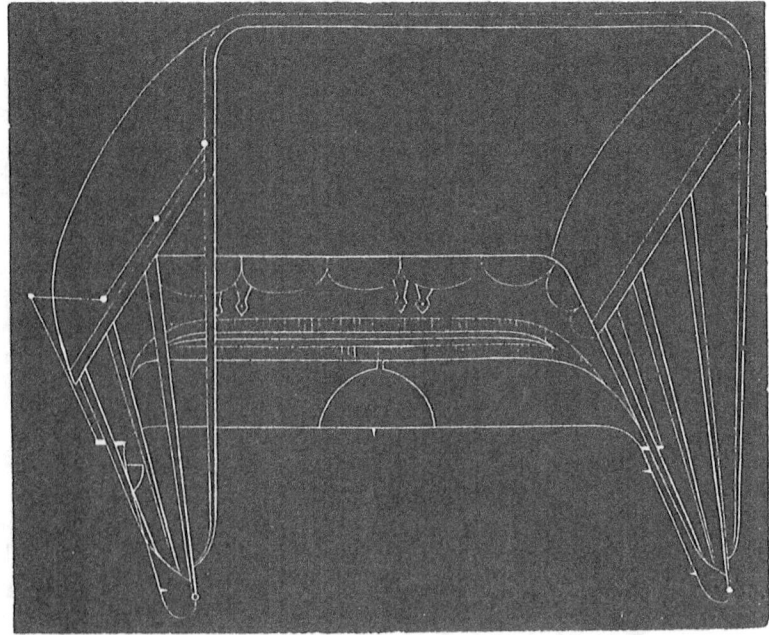

lining in, nail one end of the lace to the back of the back bow: then nail up your back valance, and cover the end of the lace with it; now paste heavily the wrong side of your lace, and also the wrong side of the unpared binding of the bottom edge; nail the front end under the front valance, straining tight; now, draw down the heap lining between the two pasted surfaces; put your hand up between the bows, and rub well together; next nail with small black nails to each bow, driving a nail in the front and back side of each bow, so as to draw the lace in slightly between each bow. This finishes the valance and lace, with the head lining in very neat style. There is no need to sew the head lining to the lace, as the paste holds it sufficiently close without.

BACK OF LIGHT TOP.

The back stays are usually made eight inches wide, the back curtain lapping on them two and a quarter inches. Back stays, made in this manner; one ply of buck ram, with muslin pasted on both sides, and a piece of enameled leather four inches wide pasted across the bottom end. When dry, paste on the leather; back curtain lined across the bottom with a piece of leather, and also at the place where the straps are stitched on. The loops shown on the back, marked thus (A) for the back-stays, are made in the following manner (or can be purchased ready made). Take a piece of harness leather, five-eighths of an inch wide and three and a half inches long shave down both ends, stitch a five-eighth buckle into it, with a few coarse stitches; then stitch the strap with buckle attached to the back stay at the point desired, but the bar of the buckle must touch the line where the back curtain is made to lap to, marked on the sketch with dots.

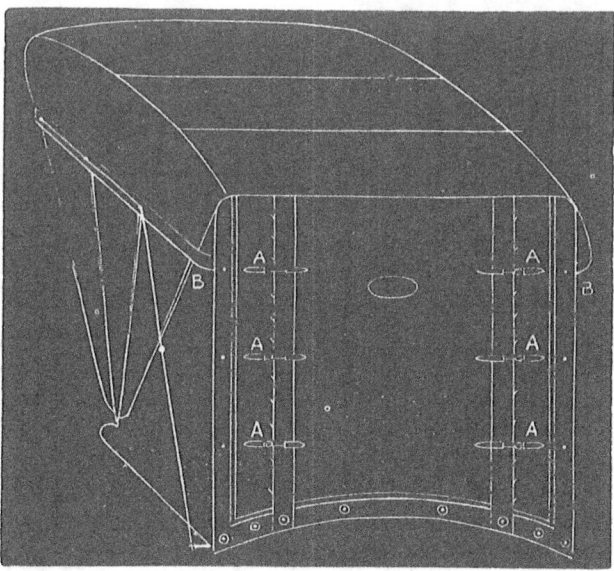

Now take a very thick piece of skirt leather, a trifle less than five-eighths of an inch wide; shape one end, as the loop marked A; let it be four inches long (one of these for each loop on the back stay, none for the curtain). Now fasten these pieces with two tacks through the wrong side of the stay, the pointed end two and a quarter inches from the curtain line (dotted). Now take a piece of light collar leather, lined with enamel, one edge of the collar turned under, and lay it on the shape while wet;

put a tack at each side, and one at the point; crease to the shape, and stitch while wet (usually in the machine). Allow the skirt leather shapes to remain in the loop until they are dry; then full them out, and they will answer for future use.

For the back curtain, you merely cut the curtain strap long enough to be a shape for the loop, and make the loop in the manner above described. The knobs on the back-strap are riveted, one opposite the end of each loop, and one and a half inches from the outer edge of the stay. The stay is also in shifting tops riveted to the rail. The lines around the stays and curtain may be stitched, or raised and stitched.

The cloth lining of the stays from the point B upward must not be trimmed off, let it project, and when stitching the outer edge, do not stitch through the cloth from this point upward, but leave the cloth loose from the leather, so that the side quarter shall pass between the stay and its cloth lining; fasten the quater to stay with a strong pop stitch, and then paste the cloth on the inside of quarter.

FITTING ON TOPS AFTER STUFFING.

Tack two strips of stay webbing across the bend of the bows; then take a piece of black muslin, eight inches wide, double it, and tack one-half just above where your props will come, then tack on white muslin, say twenty inches wide, doubled; with the open part down, and take a thin stuffing wire which will bend to the shape of the bows; put in the hair and stuff to the desired shape, then turn up the bottom edge of your black muslin, and tack it to the bows, which will prevent the white muslin from showing when the top is let down. Now proceed to fit on your top; draw on you side quarters tight (but dry); draw on your crown piece; bring the seam just above the bend; drive a tack in the front and back bows where you desire the seam to come; chalk a string and draw it across the top from tack to tack, and snap; be careful and have it straight. Cut the crown piece first, then cut the quarters by the crown. Before taking the leather off to paste, mark both the crown and quarters with a piece of chalk, each place opposite, so that when you sew them together the marks will be together, and you will have no trouble to make a smooth top.

THE BACK PART OF CLOSE TOPS

can be very much improved in appearance on *round corner* seats or jobs, by reducing the width of the back bow, say three inches less than the other bows, and then fastening a piece of wood, two inches wide, on the back of the bow outside, and as thick as it. It should run well around the bend or corners; then give it a true sweep from the center *outward;* when fitting your top *do not* have the binding for the top edge any farther out than the outer edge back stay; also have a webbing stay from the bow to the bottom center knob of back stay, for the purpose of holding the stay out, rounding when completed; let the back curtain hang very loose, and this method will give a nice round corner to the top, with the sweep continued unbroken all around, and it looks well.

"SUMMER TOPS."

We would state that there are several methods adopted and in use for putting on a "prunella" top; the most perfect, but expensive, is to proceed with it almost the same as if it were to be of leather. Set the top, cover the bows, put in the head lining the same—but, of course, it must be of "prunella;" finish the bottom edges of your head lining the same as in leather tops, and all of the inside. Put on a front valance as usual; stay and stuff and cut and fit the top as usual, with a leather top, but leave a little more goods at the seam. Put a collar-leather welt in the top, and bind the bottom edges of the side quarters with collar leather, the back stays to be pasted out the same as for leather, but use black cambric to cover the buckram on both sides; then paste on the "prunella" as with leather, but use very little paste, and it should be stiff, to prevent its squeezing through the goods. The edges of the back-stays should be bound with collar leather. Line the bottom of the back curtain with a strip of leather, as is usual in other tops, but also line both edges with a strip of leather about two inches wide; then proceed to make the curtain as usual, and line with "prunella," the same as you would line a leather curtain with cloth; stitch the knob holes in the back curtain as usual. The side curtains and just the single ply (not lined), and either bound with black worsted tape, or the edges turned over twice, then stitched; then round patches of enameled leather pasted outside over the holes, and stitched on.

The other way, for a lighter, cheaper job, is to paint or cover the bows as high as usual; then cover all the rest of each of the bows with the "prunella," nailing on top, so that the top will cover the nails; then cover a piece of webbing with "prunella," and nail it on top of the bows, *along the seam line*, so that it will hide the seam when the top is on; also another piece of webbing as a back strainer (covered also), from the middle knob (in the rail) for the back stays, to the top of the back bow.

Now in this case you use no head lining, or lining for back-stays or back curtains, but, of course, you will line the bottom edge of side quarters; then cover that lining with "prunella;" the same for the bottom of back curtain and back stays. The top should have about four inches round. You can finish the front with a valance and the back "bound off," as in the leather tops; or you can nail a welt on front and back bows, and blind-sew; the latter is preferable. Bind the edges of the curtain and stays as in side curtains of the other tops, and make all the holes the same way, except those in the bottom of the back curtain.

Cut the shanks off eight knobs, then cut eight round patches of enameled leather with a No. eight or nine punch; punch a hole in each patch; slip the head of the knob through the hole, and stitch them on the back stays (outside), two of them for the side curtain, and two for the back curtain, in place of buckles and straps; of course buckles and straps may be used.

SETTING "TUBULAR BOWS" BY DRAFT.

First prepare a draft board large enough to represent a full-size top. Draw the horizontal line A near the bottom, which represents the bottom of the seat; then draw B perpendicular to A, three feet and nine inches long, or the height desired for the top, then draw C parallel to A, forty inches for the depth of the top or longer if greater depth is wanted, then mark the points D D as much below the ends of the line C as the top is to round. Now through the points D D and E, the middle of the line C, describe the arc D E D, and space it off to correspond with the number of bows to be used. To find the point F, measure on the body of the buggy to be trimmed the distance from the bottom of the seat to the center of the pivot on which the bow sockets are to fasten. This will be the distance above the line A; then to the distance of the pivot from the back of the seat add three or four inches, or as much as it is desired the top should extend over the back of the seat, and this will be the distance of F from the line B. Now draw the lines H H H H from F to the points marked on the arc D E D, which completes the draft. This draft can be used for other buggies by simply changing the point F as may be necessary.

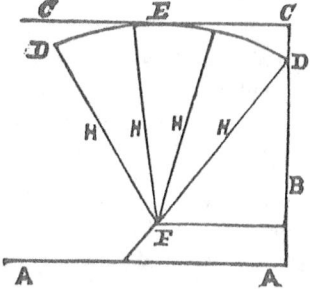

To make the top by the draft, place the bow sockets on the lines H H H H with the eye hole at F ; then place a straight-edge on the top of one of the bows; measur down the bow from the straight-edge on each side a distance equal to the distance from the top of the first bow socket to D on the arc D E D, and mark it; then from this mark measure the length you wish to have the bow run into the socket where the bow is to be cut off. Mark each bow in succession in this way, numbering them for the sockets they correspond with. The bows can run into the sockets from one to two feet, as is most convenient. It is better that the back bow should run below the point where it strikes the prop when the top is down, and the front bow should run below the points for the knobs; a foot is sufficient for the center bows.

After the bows are sawed off the proper length, run a groove along the middle of the inside of the bow, corresponding with the bead on the inside of the bow sockets. Then make the bows a true taper, so that they will go in easily to within half an inch of the marks, and then drive them to the mark.

DRAFT OF AN EXTENSION TOP.

Draw the horizontal line A, which represents the bottom of back seat; then draw line B perpendicular to line A, the height desired for top; draw line C parallel to A, the depth of top; mark D D as much below line C as you wish to have round on top;

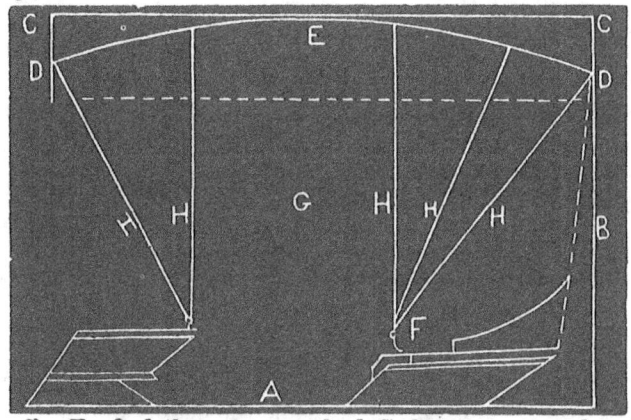

then space it off. To find the space marked G, between second and third bows, measure from pivot on back seat to eyelet hole in iron on front seat; then measure from line A to eyelet-hole K on front seat; mark on board; then draw lines H H H H H from F and K to points marked on arc D E D, and the draft is complete.

TRIMMER'S, DON'T USE STICKS.

A great many workmen still cling to the habit of using sticks for the purpose of keeping their tops stretched while trimming them. In plain English, sticks ought to be "played out." Instead of them, get the blacksmith to fit you two pieces of band iron, three-quarters wide and twenty-seven inches long, with a loop (like a strap loop), at one end of each piece, and one small hole in each of the other ends, and also a sliding loop with a small thumb-screw attached to it; then slip one piece through the stationary loop and the slip loop; screw the ends to your front and back bow, and you can regulate the size of your top without trouble, by simply sliding the irons to a greater or less distance, using the slip loop with thumb-screw to *hold it to its place.*

SOFTENING BUGGY TOPS.

Melt eight ounces of beeswax in an earthen jar, stir in two ounces of ivory or bone black, one ounce Prussian blue in oil, one ounce spirits of turpentine, and one-half ounce of copal varnish; apply with a brush, and polish with an old silk handkerchief.

MIXING AND COOKING PASTE.

First put into the pot the desired quantity of good flour, and add water in small quantities, and keep stirring until you have the mass the thickness of dough. This method prevents the flour remaining in small lumps, for the tenacity of the mass is such as to pull the lumps apart, and distribute the dry flour they contain. Afterward thin down the mass gradually until it is the consistency of cream; then stir in a tablespoonful of pulverized alum. The alum prevents the paste from souring, keeps flies away from it, and adds to its adhesive qualities. Do not cook the paste long after it gathers if you wish it to hold firmly. Your paste made, set it by until it cools.

PASTE THAT WILL KEEP.

Take one tablespoonful of flour; add gradually one pint of cold water; boil slowly, and stir well to prevent burning till it thickens. Keep it boiling till it becomes thin; then add one teaspoonful of nitro-muriatic acid, and boil till it again thickens, when it is ready for use.

HOW TO CLEAN DRAB CLOTH LININGS.

Put one pound of pearl ash in a pailful of hot water, then with a brush wash the lining all over; rub well with the brush after the lining is well wetted. Next put three ounces of oil of vitriol in a half pailful of cold water; taste it; if it bites reduce it with water until it tastes a little sour; wash the cloth with vitriol water while it is wet with the pearl ash water. The cost will be: for one pound pearl ash, 20 cents; three ounces oil vitriol, 7 cents.

STRIPS FOR CORDING.

It is simple and very handy; and if you are a piece worker it will greatly advance you. Say you want an eighth of an inch strip, and use harness leather; cut off an inch strip, and run it through your splitting knife, being careful to hold it square on its edge, and you will be surprised how nicely it is done, and am sure you will be better pleased with the time you have saved.

LEATHER BLACKING.

Three ounces extract logwood, three ounces copperas, one-half ounce bichromate potash; pulverize and add half gallon warm water. An earthen pot will be found the most convenient to keep it in.

BLACK LEATHER VARNISH.

One gallon alcohol, two ounces gum shellac; pulverize, mix and shake it well; keep it warm near the stove, or in the hot sun, until melted, which will take about one day; then add three-fourths pound Venice turpentine, three ounces good lamp-black, and keep it well corked.

BACK SUPPORTER.

Screw the two holders to the bottom of the seat, about eighteen inches apart, placing the perpendicular line A just even with the back edge of the seat. The line B shows the angle and position of the back stay. The lazy back is laid in the notch C, and tightened with a little wedge. This contrivance holds the lazy back at the proper height and distance back of the seat, and also gives it the right cant over at the top, so that all the pressure does not come on the top roll.

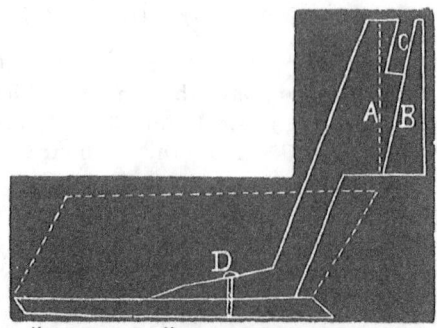

HOW TO MAKE A "BACK BOOT" FOR A WAGON WITHOUT A BACK PANEL.

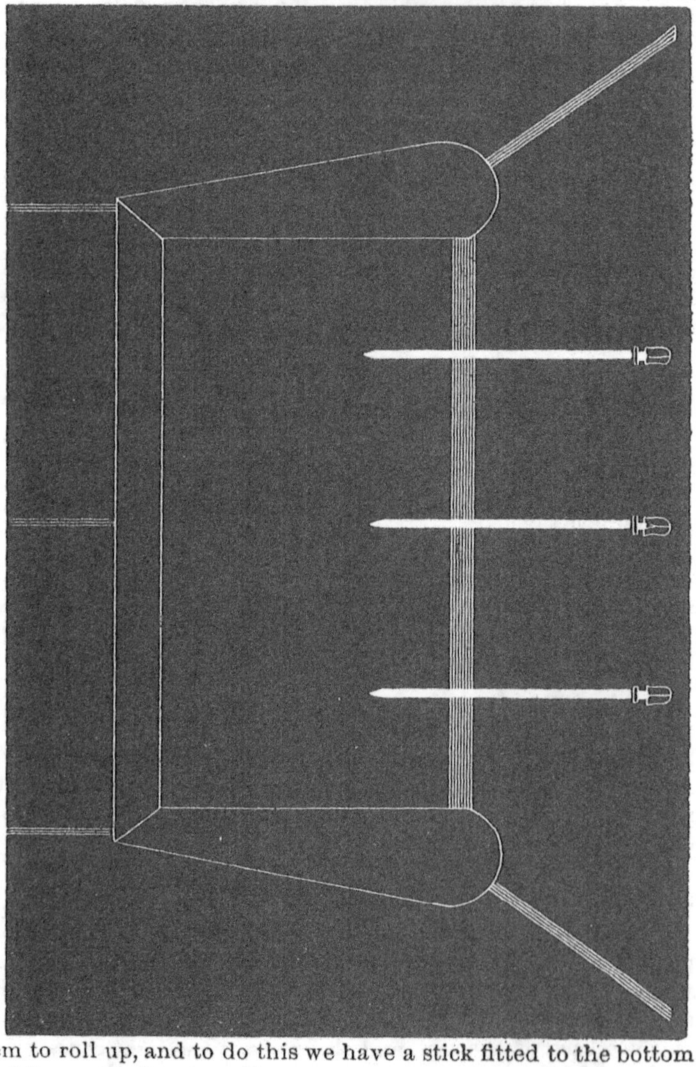

Make them to roll up, and to do this we have a stick fitted to the bottom of the seat, just inside of the seat valance, made to screw up with three screws and three notches cut in the top side of it, the same as a rockaway curtain stick; then fit a piece of enameled leather from this stick back to the edge of the back panel; then a three-inch border of enameled leather across the length of the panel, to be basted to the large piece covering the open space where the back panel would be; then a piece for side border to carry the line of the back border all around; now to the front corner of this side border stitch a strap long enough to knob on to a screw under the bottom of side panel, and three straps to the back border long enough to knob on to a screw under the *back panel;* then nail three buckled straps, five-eighths wide and thirteen inches long, on to the notches of the stick, same as rockaway roll-up straps; the borders are joined in the corners, and bound on to the main piece. Where the round corner of the side border is, it should have a piece of leather underneath as a lining to prevent its tearing.

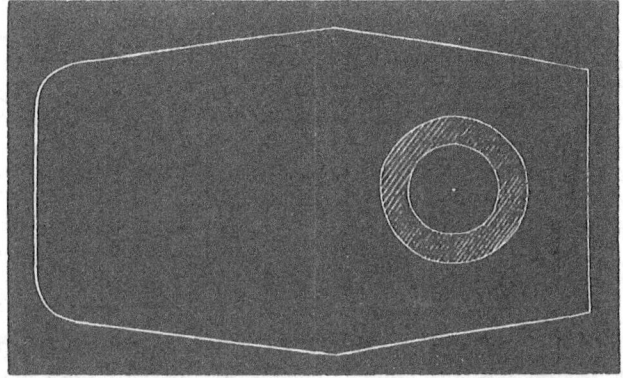

FORM TO STITCH KNOB HOLE WITH.

How to make the Stitcher.—Take two pieces of half-inch hard wood, twelve inches long seven and a half inches wide in center, and six inches at ends, diameter of hole two and three-eighths inches. Cut the hole through both pieces, and champer the hole to a sharp edge; hinge them together at the end nearest the hole with leather, fasten the lower half to the bench with a screw, spread the curtain on the bench, and insert between this clamp, placing a weight on the end, and you are ready for stitching.

A SIMPLE LITTLE MACHINE FOR CUTTING THE THREE-EDGED RAISERS.

The knife marked A can be made at any cutlery; the pointed part is the cutting edge. The standards, marked C, are made of hard wood, three inches high. The

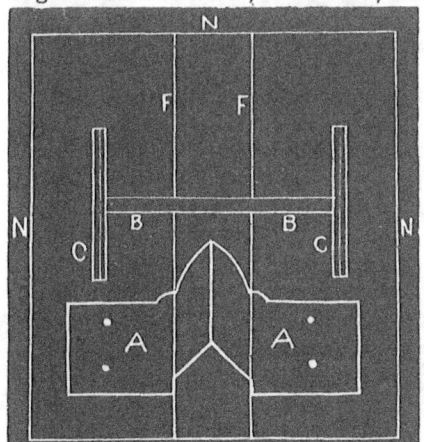

roller, marked B, is also made of hard wood, and with a shoulder then set into the standards. The lines marked F F represent a groove three-eighths of an inch deep, and three-eighths wide, cut into the block, or base, marked N. This base has a piece of wood screwed to the bottom, one and a quarter inches thick, and as long as the base, by which it can be screwed in the vise. The standards, with rollers attached, are set into the base just in front of the cutting edge of the knife, and the roller must nearly touch the surface of base. Now cut your strips of harness leather three-eighths wide; cut one end to a long point and draw them through, and you have a clear cut raiser. Be careful to have the knife made with the short perpendicular step, as shown (where it is on the base), as this gives you a chance to crease your cloth into the shape of raiser.

CUTTING STOCK.

In cutting a hide for a top, cut the quarters across the butt of the hide, and the roof in the same way, which leaves a back curtain; then cut the side curtains in the neck. It is necessary sometimes, in case of narrow necks, to turn the top front corner of the curtain down in the neck, instead of across it. Then cut out the back stays out of pieces you may have in reserve, either made then, or at some of former cutting. This is a good way, and the leather seems to work better in the quarters across the butt than in the length of the hide. But trimmers will, and do say, they do not care, as they have not time to waste in laying out stock; when, in fact, but little time is required, and they are working for their own interest in this, as well as for others. It is a credit for a man to be a close cutter, and we can give instances where this, in connection with other things, but this principally, enabled men to hold jobs in preference to better workmen in other respects. It will soon be known in a shop by the employer (in one way or another), who works for his interest, and who does not; and no employer is anxious to, or cares to keep in his employ, a man who cares altogether for himself, and none for his employer. We believe in independence—but many have a mistaken notion of true independence, which is, that while we claim our just rights, we also give to others the same.

TO MEND A BROKEN BOW.

Rip the leather off; place the broken ends of bow together; take a piece of writing paper six, eight or twelve inches long, and wrap carefully around the bow at the broken parts, and hold down well; take pocket-knife with a little straght-edge, and cut through the paper, guiding by the straight-edge; the largest paper after cutting will be the exact size of the bow. Have a tin tube made the exact size of your measure, tinner allowing for his seam, and after being soldered, hold your tube on the fractured bow with the broken parts well together; mark the bow with lead pencil at top and bottom of tube as a guide to tell when the broken ends are well together; if you get a good fit, which is not hard to do on an even trimmed bow, you will be pleased with your job. If the bow is hard to get the measure of by that plan, and is of regular taper, take a narrow strip of paper at top; lap and cut to get size; do the same at bottom where you want your tube to come.

ABOUT STOCK.

For a coach: twelve and a half yards of cloth; three yards extra if cloth lace be used.

For a four-seated phaeton, with child's seat: four and three-quarter yards of head lining cloth; three and a half yards of body lining when both seats are trimmed with cloth, and a cloth driving cushion.

For a roll-up-top buggy: four and a half to five yards of head lining cloth; piece the curtains or not; one and five-eighth yards of body lining.

For a six-seated phaeton, with dickey seat: four and three quarters to five yards of head lining; four and three-quarters to five yards of body lining.

For a four-seated coupe rockaway: eleven yards of body cloth.

For a six-seated rockaway, with partition: thirteen yards of cloth; if cloth lace be used, three yards extra.

For a brett: ten yards of cloth; if cloth lace be used, two yards extra.

For a full clarence: thirteen yards of cloth.

For a circular front coupe: eight yards of cloth; if **cloth lace be used**, **two** and a half yards extra.

PATENT LEATHER.

The "patent leather," so called, used in the manufacture of fine boots and shoes, for the skirting of saddles, for carriage tops and dash boards, for fancy styles of harness, and many other uses, is made by methods quite distinct from ordinary tanning; and it embraces some twenty-five or thirty different varieties, including those colored red, green, blue, amber, bronze and other colors.

The hides are generally tanned with hemlock, and require high liming and weak, sour liquor to secure a soft grain, and are but lightly stuffed. They are then run through splitting machines to bring them to the desired thickness. Some of the heavier hides are passed through the machine as much as five or six times. The inner and outer, that is, the grain and flesh-side splits, are commonly sold to the trunk-makers, and bring from twenty-five to upward of thirty cents per pound.

The great essential in making superior patent leather consists in properly applying the polishing substance, technically termed "sweetmeat," and which is composed for the most part of linseed oil. The hides, previously well-dried and softened by beating, with pine blocks and by what is known as "boarding," are stretched on frames provided for the purpose. The "sweetmeat" is applied in successive coats, which are fixed by exposure to a high temperature, alternating with the several applications of the material. In order to secure this, the frames are slid into properly constructed ovens, heated by steam-pipes to sixty or seventy degrees, and are left there until the "sweetmeat" or varnish is set. Any roughness that may be found at any stage of the process is polished off with pumice-stone, the dust being very carefully removed by means of a wet brush, followed by a dry one. The entire operation occupies from one to three weeks. That variety known as enameled leather requires additional manipulation, being passed through a graining machine, in which a heavy brass roller with a slightly currugated face is pressed forcibly upon the leather to perform a portion of the work of finishing the article, the balance being done by a peculiar system of "boarding" by hand.

MISCELLANEOUS--PART V.

ORDER AND DISORDER.

"Let everything be done directly and in order" is a command of Holy Writ; but we fear a great many persons have either never heard the command, or, having heard it, have cast it aside as worthless. The words used as our heading have an extensive application; but whenever applied, they present to the mind two different states or modes of conduct directly opposed to each other. Order reigns where wise and good men conduct the affairs of a nation or state, while disorder is the offspring of the jealousies and ambitions of demagogues. Contentment, profitable employment, and a steady advancement in the arts and sciences, and of morality and religion, are the fruits of peace and quiet. Disorder, on the contrary, is the parent of discontent, indolence, extravagance, and everything having a tendency to destroy the frame work of society, unlock the passions of bad men, and eat out the substance of the people. What is true with reference to a state or nation is equally true among individuals and families.

The tidy housewife generally has a smiling and agreeable husband; and, on the other hand, a man who is careful of his personal appearance, and makes it his study to provide little conveniences and comforts about his home, may rest assured he will find a loving and attentive wife.

To be a careful observer whose calling throws him in contact with men of business in different sections of the country, he observes that the thrifty merchants, manufacturers and farmers are those who do business systematically.

When we pass by a farm, and see a comfortable residence, a good barn, well-trimmed orchards, and clean fence corners, we at once decide that the owner is an honest, industrious and trustworthy farmer. When we enter a manufacturing establishment, and find a neat office, well-swept stairways and work rooms, good light and ventilation, and a studied method in every department, an evidence is before us that the proprietor will take a pride in producing first-class work.

The merchant who is deficient in taste, and careless of the interior and exterior appearance of his place of business, thoughtless in selecting suitable clerks and assistants, and slovenly in the display of his goods, may not be surprised to find buyers pass his door, and drop in at his neighbor's store to make their purchases.

But as we have to deal principally with carriage-makers, let us look in upon some of the establishments not a thousand miles away. We are sorry to record that a large proportion of them are careless about the interior appearance. How unsightly it must appear to a gentleman and lady to pass through a shop where old scrap iron, old wheels, bundles of iron, and the like, present themselves on the first floor. Shavings, saw-dust and chips, strewn over the stairway, announce the approach to the wood-shop. Plank, panel stuff, hubs, spokes, wheels, tressels, wheel-jacks, and every conceivable thing about a wood shop seem to be placed in just the position best suited to be stumbled over, and present the proper angles to catch hold of a lady's skirts, and cause her great annoyance. But they stumble along and reach the trimming room, treading on scraps of leather, clippings of rough linings, tufts of hair and moss, or may be tow, which is well calculated to give a chaotic appearance. The trimming room is generally the smallest, and easily kept clean; but in the picture before us (and it is no fancy sketch) we notice a very dirty floor.

The sweepings of several days are gathered in a mass near the stove, or in one corner. Moss, hair, old cushions, old curtains, scraps of leather, etc., lie in *splendid* disorder under the benches. On the few ready-to-tumble-down shelves overhead, leather, cloth and remnants lie as best they can, having been pitched from below by the trimmer, who, having wearied in asking the boss to put up a good cupboard to keep his stock from dust and smoke, now falls in with the ways of the shop, and helps to write disorder in his department. The paint room comes next. An artist dwelleth within. If not, who marked out that portrait on the door with dark lead, and touched it up with yellow ochre? The door-knob is glad to see you, for it clings to your hand. Within is a dark, dingy, low-ceiling room. The walls and window panes are daubed up with every color used in the shop. Bodies and carriages are mixed up together. In one corner a body is being finished; in the same room a boy is sandpapering off a gearing in putty. Confusion is written throughout the shop; and is it a wonder that customers call, look around and buy elsewhere? We would advise you then, good friends, to put on the best appearance you possibly can. Let order reign throughout the shop.

Have a place for everything, and everything in its place. Put up racks suitable for the different sized iron; keep all scraps and trash to the rear; allow no shavings to lie strewn about, especially on the stairways; put up shelving and cupboards in the trimming room, so that every article may be laid away nicely; renovate the shop from top to bottom, and be not saving of the whitewash. After cleansing the paint rooms, mark out a place in one corner of the room, and let it be kept as the only place where a brush shall be wiped. And last, but not least, straighten up the office. Tear down that old newspaper from the front window, and in its stead hang up a genteel curtain, with the firm's name and number neatly painted on it. If, after a disorderly shop has been thus renovated, the proprietor does not feel better pleased with all his surroundings, and draw a better class of customers, and get better prices, then we would say slovenliness is preferable to activity and cleanliness.

SUCCESS IN BUSINESS.

In the outset we will state that, in speaking of success in business, we do not wish to hold up to view that class of men who have amassed princely fortunes in a few years by bringing to bear, not only their superior business talent, but who have not scrupled to grasp eagerly after unlawful gains, by adding cunning deceit, and, at times, utter heartlessness, in the acquisition of wealth.

It is too often the case that we look upon success in business as that condition only in which a man has secured to himself sufficient income to *retire*, and lead a life of comparative ease and pleasure.

While we would say nothing against an individual choosing to retire from active pursuits, and enjoy the fruits of his labor, we do think that the example thus set has a tendency to create in others a desire to *speedily* arrive at a like position, careless of the means used to attain the end.

We hear occasionally of a man who, by a bold speculation, has "made a fortune" in a few months; but the majority of business men are not gifted with that keen foresight and courage which are so essential to the speculator, and must, therefore, be content with small gains, accumulating but slowly year after year.

It is well that it is so; for the cares and disappointments attendant on the conducting of any business keep down pride of heart, and secure to society a majority of that class of men who can sympathize with the unfortunate and downtrodden, and who give more liberally to the rearing of those institutions which benefit and improve the masses.

Success depends, in a great measure, on a knowledge of the business engaged in, the proper application of industry to the materials required, frugality, promptness in meeting engagements, and good moral character.

In no occupation we can name are these more essential than to those who engage in the manufacture of carriages; yet how few out of the whole number, who claim to be carriage-makers, have a good general knowledge of the business. Four distinct branches are to be looked after—wood-work, blacksmithing, painting and trimming. The material used by the respective branches are entirely dissimilar and costly, and requires the utmost vigilance on the part of the proprietor to see that there is no unneccessary waste.

We shall therefore notice, as briefly as possible, some of the principal points to be considered in successfully conducting the manufacture of carriages.

CONVENIENT AND COMFORTABLE WORK-SHOPS.

Men can, if they choose, do a greater amount of work, with more satisfaction to themselves and employers, in a clean, well-ventilated shop that they can in the opposite. How often is it the case that the smith department is in some cold shed, or in a damp, dark and poorly-ventilated cellar, with ground floor, and seldom, in either case, do we find, during a long cold winter, any fire or heat save what comes from the forge. Allowing that the forger and helper can, by constant exertion, and the little warmth they receive from the forge, keep passably warm (except the feet), with what little spirit does the filer or finisher enter upon his work on a cold morning. His tools cold, the iron he touches sticks to his fingers, and between rubbing his hands and stamping his feet many hours are actually lost to the employer during the day. Hands become dissatisfied with their situations, and when business revives obtain employment elsewhere, just at the time when the employer can least spare them and *good* hands are scarce. Much loss is occasioned by introducing new hands, supposing they can be easily obtained, which is seldom the case in busy seasons.

Much can be said respecting all the different departments, but none are so much neglected as the smith-shop. Make each department comfortable, light and well ventilated, and when you have hands that suit you change as seldom as possible.

SYSTEM.

Much loss is occasioned by a want of system in the shop. There should be in every well-regulated shop a set of hands especially for repairing; these, of course, it is better to hire by the week, and have some common jobs for them to work on when repairing is slack. But hands for new work should be kept as much as possible on the same class of work; it is better for both employer and employed. In many shops it is impossible to do so, but much dissatisfaction could be avoided by the employer studying to make as few changes from one class of work to another as the interests of his business will allow. Have a place for everything, and see that each one learns to keep things in order around him. This, of course, should be the duty of the foremen of the different departments; they should be careful, systematic men, and thoroughly acquainted with the branch in which they are engaged. If the plank pile is torn down, or a carriage is taken apart for repairs, it should be the duty of the foreman to see that each one rearranges everything after him *at once*, and not leave things go at loose ends for a week or a month, then take a general clearing up, accumulating scraps and pieces of stock that could have been used to the saving of uncut stuff.

For new work, hire by the piece; it is more satisfactory to the hands, pays better, and saves a great deal of care and watching that is necessary with many hands. You know just what the job costs you, while hands working by the week will, more or less, "take it it easy" when they are not in a working humor.

DO FIRST-CLASS WORK EMPLOY FIRST-CLASS HANDS, AND PAY FIRST-CLASS WAGES.

The curse of coach-making is the *auction shops* of our cities, of which, we are sorry to say, too many can be found. It may seem profitable to some for a time, but our experience is that very few ever accumulate any property. It is a continual drive of every one in their employ to their utmost, poor pay, poor hands, poor stock, and, consequently, poor job and poor price for it, flooding the country with carriages that are not safe to go five miles from home with. The employer is not making any reputation for himself, and in dull times his business stops. This portion of our subject is worthy ten times the space we can devote to it, for, of all things, we detest the man who will, for the sake of a few dollars profit on each job, force men to work for a small pittance, waste the stock, and turn on the community a worthless thing in the shape of a carriage, while he might, at the same time, be making for himself a reputation that would stand by him and give him his full share during dull times, and be affording a comfortable subsistence to those in his employ. Good work, good hands and good wages pay best, and is more respectable. The men should be kindly and respectfully treated by the employer, which will induce them to take a greater interest in his business, and in the production of the proper kind of work.

The best hands are invariably the cheapest, and a few dollars extra invested as wages pays much better than advertising and filling the places of competent hands with those wholly unacquainted with the shop and your manner of working.

The employer should make himself acquainted with the quality of all the material he uses, and *purchase it himself* in as large quantities as circumstances and the wants of his business may require. The habit of buying in small quantities, and of sending boys after what is wanted, is productive of no good to any one, only the dealer, who, in most instances, takes some advantage, either in quality or price of goods. Better select some honest dealer and send direct to him than trust to an incompetent person. When living at some distance from the market, time and money are often saved by sending your orders to some disinterested person in whom you have confidence in their ability to judge of the quality of the articles desired.

Practice strict economy in all departments; see that your materials are used up as closely as possible, and be careful not to accumulate a large quantity of unavailable stock. Endeavor to produce a neat, convenient class of work, that will find a ready sale.

BEWARE OF SHARPERS.

Many persons are continually on the watch for men commencing in the business. Frequently livery stable men profess to have great influence in controlling custom apart from their own, and ask for special favors on that ground; and in many instances the new beginner finds that he has, after being induced to do his work for one-third less than it was worth, lost the most of the balance, and his customer leaves him to try the same on some one else. Better do less business and be sure to get good prices, and as much as possible sell for cash; at all events settle often.

BE CAREFUL OF YOUR CREDIT.

It is a mistaken idea for any man to think it to his advantage to defer paying for his stock until he is forced to do so. It is much better to *always* buy for *cash*. A few cents per foot discount on leather, or a fraction per pound on iron, and all other kinds of material, will tell in the course of the year in the profits. There are very few that have the means always at hand, but next to paying cash fix the time for your payments when you know you can meet them, and regard your promise as sacred. Let those with whom you deal understand that you are prompt according to agreement; and should necessity require it, you will have many favors shown you that you otherwise would not receive.

A dollar saved here and there by paying cash, or meeting your notes promptly, will help to swell the profits of each succeeding year, and in time place one in easy circumstances. "Wealth is acquired by small but oft-repeated accumulations." The gross amount of these sums will be decided by your skill, industry and frugality, for, "though a man earn much, yet if he spend all he will grow no richer." There is a class of men who are prepared to pay promptly, yet ruin their credit by negligence, and assist in adding to the cares and perplexities of business life.

A man of good moral character will be mindful of his promises, and if misfortune overtake him, will find many a helping hand to assist him to rise again.

ARE THEY COMPETENT JUDGES?

Carriage-makers who seldom, if ever, take the lines in their hands, and ride out in carriages of their own manufacture—are they competent to judge of the merits or demerits of the vehicles which they with confidence recommend to others? We think not. It is one thing to oversee and pay well for the building of a fine buggy, or any other kind of vehicle, and quite another matter to experience the sensations produced by putting them into actual wear. A buggy may be handsome in general appearance, and composed of the best material, yet defective in ease of motion and comfort to the occupants. The set of the axles may cause the vehicle to run heavy, and communicate to the rider an unpleasant jarring motion, at the same time add unnecessary labor to the horse. The springs may be too stiff for their length, and fail to vibrate sufficiently under the greatest weight they may be called upon to sustain. The seat may be too low—the back placed in a position or so trimmed as to be a continual source of uneasiness; the foot-room be cramped. These and other defects may exist, while the carriage-maker, who seldom rides out, remains in total ignorance of them, in so far as his own personal experience extends. Now, an individual having purchased a buggy of such an one, might drive up to his shop door and inform him that this or that defect existed and needed to be remedied, and fail to convince the maker that such was the case. He would probably plead the skill of his workmen, the care with which every buggy was carried forward to completion, and thus fortify himself in his own opinions, through gross ignorance of what constituted comfort while seated in a vehicle carried along over roads of different degrees of smoothness.

The tendency of such a course is toward a stand-still point in the way of needed improvement, and must certainly work adversely to the carriage-maker's interests. So far as our observation extends, we are well satisfied that the builder who adopts an opposite course is by far the most successful. Becoming sensible to defects by personal experience, he is keenly sensitive and anxious to remove any cause of complaint brought to his notice by others. With such an one the customer feels that he is dealing with a manufacturer alive to his convenience and comfort, and will not be apt to go elsewhere to purchase, although he may have had occasion to point out several weak points.

The truly progressive carriage-maker tests his own work by frequently taking airings, and criticisms by those who ride a great deal he gives a careful examination. No matter who may suggest a new idea of value, he puts it away as so much gained. He gathers here a little and there a little, which, in the aggregate, when applied as little things, amount to something so important as to give to his work an indescribable *something* which marks it as superior, in short, a distinctive character.

COMPETITION.

Competition is a contest or rivalry between two individuals, companies or governments, for the purpose of profit. It is, in reality, "the life of trade," when carried on within reasonable bounds, otherwise it is ruinous to the contestants. The desire of possession is natural to every human being, and sufficient selfishness has been added to make us careful of our own, and careless to others' interests, to a certain extent.

It would be an endless task to attempt the enumeration of the forms in which competition presents itself. No trade, profession or calling that can be named but what has to be on the alert lest those who may be in the same line discover or purchase some improvement which may work injuriously, by taking money out of their pockets, and placing it in the purse of another. There was a time when men of superior skill might reap the full benefits of their productions, but the introduction of steam and labor-saving machinery have placed all nearer on a level.

The manufacturer of to-day, who does not employ the most perfect machinery, need not attempt to compete with his more enterprising rival. He may struggle along in a half-dead-and-alive manner, but must finally feel himself pushed aside to make room for those who are fully alive to their vocation. Among carriage-makers there is a strong spirit of rivalry which will not fail to show itself in the country town where there are but two shops, as well as in the metropolis, amongst the heavy manufacturers, and on down to those who run but one fire.

Builders very often allow themselves to be led away into bankruptcy, by being more anxious to sell for the sake of selling, rather than to wait and obtain a living price. They will sell at a sacrifice, rather than allow a customer to go elsewhere, especially if he should mention the name of a rival, stating that Mr. so-and-so would furnish a vehicle at the price he was now offering. This is all wrong, and, by whomsoever practiced, if persisted in, must lead to failure.

The builder should know the first cost of every carriage in his wareroom, and, having added a living per cent. to that, obtain as much more as possible, yet never take less than first cost. The prices of carriages built in different parts of the country necessarily vary, on account of the advantages one section may have over another in securing stock and labor cheaper. This fact leads to rivalry between cities; desperate struggles are made to build up a trade, where natural advantages may be all against a place, and a few years only are necessary to prove the folly of the attempt.

This kind of competition is ruinous, evidence of which is being continually brought to our notice. Prices vary so much between the large cities, both for new work and repairs, that builders in those towns, which are farther inland are often at a loss to know what value they should set on their finished work; and the journeyman is equally in the dark as to what would be a fair price for *his* labor.

In the cities we find a great many small shops struggling to get a foothold, which are compelled to take whatever may be offered, and between this class and the prince in the business, are those who lay some claim to respectability, yet who have no settled prices on their work.

In answering such questions as "What is the price for all kinds of iron or wood work, painting and trimming?" a satisfactory reply cannot be given, there being such a variety of circumstances under which the value of labor may be viewed. A comparison between two first-class shops would probably disclose the fact that one set of hands would be required to perform a certain amount of extra work in order to please the particular fancy of their employer, and, for this extra labor, would receive one-fourth more wages than those employed at a factory where equally serviceable work was built, yet deficient in certain little matters not affecting the real value of the carriage to the customer. Here the difference in wages would be balanced by the unequal amount of labor required.

Again, two shops may require of their hands an equal amount of work on each part, and yet, to piece-workers, more money could be made in one than the other, at less wages, on account of the facilities afforded in producing the work. The wood-worker, on having an abundance of seasoned stuff, kept in a dry and convenient place, the timber sawed and planed, a warm shop, etc. The blacksmith, by being provided with the sizes of iron and steel, requiring the least amount of labor to forge the parts, and not to be compelled to draw down a heavy bar, in order to obtain the size required. The trimmer, by being supplied with good stock and an abundance of it, so that he should not be compelled to waste his time for the want of a little bit of muslin, hair, or a certain kind of leather best adapted to his wants. This will apply also to the painter.

It would not be just for an employer to demand of a piece-worker an unusual amount of work for a very low price; neither would it be fair for the journeyman to claim an exorbitant price for his labor, when surrounded with every convenience for completing his work in a short time. The question of wages and piece-work prices can never be settled satisfactorily without a bill of prices, agreed upon between the employer and employee, and even then there would be conditions in which disputes would arise.

In a country town, where rents are low and living cheap, a man can afford to take a trifle less for his labor; yet, if his employer receives city prices for finished work, city prices should be paid the journeyman. Wages have much to do in the matter of competition between rival manufacturers; those who pay the least for skilled workmen being able to undersell all others. When this discovery is made by those placed at a disadvantage, their first attack will be at the wages paid their hands; the cost of the carriage must be reduced by some means, and as the material required in its construction cannot be bought at a less figure than that charged by the dealer in carriage goods, the workman must submit to a reduction, or seek other employment.

Competition should not be attempted between widely separated localities, where one section has a decided natural advantage over the other. Living profits should be looked after, regardless of what other firms may or may not obtain for their work. Cincinnati or St. Louis would be foolish to attempt competing in price with New Haven and other Eastern cities. But, by building equally as good or superior work, they may supply a large demand in their section, and obtain handsome profits.

Were every kind of business conducted on a safe basis, and less jealousy and rivalry manifested, there would be fewer failures to record.

THE PROPER CARE OF CARRIAGES.

What can be more pleasing to the eye, among the productions of the mechanic, than a carriage built by a first-class manufacturer, who has spared neither time nor money to place before the buyer a finely proportioned and elaborately finished piece of work.

Weeks and months have elapsed since the draftsman had applied his patterns to the unsightly plank, and transformed it into the graceful form presented.

The wheeler and carriage-part maker added their share of skill. The blacksmith strengthened the weak points, and by sturdy blows wrought into forms of beauty and durability the stubborn iron and steel. The trimmer, with costly fabrics of delicate texture, rich laces, carpets, etc., arranged for the comfort of the passenger; and, lastly, the painter, through months of hard and perplexing labor, finally put on the finishing stroke, and the carriage stands forth a thing of beauty.

Now this carriage which is so delicate in its construction, and so easily damaged by heat and cold, and neglect, in many ways on the part of the owner, should be properly taken care of.

But how often is it the case that the carriage is shamefully abused through ignorance or carelessness of the coachman? The owner of a fine private carriage, having provided everything necessary for keeping it in good order, may have in his employ a man who either does not understand his business, or if he does, is too heedless to attend to it properly. While every part of the carriage should receive proper attention, the surface work will require the nicest judgment.

One of the most attractive features of a carriage is the exterior finish. It is well known that the painting either adds to, or detracts from the appearance, according as it is well or poorly executed. Varnish, which forms the lustrous surface, is easily soiled. The fine gloss, for which the carriages of America are noted, is so easily dimmed by careless washing, and the coating of varnish itself ruined by scratches and hand-marks, that we have thought it would be well for manufacturers to have printed instructions ready to accompany every carriage sold, giving full instructions as to the proper care of the carriage. To the majority of city customers this might not be necessary, but to a large class of buyers it would be beneficial, and often save both parties from considerable ill-feeling.

We remember an instance to the point.

A few years ago a wealthy farmer purchased a fine carriage, in a certain city, in Kentucky. After enjoying a drive out with his family, he took the carriage to a creek, drove in, and with an old stub broom scoured the mud off, working, no doubt, in good earnest (judging from the appearance of the job when it was brought back to the shop.) The surface was ruined. The back panel looked as if some thoughtless school-boys had attempted to draw a map of the world on it, with nails and brick-bats.

The owner hurried into town, and charged Mr. ———, with having finished his carriage with inferior varnish. The scene which ensued may easily be imagined. Angry words were exchanged, but there was but one remedy; the carriage had to be re-varnished; the customer footing the bill.

A dry, clean carriage-house, entirely separated from any stable manure, should be the first thing attended to. A closet to keep the harness free from dust and dirt, should always be found in a well-planned carriage-house.

A wheel-jack, buckets, sponges, chamois, wrenches, cover, feather duster, axle grease, etc., should have each an appropriate place.

THE WASHING.

When a carriage has been run in the summer season, use the water freely, so as to remove the dust or mud before using the sponge and chamois skin. We have seen carriages ruined on the exterior by scratches in the varnish, caused by carelessness or ignorance in this particular. Mud should not be allowed to dry on the varnish if it can be avoided. The English varnishes, with which most carriages are finished, retain their elasticity for a great length of time, and mud or filth of any kind drying on them, fastens upon the body of the varnish, and the stains cannot be removed without re-varnishing.

In the winter season it is not best to wash off the mud when the water freezes while being applied. *Warm water* should *never* be used in winter time, as it will be apt to remove the varnish, or cause it to crack and peel off.

A great deal of bad feeling, and sometimes angry disputes, arise between the manufacturer and the owner of a coach, growing out of thoughtlessness on the part of the coachman who had allowed water to freeze on the panels.

GREASING.

For greasing the axles and fifth wheel, use castor-oil. It is not necessary to put on a great deal. Frequent applications and less in quantity should be the rule, for when there is an excess of oil, it oozes out, and finds its way on to the hub, and from the nub is thrown over the wheels, when the vehicle is in motion. The grease is then liable to be taken up on the sponge when washing, and also on the chamois, giving a vast amount of trouble and vexation. The fifth wheel should be looked after, and not be allowed to become entirely dry.

THE LEATHER.

Enameled leather should be kept soft and pliable, with sweet or sperm-oil. It will only be necessary while the leather is new to cleanse the top and curtains from dirt, and rub them with a greased rag. When the leather shows signs of drawing up and becoming hard and lifeless, wash it with warm water and castile soap, and with a stiff brush force the oil into the leather until the grain is filled.

SPONGES AND CHAMOIS.

Two sponges and chamois should be kept on hand, one of each for the body and gearing. The reason for this is that after the carriage has been used there is a liability to get grease on the sponge and chamois when cleansing the wheels and front axle bed. Another reason of some importance is, the gearing soon destroys a sponge, and makes it worthless for washing large panels.

WRENCHES.

Beside the axle wrench, a monkey wrench will be found of great service in looking after the nuts at those points where there is the greatest liability to accidents. Hundreds of dollars have been spent in repairs, limbs broken, and even lives lost, through neglect to inspect the carriage before starting out.

THE COVER.

When the vehicle has been washed and housed, it should be covered with an enameled cloth cover, fitted to it so as to keep it free from dust, inside and out. To preserve the wood and save expense, it should be re-painted or varnished once a year. There is no economy in saving a few dollars this year on your carriage and spending three times as much the year following.

While we have thus far spoken only of the care of the carriage after it had passed into the hands of the purchaser, a few words may not be out of place to the builder with reference to the ware-room. In cities like New York, Philadelphia, Wilmington, Cincinnati, etc., we find the ware-rooms all that could be desired; but in too many factories we find little or no pains taken to keep the finished work free from dust, smoke and the finger-marks of persons who are strolling about, buying nothing, but leaving their finger-marks, or writing their names in the dust which may have settled on some of the panels. We have in our mind a model ware room. It is well lighted, clean and neat. The doors and windows are closely fitted, excluding dust and smoke. A double floor, between which is a layer of plaster, prevents the smoke from the smith shop from finding its way into the room. The work is completed before being lowered down, and no one is admitted but those who are on business. The room is kept quiet, and no finger-marks are left on the work by mere sight-seers.

COLDNESS.

In the strife among manufacturers in the same line to excel each other in the quality, convenience, and adaptability of their products, as well as to compete in prices there is observable, too often, a coldness, a want of sociability, which is far from being pleasant to contemplate, and except in extreme cases, has no just cause for continuance. With reference to carriage-makers, where an undue advantage has been taken by one builder over another, by selling to a customer below living prices, through the customer having stated at what price he could purchase at a rival shop, allowance may be made for unfriendliness on the part of the injured party. Also, in cases where one builder speaks slightingly of a rival's work, which he knows to be equal to his own, and in those acts of doubtful honesty and good faith, the tolling off good hands, or sending spies through a shop to ascertain the secret of doing a certain portion of the work.

Where such acts have been committed, due allowance should be made for coldness between the rival firms. But, on the other hand, where nothing of that character has transpired, we must ascribe the matter to jealousy alone. We have frequently heard the remark made, that "carriage-builders are more jealous of each other, and consequently less sociable, than those engaged in any other kind of business." That there is by far too much jealousy we will admit, but with even our limited knowledge of other trades, and the professions, we stoutly deny the above assertion. We are all apt to judge hastily of, and censure severely the faults in our own, because we are not in position to ascertain the truth in regard to other trades or professions. The newspapers often make public property of quarrels among politicians, lawyers and doctors, but very rarely are they called upon to record acts of violence, as committed by rival manufacturers, and certainly the carriage-builders can show as clean a record as any that may be named. We have said there is by far too much jealousy—a word which Webster defines to be, "the fear of losing some good that another may obtain." This fear, then, when harbored and nurtured, promotes unsociability, and defeats the end aimed at by the rival parties, namely: the securing of a good run of trade at living prices. For were there no jealousies, a warm friendship might be kindled, which would result in united action on matters pertaining to the good of all, and at the same time leave the true independence of each untouched.

BE READY IN TIME.

It is by no means uncommon to hear carriage-builders regretting their lack of energy during the winter, when the spring trade opens and finds them with but a small quantity of new work finished.

We are well aware that lack of capital prevents a good many from running a full force of hands during the winter when there are but few if any sales of work. These merely regret that they are placed at a disadvantage for the want of sufficient capital to push forward their work during the dull season; and have therefore a consoling excuse. But there are builders having ample capital, or credit—which amounts to about the same—who begin croaking as soon as the leaves begin to fall. "Ah," say they, "I am fearful this will be a hard winter, and I have serious doubts about there being much call for work. Let me see, I must reduce my force of hands to the lowest number possible and have the work move on slowly." The calculation is then made as to how small a force may be retained. This having been decided on, the new order of things is inaugurated, and soon throughout the shop there is an atmosphere of dullness, which affects all who have been fortunate enough to retain their situations, and the shop does in reality move slowly, and the profits much slower. In too many instances the shop is almost wholly given over into the hands of apprentices and insight workers, and the desired result on the part of the employer is reached very naturally, namely, that of having but a small amount of work ready at the proper time; but no sooner do the cheerful spring days begin to revive trade, than we hear again the self-accusations of lack of foresight, want of confidence in the future of trade, and so on. Now it is absolutely necessary for the carriage-builder to be ready in *time*, would he supply the fullest demand that may probably be made on him, and to do this advantage must be taken of the winter season. The question should not be how small a force of hands can I get through the winter with, but after carefully considering the prospects of the coming year, ascertain how large a number of men may be kept busy. It should be understood by the employees that the winter months will be lively in preparation for the coming spring season, and if there is any lack of work, the extra time to be made up by adding a greater value to that which is in hand. The anvils should be kept ringing, for their very sound gives life and animation to the whole factory. Our highly successful carriage-builders, almost without exception, have been men of nerve and they have achieved success through a determination to be fully prepared, at the earliest possible time, for the opening of trade. If one or more customers be lost this year because they could not obtain of you what they desired, you may count your loss as double that number at least, for if they obtained good work their influence among their friends will be thrown in favor of your rival. The month of February presents a favorable time to look into the condition of work in the shop and make the calculation how much of it can be pushed forward to completion within a time which will insure its readiness for the earliest customer who may call.

FROM THE FARM WAGON TO THE LANDAU.

We desire to take a glance backward, and call to mind the gradual development of the desire for vehicles of pleasure. In this hasty review of the past, as affecting carriage-making, we shall deal only in generalities.

The farmer in former days owned his plows and farm wagons, and when he with his family prepared for church on a Sabbath morning, hitched up the team to a wagon in which chairs were placed to accommodate a portion of the family, while the sons and daughters, if he were blessed with them, mounted their horses, and the company moved off, with no visions of the latest style of buggy or stately family carriage. A journey to a distant point was accomplished in like manner. But, as the resources of the whole country became more fully developed, and the frequent transportation of freights between cities and towns became a matter of necessity, attention was directed to the laying out of turnpikes, the better to accommodate business intercourse, and to furnish more speedy and agreeable modes of travel. From the cities, turnpikes were built, radiating out to distant interior towns, and lines of stages were run regularly, to accommodate the business public.

The stage coach, in its day, was considered a very agreeable means of traveling, and doubtless induced many persons to decide on visiting from place to place; whereas, in its absence, they would have been contented to remain at home. Good roads were also conducive to travel by private conveyance, and farmers at length, one here and another there, ventured to purchase from the city a family carriage, despite the jeers of their neighbors, who looked upon the matter as a piece of foolish pride. This feeling gradually wore away, and the carriage trade increased correspondingly.

As wealth flowed in, the cities began to move in the matter of securing more speedy means of transporting goods and passengers than were afforded by rivers and canals, and the railroad was introduced. Steam carriages were made the land rivals of all other conveyances on land as well as those on the water. The stage coach gradually melted away from before this powerful rival, and sought occupation farther out on the frontier. If there had been attractions in the stage coach, which induced many persons to travel, the railroad, presenting such incomparable advantages, could not fail to increase the desire to be on the go; and thus, by degrees, the people were educated up to seeking their pleasure, or pushing their business by means of the steam carriage.

The introduction of railway carriages was looked upon by carriage-makers with rather a jealous eye, for they were fearful that persons would prefer to take the cars in going between points on the line of the railroad, and thus greatly reduce the demand for vehicles drawn by horses. To what extent this may have been true, on the firs introduction of railways, we are unable to say; but time has proven that, instead of lessening the demand for private vehicles, railways have been the means of greatly increasing it. Intercourse between the States, and the different sections of the individual States, has been quickened. The farmer who formerly was distant from a good market was brought nigh, and encouraged to lay out his best energies in order to procure large crops. By means of railroads new life was infused into every kind of business, and the material wealth of the country greatly increased.

The carriage-builder soon became convinced that in the railway he had found a friend instead of an enemy, and that the desire to own private conveyances was augmented, not lessened. During this development in the means of travel, and the trans portation of the products of the country, the farmer was being gradually weaned from his earlier prejudices, and as his children grew up to manhood and womanhood, they declared in very positive terms that to ride horseback, or in a farm wagon, was disgraceful: yes, and more than this, they began to speak slightingly of vehicles owned by their neighbors, because they were not of modern style. Now, just in the proportion in which farmers and the residents in interior towns became better judges of carriage work, did the city carriage-makers find it necessary to take advanced steps, and thus there has been going on a gradual development in the tastes of our people, until there is scarcely a village in the whole country but what has its vehicles of modern style.

Our metropolitan cities, determined on eclipsing everything in the way of style that might be attempted in the smaller cities, at length accepted European fashions, and we are now made familiar with liveried drivers and footmen in every section of our land.

There has been accomplished, then, within less than a century, a change in the tastes of the riding public, in which the farm wagon filled with chairs represents the past, and the landau in its splendor, the present degree of taste and refinement.

OUT OF WORK.

The causes that place a person in this position are the first to be looked after. Dullness of trade is the leading cause, but for this the journeyman is not responsible.

During the spring and summer months, coach-making is generally brisk, and with the large amount of repairing usual, an extra number of hands is required, and the consequence is that when fall comes the force is necessarily reduced, and many are obliged, for months, to give vent to the expression "out of work," and, as a general rule, it is that class of persons, who, during the busy season, have been improvident, spending their earnings—in many instances uselessly. Few know the extent of suffering and deprivation to those thus situated.

It is a well-known fact, that there is, and always will be, a surplus of hands in the winter season, or a deficiency in the spring and summer, and the question that interests each one, individually, is "how can I conduct myself so as to secure steady employment?"

In the first place, perfect yourself as much as possible in the branch in which you are engaged; make yourself useful to your employers.

Because hands are scarce, do not assume too much of an independent spirit, and because your employer cannot well do without you, make exactions that are not just and proper. It is not necessary for one to sacrifice manliness in order to be obliging or accommodating. A willingness to do will be remembered by the employer when he is reducing his force.

Reliability is another important trait in the character of any person. If, when he leaves the shop, his employer feels there is no dependence to be placed on seeing him the following day, and his job is delayed in consequence, it is treasured up against him, and in the end works greatly to his disadvantage,

If a man is at his post during the busy season, unless something special prevents, and his employer can safely trust him in promising a customer, he becomes an important auxiliary. Look, for instance, through the shops where you are acquainted. What is the character of the men who retain their situations winter and summer, and receive the best pay? Are they not men who take an interest in their employer's welfare? Who are obliging and anxious to please? Who have not such straight notions of independence as to believe and *act* in busy seasons as though the boss was their servant? Who are reliable and can be depended upon? Who are competent workmen?

In how many instances do we find that hands are kept on *only* because they have discharged their duties faithfully, when their services were most needed. There is, and always will be, in the minds of some, a false pride of independence; it is shown both in the employer and journeyman. In the busy season the jour feels that his services cannot be dispensed with, and he takes the opportunity to pay off some old grudge, or supposed wrong he has received when work was slack, and as the result, in many instances, it is remembered against him, and he is obliged to sit in his cheerless and cold chamber, shivering "out of work," out of money, no friends, and wishing for an opportunity to wreak vengeance on those he has forced to cast him off. How much better it would be for each one to try and make his services indispensable where he is engaged, securing the good will of all, and, as a consequence, better remuneration.

There are some employers *always* in want of hands; their names have become a regular by-word among the workmen. If a person is "out of work," his comrades say "Go to ——'s shop, he always wants hands."

A few of the causes are, false independence, overbearing disposition, using those in their employ as as though they were slaves, forcing all the manliness out of them, or obliging them to leave; poor pay—they want their work done for about half what it is worth, and are disposed to find fault in order to save a few dollars; they take advantage of the necessities of those in their employ by reducing wages in dull seasons; very slow pay, obliging their men to wait and get along as best they can, etc.

As a consequence, you very seldom find a hand stopping with them long; it affords a kind of holding-on place until there is an opening, and then he is left without hands, his customers disappointed, and his prospects of success greatly lessened; besides the many discouragements and losses attending the constant introduction of new hands, and of a class generally the poorest.

It is a mistaken policy on the part of boss or jour to act as though they were not dependent on each other. There is a certain amount of respect due from each position, and those who fail to discover it are themselves the losers.

A WORD TO APPRENTICES.

How great are the advantages of the present day for a mechanical education compared with the time when our forefathers lived, yet how few comparatively avail themselves of the opportunities surrounding them. Take, for instance, the apprenticed coach-makers, and how few of them ever carry their business farther than the shop door. When leaving their work, their minds are taken up with the momentary enjoyments of the hour; they spend their evenings at the theater or drinking saloon, and in many instances even worse sinks of iniquity, that soon drag them down to an untimely grave. If, perchance, we find one fond of reading, how often is it the case that he spends his golden moments straining his eyes over the light literature of the day, destroying every spark of taste for substantial reading that may perchance have been lingering within.

With the increased advantages for improvement come additional allurements to trifling and foolish pastime.

Go to any of our library rooms and you find works suited to the elevation of the seeker of knowledge in any branch of mechanism. See the many useful periodicals of the day, imparting knowledge invaluable to the young and future mechanics of the nation. Although the advantages for improvement are so great, we fear sometimes the tendency of mechanical ingenuity is to go backward. The chances for an apprentice who embraces every opportunity to improve himself in his particular avocation are, we think, very great. Look at it at the present time. Where you find a man well skilled in his business, he can command almost any price, while those who tell us "there is no need of their knowing more than they do, for the boss has the work done his own way." "The boss don't pay him enough wages, for he knows more *now* than the boss will let him do." This class generally are obliged to work for low wages, and so soon as their services can be dispensed with, they are set adrift to look for another job.

The apprenticeship is the best time to study and improve yourself. If you fritter away your time until you reach manhood, the attractions from study are much greater, and when the cares of a family are pressing you, but little inclination is felt to improve. Depend upon it, your employer will not be slow in finding out your ability to do, and in proportion to your usefulness will you receive compensation. Save up your little earnings, and expend it in something that you can store in your mind, and it will return to you sixty or an hundred-fold.

Where is the apprentice that could not, by a little economy or self-denial, save sufficient to procure a copy of the "*Scientific American*," "*American Artisan*," or other periodicals devoted to the dissemination of mechanical ingenuity?

Take, for instance, our own publications, devoted to the interest of coach-makers, and who will say there cannot be much instruction gained by a careful study of the productions of these very able editors and correspondents. Are you a body-maker, or learning that branch? follow, *practically*, through the explanations of the system of drafting. Are you a smith, a wheeler, a trimmer, or a painter? then carefully note the points laid down in the department you are interested in, and if you are in **earnest**, you cannot fail to be greatly benefited.

Remember, no difficulty can be overcome by simply wishing "I knew how to do it." Up and at it with a determination to conquer, and it is yours.

If this article should meet the eye of any one who has heretofore neglected to improve himself, commence at once. Cast off your slothful associates, curtail all useless expenditures, procure a share in some good library, and subscribe for those periodicals that will afford you the greatest assistance. Set your mark *high*, and strive to reach it. But remember you must put your mind on what you undertake, and then *study*.

BOXWOOD.

It is surprising to what perfection engraving on wood has been brought. A few years ago a wood cut might have been readily distinguished from all other illustrations by the coarseness of the engraving. At present we find some of the best artists devoting their time to delineating on wood; and such is the delicacy of the drawings the highest skill of the engraver is required to preserve the effect produced by the masterly use of India ink, the lead-pencil and China white. The fashion plates and smaller illustrations in the COACH-MAKERS' INTERNATIONAL JOURNAL, the pictures in the illustrated newspapers, such as Leslie's, Harper's, American Agriculturist, Appleton's Journal, Scientific American and others, are all cut in boxwood.

Boxwood, as is well known, grows in different parts of the world. The bulk, however, of that which is used in this country is imported from Turkey. The growth of the tree is slow. If it be twelve inches in diameter, its age is to be numbered by centuries, for it is above five hundred years old. Those trees which attain a diameter of eighteen inches are about one thousand years old. Block-makers prefer trees eight to ten inches in diameter. The wood is sold by the ton, is very costly, and is of such various qualities that not more than an eighth or a tenth part of a ton is suitable for the finest engravings.

The best quality of wood is of a bright canary color, the texture fine and close, and the surface free from dark markings; great care is required in preparing the blocks. After the wood in the log has been sawed up into sections of a proper thickness, and becomes thoroughly seasoned, it is ready to be cut into blocks, and here one may see what an amount of waste wood there is. Checks and other imperfections require close cutting of the sound wood, and as these pieces are necessarily small, **several must be** joined together to form a large block.

Blocks four or five inches square are composed of from four to six different pieces, fitted, doweled and glued together with such extreme nicety as to present the appearance of a solid piece. Large blocks are generally joined together with screws, so as to admit of being taken apart for engraving, and re-united when ready for the printer.

WOOD ENGRAVING—XYLOGRAPHY.

We would state, in the outset, that the wood engraver is not required to be a delineator. The block-maker furnishes the blocks of a size suitable to the design to be placed upon it—his orders from the designer or engraver always insuring the proper dimensions. The blocks are $\frac{7}{8}$ of an inch thick, which is the depth required by the printer, in order to have the face of the block to come level with the face of the type in the "form." A block that may happen to be too thick, must be planed down; therefore, it is preferable to have it scant, for it may then be raised to the proper level by attaching pasteboard to the bottom of the block. The block-maker is required to furnish the block true on the edges, and with a smooth, level surface on one of its faces. He also coats the block with a preparation composed of pumice and whiting in water, which is rubbed over the surface and then brushed, which removes all superfluous whiting, and leaves the thinnest possible film on the wood.

Should the coating be too heavy, the designer is annoyed by the particles of whiting giving away before the pencil or the inking pen. The beauty of a wood-cut depends on the skill of both the delineator and the engraver; that is to say, if a drawing is cleanly and sharply executed, and then placed in the hands of a competent engraver, the result will be a type which produces an attractive print. The delineator may perform his part well, but should the engraver be incompetent, the impressions taken from the block will be unsatisfactory. It is the engraver's duty to follow the drawing before him, and he is not held responsible for the after-appearance of the impression taken from the block, provided he has done no more than follow the tints and lines as they appeared. India-ink, the lead-pencil and china white, are all the materials required by the delineator to convey his idea to the engraver. He having faintly sketched in the pattern, or traced it by means of transparent gelatine, proceeds to wash over with India-ink those portions requiring broad tints, giving the gradation by the less or greater proportions of India-ink to that of water. The tints may be deepened by a second washing. The drawing is finished up with a No. 4 or 5 lead-pencil, where deeper shades and shadows and sharp decided lines are required; and with china-white, to indicate the high lights. The latter is not required for all drawings, the white on the face of the block being left bare for the purpose. The delineator corrects any defects in his drawing by erasing the defective part with rubber; this removes a portion of the whiting, and leaves the darker color of the wood, and unless it be again whitened, the drawing, when completed, will not be clean and sharp. A wash of water-color white is given the dark spot, which corrects the fault; the drawing may then be re-set. The block-maker, then, and the artist who puts the drawing on the block, follow distinct occupations, and are not required to be to any further extent connected with the engraver. The engraver, however, gives employment to the other two branches, as those who order a cut of any kind generally leave the whole in his hands. Having now traced the block through its most important stages, before reaching the hands of the wood engraver, we will call in at his artistically arranged and quiet little sanctum, and pass the block into his hands. On receiving it, he glances quickly over the drawing, and may signify his approbation or displeasure, according as the drawing is well or poorly defined. The gradation of tints indicates to him the kind of excision necessary to produce the proper effect, and he employs tools adapted to the different portions—these consist exclusively of gravers, small gouges and chisels. The block is placed on a circular leather cushion, filled with sand, which affords not only a firm rest to the wood, but permits it to be freely turned in all directions. The graver is held and used in a manner peculiar to this kind of engraving. The butt of the handle rests against the palm of the hand, three of the fingers closing around it while the thumb is projected forward upon the block, serving at once as a rest for the blade and a check to regulate the force in cutting—the motion of the tool being regulated by the forefinger. When an engraved block is damaged, or a serious error made, the only remedy is to drill out the part to the depth of about half the thickness of the wood, and to insert a tight-fitting plug, tapered at the bottom to insure its being driven home. The top of the plug is made level with the surface of the block, and the part re-drawn and engraved. Wood engraving is unrivaled for the production of broad, bold contrasts, and sparkling sketchy effects. The special advantage, however, which wood engraving possesses over all other forms of graphic art, is its applicability to the purposes of book illustration in the form of *text-cuts*, that is, cuts inserted and printed in the pages of type.

HISTORY OF WOOD ENGRAVING.

The earliest application of wood engraving to the production of a book, is supposed to have been in China, about the middle of the 10th century, and was probably first used for the production of playing-cards, the outlines of which were formed by impressions from wood cuts, and the coloring filled up by hand. The art made rapid progress, and the next great step was the production of books printed from wooden types, and illustrated with pictorial wood cuts. Toward the close of the 15th century, the art attained an excellence, which induced artists of celebrity and talent to select it as the means of conveying their designs to the world. From the end of the 16th century, the art to a great extent declined; but at the close of the 17th century, a certain Mr. Bewick devoted himself with enthusiasm to the art, and from that time it has continued to flourish. Originally, various kinds of wood, such as plum-tree, beach, mahogany, and pear-tree, were employed for wood engraving, and are still employed for coarse work; but there is no wood so suitable for this purpose as box, as it combines all the qualities necessary to admit of the most delicate execution.

THE ELECTROTYPE.

Having in the preceding article in this connection traced the boxwood block through the hands of draftsman, or delineator, into those of the engraver, we will now take it for granted that the engraver has completed his portion, and furnished a "proof," taken on paper from the block, which testifies that the engraving is satisfactory. To secure copies of this engraving at a comparatively cheap rate, and to preserve in metal all the delicacy of lines of the original, which would otherwise on the bare wooden block be blurred in the process of printing large editions—are advantages gained by the electrotyping process. On calling at the electrotyper's work room, we find them quite different in appearance from those of the engraver. In the latter cleanliness, quiet and an artistic air prevails, while in the former plumbago or black lead is the prevailing color of both the rooms and the workmen. But while this is true, the operations are not less interesting, and as regards the decomposition and depositing of metals, in or upon suitable moulds, through the agency of voltaic electricity, far more novel than anything connected with wood engraving. On visiting an establishment of this character, it will be noticed various machines are required to aid in producing a finished electrotype, and that a furnace must be kept glowing whereon to fuse metal and wax. The moulding press, the plumbago machine, and machines for rough cutting and planing, also circular saws for cutting both wood and metal, are made use of, in the order named. The moulding press is operated by hand, the others by steam-power.

THE FIRST OPERATION

necessary to be performed is the obtaining an impression or impressions of the block. These are taken in wax. Shallow metal pans are required to contain the wax, and benches are provided on which to place the pans so that they will lie level. The wax in a liquid form is poured into the pans, and when level full, they are allowed to remain until the wax has cooled to a certain degree; the wax when hardened presents a perfectly level surface, which is necessary in order to obtain correct impressions. The pan forms the case for the wax, and is not separated from it until the metal in solution has been deposited. In taking impressions, the pan is removed to the moulding, press. This press has a table or top which revolves, and another capable of an upward movement. The block is placed on the latter, and the pan containing the wax on the former. The revolving table is thrown forward until it lies parallel with the other table, when it is securely fastened. The block is then adjusted, and, by hand-power, the table on which it rests is forced upward, pressing the block into the wax. The top is then released and thrown back. Previous, however, to taking impressions the block is dusted with fine plumbago. The wax mould is now placed in the hands of one who devotes his whole time to trimming off the wax which projects above the level and filling, or "deepening" the blank spaces. The superfluous wax is cut off with a warm chisel, and the "blanks" deepened by using a stick of wax pressed against a hot iron tool; the wax flows down along the blank spaces forming ridges which are above the level of the mould; these are reversed in the electrotype, forming hollows, and for this reason is termed "deepening." The mould is next placed in the plumbago machine, where it moves slowly beneath brushes, which coat it thoroughly with plumbago. This operation performed, the mould has a connecting wire inserted to connect with the battery. All connection is broken with the back of the pan by passing a hot iron over the case. The mould is then washed with a strong stream of water, to remove any lumps of plumbago that possibly may remain This done, the mould is suspended in a solution of sulphate of copper; the connecting wire of the mould, connecting with the zinc plates of the battery, and the copper plate of solution is connected with the silver plate of the battery. When the coating of metal has been deposited, the mould is removed from its bath, and the "shell" or copper coating is separated from the wax by the application of hot water; the "shell" is now cleaned with hot lye, tinned on the back, and then taken and laid face down in a metal pan and liquid metal poured over it. The metal is composed of lead, tin and antimony; this process being termed "backing-up." When the metal has cooled sufficiently to be handled, the plate is cleaned, squared up, planed and mounted on wood ready for the printer.

THE GRINDING OF COLORS.

There was time, within the recollection of quite a number of coach painters, who are still following their trade, when the paint stone and muller were the only means employed in the shop for grinding colors. A large slab of marble or other stone of a suitable degree of hardness, and a large stone muller were the mill, so to speak, in which, or rather between which, the colors were, by a slow and tedious process of attrition, reduced to a proper degree of fineness. Hours then were consumed in preparing a pot of color, which at the present time would require only a corresponding number of minutes. In contemplating that slow-coach style of working, we are at a loss to conceive how any headway could have been made, in getting on with their painting. Probably they were never in a hurry with their ordered work at that time; if so, then there must have been a different class of customers from those now on the stage of action. What a sorry spectacle one of our large carriage factories would now present, with from twenty-five to fifty or more painters, and every carriage in a hurry, had we now no hand or power paint mills. It would require the time of one-third the number employed to keep the remaining two-thirds supplied with colors, and what an array of stones and mullers would there have to be along one side of the shop, and what an incessant swinging of mullers, kept up by the apprentices and other unfortunates who might be called upon to fill the position of paint grinders. But the paint stone and muller—as an exclusive means of grinding colors, were—thanks to Mr. Harris and others—set aside for something infinitely more satisfactory in results. The Harris mill was introduced some thirty years ago, and immediately found favor with all painters. It, with two or three other styles of paint mills, speedily accomplished the labor of uprooting the old method of grinding colors, and the paint stone and muller soon fell into disuse for the purpose named.

The paint mill, by hand power, has held the sway for over a quarter of a century, and with the exception of the attachment of belting in factories employing steam-power, there have been no marked changes until very recently. It now appears probable that the paint mill will soon be wholly superseded, and pass out of sight, through the introduction of coach painters' colors ready ground.

Wet colors have been furnished to house painters for several years, but all attempts to introduce any of these colors into carriage shops were signal failures. Failures, not because painters were prejudiced against them, or would not give them a fair trial, but from the fact that the colors were not adapted to carriage work, through the manner in which they were prepared. The process of grinding, and the vehicle in which they were mixed, rendered them worthless, and to this day the majority of coach painters are suspicious of all colors ground and canned ready for use. But this mistrust must soon give place to confidence in wet colors, as now prepared, for they have been tested by excellent painters, and pronounced superior in fineness, covering power and working qualities.

The paint mill, operated either by hand or steam-power, will, however, pass slowly into disuse, on account of its convenience in supplying color, at any time, of a suitable degree of fineness, and from the fact that painters have become so firmly settled in their convictions, that it is more economical to keep on hand dry colors, and the proper vehicle in which to mix them, and prepare them at the time they want to use them, and in just the quantity they require. To what degree the objections may be sustained by the facts in the case, we are not prepared to say; but judging from the improvements already made in grinding and mixing coach painters' colors, and canning them, we are firm in the belief that the time is near at hand when coach painters throughout the United States will abandon the use of the paint mill through the superiority in every particular of colors furnished in a wet state. The laborious and discouraging task of grinding colors with a stone and muller, having long since been abandoned through the introduction of the hand mill, the latter must in turn give place to that which reduces labor, and gives more satisfactory results.

TO EMPLOYERS.

Those of you who have provided no means of ventilation, suppose you call upon the finisher during extremely hot weather, and for a few moments breathe the stifling air and submit your body to the melting heat of the varnish rooms, and then notice the haggard appearance of the painter, also the dullness of the finishing coat of varnish, and whatever good resolutions may be formed at that time, put them in force at once. In short, lay out a plan for supplying the painter with a change of air.

SYSTEM IN CARRIAGE SHOPS.

Mr. Editor:—Three other persons beside myself—in all four of us, representing the four direct branches of the trade—are about forming a co-partnership for the manufacture of first-class work only, and in the conducting of our business wish to become as thoroughly systematic as possible. While I have been chosen as chief director of the whole, each of us will superintend that department to which he professionally belongs. As we are beginners, and having never had direct practice in the business, we shall willingly and thankfully accept all advice of a thorough business nature which may be given us, and are inclined to believe that from the many factories which you have visited in your connection with them as a journalist, that you may have seen and become thoroughly acquainted with many good systems, and that you would be the proper person for us to address ourselves to, is the cause of our writing you on the subject; and we would ask that you favor us with such answers as are at your command. W. T. and Associates.

We have, indeed, in our career as a journalist, visited many factories, and among them the leading ones of this country, and have never lost sight of an advantage of storing our memory with all the useful knowledge which might present itself to us. Therefore, we are at present prepared to answer the question of our correspondent readily.

Were we to be similarly placed with our correspondent, after having located the different departments advantageously, studying economy in constructing the forges, benches, etc., with a view to light, space, etc., and the comfort of the operatives, we should begin in this wise: Commencing with the wood shop—body room, wheel room and gear room—purchase a fair-sized blank book, and enter therein the number of the body, all its immediate dimensions, the dimensions of all the timbers entering into it, and by whom built. We would also pursue the same course with the wheels and gear, being careful to enter the full dimensions and shape of the gear, the number and size of spokes, size of hub-rim, and height of wheels.

After, or when the job reaches the blacksmith shop, we would adopt the same course; enter the dimensions of all irons, height from floor, height of steps from floor, size of axles and quality of same, length, width and open and grade of steel of springs, size of tire, track of axles, width of pole and shafts, size of fifth wheel, jack bolts, and all else, together with the name of the producer.

In the paint room we would keep a complete tally of how painted, number of coats, color, striping, by whom, and when done and how done, how the varnish acted, etc.

In the trimming shop the same system should prevail, the quality and color of cloth, or other material used, height of top, width of curtains, height of cushions, etc., and name of trimmer.

And lastly, when the whole was completed and ready for the salesroom or delivery, take a general outline of the whole, or a brief synopsis of what had already been recorded in the various departments, and record the weight of the same and number and, when sold, the name of the purchaser.

Our correspondent will readily see the advantage to be gained by following the above-mentioned systems and rules. It at once enables you to know the complete history of every vehicle you may build. If one of your customers, living at a distance, meets with an accident, by which one of his wheels becomes broken, you are at once able, upon the receipt of his letter, to build and completely finish his wheel without fear of being wrong, by having recourse to your history of his wagon. You can prepare him a new axle, build him a pole, send him a new spring, make a new cushion, or apron, or curtain, and be sure of that whatever is done will fit exactly, no matter if the carriage be Russia or Japan.

By this method you also acquire a knowledge of the materials entering into a carriage of a certain weight, and can easily tell how to proceed to construct a carriage of any description possible, and have the same be of the weight ordered, and without fear of having some nondescript job thrown upon your hands, that would otherwise be the case when you could illy afford it.

Again, it will afford you ample legal protection when doing business upon credit with unworthy persons that might make attempts at defrauding you of your hard-earned production. Were you forced to claim protection at the hands of the law, your complete history of the production of your handiwork in all its minute detail would prove conclusively to every fair-minded juryman or judge the justness of your claim, and your right to your own property, until such times as you shall have received an equivalent for the same.

There are many other little details which we might mention in connection with the present advice, but owing to the press of other matter will have to forego the same; but would say to our correspondent, that by strictly adhering to the rules we have offered, other rules and systems would suggest themselves at the proper period, and the whole, when interwoven into your business, will prove such a net-work of regularities, each and every one depending in a measure upon its fellow or neighbors, enough so as to force a thoroughness of your business of such a nature as to relieve you of many of the cares and perplexities that would be otherwise entailed upon you; allowing you more time for the cultivation of your mind and the development of your business interests.

CARRIAGE MATERIALS.—THE CHANGES TWENTY-FIVE YEARS HAVE WROUGHT.

Carriage materials, comprise a great variety of articles manufactured from many different substances, out of the number of which we mention iron, steel, brass, tin, lead, gold, silver, oroide, wood, leather, silk, wool, cotton, hemp, minerals, seeds, gums, ivory, bone, etc., etc. The finished carriage is required to possess strength, durability, comfort and beauty.

Iron, steel and wood contribute the necessary strength to render the vehicle strong enough to bear the load imposed, as well as to give the shape the several parts must have in order to make up in general form what is known as a carriage. Leather, silk, wool, cotton and hemp, enter into the manufacture of those materials which add comfort and beauty, without in any way contributing to the strength required for the safety of those who would ride. Gold, silver, brass, oroide, ivory and bone, add beauty and excellence of finish, and are of still less absolute value. Paints, oils and varnishes, made from lead, minerals, seeds and gums, are employed principally as preservatives of the surfaces of the exposed portions of the carriage, but may be so arranged and applied as to contribute largely to its attractiveness. But while we have in the foregoing comparisons given to iron, steel and wood, the highest value, in so far as the form, strength and durability of the vehicle is concerned, we must allow that to those, articles capable of supplying comfort, beauty and excellence, are necessary to produce a perfect whole.

But let us glance at the changes wrought in the manufacture of carriage goods. The increasing wealth of our country, during the last twenty-five years, added so greatly to the demand for vehicles for pleasure and business, that attention was naturally directed toward lessening the time and trouble of producing a finished piece of work. Formerly, there were but a few sizes of iron to be had, and as for clips, bolts, jack-clips, shaft-eyes, slat-irons, fifth wheels, steps, etc., no one ventured to make their manufacture a specialty. With reference to those parts of the carriage composed of wood, the body plank had to be sawed out by hand, and the spokes dressed out of the rough.

The supplying the carriage-maker with turned spokes was the first innovation in the wood department. Factory-made wheels followed, and still later the carriage parts were made a distinct branch of manufacture. But the divisions and sub-divisions in the manufacture of the parts composed of wood appeared to be a more natural one than that of those parts composed of iron and steel, hence the manufacture of tires of various sizes, to suit the different grades of work, followed more slowly, as did also that of bar iron of sizes exactly adapted for stays, dash rods and steps. But even these were in advance of what we are now familiar with, viz., bolts, clips, fifth wheels, shaft-eyes, slat-irons, etc., etc., made in the best manner of the finest quality of iron.

On entering a carriage-shop at present, and examining the stock of materials on hand, we find that but little more comparatively is required of the builder than to place in the hands of his workmen the several parts to be joined together. For the builder may now order the body, wheels, carriage part, tires, springs, axles, exact sizes of iron and steel, the fifth wheel, clips suitable to every part, steps, clip-bars, shaft-eyes, slat-irons, top-props, and joints, and those articles which have been supplied a greater length of time, as whip sockets, bands, door-handles, locks, hinges, buckles, buttons, laces, cord, etc., etc., thus almost making up a carriage in detached parts, many of them being complete in themselves, and requiring only to be joined together, and the carriage painted and trimmed.

The hasty glance given to the changes wrought in the manufacture of carriage materials, may serve to awaken a passing thought concerning the advantages possessed by the carriage builder, over those in the same line in times past, and pride, as he remembers that the United States offers these advantages to an extent far exceeding other countries.

GODDARD METHOD OF PAINTING.

Thos. Goddard, of Boston, well known to our readers as a builder of first-class carriages, carries on his shop after a fashion of his own, and each department presents singular methods of working. The paint shop is the only one we will notice at present. On entering the paint rooms nothing remarkable presents itself; in fact one is disappointed at finding so little in the arrangements and conveniences of the rooms worthy of noting down.

But in the method of carrying forward a body, there is much to interest painters unacquainted with a slow process of painting. Lead, rough-stuff, and American and English varnishes, are used. Japan and turpentine also. But a certain make and grade of each must be had; for instance, English lead alone is counted worthy of a place in the shop; then follows English filling and English varnish, Minett's American rubbing, and Noble & Hoare's English, are the varnishes in favor, these having been tested and pronounced reliable by the head of the concern.

English filling also having proven to the proprietor's satisfaction to be good enough, holds sway against all competitors; the old time-honored yellow ochre sharing its honors.

Not a gill of japan is bought from any maker of japan; on the contrary, the article used is cooked by the foreman of paint shop after a method of his own. This article is comparatively pale, and possesses sufficient oil to avoid the necessity of adding any more oil when paints are being mixed.

When a body is ready for priming, it receives a coat of English lead, mixed to dry slowly. The body, after standing until hard, receives thereafter two coats more of lead; is puttied and rough-stuffed. The rough-stuff is mixed in about the proportions of $\frac{1}{3}$ English lead, dry, $\frac{1}{3}$ English filling, and $\frac{1}{3}$ yellow ochre, in japan and a good quality of rubbing varnish, and reduced to the [proper working consistency with turpentine. Lead is used in larger proportion in the first coats, while the ocher predominates in the last coats. There is no such word as hurry known in the shop, therefore each coating remains a week, if necessary, to insure its being hard. When the varnishing process is reached, there is more novelty in store for the painter from a distance.

Instead of employing American rubbing varnish, which will dry firmly and cut down pleasantly, American and English rubbing are mixed half and half, and from a week to ten days allowed between each coat for hardening. The idea entertained is that each coat of varnish should be so mixed as to retain considerable tackiness, which, while it may sacrifice a full silky luster, fully compensates for this deficiency on the new carriage, by wearing much better when put into hard service.

The rubbing coats cut down very unpleasantly on account of their tackiness, and the painter who stands by and witnesses the operation grows nervous at the tedious process. After two or three coats of rubbing varnish have been applied, the finish is given with Noble & Hoare's wearing body.

The carriage part is carried forward with similar care, the whole when finished presenting a good appearance; but by far inferior in luster to the opposite method of painting. The Goddard painting has the reputation of wearing exceedingly well, not being liable to crack, peel or blister, thus apparently setting at nought the theory that the under coats of varnish should be free from tackiness in order to insure a durable finish.

THE REINS, AS HELD BY THE FAIR SEX.

The introduction of that cosy, convenient and showy little article known as the Pony Phaeton, has given to the fashionable class of females a comparatively new, and, we suppose, delightful means of taking their airings and attracting the attention of the dashing fellows who are courageous enough to drive their own high-mettled animals, and the envy of those gents who invariably walk, through fear of sitting astride or behind a horse. Ladies reared in the city, who, as a general thing, have not had an opportunity to become proficient horse-women, and yet desire to follow closely the fashions as they are presented, must find it very difficult to overcome their natural dread of taking charge of a horse with no masculine hand and strong arm near by to check any ugliness that may be exhibited. But we suppose that it is with this, as with all other demands of fashion, which require of its votaries the sacrifice of health, and even life, if they would prove themselves true worshipers. At any rate, it is quite certain that females drive out in our parks, who are not capable of managing a horse when he becomes in the least frightened or excited, and they risk their limbs and life in the attempt to appear in the drive in the latest approved mode. We must give them credit for their spirit of independence, and would by no means say aught against the custom of ladies enjoying the fresh air and the beauties of nature and art, as exhibited at the parks. But those who drive out for mere fashion sake, and are incapable of managing a horse, endanger not only their own lives, but those of others. It is proverbial that women are reckless drivers, and no one doubts it who has seen a buxom Indiana or Kentucky farmer's daughter start off alone on a visit to a friend living a few miles distant. "Go long," is the word with her, and the family horse must reach out to the full extent of his metal. Women favorably situated to acquire control of and management of the horse, display all the courage of man, and many of them add to courage great recklessness.

We remember well a case in point: In a western city a carriage-maker sought and won the hand of a beautiful country girl, who resided some eight or ten miles from the city. She had been familiar with the management of horses from childhood, and, country girl fashion, had taken her first lesson in riding by mounting old "Tom" or "Pete" bare-backed, and putting him through over the fields. After marrying, she, of course, moved into the city, and occasionally felt anxious to pay a short visit to her old home, and as men of business cannot always lay down their cares and be ready to accompany their wives at just the time they "*must go or die,*" Mrs. ——— at such times had a horse and buggy ordered for her, and snapped her fingers at the idea of being disappointed. Through the city she drove at a lively pace, but having reached the country road she gave her horse the reins and dashed away at a frightful pace, causing persons to rush out and attempt to stop what they supposed to be a runaway horse.

The same spirit is occasionally exhibited in our parks, resulting at times in pretty serious damage to horse and driver.

Recently an accident occurred in Fairmount Park. Two young ladies had been taking an airing, and on returning from the park the horse took fright, and plunged forward down the road, regardless of consequences. The driveress headed for the Lincoln Monument—a pile of granite surmounted by a hideous bronze statue of Lincoln—and " brought the horse up standing," as we say. The horse and the fair occupants of the phaeton were badly injured, and the whole was doubtless due to reckless driving, or lack of skill on the part of the fair hands that held the reins.

It would be well for city ladies who lack courage and presence of mind at a time when it is most needed, to ascertain beyond a doubt, if possible, the reputation or character of the horse they venture to drive. A lady venturing to drive to a large park must calculate on falling in company with a great variety of persons, among them no small share of a class who are careless of the convenience of others, and who, to make a great display on the road, cause an otherwise gentle horse to become restive.

There should be driving schools, where city ladies could be taught to manage the horse when he is quiet and obedient, and also when frightened and in a bad humor. A few simple rules to guide them, and the experience and confidence gained by holding the reins and driving alone in a safe enclosure would soon give courage to the most timid. We are always pleased to meet the jaunty pony phaetons at the park, with their burden of fair occupants, but at the same time have an instinctive fear that some accident may overtake them.

THE SARVEN PATENT WHEEL.
AN IMPORTANT DECISION IN ITS FAVOR.

On April 23d, 1872, the United States District Court sat in New Haven, Conn., and in the case of James D. Sarven *vs.* Elihu Hall & Co., which was a bill in equity, to restrain the defendants from using certain improvements, patented to the plaintiff, for making wheels. Judge Woodruff read an opinion deciding in favor of the plaintiff, with an order for an injunction.

"The New Haven Wheel Company appear as one of the principal plaintiffs against Elihu Hall & Co., who are manufacturing the so-called Warner patent. The decision of the court sustained the Sarven patent as against many other so-called patent wheels, and was therefore a test case. All infringing patents are, therefore, enjoined from further manufacturing. For a considerable time the plaintiffs have been injured by a large number of infringements which the success of the wheel has induced; some twenty-five or thirty have sprung up in the course of three or four years. The matter was such an important one that a special term of the United States District Court, Judges Woodruff and Shipman sitting, was held in Hartford last December, to try the case, and the trial excited considerable attention, on account of the extensive interests involved. The ablest patent lawyers in the country, including Judge Fisher, late Commissioner of Patents, Keller and Blake of New York, Thurston, of Providence, Beach, Ingersoll and Earle, of New Haven, were engaged. Such experts as the Rennicks and Treadwell, of New York, and Waters, of Boston, were also employed. The case has been a long one, but it is probable that the decision given will be final."

By private letter, we learn that the court has given E. Hall & Co. until September, 1872, close up, and has appointed a master to receive statement of all wheels made by them, assess damages, and report same at next term of the court. E. Hall & Co. go under bonds until September, when the injunction takes effect. It is expected that all other infringements will be enjoined immediately.

BLACKSMITHING IN GERMANY.

The following may prove amusing to some of our vulcanites, and is furnished us by one of our German subscribers.

In the interior towns and villages of Germany, it has been the **custom**, for many years, for the farmer to purchase the iron for his tire and horse-shoes, and in some instances, when having a new wagon built, to purchase all the iron entering into the same, the lengths of every piece being furnished him by the smith. One part of the contract is, that the smith shall return to the farmer all ends and cuttings from the iron, and it frequently occurs that the farmer remains at the shop until the iron is all cut up, in order that the smith shall not indulge in too much cabbage. Each smith-shop has what is termed "the hell," and in cutting off a set of tires, if the farmer be not present, the largest half of the end cut off finds its way to "the hell;" the duty in putting it there devolves upon the youngest apprentice. From this always plentiful store, the smith furnishes his material for the manufacture of bolts, horseshoes, etc., for transient customers.

The horse-shoeing part is also a feature. The farmer will bring with him the end of some piece of iron or tire, with which to make the shoes, or perhaps a dozen or more old horse-shoes to be converted into new ones. The farmer must blow the bellows until the work is forged or the shoes all made, and must then hold up the horse's foot while the shoes are being driven on, or fitted, or taken off, and invariably carries the old shoes home with him, or if he prefers, he can give the old shoes in payment for the apprentices' services in holding up the feet.

WHERE CARRIAGES SHOULD BE KEPT.

It is a common and very vexatious complaint from parties who store their carriages in the same building with their horses, or in damp and close brick carriage-houses, that the varnish becomes soft, and cracks. This *will* happen, even after the very best material and varnish have been used. Then the owner blames the poor painter, "who has endured the stifling fumes of a close varnish room trying his best," and also informs the builder that he is using vile trash instead of good serviceable varnish, when in most cases they are both innocent, *for no varnish ever was made, or ever can be made*, that will stand the steam arising in a stable where horses are kept.

Oils, by contact with alkalies, are more or less readily converted into soaps soluble in water (among the most easily saponified oils is linseed, used largely in the manufacture of varnish), which when shaken up with a solution of ammonia unites with the alkali; forming a thick solution of soft soap. Ammonia is a gas, and occurs in the air wherever organic fermentation is in progress. When a varnished carriage is exposed to an atmosphere of ammonia, arising from manure or decaying vegetable matter, the alkali unites with the oil of the varnish, forming an almost imperceptible filament of soap, which, when the carriage is washed, dissolves in the water, and is removed, leaving a fresh surface to be again acted on by the ammonia, so that the oil is gradually removed from the varnish, leaving the brittle gum to look like rosin and crumble away.

Therefore, a *dry, clean* carriage-house, *entirely separated from the stable or stable manure*, is the first thing required; and, secondly, to have it revarnished when it is required, "not the house but the carriage," for there is nothing saved by letting a carriage run three or four years without varnishing and then spending five times the amount in having it repainted. In fact, a carriage that is used continually should be varnished once a year.

WATER-PROOF GLUE.

It is often found that joints glued together, will allow water to dissolve the glue, and thereby destroy its adhesive power. It may have been well painted, and every care taken to it make impervious to water, but owing to its exposed position, water has managed to get in. Often where screws are put in, the glue around them will be dissolved, caused by the screws sweating; and we have almost always found that where the screws are inserted in a panel, that the glue loses its strength and allows the joint to open, and the wood is clear of all glue, which shows that moisture has absorbed the glue. Ordinary glue can be rendered insoluble by water, by adding to the water with which glue is mixed, when required for use, a small quantity of bichromate of potash. Chromic acid has the property of rendering glue or gelatine insoluble; and as the operation of heating the glue-pot is conducted in the light, no special exposure of the pieces joined is necessary.

Glue prepared in this manner is preferable in gluing the panels on bodies where there is danger of water affecting the glue. The strength of glue is not diminished by the addition of the potash.

A SAFE AND PROFITABLE INVESTMENT

Is always sought by the capitalist, and so alive is he to his own interests that he will not suffer his money to remain idle a moment. If a high rate of interest cannot be realized, he takes a lesser rate, and thus ever insures to himself some return from money which is dead unless kept in motion. Those who possess very large means seek out those investments which require heavy expenditures of money to set them into profitable operation, and which, at the same time, promise the handsomest returns. Men of small capital are compelled to move more cautiously, and be satisfied with small returns. But no matter how large or small the amount of money one possesses, there can be nothing added to it without it is kept in motion. Now, suppose one has no capital except his knowledge of a good trade, what shall he do? Sit down and repine, and determine to shut out all further knowledge of his trade, or, to say the least, refuse to search out hidden things, or by a small outlay of his money receive the fruits of other men's labors? Alas, we fear—we may speak more positively—we *know* a large number of mechanics whose only capital is in their knowledge of a trade, who are ever ready to pour out their complaints in one's ear, and bemoan their lot, who have not yet been faithful in that which is least. They have been satisfied with a partial knowledge of their vocation, and that, too, when the means were just at hand, which when supplemented with a little energy on their part, would add materially to their capital; and through persevering effort, industry and economy, eventually place in their hand capital in the more potent form of money. The carriage artisan of to-day, who stands still and gazes at empty space, while the sound of printing presses is thundering in his ears, and printed sheets filled with carriage literature are falling cheaply at his side, is not investing either his muscular or brain capital, and should not be disappointed if he grows pooror as the years roll past. There are to-day hundreds of journeymen carriage-makers in the States, and a large number of employers, who are wrapped up in a thick cloak of self-conceit, and many others who appear to be dreaming away their time. The first class named, look out from beneath their mantle with disdain upon one who should present to them for inspection the thoughts of fellow-craftsmen, arranged in the form of a periodical. They have no need of books to instruct them, their styles are all original, and they mean to keep in advance of rivals by continuing to do their own thinking and drafting. The second class are equally indifferent to their own best interests, but receive you without a dissenting word, and hear without heeding; in fact, assent to a proposition through lack of spirit to reply. As opposite to these, we could name several carriage artisans, if we chose to do so, who some four or five years ago invested their brain and muscular capital in the right way, that to-day are receiving as interest, not only honors for their skill, but salaries which richly repay them for their toils. The way lies open for others to enter, and although all cannot occupy the best paying positions, each one, without exception, may increase his capital by using the means at hand for self-improvement.

AMERICAN TIMBER INTERESTS.

We are threatened with a want of sufficient quantity of timber to meet the actual necessities of life. Twenty million of people are living in dwellings chiefly constructed of wood. Their barns and out-buildings are of wood; and the fencing of their farms, more expensive than their other improvements, is of wood; and all these are perishable with time. Moreover, our sixty thousand miles of railroad consume annually immense quantities of timber. Twenty-one thousand six hundred cords of wood are daily consumed in running railway trains three hundred and twenty thousand miles each twenty-four hours. Sixty thousand miles of road require twenty-five hundred ties to the mile, and as they must be replaced every five years, an annual consumption of thirty million ties is required. We will soon construct each year ten thousand miles of new road, requiring twenty-five million more ties, and, when we add to all these sources of forest destruction, the wood required in the fencing of these railroads, the half million telegraph poles which each year will be required, and the vast amount of the destruction of forests by flood or fire, we must be absolutely startled with the conviction that whole provinces of woods which have required a hundred years to grow, are each year being swept away, while nothing is done by either public authority or private zeal to supply the place of that which is destroyed, or protect in any measure that which exists. These are "hard facts," and whether people mind them or not to-day, they will give them some thought hereafter. In France, the forests were cut down with the utmost recklessness, and for the last thirty years her fertile valleys have been swept by terrible floods, carrying away all kinds of property, and covering the rich soil with gravel and sand. In Russia, the forests are beginning to disappear, and a law is now in force making it illegal to use anything but coal for fuel on the railroads. The timber lands of Germany are under the special protection of the government, while in Japan every one who cuts down a tree is compelled at once to plant another. The experience of these countries foreshadows that of our thoughtless men and reckless corporations may go on stripping the land of its forests, but at last every one will be convinced of the necessity of a change."

www.ingramcontent.com/pod-product-compliance
Lightning Source LLC
Chambersburg PA
CBHW071215080526
44587CB00013BA/1380